W9-DEP-145

HARCOURT BRACE JOVANOVICH COLLEGE OUTLINE SERIES

DISCRETE STRUCTURES

Essential Computer Mathematics

John L. Van Iwaarden

Department of Mathematics
Hope College
Holland, Michigan

Consulting Editor

Herbert Dershem

Department of Computer Sciences
Hope College
Holland, Michigan

Books for Professionals
Harcourt Brace Jovanovich, Publishers
San Diego   New York   London

Copyright © 1988 by Books for Professionals, Inc.

All rights reserved. No part of this publication may be reproduced or transmitted in any form or by any means, electronic or mechanical, including photocopy, recording, or any information storage and retrieval system, without permission in writing from the publisher.

Requests for permission to make copies of any part of the work should be mailed to:
    Permissions
    Harcourt Brace Jovanovich, Publishers
    Orlando, Florida 32887

Library of Congress Cataloging-in-Publication Data

Van Iwaarden, John L.
    Discrete structures: essential computer mathematics/John L. Van Iwaarden; consulting editor, Herbert Dershem.

    p. cm.—(Harcourt Brace Jovanovich college outline series) (Books for professionals)
    Includes index.
    ISBN 0-15-601578-1
    1. Electronic data processing—Mathematics.   I. Dershem, Herbert.   II. Title.   III. Series.
    IV. Series: Books for professionals.
QA76.9.M35V36   1988
004′.01′51—dc19

Printed in the United States of America

First Edition   88-21201

A   B   C   D   E

PREFACE

Do not *read* this Outline—**use** it. You can't learn computer-related mathematics simply by reading it: you have to *do* it. Solving specific, practical problems is the best way to master—and to demonstrate your mastery of—the theories, laws, and definitions upon which the study and use of mathematics are based. Outside the classroom, you need three tools to do discrete mathematics: a pencil, paper, and a calculator. Add a fourth tool, this Outline, and you're all set.

This HBJ College Outline has been designed as a tool to help you sharpen your problem-solving skills in college-level discrete structures mathematics. Each chapter covers a unit of material, whose fundamental principles are broken down in outline form for easy reference. The outline text is heavily interspersed with worked-out examples, so you can see immediately how each new idea is applied in problem form. Each chapter also contains a Summary and a Raise Your Grades section, which (taken together) give you an opportunity to review the primary principles of a topic and the problem-solving techniques implicit in those principles.

Most important, this Outline gives you plenty of problems to practice on. Work the Solved Problems, and check yourself against the step-by-step solutions provided. Test your mastery of the material in each chapter by doing the Supplementary Exercises. (In the Supplementary Exercises, you're given answers only—the details of the solution are up to you.) Finally, you can review all the topics covered in the Outline by working the problems in Exams 1, 2, and 3 and the Final Exam. (Use the exam problems to diagnose your own strengths and weaknesses.)

Having the tools is one thing; knowing how to use them is another. The solution to any problem in mathematics requires six procedures: (1) UNDERSTANDING, (2) ANALYZING, (3) PLANNING, (4) EXECUTING, (5) CHECKING, (6) REPORTING. Let's look at each of these procedures in more detail.

1. **UNDERSTANDING**   Read over the problem carefully and be sure you understand every part of it. If you have difficulty with any of the terms or ideas in the problem, reread the text material on which the problem is based. (In this Outline, important ideas, principles, laws, and terms are printed in boldface type, so they will be easy to find.) Make certain that you understand what kind of answer will be required.

2. **ANALYZING**   Break the problem down into its components. Ask yourself
   - What are the data?
   - What is (are) the unknown(s)?
   - What equation, law, or definition connects the data to the unknowns?

3. **PLANNING**   Trace a connection between the data and the unknowns as a series of discrete operations (steps). This often involves manipulating one or more mathematical expressions to isolate unknown quantities. Once you have a clear, stepwise path between data and solution, take note of any steps that require ancillary operations, such as using special symbols or converting units. (Keep a sharp watch on units—they are often useful clues.)

4. **EXECUTION**   Follow your plan and execute the mathematical operations. It helps to work with symbols whenever possible: substituting data for variables should be the *last* thing you do. Make sure you've used the correct signs, exponents, and units.

5. **CHECKING**   Never consider a problem solved until you have checked your work. Does your answer—
   - make sense?
   - have the right units?
   - answer the question?

6. REPORTING   Make sure you have shown your reasoning and method clearly, and that your answer is readable. (It can't hurt to write the word "Answer" in front of your answer. That way, you—and your instructor—can find it at a glance, saving time and trouble all round.)

The path from concept to bound book is a long and time-consuming one and I have not traveled it alone. I thank the editorial staff at Harcourt Brace Jovanovich for their expert work and helpful suggestions. I also thank Hope College for granting me the necessary time and facilities to complete this work. Finally, I thank my friends and colleagues for their patient support during these many months.

Hope College,   JOHN L. VAN IWAARDEN
  Holland, Michigan

CONTENTS

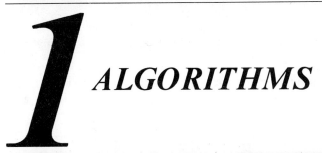

1 ALGORITHMS

THIS CHAPTER IS ABOUT

☑ **Algorithms Defined**
☑ **Characteristics of an Algorithm**
☑ **The Language of Algorithms**
☑ **Types of Algorithms**
☑ **Top-Down Design**

1-1. Algorithms Defined

For centuries mathematicians have sought ways to short-cut time-consuming calculations when solving complicated problems. With the advent of computers, we have been able to reduce many otherwise tedious mathematical manipulations to a series of mechanical, repetitive, routine processes called algorithms. Simply put, an "algorithm" is a step-by-step procedure that tells you how to perform an activity, be it cooking a steak, building a model airplane, filling out tax forms, sorting names into alphabetical order, or finding the sum of the first fifteen prime numbers. Specifically, an **algorithm** is

- a complete set of unambiguous instructions needed to carry out a task or to solve a problem; or
- a clear description of how a computer is to proceed in an organized way through a sequence of elementary steps.

The method of forming a routine solution to a mathematical problem is called the *algorithmic method*.

EXAMPLE 1-1 Supply an efficient algorithm for sharpening a new pencil (using a mechanical sharpener).

Solution

1. Insert noneraser end of pencil into hole in side of sharpener.
2. Hold pencil firmly in place with one hand.
3. Turn crank of sharpener with other hand.
4. Remove pencil from sharpener to check point of pencil.
5. If point is sharp enough, then stop.
6. Else repeat steps 1–4 until pencil point is sharp.

Most tasks, like sharpening a pencil, can be reduced to a series of discrete, individual steps or motions that are usually repeated, routine processes. How detailed an algorithm becomes is governed by how detailed it needs to be. For example, the shampoo label that reads "apply, lather, rinse, repeat" offers a general algorithm that explains how to use the product; it assumes you already know the other motions required to wash your hair—"wet hair, open bottle, pour shampoo into palm," etc.

Algorithms are particularly useful in mathematics. In many cases they describe to a person—or a computer—how to solve a mathematical problem (essentially, most computer programs are algorithms put into action). We write an algorithm by reducing a set of complex mathematical manipulations to a sequence of mechanical, repetitive, routine processes. Let's walk through the creation of an algorithm that will solve for the greatest common divisor (GCD) of two numbers.

recall: The greatest common divisor of two positive integers A and B is the largest integer that divides both A and B.

EXAMPLE 1-2   Write an algorithm that will find the GCD of two given positive integers A and B.

Solution

1. Write the numbers A and B, in that order, on a line.
2. If A and B are equal, stop, for each is the desired result. If not, go on.
3. If A is less than B, then interchange A and B on the line and proceed.
4. Subtract the number on the right from the one on the left and replace the number on the left by the remainder of the subtraction; rename these numbers as new A and B, respectively.
5. Repeat steps 2–4.

note: Steps 2–5 make up what's called a **loop**: the algorithm tells you (or a computer) to circle around and go back to perform the same task you've already done, only using new numbers.

Test: Let $A = 30$, $B = 75$.

1. $A = 30 \qquad B = 75$
2. $A \neq B$
3. $A < B$, so interchange: 75   30
4. $75 - 30 = 45$; new $A = 45$ and new $B = 30$
2′. $A \neq B$
3′. $A \not< B$; no change
4′. $45 - 30 = 15$; new $A = 15$ and new $B = 30$

2″. $A \neq B$
3″. $A < B$, so interchange: 30   15
4″. $30 - 15 = 15$; new $A = 15$ and new $B = 15$
2‴. $A = B$, so we stop
    The GCD is 15.

note: The *prime* ($'$), *double prime* ($''$), and *triple prime* ($'''$) on the step numbers indicate successive passes through the same loop.

The algorithm in Example 1-2, though successful, is unwieldy and inefficient because it requires many return visits to the loop. The process in Example 1-3 is much more practical.

EXAMPLE 1-3   Devise a division-based algorithm for finding the GCD of two given positive integers A and B.

Solution

1. Write down A and B, making A the larger integer and B the smaller.
2. Divide A by B and let R1 be the remainder.
3. If R1 = 0, then stop; the GCD is B.

4. Else, rename *B* and R1 as new *A* and new *B*, respectively.

5. Repeat steps 2–4.

Test: Let *A* = 30, *B* = 75.

1. *A* = 75   *B* = 30
2. 75 ÷ 30 = 2 with R1 = 15
3. R1 ≠ 0
4. *B* = 30   R1 = 15, renamed as *A* = 30   *B* = 15

2′. 30 ÷ 15 = 2 with R1 = 0
3′. R1 = 0; stop
    The GCD is *B* = 15.

Notice that the algorithm in Example 1-3 required only six steps and one visit back through the loop, while it took the algorithm in Example 1-2 eleven steps and three visits back through the loop to accomplish the same task.

1-2. Characteristics of an Algorithm

To be useful, an algorithm must be

- *unambiguous*—the instructions must be clear; avoid using words like "except," "unless," "providing," etc., which muddle the message with two many exceptions.
- *precise*—the instructions must be explicit, and methods of arithmetic computation must be concise.
- *finite* in description and execution—the algorithmic process must yield a finite answer within a finite sequence of steps.
- *effective*—the algorithm must produce the desired result in a reasonable number of steps.

Let's examine another algorithm in light of these characteristics.

EXAMPLE 1-4   Write an algorithm to add two integers.

Solution

1. Write the integers one below the other, aligning their rightmost digits.
2. If one number has more digits than the other, then
   a. Insert zeros at the beginning of the smaller number until the integers are of the same length.
3. Draw a line under the lower number.
4. Start with the rightmost column, called the current column.
5. Add the digits in the current column.
   a. If the sum is less than ten, then
      1. record the sum in the current column, below the line.
   b. Else
      1. Subtract 10 from the sum value.
      2. Record the difference below the line.
      3. Record the value 1 in a position above the upper number in the column to the left of the current column.
6. Designate the column immediately to the left of the current column as the new current column.
7. If the current column is empty, then
   a. Stop.
   b. The result is below the line.
8. Else
   a. Go to step 5.

It is important to note the following characteristics of the Example 1-4 algorithm:

- The instructions are clear and unambiguous; note the imperatives "write," "add," "record," and "subtract."
- The algorithm is precise; its instructions cover all potential "what-if's."
- Only these eight steps are needed to describe the algorithm, regardless of the size of the integers given; this algorithm will always yield an answer in a finite set of steps because the input values are finite integers.
- The algorithm is effective because a result is readily obtainable if the steps are executed properly.

In addition, this algorithm displays *generality*; it can solve many types of addition problems since the form of the input values remains general, not specific to the problem. (For example, with just a few changes, this algorithm could be used to add several integers or to add decimal numbers.)

1-3. The Language of Algorithms

For an algorithm to be useful and workable, we must have a language that will allow us to express it clearly. There are four alternatives:

(1) natural language (such as English)
(2) computer programming languages
(3) flowcharts
(4) pseudocode

Let's look at the merits and/or drawbacks of each.

A. Natural language

Although we normally converse in English and describe problems to one another in English, such natural language descriptions often contain ambiguities (clichés, idiomatic expressions, and complicated qualifications such as "except" or "provided that"). In addition, word descriptions of solutions—let alone of problems—can sometimes occupy reams of paper better used for paper airplanes and shelf lining. As useful as it is in normal conversation, natural language is severely limited in its capacity to describe mathematical computations and manipulations, so it can be readily dismissed as a way to express algorithms.

B. Computer programming languages

Unlike standard English, computer languages can handle various mathematical requirements with relative ease. Unfortunately, every computer language also has built-in details specific to its type that tend to hide the main structure of an underlying algorithm. These built-in nuances can make it very difficult for someone who knows one computer language to interpret the algorithm of another computer language. Because we want algorithms to be free from programming details such as punctuation, looping structures, and format, writing a general algorithm in a specific computer language is often too limiting. We must therefore also dismiss this method of representing algorithms.

C. Flowcharts

A **flowchart** is a diagram of directed lines that connect variously shaped boxes. To write an algorithm in flowchart form, you place each step of the algorithm into a specific type of box (see Table 1-1). The boxes are then joined by a series of lines to determine the order in which the processes are executed.

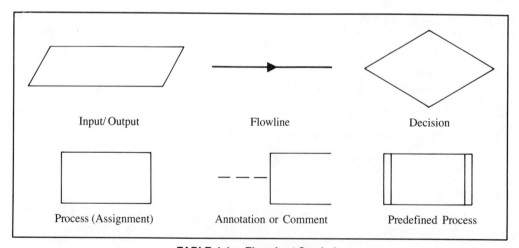

TABLE 1-1. Flowchart Symbols

A flowchart is a good method for giving a pictorial representation of an algorithm. And, unlike a specific computer programming language, a flowchart allows the algorithm to be universally applied. Its drawback, however, is its tendency to become so complex that you need a map, compass, or live-in navigator to find your way through it. Though useful as a first attempt to organize an algorithm, the flowchart is usually not effective as an algorithm's final form.

D. Pseudocode

The most effective way to express an algorithm is in **pseudocode** ("pseudo" means false). A "false system" of language, pseudocode is neither English nor math nor a specific computer language. Instead, it's a middleman, a sort of go-between, that expresses an algorithm precisely, efficiently, and unambiguously.

Pseudocode works as a series of statements, much like a computer program. Its typical structure is:

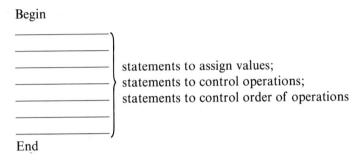

There are three basic types of pseudocode statements:

(1) *assignment statements*—statements that tell the computer to rename something or to replace one variable with a new value, variable, or expression. For example:

- 1. variable name ← value $x \leftarrow 3$
- 1. variable name ← expression $x \leftarrow x + 1$

    note: The equals sign is NOT used in algorithmic assignment statements, although you will see it used in computer programming. The left-facing arrow ($\leftarrow$) in $x \leftarrow x + 1$ means to assign the value of $x + 1$ to be the *new* value of x. The programming statement $x = x + 1$ means the same thing to the computer, but it is rather confusing mathematically for pseudocode.

(2) *condition statements*—statements that tell the computer to select options when certain conditions exist or do not exist. For example:

- 1. IF (condition) THEN
      a. (statement)

        1. IF $x = 4$ THEN
             a. $y \leftarrow x$

- 1. IF (condition) THEN
      a. (statement A)
   2. ELSE
      a. (statement B)

        1. IF $x = 4$ THEN
             a. $y \leftarrow x$
           2. ELSE
             a. $y \leftarrow x + 1$

(3) *iterative statements*—statements that tell the computer to repeat a certain series of calculations until a given condition is met. For example:

- 1. FOR (all set elements) TO (final value) DO
      a. (statement)

        1. FOR $K = 1$ TO N DO
             a. $A(K) \leftarrow 3K + 5$

    note: The phrase "all set elements" stands for something like a set $A(1), A(2), \ldots, A(N)$; $A(K)$ is the Kth term in a given or stored set of N terms.

- 1. FOR (initial value of variable) TO (final value) DO
      a. (statement)

        1. FOR $x = 2$ TO 17 DO
             a. $y \leftarrow 3x + 5$

- 1. WHILE (expression) DO
      a. (statement(s))

        1. WHILE $x \leq 17$ DO
             a. $y \leftarrow 3x + 5$

• 1. REPEAT (statement collection) UNTIL
  a. (condition)

1. REPEAT $y \leftarrow 3x + 5$
   UNTIL
   a. $x \geq 17$

Now that we've introduced the basic types of statements, let's look at some examples of how pseudocode is used and written. For a more visual approach, we compare pseudocode to the corresponding flowchart of an algorithm.

EXAMPLE 1-5 Use the IF-THEN-ELSE conditional statement to create an algorithm in flowchart and pseudocode that will find the larger of two given values.

Solution See Figure 1-1.

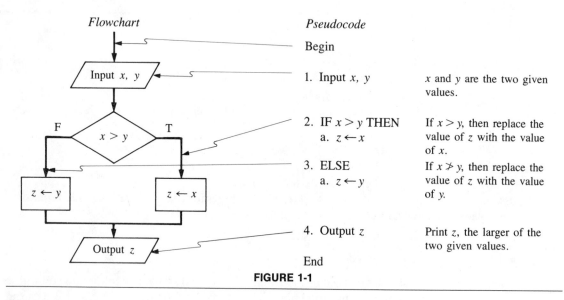

Flowchart   *Pseudocode*

Begin

1. Input x, y x and y are the two given values.

2. IF $x > y$ THEN   If $x > y$, then replace the
   a. $z \leftarrow x$ value of z with the value of x.

3. ELSE   If $x \not> y$, then replace the
   a. $z \leftarrow y$ value of z with the value of y.

4. Output z Print z, the larger of the two given values.

End

FIGURE 1-1

EXAMPLE 1-6 Using the WHILE-DO iterative statement, create an algorithm in flowchart and pseudocode to find the sum of the first n positive integers.

Solution See Figure 1-2.

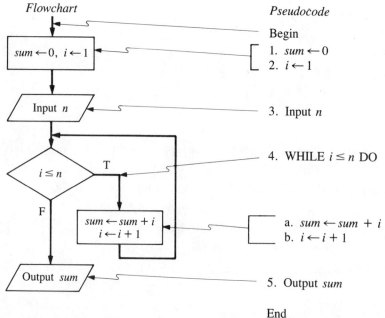

Flowchart   *Pseudocode*

Begin

1. $sum \leftarrow 0$ Initialize sum and i. (sum
2. $i \leftarrow 1$ is the accumulated value of the addition; i is the most current integer.)

3. Input n n is the number of integers you want added together.

4. WHILE $i \leq n$ DO   While the number of integers already added (i) is less than or equal to the number of integers requested (n), add the most
   a. $sum \leftarrow sum + i$ current value of i to the
   b. $i \leftarrow i + 1$ cumulative total sum. Increase i by one and go through the loop again.

5. Output sum Print sum, which is now the final total.

End

FIGURE 1-2

note: To initialize means to assign what the first (initial) values of the variables will be.

By examining Examples 1-5 and 1-6, we can glean several rules for writing pseudocode:

- always have only one statement or condition per line of code
- always label successive steps consecutively
- always indent subordinate lines, or loops, and alternately letter and number them according to their level, or degree of succession:

```
Begin
1. (line 1)
2. (line 2)
   a. (line 3)
      1. (line 4)
      2. (line 5)
   b. (line 6)
3. (line 7)
   a. (line 8)
4. (line 9)
5. (line 10)
End
```

- always begin with the word "Begin" and end with the word "End"; these do not receive statement numbers
- when necessary, have an *input statement* to input your givens (the computer needs to know what values to work with)
- when necessary, include an *output statement* to make the computer print the answer it was working so hard to get

Now that we've discussed the "syntax" of the pseudocode language, let's practice creating more algorithms in this new code.

EXAMPLE 1-7 Create an algorithm to find the largest (MAX) value in a set of input values $v(1), v(2), \ldots, v(n)$. Test it.

Solution

Begin
1. Input n n is the number of values you want checked.

2. Input $v(1), v(2), \ldots, v(n)$
3. $i \leftarrow 1$ i is the location of the value currently under examination (e.g., when $i = 3$ we have $v(i) = v(3)$ = the third value in the list).

4. MAX $\leftarrow v(1)$ Let MAX be the first value.
5. FOR $i = 2$ TO n DO   This loop tests each subsequent
     a. IF $v(i) >$ MAX THEN   value against the current value of
       1. MAX $\leftarrow v(i)$ MAX and replaces it if a new, larger MAX is found.

6. Output MAX
End

Test: Let $v(1) = 15$, $v(2) = 19$, $v(3) = 4$, $v(4) = 8$, $v(5) = 27$, and $v(6) = 3$; $n = 6$.

i	v(i)	MAX	
initialize 1	15	15	Read this line as: "When $i = 1$, $v(i) = 15$, and MAX $= 15$."
2	19 ($>$MAX)	19	
3	4 ($\not>$MAX)	19	
4	8 ($\not>$MAX)	19	
5	27 ($>$MAX)	27	
6 ($i = n$)	3 ($\not>$MAX)	27	Output MAX: 27

EXAMPLE 1-8 Form the algorithm to generate the Fibonacci sequence of numbers, 1, 1, 2, 3, 5, 8,..., up to N terms. Test it.

Solution The Fibonacci sequence begins with the numbers 1 and 1, and is defined thereafter by the recursion formula $a_n = a_{n-1} + a_{n-2}$ (i.e., each member is the sum of the previous two values). We will form two versions—the first as a flowchart and the second in pseudocode (see Figure 1-3).

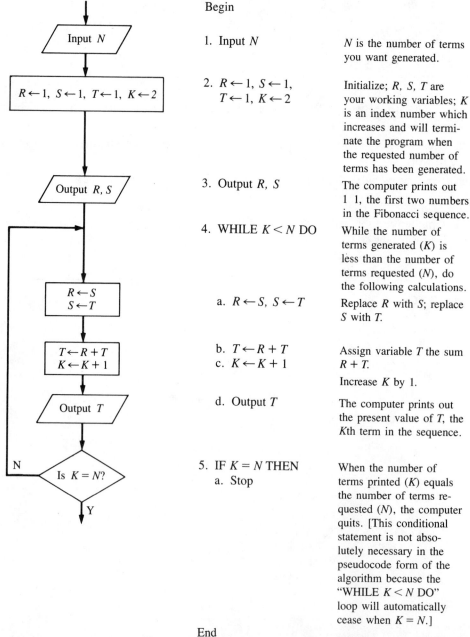

Begin

1. Input N N is the number of terms you want generated.

2. $R \leftarrow 1$, $S \leftarrow 1$, $T \leftarrow 1$, $K \leftarrow 2$ Initialize; R, S, T are your working variables; K is an index number which increases and will terminate the program when the requested number of terms has been generated.

3. Output R, S The computer prints out 1 1, the first two numbers in the Fibonacci sequence.

4. WHILE $K < N$ DO   While the number of terms generated (K) is less than the number of terms requested (N), do the following calculations.

     a. $R \leftarrow S$, $S \leftarrow T$ Replace R with S; replace S with T.

     b. $T \leftarrow R + T$ Assign variable T the sum $R + T$.
     c. $K \leftarrow K + 1$ Increase K by 1.

     d. Output T The computer prints out the present value of T, the Kth term in the sequence.

5. IF $K = N$ THEN   When the number of
     a. Stop   terms printed (K) equals the number of terms requested (N), the computer quits. [This conditional statement is not absolutely necessary in the pseudocode form of the algorithm because the "WHILE $K < N$ DO" loop will automatically cease when $K = N$.]

End

FIGURE 1-3

Test: We run the algorithm for $N = 6$ terms.

	R	S	T	K	output
initialize	1	1	1	2	1 1
pass 1	1	1	2	3	(1 1) 2
pass 2	1	2	3	4	(1 1 2) 3
pass 3	2	3	5	5	(1 1 2 3) 5
pass 4	3	5	8	6	(1 1 2 3 5) 8

$K = N$, stop;
final output: 1 1 2 3 5 8

EXAMPLE 1-9   Create a more efficient version of the GCD algorithm of Example 1-3 and draw its flowchart. Test it first with $A = 75$ and $B = 30$ and again with $A = 75$ and $B = 17$.

Solution   See Figure 1-4.

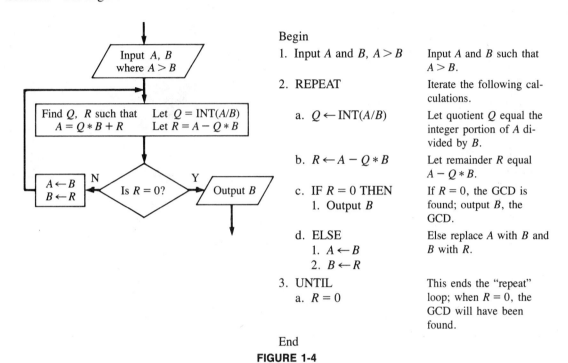

Begin
1. Input A and B, $A > B$ — Input A and B such that $A > B$.

2. REPEAT — Iterate the following calculations.

   a. $Q \leftarrow \text{INT}(A/B)$ — Let quotient Q equal the integer portion of A divided by B.

   b. $R \leftarrow A - Q * B$ — Let remainder R equal $A - Q * B$.

   c. IF $R = 0$ THEN
     1. Output B — If $R = 0$, the GCD is found; output B, the GCD.

   d. ELSE
     1. $A \leftarrow B$
     2. $B \leftarrow R$ — Else replace A with B and B with R.

3. UNTIL
   a. $R = 0$ — This ends the "repeat" loop; when $R = 0$, the GCD will have been found.

End

FIGURE 1-4

note: The term $\text{INT}(A/B)$ (statement 2) tells the computer to calculate the division in parentheses and to return only the integer portion of the quotient; that is, it simply drops the decimal portion and gives the integer, without rounding up or down.

agreement: We will adopt certain conventions from computer programming in regard to arithmetic operations. In this example and elsewhere, the "*" is used for multiplication. We will restrict our use of the "×" for scientific notation.

Test 1: Let $A = 75$, $B = 30$.

A	B	Q	R
75	30	2	15
30	15	2	0

Output B: 15

Test 2: Let $A = 75$, $B = 17$.

A	B	Q	R
75	17	4	7
17	7	2	3
7	3	2	1
3	1	3	0

Output B: 1

EXAMPLE 1-10   One way to find the square root of a positive number A is to use the recursion formula

$$x_{i+1} = \frac{1}{2}\left(x_i + \frac{A}{x_i}\right)$$

where x_i is the ith estimate of the square root.

Write an algorithm in pseudocode that uses the recursion formula to obtain the value of $\sqrt{A}$, correct to p decimal places, when you input an estimated value.

Solution

Begin

1. Input A A = the number whose square root you want.

2. Input x x = your estimated value of $\sqrt{A}$.

3. Input p p = the number of decimal places to compute to.

4. $y \leftarrow 0.5(x + A/x)$ Initialize the expression for y (using the recursion formula).

5. IF $|y \uparrow 2 - A| \leq 0.5 * 10 \uparrow (-p)$ THEN   Test the current estimate: if it's close to correct, $|y^2 - A|$ will be less than or equal to 0.5×10^{-p} (this carries the computations to the requested decimal place) and y will be output.
   a. Output y

6. ELSE   If the current estimate isn't close enough, replace x with the closer estimate y and run through the loop again.
   a. $x \leftarrow y$

7. Return to step 4

End

---

note: The upward arrow ($\uparrow$) indicates the exponentiation; e.g., $x \uparrow 2$ is the code for x^2 and $x \uparrow (n-1)$ is the code for x^{n-1}. Some computer languages use $**$ for exponentiation.

EXAMPLE 1-11 Use the algorithm of Example 1-10 to compute $\sqrt{2}$, correct to four decimal places.

Solution We have $A = 2$ and $p = 4$, and we will use the crude estimate $x = 1.5$ (the computations are numbered to match the original algorithm steps in Example 1-10):

4. $y = 0.5(1.5 + 2/1.5) = 0.5(1.5 + 1.3333) \cong 1.4167$
5. $|y^2 - A| = |1.4167^2 - 2| = |0.0070| \nleq 0.5 \times 10^{-4}$ $(0.0070 > 0.0005)$
6. new $x = 1.4167$
4′. $y = 0.5(1.4167 + 2/1.4167) = 0.5(1.4167 + 1.4117) \cong 1.4142$
5′. $|y^2 - A| = |1.4142^2 - 2| = |-0.000\,038\,36| < 0.5 \times 10^{-4}$, so $y = 1.4142$ is correct to four decimal places.

Output y: 1.4142

1-4. Types of Algorithms

The essential mathematics for computer science work uses six common types of algorithms: *direct computation, enumeration, divide-and-conquer, iteration, trial-and-error,* and *recursion.*

A. Direct computation

With **direct computation** a computer obtains an exact answer by using a well-defined sequence of elementary operations. For example, an algorithm that finds the roots of a quadratic equation by applying the quadratic formula would be a direct computation algorithm. So would your using a given set of instructions to calculate your income tax.

B. Enumeration

Sometimes the solution to a mathematical problem is one element of a given set of elements, and you need to find that single element. By using an **enumeration** algorithm, the computer will try all possible solutions until it identifies the one element that solves the problem. For instance, the algorithm in Example 1-7, in which the maximal element of a set is determined, uses the enumeration approach. In that algorithm, the first element was initially called MAX. The computer compared succeeding elements against MAX and replaced it with a new MAX when a larger element was found. The computer had to scan the *entire list* to find the largest value.

C. Divide-and-conquer

Sometimes the enumeration technique is so long and cumbersome (remember that *each* element must be directly checked) that it is easier to subdivide the task for increased efficiency. The **divide-and-conquer** method divides a large problem into smaller but similar problems that can be either solved directly and more easily or subdivided further by the same technique. A practical example would be the search for a particular name in a telephone book. With the enumeration method, you would have to look at *each* entry to determine if it is the name you want. Obviously, you'd greet your old age before you found "Zbaygen" in something like a New York City directory. If you divided the book in the middle, however, and determined in which half the desired name resides, you would have already cut your task in half. Then you'd simply apply the same technique on progressively smaller lists. For example, say your directory only had 64 names in it, and you wanted to find the name "Zbaygen." After the first comparison you would have only 32 names left. After two comparisons you would retain 16, after the third you'd have 8, the fourth 4, the fifth 2, and the sixth one. This final list of one entry is *very easy* to search! This particular divide-and-conquer algorithm—continually dividing the task in half—is called a *binary search*. It allows you to find "Zbaygen," or any name for that matter, in a 64-entry list by using only 6 comparisons, instead of the 64 the enumeration method would require.

D. Iteration

In an **iteration** algorithm, a series of increasingly precise approximate solution values are computed until a value is obtained that is "sufficiently close." Usually the "exact answer" requires an infinite number of steps.

recall: An algorithm must always yield a finite answer in a finite number of steps. By identifying a value that's "sufficiently close," you thankfully tell the computer when to stop, else it would gamely continue calculating the "exact" answer until the poor thing suffered a capacitor breakdown (known as being "incapacitated").

Iteration techniques usually apply to mathematical problems in which power series are used for computational purposes. For example, one of the function expressions derived from the MacLaurin Expansion Theorem of calculus is

$$\cos(x) = 1 - \frac{x^2}{2!} + \frac{x^4}{4!} - \frac{x^6}{6!} + \cdots$$

To exactly evaluate this expression at a given value of x would require an infinite number of computations. We may, however, limit ourselves to any finite number of terms to arrive at a result that is as precise as we might desire.

E. Trial-and-error

The trial-and-error method is a form of iteration in which each successive approximation is based on the amount of error obtained in a previous approximation. In driving a car through a curve in the highway, for example, you apply a trial amount of steering wheel shift. When you observe the result of this action, you make a correction to it; if this correction is still not adequate, you make a new correction based on the previous one, etc. The iteration algorithm we used in Example 1–10 to find the square root of a number also used trial and error: the value for A was tested at each new iteration for its error amount, and the next iteration was based on the previous value.

F. Recursion

For certain kinds of problems, an algorithm might need to call itself in the process of executing its steps. Being able to reference itself makes a **recursion** algorithm economical, and it often becomes a very effective tool for problem solving.

A relatively simple example is the computation of the factorial of a positive integer N. Remember that $N! = 1 * 2 * 3 * \cdots * N$. But we can also write this as follows:

$$1! = 1$$

$$N! = N * (N - 1)! \quad \text{for} \quad N > 1$$

Here, computing $N!$ depends on already having computed $(N - 1)!$. But computing $(N - 1)!$ depends on already having computed $(N - 2)!$, which depends on already having computed $(N - 3)!$, etc. This dependence on a previous computation is passed all the way back to the point where we need the value of $1!$. But that value is part of the given, so from the fact that $1! = 1$, we get

$$2! = 2(1!) = 2$$

$$3! = 3(2!) = 6$$

$$4! = 4(3!) = 24$$

$$\vdots$$

$$N! = N(N - 1)!$$

EXAMPLE 1-12   Write the factorial algorithm as a recursive algorithm in pseudocode.

Solution

Begin
1. Input N
2. IF $N = 1$, THEN
   a. $N! \leftarrow 1$ Return the factorial value as 1.
3. ELSE
   a. $N! \leftarrow N * (N - 1)!$ Return the factorial value as N times
End   factorial $(N - 1)$. At this point, the computer can't calculate $N!$ until it has calculated $(N - 1)!$, so it puts $N!$ on "hold" and automatically calls itself (goes back to step 2) to compute $(N - 1)!$. Since it can't compute $(N - 1)!$ until it knows $(N - 2)!$, it puts $(N - 1)!$ on hold and calls itself again to compute $(N - 2)!$. It keeps calling itself until the value of $1!$ is called for. Then the value 1 is returned to the last call, $1 * 2$ is returned to the next to the last call, then $1 * 2 * 3$ is returned, etc., until $N!$ is found.

---

note: A factorial algorithm like that of Example 1-12 is usually built into larger computers (more than your \$66 brand) and it is treated as a *subroutine*. That is, when a program you write needs $N!$, the computer pauses to call up the factorial subroutine, uses it to calculate $N!$, returns from the subroutine with the value of $N!$, plugs it into your program at the appropriate point, then keeps on truckin' with the remaining part of your program. For this purpose, there is no need for an output line because the computer simply returns the computed value of $N!$ to the main program.

1-5. Top-Down Design

Sometimes the process of creating an algorithm is so complex that constructing its details in one continuous collection of ideas is impossible. Knowing this, you might first outline the major jobs required to finish the task and then put them into an operational sequence. Then you might look at each job independently and outline its major subjobs and their flow, going deeper and deeper into the structure until the individual jobs at the bottom level are elementary. When all of these components are correctly "glued together," the overall algorithm will be complete. This modular approach to algorithm construction is called **top-down design**, because you work downward from the "top" of the problem and keep spreading it out, branching down like the roots of a tree, to make each step increasingly simpler.

EXAMPLE 1-13   Create a generalized top-down flowchart to calculate the paycheck of an employee of a small manufacturing company and establish a general format algorithm in natural language.

Solution   See Figure 1-5.

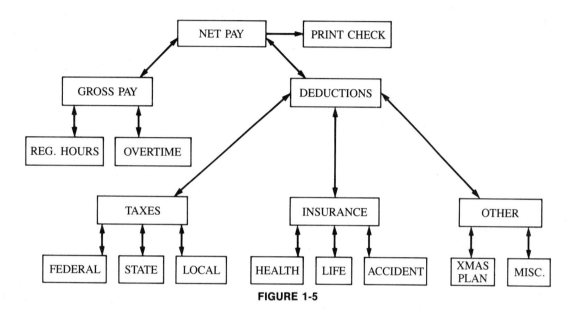

FIGURE 1-5

Looking at the "flowchart" shown in Figure 1-5, you can see that each subdivision would require a loop or a *subroutine* to make the intermediate calculations of the various deductions.

note: A **subroutine** is simply a subordinate group of related steps that add more detail to an algorithm; it serves the same function as a loop, telling the computer to hold on to the current calculations for a moment while it travels through the subroutine or loop. Because the conditional statement loops (such as WHILE–DO and IF–THEN–ELSE) can usually replace subroutines, and because subroutines are often specific to a particular computer language, they are not used in pseudocode except to refer to an algorithm already established elsewhere.

Using the top-down flowchart here, you could set up the general format of the algorithm to be something like:

1. Gross pay = regular hours * rate of pay + overtime hrs * rate of pay
2. Net pay = gross pay − deductions
3. Deductions = deduction tax + deduction insurance + deduction other
   a. Deductions tax = tax federal + tax state + tax local
      1. tax federal = gross pay * %
      2. tax state = gross pay * %
      3. tax local = gross pay * %
   b. Deductions insurance = deduction health + deduction life + deduction accident
      1. deduction health = gross pay * %   (or * steady amount)
      2. deduction life = gross pay * %   (or * steady amount)
      3. deduction accident = gross pay * %   (or * steady amount)
   c. Deductions other = deductions Christmas + deductions misc. = gross pay * %   (or * steady amount)
4. Calculate net pay
5. Print check for net pay

EXAMPLE 1-14 Create a top-down flowchart to find the roots of the cubic polynomial $ax^3 + bx^2 + cx + d = 0$.

Solution The cubic polynomial will have either three real roots or one real root and two complex roots. See Figure 1-6.

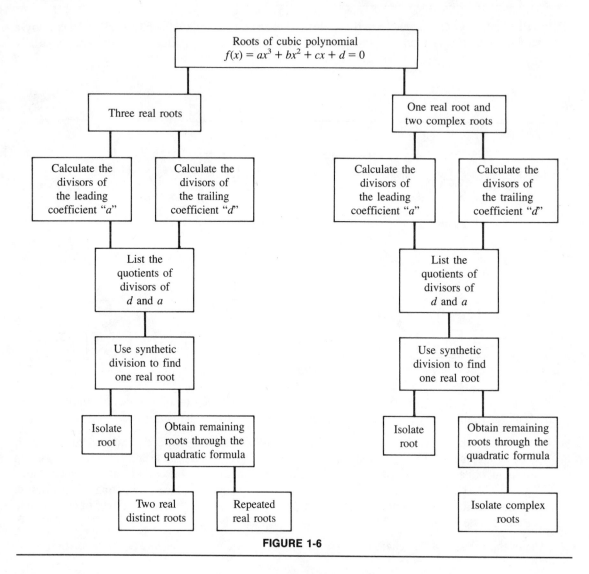

FIGURE 1-6

SUMMARY

1. An algorithm is an unambiguous set of precisely stated step-by-step instructions that lead to the solution of a problem.
2. A correctly posed algorithm is unambiguous, precise, finite in description and execution, and effective.
3. Algorithms may be represented in natural language, computer programming language, flowcharts, or pseudocode. The latter two are the most effective for planning and expressing an algorithm.
4. The six common types of algorithms are direct computation, enumeration, divide-and-conquer, iteration, trial-and-error, and recursion.
5. The process of top-down design is often very useful in structuring the algorithm of a complicated problem.

RAISE YOUR GRADES

Can you...?

☑ give more than one definition of an algorithm
☑ define each of the four characteristics of an algorithm
☑ present arguments for the merits of each of the four ways to express algorithms
☑ list the six common types of algorithms
☑ give an example of each of the six types of algorithms
☑ explain the concept of top-down design

SOLVED PROBLEMS

PROBLEM 1-1   Design an algorithm for changing a flat tire.

Solution   Use natural language to write down each step you'd take to change a tire:

1. Set emergency brake.
2. Open trunk.
3. Check condition of spare tire.
4. Remove spare from trunk.
5. Remove jack from trunk.
6. Block wheels.
7. Remove hub cap.
8. Loosen lug nuts slightly.
9. Jack up car.
10. Remove lug nuts.
11. Remove flat tire.
12. Install spare tire.
13. Tightly replace lug nuts.
14. Lower car and remove jack.
15. Put all remaining paraphernalia in trunk.
16. Head for service station.

note: Order is important; details are necessary; additional subdetails could be added.

PROBLEM 1-2   Design an algorithm to accept an arbitrary number x and return the absolute value of x.

Solution   Since this is a mathematical problem, pseudocode is the most useful language for expressing the algorithm:

Begin
1. IF $x < 0$ THEN
      a. $a \leftarrow -x$
2. ELSE
      a. $a \leftarrow x$
3. Output a
End

PROBLEM 1-3   In the country Nizpan the tax structure is as follows: A person who earns more than $50,000 pays $7,000 tax; someone who gets paid from $30,000 through $50,000 pays $4,000 tax; and someone who makes less than $30,000 pays a 9% tax. Write an algorithm to compute the tax owed for an arbitrary income I.

Solution

Begin
1. Input I
2. IF $I > 50\,000$ THEN
   a. $T \leftarrow 7000$
3. IF $I < 30\,000$ THEN
   a. $T \leftarrow 0.09 * I$
4. ELSE
   a. $T \leftarrow 4000$
5. Output T
End

PROBLEM 1-4   Create an algorithm to find the sum (SUM) of the squares of the first n positive integers. Test it.

Solution

Begin
1. SUM $\leftarrow 0$ Initialize.
2. $i \leftarrow 1$
3. Input n
4. REPEAT   Steps 4 and 5 make up a conditional
   a. SUM $\leftarrow$ SUM $+ i * i$ loop; the computer will iterate the
   b. $i \leftarrow i + 1$ calculation until the value of the
                                 counter variable i is greater than n. It
                                 has then added n quantities.
5. UNTIL
   a. $i > n$
6. Output SUM
End

Test: Let $n = 5$; we are to compute $1^2 + 2^2 + 3^2 + 4^2 + 5^2$.

	SUM		i	n		
initialize	0		1	5		
pass 1	1	$(0 + 1^2)$	2	5	Is $i > n$? No	
pass 2	5	$(1 + 2^2)$	3	5	Is $i > n$? No	
pass 3	14	$(5 + 3^2)$	4	5	Is $i > n$? No	
pass 4	30	$(14 + 4^2)$	5	5	Is $i > n$? No	
pass 5	55	$(30 + 5^2)$	6	5	Is $i > n$? Yes	Output SUM: 55

PROBLEM 1-5   Create an algorithm that will find all the two-digit numbers that are equal to the sum of the squares of their digits.

Solution   Set up m and n as the left- and right-hand digits, respectively, of the integer (for example, integer 87 has $m = 8$, $n = 7$). Using the enumeration method, test each number from 10 to 99 to see if

$$\underset{\substack{\text{(the tens} \\ \text{column)}}}{10 * m} + \underset{\substack{\text{(the ones} \\ \text{column)}}}{n} = m^2 + n^2$$

which will satisfy the requirements of the problem.

Begin
1. $m \leftarrow 0$
2. REPEAT
   a. $m \leftarrow m + 1$ Increase m by 1; this will eventually
                              carry us through the 10's, 20's, 30's,
                              etc.

b. IF $m > 9$ THEN
    1. Stop
c. ELSE
    1. $n \leftarrow 0$
d. WHILE $n \leq 9$ DO

The tens column (m) is limited to digits 1 to 9.

Set n, the right-hand digit, to 0.
Like m, n cannot exceed the digit 9. When $n = 10$, the computer skips down to step 3a; if this test in 3a fails, it goes up to step 2a and increases m by one and resets n to 0 in step 2c; "19" becomes "20."

    1. IF $10 * m + n = m \uparrow 2 + n \uparrow 2$ THEN
        a. Output m, n
        b. $n \leftarrow n + 1$
    2. ELSE
        a. $n \leftarrow n + 1$
3. UNTIL
    a. $10 * m + n = 100$

Test the number, print it if it meets the qualifications, and then move on to the next two-digit integer.

The computer will stop once it has tested all integers up to 99.

End

PROBLEM 1-6 Write an algorithm for finding all the factors of a given integer N.

Solution Because the largest number that can divide another number is itself, you can use a WHILE–DO loop to make the computer try all possible factors up to the integer N.

Begin
1. Input N
2. $A \leftarrow 0$
3. WHILE $A \leq N$ DO
    a. $A \leftarrow A + 1$
    b. IF $INT(N/A) = N/A$
        1. Output N/A
4. Stop
End

recall: The integer function INT analyzes its argument and deletes the portion after the decimal point.

PROBLEM 1-7 Form an algorithm to take any three integers written on a line and rewrite them from smallest to largest. Test it.

Solution This algorithm can be made very efficient if we first create a subalgorithm called SWAP. Define $SWAP(A, B)$ as follows:

Begin
1. $T \leftarrow A$
2. $A \leftarrow B$
3. $B \leftarrow T$
End

This will interchange the order of two values A and B, regardless of their values; e.g., if we start with $A = 2$, $B = 7$, we end up with $A = 7$, $B = 2$.

Now the algorithm for three given input values P, Q, R, in arbitrary order:

Begin
1. Input P, Q, R
2. IF $P > Q$ THEN
    a. SWAP(P, Q)
3. IF $Q > R$ THEN
    a. SWAP(Q, R)

4. IF $P > Q$ THEN
   a. SWAP(P, Q)
5. Output P, Q, R
End

Test: Let the given numbers be 5, 8, 2.

	P	Q	R
Original order	5	8	2
after step 2	5	8	2
after step 3	5	2	8
after step 4	2	5	8

PROBLEM 1-8   Design an algorithm to determine if a given number N is prime.

Solution   Begin with some facts and observations: A number N is prime if the smallest integer greater than 1 that divides N is N itself. We could use the enumeration method, beginning with 2, and divide every value through N into N to see if it leaves a zero remainder; unfortunately, this method would be incredibly time-consuming and inefficient. A more practical way to shorten the list of test values is to use only those values through $\sqrt{N}$, since if N is *not* prime, then one of its divisors will be smaller than or equal to $\sqrt{N}$.

    When dividing an integer N by another integer D, we say that D divides N, denoted $D \mid N$, if there is a zero remainder. Else we say that D does not divide N and denote it by $D \nmid N$. Using the above observations, we obtain the algorithm.

Begin
1. Input N — Input integer to be tested.
2. $D \leftarrow 2$ — Initialize D as 2, the first number (after 1) that might divide evenly into N.

3. IF $D * D \leq N$ and $D \nmid N$ THEN — Test to see if $D \leq \sqrt{N}$ and if D does
   a. $D \leftarrow D + 1$ and return to step (3) — not divide N exactly.
                                     If statement 3 is true, then try the next higher value of D. Continue doing this until statement (3) is false.

4. ELSE — If part of statement 3 is false and 4a
   a. IF $D * D > N$ THEN — is true, then N is prime; replace D
      1. $D \leftarrow N$ — with N and output D.
5. Output D — D will be a divisor of N; if it is not N, it divides N, so N is not prime.

End

PROBLEM 1-9   Draw a flowchart and design an algorithm for finding the average of N numbers, $A(1)$, $A(2)$, $A(3), \ldots, A(N)$.

Solution   Assume the numbers are already stored in locations $A(1)$, $A(2), \ldots, A(N)$. See Figure 1-7.

PROBLEM 1-10   Draw a flowchart and design an algorithm for the following problem. A salesperson normally makes a 20% commission, but if sales are greater than $2,000, then a $100 bonus is added to the commission. How do we compute the gross pay?

Solution   See Figure 1-8.

PROBLEM 1-11   To calculate the value of a certificate savings account at the end of its term, we need the following information: the initial deposit D, the number of years the money is on deposit Y, the annual interest rate I, and the number of pay periods per year P. Form an algorithm to calculate the savings account balance when the account matures.

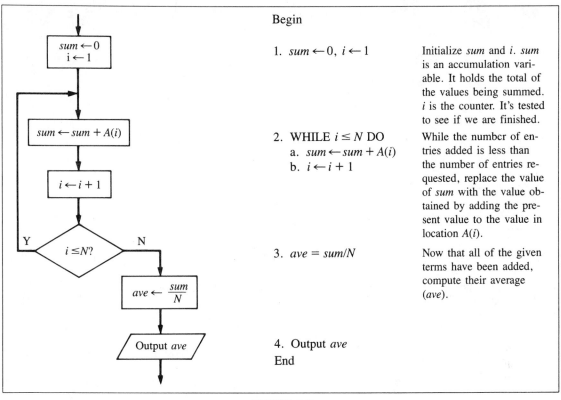

Begin

1. $sum \leftarrow 0$, $i \leftarrow 1$ — Initialize *sum* and *i*. *sum* is an accumulation variable. It holds the total of the values being summed. *i* is the counter. It's tested to see if we are finished.

2. WHILE $i \leq N$ DO
   a. $sum \leftarrow sum + A(i)$
   b. $i \leftarrow i + 1$

   While the number of entries added is less than the number of entries requested, replace the value of *sum* with the value obtained by adding the present value to the value in location $A(i)$.

3. $ave = sum/N$ — Now that all of the given terms have been added, compute their average (*ave*).

4. Output *ave*

End

FIGURE 1-7

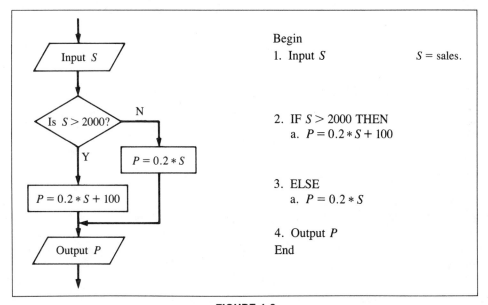

Begin

1. Input S — S = sales.

2. IF $S > 2000$ THEN
   a. $P = 0.2 * S + 100$

3. ELSE
   a. $P = 0.2 * S$

4. Output P

End

FIGURE 1-8

Solution Recognizing that the new balance per period = previous balance + (previous balance * interest per pay period), we must first calculate the interest per pay period.

Begin

1. Input D, Y, I, P

2. PERIODS $\leftarrow Y * P$ — Find the number of pay periods.

3. INTEREST $\leftarrow I/P$ — Find the interest rate per pay period.

4. BALANCE $\leftarrow D$ — Initialize BALANCE.

5. FOR $j = 1$ TO PERIODS DO
   a. BALANCE $\leftarrow$ BALANCE + (BALANCE * INTEREST)

6. Output BALANCE

End

PROBLEM 1-12   Trace through the savings account algorithm of Problem 1-11 using the values $D = \$100$, $Y = 2$ years, $I = 10\%$ (0.10), $P = 4$ times yearly.

Solution

periods	interest	j	BALANCE	
8	0.025	—	100.00	
		1	102.50	
		2	105.06	
		3	107.69	
		4	110.38	
		5	113.14	
		6	115.97	
		7	118.87	
		8	121.84	Output BALANCE: 121.84

PROBLEM 1-13   Suppose we have a linear *array* (list) of N data items $D(1), D(2), \ldots, D(N)$ stored in computer memory and a specific item I contained therein. Write an algorithm to search through the array and find the location L of item I and to report that location. If I is not found, report this fact.

Solution

Begin
1. $J \leftarrow 1, L \leftarrow 0$
2. WHILE $L = 0$ and $J \leq N$ DO $L = 0$ means we are at the beginning
                                                         of the problem and the item is not
                                                         yet found.
   a. IF $I = D(J)$ THEN   If requested, item I is the Jth item,
      1. $L \leftarrow J$ $D(J)$, then location $L = J$.
   b. ELSE
      1. $J \leftarrow J + 1$
3. IF $L = 0$ THEN   If L still equals 0 when $J = N$, item I
   a. OUTPUT "ITEM NOT FOUND"   was never found in the list.
4. ELSE
   a. Output L
End

note: This type of algorithm would most likely be a subroutine, perhaps named SEARCH, that we would call on when a search is warranted. For example, say we have a list of subscribers to a pet store newsletter, and we want to find subscriber Jones. We might call the SEARCH subroutine by saying SEARCH(JONES). The word SEARCH activates the subroutine, and the parameter JONES provides the item I to be sought. This algorithm therefore doesn't require a line saying "Input I," for it automatically sets I to be "Jones."

PROBLEM 1-14   *Mind Reading Trick.* A "mind reader" (computer) asks a person in the audience to think of a number, multiply it by 5, add 6, multiply by 4, add 9, multiply by 5, and then state the result. The person chooses the number 15 and calculates as follows: $15 \rightarrow 75 \rightarrow 81 \rightarrow 324 \rightarrow 333 \rightarrow 1665$. She announces 1665. The mind reader then subtracts 165 from the result, knocks off two zeros, and informs the person that her original number was 15.

   Create an algorithm for a computer to "read minds" in this way.

Solution

Begin
1. Input A A is the original number.
2. $R1 \leftarrow 5 * A$ $5A$ (Multiply by 5)
3. $R2 \leftarrow R1 + 6$ $5A + 6$ (Add 6)

4. $R3 \leftarrow R2 * 4$ $4(5A + 6) = 20A + 24$ (Multiply by 4)

5. $R4 \leftarrow R3 + 9$ $20A + 24 + 9 = 20A + 33$ (Add 9)

6. $R5 \leftarrow R4 * 5$ $5(20A + 33) = 100A + 165$ (Multiply by 5)

7. $R \leftarrow (R5 - 165)/100$ $R = A$

8. Output R

End

PROBLEM 1-15 A straightforward mathematical problem for using recursion is the calculation of x^n for any given x and positive integer n. Observe that the formula $x^n = x * x^{n-1}$ displays the recursive call. Write the algorithm, denoted POWER(x, n), for this computation.

Solution

Begin
1. Input x, n
2. IF $n = 0$ THEN
   a. POWER $\leftarrow 1$
3. ELSE
   a. POWER $\leftarrow x *$ POWER$(x, n - 1)$
End

Return the POWER value as x times POWER$(x, n - 1)$. At this point, the computer can't calculate POWER until it has calculated POWER$(x, n - 1)$, so it puts the first POWER on "hold" and automatically calls itself (goes back to step 2) to compute POWER$(x, n - 1)$. Since it can't compute POWER$(x, n - 1)$ until it knows POWER$(x, n - 2)$, however, it puts POWER$(x, n - 1)$ on hold and calls itself again to compute POWER$(x, n - 2)$. It keeps calling itself in this manner until the value of POWER$(x, 0)$ is called for. Then the value 1 is returned to the last call, $1 * x$ is returned to the next to the last call, then $1 * x * x$ is returned, etc., until POWER(x, n) is found.

PROBLEM 1-16 Trace through the recursive algorithm POWER(x, n) of Problem 1-15 using values $x = 5, n = 3$ (for finding 5^3).

Solution

$$
\begin{aligned}
\text{POWER}(5, 3) &= 5 \cdot \text{POWER}(5, 2) \\
&= 5 \cdot (5 \cdot \text{POWER}(5, 1)) \\
&= 5 \cdot (5 \cdot (5 \cdot \text{POWER}(5, 0))) \\
&= 5 \cdot (5 \cdot (5 \cdot 1)) \\
&= 5 \cdot (5 \cdot (5)) \\
&= 5 \cdot (25) \\
&= 125
\end{aligned}
$$

PROBLEM 1-17 The developer of a newly constructed, three-unit office building (Figure 1-9) wants to cover all of its outside walls with an expensive polished aluminum siding and the roof with crushed marble. The developer has three grades of siding available, SA, SB, and SC, and wants to know the cost comparisons of using each grade. For the roof's price, the developer must consider not only the cost of three grades of marble, MA, MB, and MC, but also the cost of the underlying roofing materials and the overall roofing thickness desired (two thicknesses are available, TA and TB). Use a top-down design approach to create a general flow chart of the steps needed to make an algorithm that will accurately predict the final cost of materials to be used, depending upon the various grades of siding and roofing marble and the various thicknesses of roof covering.

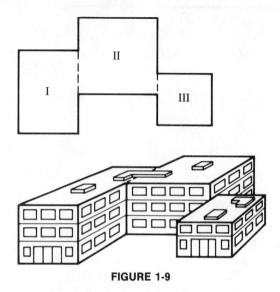

FIGURE 1-9

Solution   Though you'd eventually want to output the final costs based on the eighteen various combinations available (e.g., siding SA with marble MB and roofing thickness TB), the general computations you'd have to consider would be as shown in Figure 1-10.

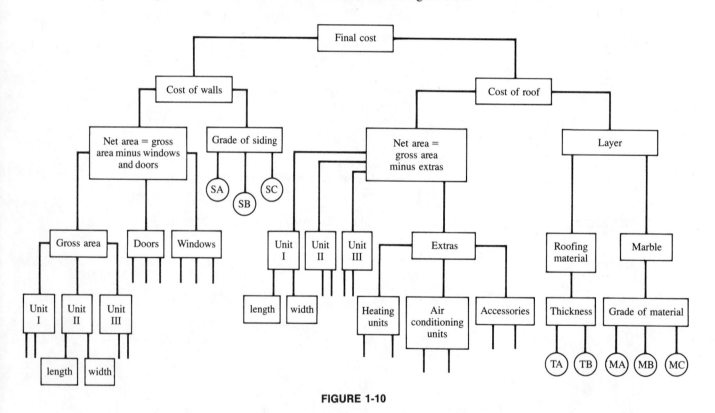

FIGURE 1-10

Supplementary Exercises

PROBLEM 1-18   Create an algorithm to boil an egg.

PROBLEM 1-19   Design an algorithm to find the value of the hypotenuse of a right triangle given the lengths of the legs. (*Hint:* Use SQR(c) to denote the square root of c.)

PROBLEM 1-20 Design an algorithm to test if a given year is a leap year.

PROBLEM 1-21 Design an algorithm to accept a measurement in feet and inches and produce that measurement in centimeters.

PROBLEM 1-22 Your favorite jewelry store has a sale on diamonds. You will get a 30% discount off the list price but must pay 13% tax on the discounted price. Write an algorithm to calculate your final price for any given list price.

PROBLEM 1-23 Numbers like 4, 9, 16, 25, ... are called *perfect squares*. Certain numbers, when added and subtracted, give a perfect square, for example $6 + 10 = 16$ (a perfect square); $10 - 6 = 4$ (a perfect square). Create an algorithm to find all pairs of numbers less than 100 that give a perfect square when added and when subtracted. (There are 31 such pairs.)

PROBLEM 1-24 Write an algorithm to convert a given temperature in Celsius to the corresponding temperature in Fahrenheit.

PROBLEM 1-25 The postal service will allow a rectangular box to be sent first-class if its length-plus-girth measurement is less than 100 inches. The length is the box's longest dimension; the girth is the distance around it, measured perpendicular to the length. Write an algorithm to accept the three unidentified dimensions of a box, to output its length-plus-girth measurement, and to denote whether or not the box may be mailed first-class. (*Hint:* Use the SWAP algorithm, Problem 1-7.)

PROBLEM 1-26 A circular pool has diameter D and depth h. It is to be filled to the top through a hose that pumps v gallons of water per minute. Write an algorithm to calculate how many minutes it will take to fill a pool that has arbitrary given dimensions. (*Hint:* One cubic foot = 7.48 gallons.)

PROBLEM 1-27 The number 153 is an interesting number because $153 = 1^3 + 5^3 + 3^3$. Any N-digit number is called an *Armstrong number* if the sum of the Nth power of its digits is equal to the original number. Write an algorithm to find other three-digit Armstrong numbers. (If you have a computer available, write the program to find the others: 370, 371, 407.)

PROBLEM 1-28 A salesperson earns a 7% monthly commission on her first $1500 of sales income, a 10% commission on sales income over $1500 but less than or equal to $5000, and a 15% commission on sales income above $5000 (e.g., $6000 total sales = $1500 at 7%, $3500 at 10%, and $1000 at 15%,). Write an algorithm that will take the total sales amount for a month and compute the salesperson's commission and the net income the company receives after the commission has been deducted.

PROBLEM 1-29 The factorial of an integer N greater than zero is defined as the product of all the integers from 1 through N. Without using recursion, write an algorithm to calculate the factorial of a given integer.

PROBLEM 1-30 A lad and his dad were bargaining to determine the boy's pay for doing yard work. Rather than accept his dad's offer of $20 for two weeks of work, the son proposed the following: Pay me 1¢ for the first day's work, 2¢ for the second, 4¢ for the third, etc., doubling the pay each day for the 14-day period. The father agreed. Write an algorithm to calculate the boy's total pay. Who made the better deal?

PROBLEM 1-31 Write an algorithm that will take the input coefficients A, B, C of the quadratic equation $Ax^2 + Bx + C = 0$ and output the real solutions, if any.

PROBLEM 1-32 The four-digit number 3025 is special. The sum of the first two digits (30) and the last two digits (25) is 55. If you square 55, the original number is obtained ($55^2 = 3025$). Create an algorithm to find all other four-digit numbers that have this property. (If you have a computer available, create the program to find the others: 2025, 9801.)

PROBLEM 1-33 Suppose we are given a long list of n numbers already in computer memory. Write an algorithm to find the sum of the squares of these numbers.

PROBLEM 1-34   Write the algorithm for computing the value of x^{-n} for any nonzero value x and any negative integer n.

PROBLEM 1-35   Suppose you draw N straight lines on a sheet of paper so that every pair of lines intersect but no three lines intersect at one point. Find the recurrence formula to calculate the number of regions into which the plane is divided by these lines and create the corresponding algorithms.

Answers to Supplementary Exercises

1-18   Answers will vary according to personal procedure and degree of detail. Be sure to be ordered and complete.

Begin
1. Fill small kettle with water.
2. Place raw egg—still in shell—in kettle.
3. Place kettle on stove burner.
4. Turn on stove to medium heat.
5. When water boils, cook egg for 5 minutes.
6. Remove kettle from heat.
7. Turn off stove.
8. Pour off water.
9. Let egg stand until cool.
End

1-19   Begin
1. Input a, b
2. $c \leftarrow a \uparrow 2 + b \uparrow 2$
3. $H \leftarrow$ SQR(c)
4. Output H
End

1-20   Leap year occurs in every year that is divisible by 4 except the years that mark the even hundreds; if a century year is divisible by 400, however, it is also a leap year.

Begin
1. Input Y
2. IF $100 \mid Y$ and $400 \nmid Y$ THEN
   a. Output "NOT A LEAP YEAR"
3. IF $4 \mid Y$ THEN
   a. Output "LEAP YEAR"
4. ELSE
   a. Output "NOT A LEAP YEAR"
End

1-21   Begin
1. Input F, I
2. $IN \leftarrow F * 12 + I$
3. $CM \leftarrow 2.54 * IN$
4. Output CM
End

1-22   Begin
1. Input L
2. $P \leftarrow 0.30L * 1.13$
3. Output P
End

1-23   Begin
1. FOR $N = 1$ TO 100 DO
   a. FOR $P = N + 1$ TO 100 DO
      1. IF SQR($N + P$) = INT(SQR $(N + P)$) and IF SQR($P - N$) = INT(SQR($P - N$)) THEN
         a. Output $N, P, N + P, P - N$
End

1-24   Begin
1. Input C
2. $F \leftarrow (9/5) * C + 32$
3. Output F
End

1-25   Begin
1. Input A, B, C
2. IF $A > B$ THEN
   a. SWAP(A, B)
3. IF $B > C$ THEN
   a. SWAP(B, C)
4. IF $A > B$ THEN
   a. SWAP(A, B)
5. $L \leftarrow C, W \leftarrow B, H \leftarrow A$
6. *Measurement* $\leftarrow L + 2W + 2H$
7. Output *Measurement*
8. IF *Measurement* < 100 THEN
   a. Output "OK"
9. ELSE
   a. Output "NOT OK"
End

1-26   Begin
1. Input D, h, v
2. *Volume* $\leftarrow \pi * (D/2) \uparrow 2 * h$
3. *Gallons* $\leftarrow$ *Volume* $* 7.48$
4. *Minutes* $\leftarrow$ *Gallons*$/v$
5. Output *Minutes*
End

1-27   Begin
1. FOR $N = 100$ TO 999 DO
   a. $A \leftarrow$ INT($N/100$)
   b. $B \leftarrow$ INT($N/10$) $- 10 * A$
   c. $C \leftarrow N - 100 * A - 10 * B$
   d. IF $N \neq A \uparrow 3 + B \uparrow 3 + C \uparrow 3$ THEN
      1. Next N
   e. ELSE
      1. Output N
End

1-28 Begin
   1. Input S
   2. IF $S \leq 1500$ THEN
     a. $C = 0.07 * S$
   3. IF $S \geq 5000$ THEN
     a. $C = 0.07 * 1500 + 0.10 * 3500$
       $+ 0.15 * S - 5000$
   4. ELSE
     a. $C = 0.07 * 1500 + 0.10 * S - 1500$
   5. $I = S - C$
   6. Output C, I
   End

1-29 Begin
   1. Input N
   2. $fac \leftarrow 1$
   3. FOR $i = 1$ TO N DO
     $fac \leftarrow fac * i$
   4. Output fac
   End

1-30 Begin
   1. $pay \leftarrow 1$
   2. FOR $day = 1$ TO 13 DO
     a. $pay = (pay + 2\uparrow day)/100$
   3. Output pay
   4. IF $pay > 20$ THEN
     a. Output "GOOD DEAL!"
   End
   Actual pay is $163.83

1-31 Begin
   1. Input A, B, C
   2. $D \leftarrow B\uparrow 2 - 4 * A * C$
   3. IF $D > 0$ THEN
     a. $R1 \leftarrow (-B + SQR(D))/2 * A$
     b. $R2 \leftarrow (-B - SQR(D))/2 * A$
     c. Output $R1, R2$
   4. ELSE
     a. IF $D = 0$ THEN
       1. $R \leftarrow -B/2 * A$
       2. Output R
     b. ELSE
       1. Output "NO REAL SOLUTIONS"
   End

1-32 Begin
   1. FOR $N = 1000$ TO 9999 DO
     a. $F \leftarrow INT(N/100)$
     b. $L \leftarrow N - 100 * F$
     c. IF $(F + L)\uparrow 2 \neq N$ THEN
       1. Next N
     d. ELSE
       1. Output $N, F + L, (F + L)\uparrow 2$
   2. Next N
   End

1-33 Begin
   1. Input n
   2. $sum \leftarrow 0$
   3. FOR $i = 1, 2, \ldots, n$
     a. $sum \leftarrow sum + x(i) * x(i)$
   4. Output sum
   End

1-34 Begin
   1. Input x, n
   2. $x \leftarrow 1/x$
   3. $n \leftarrow -n$
   4. $power \leftarrow 1$
   5. FOR $j = 1$ TO n DO
     a. $power = power * x$
   6. Output $power$
   End

1-35 $R_n = R_{n-1} + n$; let the value calculated recursively be designated as $regions(n)$.

   Begin
   1. Input n
   2. IF $n = 3$ THEN
     a. $regions(n) \leftarrow 7$
   3. $regions(n) \leftarrow regions(n - 1) + n$
   End

note: This works in the following way: let $n = 6$.

$$
\begin{aligned}
regions(6) &= regions(5) + 6 \\
&= (regions(4) + 5) + 6 \\
&= ((regions(3) + 4) + 5) + 6 \\
&= ((7 + 4) + 5) + 6 \\
&= (11 + 5) + 6 \\
&= 16 + 6 \\
&= 22
\end{aligned}
$$

2 NUMBERS AND BASES

THIS CHAPTER IS ABOUT

☑ **Number Classification**
☑ **Decimal Representation**
☑ **Binary Representation**
☑ **Octal and Hexadecimal Representation**
☑ **Complements**
☑ **Modular Arithmetic**

2-1. Number Classification

The basic building block of all mathematics is the *number*. As with many other entities in this world, numbers come in many varieties. First, there's the set of numbers we've learned from infancy, the set of **natural numbers**:

NATURAL NUMBERS $$\mathbb{N} = \{1, 2, 3, 4, 5, \ldots\}$$

This set allows us to count objects such as fingers and toes and to compare collections of objects like apples and oranges. This simple set of building blocks, however, is too limited to use by itself in mathematics, for we can't solve something as simple as $6 + x = 3$ with only the natural numbers. For this, we need the option of negative numbers, so we rely on the extended set of numbers called the **integers**:

INTEGERS $$\mathbb{Z} = \{\ldots, -4, -3, -2, -1, 0, 1, 2, 3, 4, \ldots\}$$

With the new set of negative numbers, zero, and positive numbers, we can perform all manner of addition, subtraction, and multiplication and still come up with a member of the integer set for an answer. Until, that is, we stumble upon something like $3y = 7$, for which no integer solves the equation.

Undaunted, we cross another mathematical threshold and meet up with fractions and decimals in the world of the **rational numbers**:

RATIONAL NUMBERS $$\mathbb{Q} = \left\{ \frac{p}{q} \,\middle|\, p \text{ and } q \text{ are integers}, q \neq 0 \right\}$$

note: Read the vertical bar as "such that." The whole phrase above reads as: "Q is the set of numbers p divided by q such that p and q are integers and q is not equal to 0."

The expanded set of rational numbers lets us turn the equation $ay = b$ into $y = b/a$ so that we can solve it for *any* integers a, b, and y (with $a \neq 0$). In the set of rationals, we find integers, fractions, terminating decimals ($3/4 = 0.75$), and repeating decimals ($1/3 = 0.\overline{3}$).

note: The overbar on a repeating decimal indicates the portion that repeats. Thus $2.57\overline{676}$ means that 76 repeats infinitely.

Because the rationals can fulfill almost any mathematical computation, they are used extensively in measurement. For instance, with rationals we can determine the length of our fingers and toes and the height or weight of apples or oranges. Until, that is, science comes in to spoil our fun by demanding *exact* measurements: the exact length of the diagonal of a square of unit length; the exact length of the invisible hypotenuse formed between the tip of your thumb and the tip of your index finger when you create a right angle with them; the exact ratio of the circumference of an orange to its diameter.

So, obligingly, we take the quantum leap from the confining world of rational thinking and numbers to the boundless realm of the infinitely continuing decimal, that of the **irrational numbers**:

IRRATIONAL NUMBERS $\qquad\qquad \mathbb{I} = \{x \mid x \text{ is not rational}\}$

The set of irrational numbers is where we put such values as $\sqrt{2}$ and π.

So, from the humble beginnings of the natural numbers, we have four basic sets of numbers at our fingertips. Given these sets, we find that *any* number can be categorized as either:

- rational (terminating *or* repeating), or
- irrational (nonterminating *and* nonrepeating)

To make it easier to talk about all numbers as a group, mathematicians lump all the rationals and irrationals together into a fifth, single collection called the set of **real numbers**:

REAL NUMBERS $\qquad\qquad \mathbb{R} = \mathbb{Q} \cup \mathbb{I}$

The real numbers $\mathbb{R}$ consist of all the points on a continuous number line.

Okay. Now that we've traced down the path of the various types of number sets, let's take a look at how they're all related. As for Rome, all roads lead to Real:

$$\mathbb{N} \subset \mathbb{Z} \subset \mathbb{Q} \subset \mathbb{R}$$
$$\mathbb{I} \subset \mathbb{R}$$
$$\mathbb{Q} \cup \mathbb{I} = \mathbb{R}$$

EXAMPLE 2-1 Given the set $\{-4/5, 9, \sqrt{3}, 3.4, \pi/2, -5.\overline{3}, 0, \sqrt{2}, -8\}$, find its subset of **(a)** natural numbers, **(b)** integers, **(c)** rational numbers, and **(d)** irrational numbers.

Solution

(a) $\{9\}$ **(b)** $\{-8, 0, 2, 9\}$ **(c)** $\{-8, -5.\overline{3}, -4/5, 0, 2, 3.4, 9\}$ **(d)** $\{\sqrt{2}, \pi/2, \sqrt{3}\}$

2-2. Decimal Representation

You are probably so thoroughly familiar with the idea of decimal notation that you give little or no thought to what it really means: that something like 356 stands for 3 hundreds, 5 tens, and 6 units. We can rewrite this as:

$$356 = 3(10^2) + 5(10^1) + 6(10^0) \qquad\qquad \textbf{(2-1)}$$

In equation (2-1), 356 is said to be written in **expanded decimal notation**, that is, as the sum of the digits multiplied by the powers of 10. In an integer, the position of a digit, counting from the right, determines its *power value*; that is, it determines the exponent for the base number 10.

In the decimal—or base 10—system, we are limited to the ten digits 0, 1, 2, 3, 4, 5, 6, 7, 8, 9. This system is, however, not the only numbering system available. For instance, computers use the binary system (base 2), the octal system (base 8), or the hexadecimal system (base 16); you even work daily on a base 60 system every time you use a clock or a stopwatch (60 minutes to the hour, 60 seconds to the minute).

Understanding how our decimal system works is the key to understanding any other number system, so let's go through a quick review of the decimal system before we move on to those other bases that computers use.

EXAMPLE 2-2 Write 2074 in expanded decimal notation.

Solution The decimal number 2074 means that we have 2 thousands, 0 hundreds, 7 tens, and 4 units.

Thus we *expand* the integer to be a sum of its component parts:

$$2074 = 2(1000) + 0(100) + 7(10) + 4(1)$$

and then express this as a sum of the powers of 10:

$$2074 = 2(10^3) + 0(10^2) + 7(10^1) + 4(10^0)$$
$$= 2(10^3) + 7(10^1) + 4(10^0) \longleftarrow \text{expanded decimal notation}$$

note: The *decimal point* is a mark that puts the magnitudes larger than or equal to zero to the left of the mark and the magnitudes smaller than zero to the right of the mark.

EXAMPLE 2-3   Write the number 21.78 in its expanded notation.

Solution   The decimal number 21.78 has 2 tens, 1 unit, 7 tenths, and 8 hundredths and the decimal point is to the right of the units digit. So we write:

$$21.78 = 2(10) + 1(1) + 7(\tfrac{1}{10}) + 8(\tfrac{1}{100})$$
$$= 2(10^1) + 1(10^0) + 7(10^{-1}) + 8(10^{-2})$$

To write a fraction in expanded notation, you must first convert it to a decimal.

EXAMPLE 2-4   Write the number $\frac{3}{8}$ in expanded decimal notation.

Solution   First plug $\frac{3}{8}$ into your handy-dandy calculator (or use the old-fashioned method of pencil division if you're a purist), and you get:

$$\frac{3}{8} = 0.375$$
$$= 0 \text{ units} + 3 \text{ tenths} + 7 \text{ hundredths} + 5 \text{ thousandths}$$
$$= 0(1) + 3(\tfrac{1}{10}) + 7(\tfrac{1}{100}) + 5(\tfrac{1}{1000})$$
$$= 3(10^{-1}) + 7(10^{-2}) + 5(10^{-3})$$

Now that we've spread them all out, it's time to pack up your decimal numbers for a while, because now we'll consider other bases.

2-3. Binary Representation

It is generally accepted that base 10 arithmetic was first established because most people come equipped with ten fingers (digits) and that they originally counted up to ten then started over. Unlike people, however, computers are not blessed with phalanges; instead have only a simple two-choice, on/off approach of electronic circuits, so researchers and computer architects took up the *binary number system* ("bi" = two) for computers. The main reasons for having binary in computers are that:

(1) Storage devices inside a computer are usually activated by a two-choice system: yes or no, on or off.
(2) Two-choice devices are inherently more reliable than devices with more choices.
(3) Binary arithmetic is simple and quick and doesn't require complicated circuitry.

In a *binary representation* of numbers—that is, in **base 2**—we are limited to two digits, 0 and 1, to use in representing any number.

Because the binary system is based upon the powers of two instead of upon the powers of ten, a **number in base 2** is a sequence of 0's and 1's, with or without an embedded **binary point** (like a decimal point, only for binary numbers). The integral part (to the left of the binary point) uses nonnegative powers of 2, while the fractional part (to the right of the binary point) uses negative powers of 2. The location of each binary digit (called a **bit**) to the right or left of the binary point determines its magnitude, or power of 2:

$$1011.011 = 1(2^3) + 0(2^2) + 1(2^1) + 1(2^0) + 0(2^{-1}) + 1(2^{-2}) + 1(2^{-3}) \qquad \textbf{(2-2)}$$

$$11011 = 1(2^4) + 1(2^3) + 0(2^2) + 1(2^1) + 1(2^0) \qquad \textbf{(2-3)}$$

note: The binary numbers in eqs. (2-2) and (2-3) are said to be written in **expanded binary notation**.

Whenever you use a computer or calculator, you input data in the decimal system and get back an answer in the decimal system. Internally, however, the machine converts the information into another base—usually base 2—reconverts it when it's done, and displays the answer in decimal form. This conversion is a straightforward process.

A. Converting from binary to decimal

The binary (base 2) to decimal (base 10) conversion is a matter of simple multiplication and addition. First you expand the binary number, then convert the powers of two to their respective decimal equivalents, and then add everything together.

EXAMPLE 2-5   Find the decimal equivalent of the binary numbers (a) 1011.011 and (b) 10.101.

Solution

(a) $1011.011_2 = 1(2^3) + 0(2^2) + 1(2^1) + 1(2^0) + 0(2^{-1}) + 1(2^{-2}) + 1(2^{-3})$
$= 1(8) + 0(4) + 1(2) + 1(1) + 0(\frac{1}{2}) + 1(\frac{1}{4}) + 1(\frac{1}{8})$
$= 8 + 2 + 1 + 0.25 + 0.125$
$= 11.375_{10}$

(b) $10.101_2 = 1(2^1) + 0(2^0) + 1(2^{-1}) + 0(2^{-2}) + 1(2^{-3})$
$= 1(2) + 0(1) + 1(\frac{1}{2}) + 0(\frac{1}{4}) + 1(\frac{1}{8})$
$= 2 + 0.5 + 0.125$
$= 2.625_{10}$

---

agreement: To avoid confusion in the rest of the chapter, we identify the base of a given number by writing the base as a subscript notation to the right of the number. If there is no subscript base, it will be in base 10.

B. Converting from decimal to binary

There are two general methods for converting from decimal to binary: the *power-2 deduction method* and the *algorithm method*.

1. The power-two deduction method

In this method, you write the decimal number as the sum of the powers of 2 and then perform binary expansion backwards (instead of expanding, you condense). To do this, first make note of the positive powers of 2:

$$2^0 = 1 \qquad 2^4 = 16 \qquad 2^8 = 256$$
$$2^1 = 2 \qquad 2^5 = 32 \qquad 2^9 = 512$$
$$2^2 = 4 \qquad 2^6 = 64 \qquad 2^{10} = 1024$$
$$2^3 = 8 \qquad 2^7 = 128 \qquad \vdots$$

Then examine your decimal integer and subtract out the *largest* power of 2 that you can and record that power; take the remainder and subtract the *largest* power of 2 that you can from that and record that power. Continue down until you have a zero remainder; you will have written your decimal number as the sum of the powers of 2.

EXAMPLE 2-6   Convert 149_{10} to base 2 using the power-two deduction method.

Solution   The largest power of 2 in 149_{10} is $2^7 = 128$.

$$
\begin{array}{r}
149 \\
-128 \longrightarrow \text{Record } 2^7 \\
\hline
21
\end{array}
$$

$$
\begin{array}{r}
21 \\
-\ 16 \\
\hline
5
\end{array}
\longrightarrow \text{Record } 2^4 \qquad \text{(the largest power of 2 in remainder 21)}
$$

$$
\begin{array}{r}
-\ 4 \\
\hline
1
\end{array}
\longrightarrow \text{Record } 2^2 \qquad \text{(the largest power of 2 in remainder 5)}
$$

$$
\begin{array}{r}
-\ 1 \\
\hline
0
\end{array}
\longrightarrow \text{Record } 2^0 \qquad \text{(the largest power of 2 in remainder 1)}
$$

So now we have

$$149_{10} = 2^7 + 2^4 + 2^2 + 2^0 \tag{2-4}$$

Using a multiplier of zero, insert the missing powers of 2 and we get:

$$149_{10} = 1(2^7) + 0(2^6) + 0(2^5) + 1(2^4) + 0(2^3) + 1(2^2) + 0(2^1) + 1(2^0) \tag{2-5}$$
$$= 1 + 0 + 0 + 1 + 0 + 1 + 0 + 1$$
$$= 10010101_2$$

shortcut: You can skip the full expansion step of eq. (2-5) by using only eq. (2-4): Place a 1 for every power of 2 magnitude you *do* have in your sum, and place a 0 for every magnitude you *don't* have. Thus,

$$
\begin{array}{c}
149_{10} = 2^7 \qquad + \qquad 2^4 + 2^2 + 2^0 \\
\downarrow \qquad\qquad \downarrow \quad\ \downarrow \quad\ \downarrow \\
= 1 \quad\ 0 \quad\ 0 \quad\ 1 \quad 0 \quad 1 \quad 0 \quad 1 \\
\uparrow \quad\ \uparrow \quad\quad \uparrow \quad\ \uparrow \\
(2^6) \ (2^5) \quad (2^3) \ (2^1) \\
= 10010101_2
\end{array}
$$

EXAMPLE 2-7   Convert 509_{10} to base 2 using the power-two deduction method.

Solution   The largest power of 2 in 509_{10} is 256.

$$
\begin{array}{r}
509 \\
-256 \\
\hline
253
\end{array}
\longrightarrow \text{Record } 2^8
\qquad\qquad
\begin{array}{r}
29 \\
-\ 16 \\
\hline
13
\end{array}
\longrightarrow \text{Record } 2^4
$$

$$
\begin{array}{r}
-128 \\
\hline
125
\end{array}
\longrightarrow \text{Record } 2^7
\qquad\qquad
\begin{array}{r}
-\ 8 \\
\hline
5
\end{array}
\longrightarrow \text{Record } 2^3
$$

$$
\begin{array}{r}
-\ 64 \\
\hline
61
\end{array}
\longrightarrow \text{Record } 2^6
\qquad\qquad
\begin{array}{r}
-\ 4 \\
\hline
1
\end{array}
\longrightarrow \text{Record } 2^2
$$

$$
\begin{array}{r}
-\ 32 \\
\hline
29
\end{array}
\longrightarrow \text{Record } 2^5
\qquad\qquad
\begin{array}{r}
-\ 1 \\
\hline
0
\end{array}
\longrightarrow \text{Record } 2^0
$$

$$509_{10} = 2^8 + 2^7 + 2^6 + 2^5 + 2^4 + 2^3 + 2^2 + 2^0$$
$$= 1 \quad 1 \quad 1 \quad 1 \quad 1 \quad 1 \quad 1 \quad 0 \quad 1$$
$$= 111111101_2$$

---

2. The algorithm method

Mathematicians have created an algorithm to convert an integer from decimal to binary.

CONVERSION ALGORITHM—DECIMAL INTEGER TO BINARY

1. Divide the integer by 2 and record the remainder.
2. Divide the quotient by 2 and record the remainder.

3. Repeat step 2 until the final quotient is 0.

4. Write the recorded remainders in reverse order.

The remainders from this process will always be either a 0 or a 1 (since we are dividing by 2), and the sequence of these digits in *reverse order* is the correct binary representation.

EXAMPLE 2-8   Convert the number 115_{10} to binary form using the algorithm method. Check it.

Solution

$$\begin{array}{c} 57 \\ 2\overline{)115}, \quad R = 1 \end{array}$$

$$\begin{array}{c} 28 \\ 2\overline{)57}, \quad R = 1 \end{array}$$

$$\begin{array}{c} 14 \\ 2\overline{)28}, \quad R = 0 \end{array}$$

$$\begin{array}{c} 7 \\ 2\overline{)14}, \quad R = 0 \end{array} \qquad 1 \quad 1 \quad 1 \quad 0 \quad 0 \quad 1 \quad 1$$

$$\begin{array}{c} 3 \\ 2\overline{)7}, \quad R = 1 \end{array}$$

$$\begin{array}{c} 1 \\ 2\overline{)3}, \quad R = 1 \end{array}$$

$$\begin{array}{c} 0 \\ 2\overline{)1}, \quad R = 1 \end{array}$$

Thus $115_{10} = 1110011_2$.

Check:
$$\begin{aligned} 1110011_2 &= 1(2^6) + 1(2^5) + 1(2^4) + 0(2^3) + 0(2^2) + 1(2^1) + 1(2^0) \\ &= \ \ 64 \ + \ 32 \ + \ 16 \ + \ 0 \ + \ 0 \ + \ 2 \ + \ 1 \\ &= 115_{10} \end{aligned}$$

If the base 10 number is a decimal fraction, the algorithm changes to accommodate them:

CONVERSION ALGORITHM—DECIMAL FRACTION TO BINARY

1. Multiply the number by 2 and record the integral part of the result.

2. Using only the fractional part of the result, multiply by 2 and again record the integral part.

3. Repeat steps 1 and 2 until the fractional part is zero.

4. Write the integral parts in order of their appearance.

The sequence of the integral-part digits, *in calculated order*, is the binary representation.

EXAMPLE 2-9   Convert 0.828125_{10} to binary form using the algorithm method.

Solution

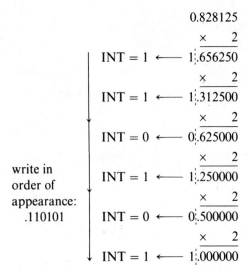

Thus $0.828125_{10} = 0.110101_2$

caution: Remember to add the binary point when converting a decimal fraction to binary.

If the decimal number you want to convert has both an integral part and a fractional part, make the conversions separately.

EXAMPLE 2-10   Convert 12.3125_{10} to binary form.

Solution $12.3125 = 12 + 0.3125$.

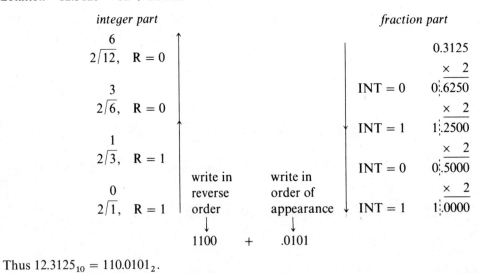

Thus $12.3125_{10} = 110.0101_2$.

In Examples 2-9 and 2-10, the binary representation of the decimal fractional part terminated. This will not always be the case.

EXAMPLE 2-11   Convert the decimal value 0.3 to binary.

Solution

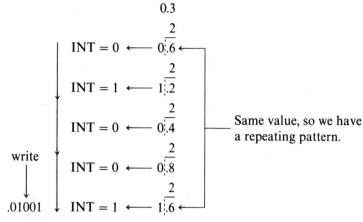

Therefore, $0.3_{10} = 0.0\overline{1001}_2$.

2-4. Octal and Hexadecimal Representation

Although binary numbers are suitable inside a computer, they often give people problems because of their length and unwieldy nature. For instance, could you copy down the 24-bit binary number 110010111001011001101101 without making a mistake? Or compare it with another 24-bit binary number "bit by bit"? (Or how would you like to make a check out for $101101001 instead of $361?)

Handling such long number strings is something like trying to pick up a dozen raw eggs in your hands without the benefit of an egg carton: you're sure to make a mess of it. To avoid the difficulty of juggling long binary numbers, computer programmers often use one of two alternate number systems: octal and hexadecimal. Both systems are closely related to binary, and conversion from one base to another is relatively simple.

A. The octal system

The **octal representation** of a number uses the base value 8 and the digits 0, 1, 2, 3, 4, 5, 6, 7. The place values in this system are the powers of 8:

$$8^0 = 1 \qquad 8^{-1} = \frac{1}{8}$$

$$8^1 = 8 \qquad 8^{-2} = \frac{1}{64}$$

$$8^2 = 64 \qquad 8^{-3} = \frac{1}{512}$$

$$8^3 = 512 \qquad \vdots$$

$$8^4 = 4096$$

$$\vdots$$

For example,

$$347.02_8 = 3(8^2) + 4(8^1) + 7(8^0) + 0(8^{-1}) + 2(8^{-2})$$

1. Converting from octal to decimal

To convert from octal to decimal you simply expand the octal number to be the sum of the powers of eight and then multiply it all out.

EXAMPLE 2-12   Convert 4126_8 to base 10.

Solution

$$4126_8 = 4(8^3) + 1(8^2) + 2(8^1) + 6(8^0)$$
$$= 4(512) + 1(64) + 2(8) + 6(1)$$
$$= 2134_{10}$$

EXAMPLE 2-13   Convert 35.14_8 to base 10.

Solution

$$35.14_8 = 3(8^1) + 5(8^0) + 1(8^{-1}) + 4(8^{-2})$$
$$= 3(8) + 5(1) + 1(\tfrac{1}{8}) + 4(\tfrac{1}{64})$$
$$= 24 + 5 + 0.125 + 0.0625$$
$$= 29.1875_{10}$$

2. Converting from decimal to octal

To convert from decimal to octal follows the same general pattern as the conversion from decimal to binary.

EXAMPLE 2-14   Convert 315_{10} to base 8.

Solution   We divide the base 10 value by 8 and record the remainder, then divide each new quotient by 8 and record its remainder, respectively, until we obtain a zero quotient. The remainders in *reverse order* give the octal representation.

$$\begin{array}{l} \phantom{8\overline{)}}39 \\ 8\overline{)315}, \quad R = 3 \\[2mm] \phantom{8\overline{)}}4 \\ 8\overline{)39}, \quad R = 7 \\[2mm] \phantom{8\overline{)}}0 \\ 8\overline{)4}, \quad R = 4 \end{array}$$

write in reverse order: 473

Thus $315_{10} = 473_8$.

EXAMPLE 2-15   Convert 0.34375_{10} to base 8.

Solution   Multiply the base 10 fractional part by 8 and record the integer part. Using only the fractional part, multiply by 8 again and record the integral part. Continue until the fractional part is zero. The integer parts *in the order obtained* are the digits of the octal representation.

$$0.34375$$

write in order of appearance: .26

$$\text{INT} = 2 \longleftarrow \quad 2.75000 \quad \times 8$$
$$\text{INT} = 6 \longleftarrow \quad 6.00000 \quad \times 8$$

Thus $0.34375_{10} = 0.26_8$.

caution: Always remember to place the *octal point* (like a decimal point, only for octal numbers) when converting a fraction.

As with decimal to binary conversion, if the decimal number you want to convert to octal has both an integral and a fraction part, you must do both parts of the conversion separately.

EXAMPLE 2-16   Convert 142.15625_{10} to base 8.

Solution   Work in two parts.

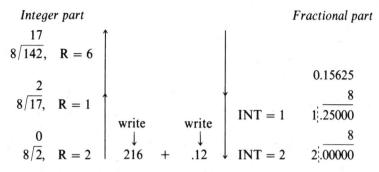

Thus $142.15625_{10} = 216.12_8$.

3. Converting from binary to octal

To convert from the binary to octal base is particularly simple because for each digit in octal—0, 1, 2, 3, 4, 5, 6, 7—there exists a corresponding binary "triple"—a 3-bit group—as shown in Table 2-1.

TABLE 2-1

octal digit	3-bit binary	octal digit	3-bit binary
0	000	4	100
1	001	5	101
2	010	6	110
3	011	7	111

If we wish to convert from binary to octal, we merely count off groups of three bits on our binary number, starting from the right, and substitute the corresponding octal digit for each binary triple.

EXAMPLE 2-17   Convert $11001011100101100110 1101_2$ to octal.

Solution   Using Table 2-1 we find that the binary number can be broken down as follows:

$$11001011100101100110 1101_2 = \underline{110}\ \underline{010}\ \underline{111}\ \underline{001}\ \underline{011}\ \underline{001}\ \underline{101}\ \underline{101} = 62713155_8$$
$$\quad\quad 6 \quad 2 \quad 7 \quad 1 \quad 3 \quad 1 \quad 5 \quad 5$$

note: Always start counting your 3-bit groups from the *right* side of the binary number. Add extra (meaningless) zeros to the left to come up with a complete group of three digits.

EXAMPLE 2-18   Convert 1101110_2 to octal.

Solution

add extra 0's to pad

$$1101110_2 = \overset{\downarrow}{\overline{001}}\ \underline{101}\ \underline{110} = 156_8$$
$$\quad\quad 1 \quad\ 5 \quad\ 6$$

4. Converting from octal to binary

To convert this way is as easy as to convert from binary to octal because you once again refer to Table 2-1, only this time you replace each octal digit with its corresponding binary triple.

EXAMPLE 2-19   Find the binary representation for 31426_8.

Solution

$$\begin{array}{ccccc} 3 & 1 & 4 & 2 & 6 \\ \end{array}$$

$$31426_8 = \overbrace{011} \ \overbrace{001} \ \overbrace{100} \ \overbrace{010} \ \overbrace{110} = 11001100010110_2$$

↑
drop this first 0
in the answer

B. The hexadecimal system

In the hexadecimal system of numbers, we are given 16 digits ("hexa" = 6 + "deci" = 10) to work with. The digits are 0, 1, 2, 3, 4, 5, 6, 7, 8, 9, A, B, C, D, E, F. The letters act as the "digits" 10, 11, 12, 13, 14, 15, respectively (much like in a deck of cards, in which the Jack can be 11, the Queen 12, and the King 13).

In hexadecimal, the base is 16, so the hexadecimal number $D436_{16} = D(16^3) + 4(16^2) + 3(16^1) + 6(16^0)$ in expanded notation. You can see how much less space it takes to write a large number using the hexadecimal system.

1. Converting from hexadecimal to decimal

As with conversions from other bases, converting to decimal from hexadecimal involves expanding the number and multiplying out the figures.

EXAMPLE 2-20   Convert $B37_{16}$ to decimal.

Solution

$$\begin{aligned} B37_{16} &= B(16^2) + 3(16^1) + 7(16^0) \\ &= 11(256) + 3(16) + 7(1) \\ &= 2816 + 48 + 7 \\ &= 2871_{10} \end{aligned}$$

---

2. Converting from decimal to hexadecimal

To convert from decimal to hexadecimal, you need to divide the base 10 value by 16 and record the remainder, then divide each new quotient by 16 and record its remainder, respectively, until you obtain a zero quotient. The remainders in reverse order give the hexadecimal representation.

EXAMPLE 2-21   Convert 421_{10} to hexadecimal.

Solution   Keep dividing the number by 16 and record the remainder until you have a zero quotient. Be sure to use the digit and letter representations.

$$
\begin{array}{r}
26 \\
16\overline{)421} \\
32 \\
\hline
101 \\
96 \\
\hline
5
\end{array}
\longrightarrow R = 5
$$

$$
\begin{array}{r}
1 \\
16\overline{)26} \\
16 \\
\hline
10
\end{array}
\longrightarrow R = A
$$

$$
\begin{array}{r}
0 \\
16\overline{)1} \\
0 \\
\hline
1
\end{array}
\longrightarrow R = 1
$$

write in reverse order: 1A5

Thus $421_{10} = 1A5_{16}$.

3. Converting from binary to hexadecimal

Like converting from binary to octal and back, moving between hexadecimal and binary is a simple procedure. This time, we mark off the binary digits into groups of four that correspond to each digit in the hexadecimal system, as Table 2-2 illustrates.

TABLE 2-2

hexadecimal digit	4-bit binary	hexadecimal digit	4-bit binary
0	0000	8	1000
1	0001	9	1001
2	0010	A	1010
3	0011	B	1011
4	0100	C	1100
5	0101	D	1101
6	0110	E	1110
7	0111	F	1111

EXAMPLE 2-22   Using the same binary number as in Example 2-17, convert it to hexadecimal using Table 2-2.

Solution   Start from the right and create 4-bit groups;

$$
110010111001011001101101_2 = \underline{1100}\ \underline{1011}\ \underline{1001}\ \underline{0110}\ \underline{0110}\ \underline{1101} = CB966D_{16}
$$
$$
\ \ C\ \ \ \ \ B\ \ \ \ \ 9\ \ \ \ \ \ 6\ \ \ \ \ \ 6\ \ \ \ \ \ D
$$

As with binary to octal conversion, it may be necessary to pad with zeros to the left to come up with a complete 4-bit group. Always start counting the 4-bit groups from the *right* of the number.

EXAMPLE 2-23   Convert the 18-bit binary number 101101010000110110 to hexadecimal representation.

Solution

$$101101010000110110_2 = \underbrace{0010}_{2} \quad \underbrace{1101}_{D} \quad \underbrace{0100}_{4} \quad \underbrace{0011}_{3} \quad \underbrace{0110}_{6} = 2D436_{16}$$

4. Converting from hexadecimal to binary

Once again use Table 2.2, only this time substitute the corresponding binary 4-bit group for each hexadecimal digit.

EXAMPLE 2-24   Convert $A2E47_{16}$ to binary.

Solution

$$A2E47_{16} = \overset{A}{\overbrace{1010}} \quad \overset{2}{\overbrace{0010}} \quad \overset{E}{\overbrace{1110}} \quad \overset{4}{\overbrace{0100}} \quad \overset{7}{\overbrace{0111}} = 10100010111001000111_2$$

Remember to drop any extra zeros to the left of the new binary number.

EXAMPLE 2-25   Convert $1D7_{16}$ to binary.

Solution

$$1D7_{16} = \overset{1}{\overbrace{0001}} \quad \overset{D}{\overbrace{1101}} \quad \overset{7}{\overbrace{0111}} = 111010111_2$$

drop these extra 0's
in final answer

2-5. Complements

A. Binary arithmetic and complements

We all know how to add and subtract in decimals. Binaries are a tad trickier, however, until you completely let go of the secure decimal patterns you're familiar with and wholeheartedly grasp the basic rules of **binary addition and subtraction**:

BINARY ADDITION

$$0 + 0 = 0$$
$$1 + 0 = 1$$
$$1 + 1 = 0, \text{ carry } 1$$
$$1 + 1 + 1 = 1, \text{ carry } 1$$
$$1 + 1 + 1 + 1 = 0, \text{ carry two 1's}$$

Examples:

$$
\begin{array}{cccc}
 & & \overset{\scriptscriptstyle 1}{} & \\
 & & \overset{\scriptscriptstyle 1\,1}{10} & \\
\overset{\scriptscriptstyle 1}{10} & 10 & 10 & \\
+\ 10 & +\ 10 & 10 & \\
\hline
100 & \overline{110} & +\ \underline{10} & \\
 & & 1000 &
\end{array}
$$

$$
\begin{array}{cc}
\overset{\scriptscriptstyle 1}{101} & \overset{\scriptscriptstyle 1\,1\,1}{111} \\
+\ 100 & 100 \\
\hline
1001 & +\ \underline{11} \\
 & 1110
\end{array}
$$

**BINARY
SUBTRACTION**

$$0 - 0 = 0$$
$$1 - 0 = 1$$
$$1 - 1 = 0$$
$$10 - 1 = 1 \text{ (by "borrowing")}$$

Examples:

$$
\begin{array}{r}
{}^{0\ 10}\\
\cancel{1}\cancel{0}1\\
-\ \ 1\ 0\\
\hline
1\ 1
\end{array}
\qquad
\begin{array}{r}
1\ 1\ 1\\
-\ \ 1\ 1\\
\hline
1\ 0\ 0
\end{array}
\qquad
\begin{array}{r}
{}^{1}_{0\ \cancel{1}0\ 10}\\
\cancel{1}\cancel{0}\ \cancel{0}\\
-\ \ 1\ 1\\
\hline
1
\end{array}
\quad
\left(
\begin{array}{r}
Check:\\
{}^{1\ 1}\\
1\ 1\\
+\ \ \ 1\\
\hline
1\ 0\ 0
\end{array}
\right)
\qquad
\begin{array}{r}
1\ 1\ 1\\
{}^{0\ \cancel{1}0\ \cancel{1}0\ \cancel{1}0\ 10}\\
\cancel{1}\ \cancel{0}\ \cancel{0}\ \cancel{0}\ \cancel{0}\\
-\ \ \ \ 1\ 1\ 0\ 1\\
\hline
1\ 1
\end{array}
\quad
\left(
\begin{array}{r}
Check:\\
{}^{1\ 1\ 1\ 1}\\
1\ 1\ 0\ 1\\
+\ \ \ \ \ 1\ 1\\
\hline
1\ 0\ 0\ 0\ 0
\end{array}
\right)
$$

And just for good measure, we throw in the following fact, also related to binary arithmetic:

$$\text{base } 2^p = \underbrace{1000\cdots0}_{p \text{ zeros}}$$

Now that we've done those mental gymnastics, put binary arithmetic into a pocket until we need it again, and let's talk computers and complements for a moment.

You may have already discovered that computers, as smart as they are, are relatively stupid machines. They usually can't count above 2, and technically they can't subtract; they can only add. To perform a subtraction problem, a computer must transform it into an addition problem by something mathematicians call a **complement**. Given a real number A in base r and with p digits in its integral part, we have

COMPLEMENT
$$A^{**} = r^p - A \qquad\qquad (2\text{-}6)$$

EXAMPLE 2-26   Find the complement of 345_{10}.

Solution   In base 10 we have $r = 10$ and $p = 3$. So

$$A^{**} = r^p - A$$
$$= 10^3 - 345 = 1000 - 345$$
$$= 655_{10}$$

EXAMPLE 2-27   Find the complement of 21.73_{10}.

Solution   Again $r = 10$ but now $p = 2$. So

$$A^{**} = 10^2 - 21.73 = 100 - 21.73$$
$$= 78.27_{10}$$

Now, reach into your pocket and pull out that binary arithmetic you put there, because it's time to find the complement of a binary number.

EXAMPLE 2-28   Find the complement of (a) 1101_2 and (b) 10101_2.

Solution

(a) Base $r = 2$, so here the complement is $2^p - A$. With $p = 4$ we have

$$A^{**} = 2^4 - 1101_2$$
$$= 10000_2 - 1101_2 \longrightarrow
\begin{array}{r}
1\ 1\ 1\\
{}^{0\ \cancel{1}0\ \cancel{1}0\ \cancel{1}0\ 10}\\
\cancel{1}\ \cancel{0}\ \cancel{0}\ \cancel{0}\ \cancel{0}\\
-\ \ \ \ 1\ 1\ 0\ 1\\
\hline
0\ 0\ 1\ 1
\end{array}$$
$$= 0011_2 \longleftarrow$$

Check: Using base 10 arithmetic: $2^4 - 1101_2 = 16_{10} - 13_{10} = 3_{10} = 0011_2$.

(b) We have $r = 2$ and $p = 5$, so

$$A^{**} = 2^5 - 10101_2$$
$$= 100000_2 - 10101_2 \longrightarrow$$
$$= 01011_2 \longleftarrow$$

$$\begin{array}{c} \overset{1}{1}\,\overset{1}{0}\,\overset{1}{0}\,\overset{1}{0}\,\overset{1}{0}\,0 \\ 1\,0\,0\,0\,0\,0 \\ -\quad 1\,0\,1\,0\,1 \\ \hline 0\,1\,0\,1\,1 \end{array}$$

shortcut: Carefully examining the results of Example 2-28 will reveal the following fact: To get the complement of a *binary number*, we need only to change the original number A as follows:

1. Replace its 0's with 1's.
2. Replace its 1's with 0's.
3. Add 1 to the result.

EXAMPLE 2-29 Obtain the complement of 110_2 **(a)** by the definition (eq. 2-6) and **(b)** by the shortcut.

Solution

(a) By the definition of complement we plug in $p = 3$ and $r = 2$:

$$A^{**} = 2^3 - 110_2$$
$$= 1000_2 - 110_2 \longrightarrow$$
$$= 0010_2 \longleftarrow$$

$$\begin{array}{c} \overset{1}{0}\,\overset{}{0}\,0 \\ 1\,0\,0\,0 \\ -\quad 1\,1\,0 \\ \hline 0\,0\,1\,0 \end{array}$$

note: Drop the extra zeros preceding the number:

$$A^{**} = 10_2$$

(b) Using the shortcut we swap the 0's and 1's of 110_2 and get 001_2. We then add 1 to obtain:

$$A^{**} = 001_2 + 1_2 = 10_2$$

EXAMPLE 2-30 Use the shortcut to find the complement of 10111_2.

Solution Swapping the 0's and 1's, we get 01000_2; then we add 1 to obtain:

$$A^{**} = 01000_2 + 1_2 = 1001_2$$

Check: Using decimal conversion we find our original number $10111_2 = 23_{10}$. The five-digit number (in base 2) makes $p = 5$, so we are in 2^5 complementation. But $2^5 = 32_{10}$, and the complement of $23_{10} = 32_{10} - 23_{10} = 9_{10}$ which, converted back to base 2, is 1001_2.

B. Applications of complements

Recall that in the previous section we said that computers can't subtract. Instead, whenever a minus sign appears on an arithmetic application, the computer uses complementation to turn the subtraction problem into an equivalent addition problem. For example, for $A = 514$ and $B = 346$, we have $A - B$ (watch the signs):

$$\begin{array}{c} \text{complement of 346:} \\ B^{**} = 10^3 - B \\ \overbrace{} \\ 514 - 346 = 514 + \underbrace{(1000 - 346)}_{\substack{\text{complement} \quad -346 \\ \downarrow}} - 1000 \end{array}$$

$$= 514 + 654 - 1000$$
$$= 1168 - 1000$$
$$= 168$$

(*note:* This may look like "subtraction," but it is actually just a "digit drop" in base 10 arithmetic.)

Check: $514 - 346 = 168$

Generically, the above procedure converts to:

$$\text{complement of } B$$
$$\downarrow$$
$$A - B = A + \overbrace{(r^p - B)}^{} - r^p$$
$$\underbrace{}_{-B}$$

or

$$A - B = A + B^{**} - r^p \qquad\qquad (2\text{-}7)$$

Eq. (2-7) will help us to evaluate any subtraction $A - B$ where $A > B$.

EXAMPLE 2-31   Use complementation to find $46_{10} - 19_{10}$ by writing it as an addition problem.

Solution   According to eq. (2-7) we must find the complement B^{**} of B. Referring to eq. (2-6) we identify $r = 10$ since we are base 10; $p = 2$ because B is a two-digit number. The difference of 46 and 19 will be the sum of 46 and the complement of 19, minus 10^2.

$$\begin{aligned}
46_{10} - 19_{10} &= 46 + (10^2 - 19) - 10^2 \\
&= 46 + 81 - 100 \\
&= 127 - 100 \\
&= 27_{10}
\end{aligned} \qquad\qquad (2\text{-}8)$$

---

note: To get around having the computer make the subtraction in eq. (2-8), we simply tell it to examine the number 127 and to *ignore* (that is, drop) the initial 1 (which is in the 10^2, or r^p, column). This maneuver essentially subtracts 100 (r^p) from the value (per the requirements of eq. 2-7) and gives us 27, the correct answer for Example 2-31. Dropping this 1 is an essential part of using complements for subtraction in computers.

Eq. (2-7) and the manipulations of Example 2-31 demonstrate a general algorithm for subtraction using complementation. For $A - B \, (A > B)$,

SUBTRACTION ALGORITHM

1. If A has more digits than B, add zeros before B to even them up.
2. Compute the complement of B (count any extra zeros you've added as part of the value of p).
3. Add the complement of B to A.
4. Delete the digit 1 in the r^p-th column of the sum.

EXAMPLE 2-32   Use the subtraction algorithm to perform $523_{10} - 79_{10}$.

Solution   We have $A = 523$ and $B = 79$. B has fewer digits than A, so we add a meaningless zero to make $B = 079$. Now, with $r = 10$ and $p = 3$, we solve for B^{**}:

$$B^{**} = 1000 - 079 = 921$$

Now add B^{**} to A and delete the first 1.

$$\begin{aligned}
A + B^{**} &= 523 + 921 \\
&= 1444 \\
&= 444_{10}
\end{aligned}$$

---

note: Here's why we need to delete that extra 1 (the value of r^p) in Example 2-32: Say we want to perform $A - B$. The complement of $B = B^{**} = r^p - B$. According to the algorithm, we must add B^{**} to A, which gives us:

$$A - B = A + B^{**}$$
$$= A + (r^p - B)$$
$$= (A - B) + r^p \qquad \text{(2-9)}$$

Unfortunately, eq. (2-9) is *false* because:

$$A - B \neq (A - B) + r^p$$

For eq. (2-9) to be true, we will have to deduct the value of r^p from the right-hand side. Therefore we have

$$A - B = A + B^{**} - r^p$$
$$= A + (r^p - B) - r^p$$
$$= (A - B) + r^p - r^p$$
$$= (A - B) \qquad \text{(2-10)}$$

which is true. (Notice that eqs. (2-10) and (2-7) are the same, although we derived them from different procedures.)

But, Aha!, you say, using the subtraction algorithm still means that the computer must *subtract* to perform the complement function, so this silly backwards method doesn't really turn a subtraction problem into an addition problem.

Wrong. This is where the computer outsmarts us. By converting to the binary number system, the computer does not have to subtract to get a complement (remember, it has only to swap 1's and 0's and *add* 1). In addition, once we tell the computer to add zeros to make the numbers the same bit length, it *always* provides its own extra 1 in the r^p-th place when it carries out the subtraction algorithm.

EXAMPLE 2-33   Use the algorithm method to calculate $10110_2 - 1101_2$.

Solution   First add zeros to the front of B so that the two numbers have an equal number of bits. Then perform the subtraction algorithm.

Check: Converting to base 10, we have: $10110_2 - 1101_2 = 22_{10} - 13_{10} = 9_{10} = 1001_2$.
We have just successfully performed subtraction by using only addition.

EXAMPLE 2-34   Pretend you're a computer and subtract 14_{10} from 152_{10}.

Solution   A computer would take these numbers, convert them to binary (see Section 2-2), perform the subtraction via complementation, and then reconvert the answer to decimal.

$$A = 152_{10} = 10011000_2 \qquad B = 14_{10} = 1110_2$$

In the upcoming section you'll learn how to subtract using complementation when $A < B$.

C. Additional terms

In some textbooks you will find that the result of each step of the way toward a complement receives a separate name. For binary numbers we have:

- The number obtained by swapping the 0's and 1's in a binary number A is called the **one's-complement**. It is denoted A^*. The one's-complement of $101_2 = A^* = 010_2$.

- The number obtained by adding 1 to the one's-complement is called the **two's-complement**. It is denoted A^{**}.

  note: The two's-complement is the complement you've been finding all along for binary numbers and is the one the computer uses in subtraction.

EXAMPLE 2-35   Find the one's-complement and two's-complement of (a) 1011010, (b) 11011101, and (c) 111011111.

Solution

	(a)	**(b)**	**(c)**
Binary Number	1011010	11011101	111011111
A^* (one's-complement)	100101	100010	100000
A^{**} (two's-complement)	100110	100011	100001

For the decimal system, we have what are called the nine's-complement and the ten's-complement:

- The **nine's-complement** of a decimal number A is the number obtained by subtracting each digit of A from 9. It is denoted A^*.

$$\begin{array}{r} 999 \\ -314 \\ \hline \end{array}$$
$$\text{nine's-complement of } 314_{10} = A^* = 685_{10}$$

- The **ten's-complement** of a decimal number A is the nine's-complement plus 1. It is denoted A^{**}.

  note: The ten's-complement is the complement you've been finding all along for decimal numbers.

EXAMPLE 2-36   Find the nine's-complement and ten's-complement of (a) 43612_{10}, (b) 5943_{10}, and (c) 10900_{10}.

Solution

	(a)	**(b)**	**(c)**
Decimal Number	43612	5943	10900
A^* (nine's-complement)	56387	4056	89099
A^{**} (ten's-complement)	56388	4057	89100

Check: For the ten's-complement we have $A^{**} = 10^p - A$.

(a) $A^{**} = 10^5 - 43612 = 56388$
(b) $A^{**} = 10^4 - 5943 = 4057$
(c) $A^{**} = 10^5 - 10900 = 89100$

In the subtraction of $A - B$, when $A < B$ the answer will be a negative number, but you can still use complementation to perform the operation. You must, however, tell the computer to do something special when $A < B$ and to insert a minus sign in front of the final result.

EXAMPLE 2-37   Find $149_{10} - 215_{10}$ using complementation.

Solution

149
-215 $\xrightarrow[\text{complement}]{\text{get}}$ 784 $\longleftarrow$ (nine's-complement)
B^{**} $+\quad 1$ 149
                $\overline{785}$ $\longleftarrow$ (ten's-complement) $\longrightarrow$ $+785$
                                                                        $\overline{934}$

Notice that the 10^3 (r^p) column is empty; according to eq. (2-7), however, we must still subtract 10^3 from 934. But: $934 - 1000 = -66$, which is the *opposite of the complement of 934* and is the answer we're looking for. To obtain that number, then, we must take the complement of 934 and stick a minus sign in front of it.

Ten's-complement of $934 = 65 + 1 = 66 \rightarrow -66$. Voilà, the correct answer.
Check: $149 - 215 = -(215 - 149) = -66$

Use the same procedure as in Example 2-37 every time $A < B$ in a subtraction problem.

EXAMPLE 2-38   Perform the arithmetic base 2 operation $100101_2 - 101110_2$.

Solution   The result will be a negative number since $A < B$.

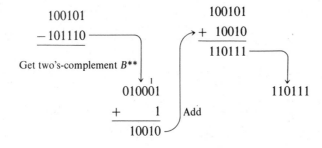

There is no extra 1 in the 2^6 column because the answer is negative. To get the result, we now take the two's-complement of this value and affix the negative sign in front:

001000
$+\qquad 1$
$\overline{-001001}$ $\leftarrow$ Result

Check: In base 10: $100101_2 - 101110_2 = 37_{10} - 46_{10} = -9_{10} = -1001_2$.

> *note:* Besides using them for subtraction, computers use complements for number storage; some machines store negative numbers in the form of their complements.

2-6. Modular Arithmetic

Another operation with numbers that is valuable in computer work is modular arithmetic. Our most familiar use of it in everyday living is the clock. In the world of time, we have a 12-hour clockface and the following system:

real time	modular math
2 o'clock + 3 hrs = 5 o'clock	2 + 3 = 5 mod 12
10 o'clock + 3 hrs = 1 o'clock	10 + 3 = 1 mod 12
11 o'clock + 4 hrs = 3 o'clock	11 + 4 = 3 mod 12

That is, if the result of our addition is greater than 12, we subtract 12 (or a multiple of 12) to find out what time it really is. We call the clock system of 12 hrs a **modulo 12** system; adding, subtracting, multiplying, and dividing in modulo systems is known as **modular arithmetic**.

note: In modular arithmetic you write

$$15 \quad = \quad 3 \quad \text{mod} \quad 12$$

the nonmodular value	the modular value	identifies that you're in modular	identifies your modular "base"

Now, say you make a machine that has a gear connected to a readout that will display only the five digits 0, 1, 2, 3, 4. Begin at 0. With one click, the readout goes to 1, with two clicks, it reads 2, with three, 3, and so on. When it counts five clicks, the readout sets itself to 0 again; for six clicks, it reads 1, for seven clicks 2, and so on; for ten clicks, it resets to 0, eleven clicks reads 1, and so it goes.

We can call this gear's routine a **modulo 5** system: we have only five digits, and all integers (number of clicks) can be recorded as one of these five digits, though you may not know which integer the digit is standing for. (For example, if the machine reads 2, it could have registered 2 clicks, 7 clicks, 12 clicks, or whatever.)

Mathematically, modulo is defined as: If $a = b$ modulo n, then

MODULO $$a = k \cdot n + b \quad \text{for some integer } b \tag{2-11}$$

In modulo n, we are limited to n digits.

note: $k =$ the number of times that n (the modular base) goes into a (the number you want written in a specific modular base), and b (the value of a in the specific modular base) is the remainder of the division of a/n.

EXAMPLE 2-39   Using eq. (2-11) find the following values in the modular arithmetic system: **(a)** In modulo 7 what is 25, 49, −12, 3? **(b)** In modulo 10 what is 37, 176, −18?

Solution

(a) In the modulo 7 system, we have only the digits 0, 1, 2, 3, 4, 5, 6, so from each number we extract all multiples of 7:

$$25 = \quad 3 \cdot 7 + 4 = 4 \text{ mod } 7$$
$$49 = \quad 7 \cdot 7 + 0 = 0 \text{ mod } 7$$
$$-12 = -2 \cdot 7 + 2 = 2 \text{ mod } 7$$
$$3 = \quad 0 \cdot 7 + 3 = 3 \text{ mod } 7$$

(b) Extract all multiples of 10:

$$37 = \quad 3 \cdot 10 + 7 = 7 \text{ mod } 10$$
$$176 = \quad 17 \cdot 10 + 6 = 6 \text{ mod } 10$$
$$-18 = -2 \cdot 10 + 2 = 2 \text{ mod } 10$$

Because we are limited to a certain number of digits in a specific modulo, the value of a modulo can represent any number of integers. For example:

$$0 \bmod 4 = \{\ldots -12, -8, -4, 0, 4, 8, 12, \ldots\}$$
$$1 \bmod 4 = \{\ldots -11, -7, -3, 1, 5, 9, 13, \ldots\}$$
$$2 \bmod 4 = \{\ldots -10, -6, -2, 2, 6, 10, 14, \ldots\}$$
$$3 \bmod 4 = \{\ldots -9, -5, -1, 3, 7, 11, 15, \ldots\}$$

To perform modular arithmetic, simply make the required computations then convert to your modular base;

$$11 + 3 = 14 = 2 \bmod 12 \qquad 6 + 7 = 13 = 6 \bmod 7$$
$$9 - 3 = 6 = 1 \bmod 5 \qquad 13 \cdot 3 = 39 = 3 \bmod 4$$

EXAMPLE 2-40   Create the modular arithmetic table for a modulo 5 system.

Solution   Form addition for the nonnegative integers.

note: Read this like a multiplication table, only for addition: in mod 5, $0 + 0 = 0 \bmod 5$; $1 + 2 = 3 \bmod 5$, $6 + 5 = 1 \bmod 5$, etc. Notice where the pattern starts repeating.

Addition

mod 5	0	1	2	3	4	5	6	$\cdots$
0	0	1	2	3	4	0	1	$\cdots$
1	1	2	3	4	0	1	2	$\cdots$
2	2	3	4	0	1	2	3	$\cdots$
3	3	4	0	1	2	3	4	$\cdots$
4	4	0	1	2	3	4	0	$\cdots$
5	0	1	2	3	4	0	1	$\cdots$
6	1	2	3	4	0	1	2	$\cdots$
$\vdots$	$\vdots$	$\vdots$	$\vdots$	$\vdots$	$\vdots$	$\vdots$	$\vdots$	

SUMMARY

1. Numbers may be classified into a variety of interrelated categories; the real numbers contain the rationals, which contain the integers, which contain the natural numbers. Irrational numbers are also real numbers.

2. Our normal counting patterns are based on the decimal system; each number is written as a sum of digits multiplied by a power of 10.

3. For computer work, the base 2 numeration system is more practical than the decimal, base 10, system.

4. Other bases, such as base 8 and base 16, are popular and necessary number systems used by computer programmers.

5. Well-defined algorithms exist for converting a number from one given base to the equivalent number in another base. (Table 2-3 displays a few equivalence relationships among bases 2, 8, 10, and 16.)

6. Complements of numbers are helpful in the inner construction of many computers.

7. The modular arithmetic system allows addition within a restricted framework. In modulo n we have exactly n digits to work with.

TABLE 2-3 Binary, Octal, and Hexadecimal Numbers between 0 and 18

Decimal	Binary	Octal	Hexadecimal	Decimal	Binary	Octal	Hexadecimal
0	0	0	0	10	1010	12	A
1	1	1	1	11	1011	13	B
2	10	2	2	12	1100	14	C
3	11	3	3	13	1101	15	D
4	100	4	4	14	1110	16	E
5	101	5	5	15	1111	17	F
6	110	6	6	16	10000	20	10
7	111	7	7	17	10001	21	11
8	1000	10	8	18	10010	22	12
9	1001	11	9				

RAISE YOUR GRADES

Can you...?

☑ explain the five different number classifications
☑ give examples of each of the number classifications
☑ explain the expanded notation for a number in decimal or binary representation
☑ give three reasons why binary numbers are used in computers
☑ convert a number in a given base to the equivalent number in another base
☑ find the complement of a given number and use it in applications
☑ use modular arithmetic to reduce a given number to a particular modulo base

SOLVED PROBLEMS

Number Classification

PROBLEM 2-1 Create a pictorial representation of the various number sets.

Solution

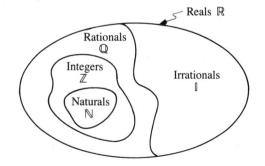
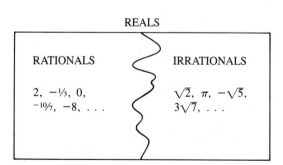

PROBLEM 2-2 Give a precise classification of each of the following numbers first by stating if it is rational or irrational, then by identifying the *smallest* of the subsystems to which it belongs:

(a) -4 **(b)** $\sqrt{11}$ **(c)** $\dfrac{\pi}{5}$ **(d)** $\dfrac{1}{8}$ **(e)** $\dfrac{27}{3}$ **(f)** $\dfrac{1}{1-\sqrt{3}}$ **(g)** -49 **(h)** $(-2)(-4)$

Solution

(a) -4 is rational and is a negative integer   **(b)** $\sqrt{11}$ is irrational

(c) $\dfrac{\pi}{5}$ is irrational   **(d)** $\dfrac{1}{8}$ is rational

(e) $\dfrac{27}{3}$ is rational but equals 9 and thus is also a natural number

(f) $\dfrac{1}{1 - \sqrt{3}} = \dfrac{1 + \sqrt{3}}{(1 - \sqrt{3})(1 + \sqrt{3})} = \dfrac{1 + \sqrt{3}}{1 - 3} = \dfrac{1 + \sqrt{3}}{-2} = -\dfrac{1}{2}(1 + \sqrt{3})$ is irrational

(g) -49 is rational and is an integer
(h) $(-2)(-4) = 8$, which is rational and is a natural number

PROBLEM 2-3 Rational numbers have decimal expansions that terminate or repeat. Convert the following rational numbers to one of these two forms: **(a)** $\frac{4}{5}$, **(b)** $\frac{2}{3}$, **(c)** $\frac{47}{2}$, **(d)** $\frac{1}{8}$, **(e)** $\frac{1}{11}$.

Solution

(a) $\frac{4}{5} = \frac{8}{10} = 0.8$ (terminating)   Multiply by n/n to get the denominator to be a power of 10.

$$0.666\ldots$$
$$3\overline{)2.0^20^20\ldots}$$

(b) $\frac{2}{3} = 0.\overline{6}$ (repeating)

(c) $\frac{47}{2} = \frac{46}{2} + \frac{1}{2} = 23 + 0.5 = 23.5$ (terminating)

(d) $\frac{1}{8} = \frac{1}{2 \cdot 2 \cdot 2} = (\frac{1}{2})(\frac{1}{2})(\frac{1}{2}) = (0.5)(0.5)(0.5) = 0.125$ (terminating)

$$0.0909\ldots$$
$$11\overline{)1.00^100^10\ldots}$$

(e) $\frac{1}{11} = 0.\overline{09}$ (repeating)

PROBLEM 2-4 Convert the following rational numbers into the form of the quotient of two integers: **(a)** $0.\overline{1}$, **(b)** $0.\overline{31}$, **(c)** $0.\overline{9}$ **(d)** $3.\overline{261}$.

Solution The technique is to set the given number equal to x (or any variable) and to multiply both sides by 10^p, where p is the number of digits in the repetition pattern. Then by subtraction, the repeating portion may be eliminated.

(a) Let $x = 0.1111\ldots$, so $10x = 1.111\ldots$

$$10x - x = (1.1111\ldots) - (0.1111\ldots)$$

$$9x = 1.000\ldots$$

$$x = \frac{1}{9}$$

(b) Let $y = 0.313131\ldots$, so $100y = 31.3131\ldots$

$$100y - y = (31.3131\ldots) - (0.3131\ldots)$$

$$99y = 31$$

$$y = \frac{31}{99}$$

(c) Let $z = 0.9999\ldots$, so $10z = 9.9999\ldots$

$$10z - z = (9.9999\ldots) - (0.9999..)$$

$$9z = 9$$

$$z = 1$$

    note: z is exactly nine times larger than x, which is reasonable since the decimal in **(c)** is nine times larger than that in **(a)**.

(**d**) Let $w = 3.261261...$, so $1000w = 3261.261261...$

$$1000w - w = (3261.261261...) - (3.261261...)$$

$$999w = 3258$$

$$w = \frac{3258}{999}$$

PROBLEM 2-5 Prove that $\sqrt{2}$ is irrational.

Solution Many methods of proof are available and will be discussed in detail later (see Chapter 7). Here we use the method of contradiction.

(**1**) Suppose that $\sqrt{2}$ is rational.

(**2**) Then $\sqrt{2} = \dfrac{p}{q}$ for some integers p and q that have no common factor.

(**3**) Then $2 = \dfrac{p^2}{q^2}$ so that $p^2 = 2q^2$.

(**4**) But it is a fact that the square of any number, when written in its prime factorization, has each prime occurring as a factor an even number of times.

   note: Check this fact:

$$12^2 = 12 \cdot 12 \qquad\qquad 15^2 = 15 \cdot 15$$
$$= 2^2 \cdot 3 \cdot 2^2 \cdot 3 \qquad\qquad = 3 \cdot 5 \cdot 3 \cdot 5$$
$$= 2^4 \cdot 3^2 \qquad\qquad\quad = 3^2 \cdot 5^2$$

(**5**) Hence p^2 has the factor 2 an even number of times, while $2q^2$ has the factor 2 an odd number of times.

(**6**) This condition is impossible, so there is an error in our supposition (1). Therefore $\sqrt{2}$ is irrational.

Decimal Representation

PROBLEM 2-6 Write 3482_{10} in expanded decimal notation.

Solution You have 3 thousands, 4 hundreds, 8 tens, and 2 units. Thus

$$3482 = 3(1000) + 4(100) + 8(10) + 2(1)$$
$$= 3(10^3) + 4(10^2) + 8(10^1) + 2(10^0)$$

PROBLEM 2-7 Write 44.2875 in expanded decimal notation.

Solution

$$44.2875 = 40 + 4 + \frac{2}{10} + \frac{8}{100} + \frac{7}{1000} + \frac{5}{10000}$$

$$= 4(10^1) + 4(10^0) + 2(10^{-1}) + 8(10^{-2}) + 7(10^{-3}) + 5(10^{-4})$$

Binary Representation

PROBLEM 2-8 Write the binary number 110.1001 in expanded binary notation.

Solution Since the base value is 2, the integral part (110) will use positive powers of 2 and the fractional part (0.1001) will use negative powers of 2. Thus

$$110.1001 = 1(2^2) + 1(2^1) + 0(2^0) + 1(2^{-1}) + 0(2^{-2}) + 0(2^{-3}) + 1(2^{-4})$$

PROBLEM 2-9 Convert 110.1001_2 to decimal.

Solution

$110.1001_2 = 2^2 + 2^1 + 2^{-1} + 2^{-4}$ (*shortcut:* Omit the zero terms.)

$$= 4 + 2 + \frac{1}{2} + \frac{1}{16}$$

$$= 6 + \frac{9}{16}$$

$$= 6.5625_{10}$$

PROBLEM 2-10 Convert 173_{10} to binary using the power-two deduction method.

Solution Noting the various powers of 2 ($2^0 = 1$, $2^1 = 2$, $2^2 = 4$, $2^3 = 8$, $2^4 = 16$, $2^5 = 32$, $2^6 = 64$, $2^7 = 128$, $2^8 = 256$, etc.), we subtract out the largest power of two possible. With the remaining value, repeat this. Continue until the remaining value is 0 or 1.

$$
\begin{array}{r}
173 \\
-128 \longrightarrow \text{Record } 2^7 \\
\hline
45 \\
- 32 \longrightarrow \text{Record } 2^5 \\
\hline
13 \\
- 8 \longrightarrow \text{Record } 2^3 \\
\hline
5 \\
- 4 \longrightarrow \text{Record } 2^2 \\
\hline
1 \\
- 1 \longrightarrow \text{Record } 2^0 \\
\hline
0
\end{array}
$$

The right side is now an expanded binary representation. Put the digit one as the place value for each given power:

$$173_{10} = 2^7 + 2^5 + 2^3 + 2^2 + 2^0$$
$$= 1 \quad 0 \ 1 \quad 0 \ 1 \quad 1 \quad 0 \ 1$$
$$ \uparrow \qquad \uparrow \qquad \uparrow \ \uparrow \qquad \uparrow$$
$$= 2^7 \quad 2^5 \quad 2^3 \, 2^2 \quad 2^0$$

PROBLEM 2-11 Using the algorithm method, convert the decimal number 173 to binary representation.

Solution

recall: The algorithm for this conversion is:

1. Divide the integer by 2 and record the remainder.
2. Divide the quotient by 2 and record the remainder.
3. Repeat step 2 until the final quotient is zero.

4. Write the recorded remainders in reverse order.

$$
\begin{array}{l}
\quad\ 86 \\
2\overline{\smash{\big)}173}, \quad R = 1 \\[4pt]
\quad\ 43 \\
2\overline{\smash{\big)}86}, \quad R = 0 \\[4pt]
\quad\ 21 \\
2\overline{\smash{\big)}43}, \quad R = 1 \\[4pt]
\quad\ 10 \\
2\overline{\smash{\big)}21}, \quad R = 1 \\[4pt]
\quad\ 5 \\
2\overline{\smash{\big)}10}, \quad R = 0 \\[4pt]
\quad\ 2 \\
2\overline{\smash{\big)}5}, \quad R = 1 \\[4pt]
\quad\ 1 \\
2\overline{\smash{\big)}2}, \quad R = 0 \\[4pt]
\quad\ 0 \\
2\overline{\smash{\big)}1}, \quad R = 1
\end{array}
$$

write in reverse order:

1 0 1 0 1 1 0 1

Thus $173_{10} = 10101101_2$

PROBLEM 2-12   Using the power-two deduction method, convert 0.734375_{10} to binary.

Solution   The negative powers of two are:

$$2^{-1} = \frac{1}{2} = 0.5 \qquad 2^{-3} = \frac{1}{8} = 0.125 \qquad 2^{-5} = \frac{1}{32} = 0.03125$$

$$2^{-2} = \frac{1}{4} = 0.25 \qquad 2^{-4} = \frac{1}{16} = 0.0625 \qquad 2^{-6} = \frac{1}{64} = 0.015625$$

Thus:

$$
\begin{array}{rl}
0.734375 & \\
-0.5 & \longrightarrow \ \text{Record } 2^{-1} \\
\hline
0.234375 & \\
-0.125 & \longrightarrow \ \text{Record } 2^{-3} \\
\hline
0.109375 & \\
-0.0625 & \longrightarrow \ \text{Record } 2^{-4} \\
\hline
0.046875 & \\
-0.03125 & \longrightarrow \ \text{Record } 2^{-5} \\
\hline
0.015625 & \\
-0.015625 & \longrightarrow \ \text{Record } 2^{-6} \\
\hline
0 &
\end{array}
$$

We therefore have

$$
\begin{aligned}
0.734375 &= 0.5 + 0.125 + 0.0625 + 0.03125 + 0.015625 \\
&= 2^{-1} + 2^{-3} + 2^{-4} + 2^{-5} + 2^{-6} \\
&= 0.101111_2
\end{aligned}
$$

PROBLEM 2-13 Convert the fractional decimal number 0.734375_{10} to binary representation.

Solution

recall: The algorithm for this conversion is:

1. Multiply the number by 2 and record the integral part of the result.
2. Using only the fractional part of the result, multiply by 2 and again record the integral part.
3. Repeat steps 1 and 2 until the fractional part is zero.
4. Write the integral parts in order of appearance.

$$0.734375$$
$$\times \quad 2$$

INT = 1 ⟵ 1.468750

$$\times \quad 2$$

INT = 0 ⟵ 0.937500

$$\times \quad 2$$

INT = 1 ⟵ 1.875000

$$\times \quad 2$$

INT = 1 ⟵ 1.750000

record in order of appearance: 101111

$$\times \quad 2$$

INT = 1 ⟵ 1.500000

$$\times \quad 2$$

INT = 1 ⟵ 1.000000

Thus $0.734375_{10} = 0.101111_2$

PROBLEM 2-14 Convert 0.85_{10} to binary.

Solution Not all conversions terminate. An arbitrarily chosen decimal fractional value may have a monstrous-looking binary expansion, so in the process of multiplying by 2, retaining the fractional part, and recording the integral part, we watch for the repeat appearance of any fractional value.

$$0.85$$
$$\times 2$$

INT = 1 ⟵ 1.70

$$\times 2$$

INT = 1 ⟵ 1.40

$$\times 2$$

INT = 0 ⟵ 0.80

$$\times 2$$

INT = 1 ⟵ 1.60

read

.110110

INT = 1 ⟵ 1.20

$$\times 2$$

INT = 0 ⟵ 0.40

Same value; the next four integral values obtained when the process is repeated will be 0110. This pattern will occur over and over again, so we have a repeating binary.

Thus, $0.85_{10} = 0.1\overline{10110}_2$.

PROBLEM 2-15 Your company has produced 1100 photographic whimmybobs that self-destruct if they are exposed to light. They are manufactured and to be packed into boxes in a darkroom. You, as chief whimmybob packer, are told to put them all into as few boxes as possible but in such a way that, if your main (and only) customer requests any number of whimmybobs from 1 to 1100, you can give the customer the *exact* number of whimmybobs requested without inspecting or opening any box. How do you pack them?

Solution Having just finished learning about binaries, you might recognize that you can make your task easier if you apply the binary numbering system. So, moving carefully in your cramped little darkroom, you begin to box your whimmybobs.

You put 1 whimmy in the first box and label it your 0 box;
You put 2 whimmies in the next box and label it your 1 box;
You put 4 whimmies in the next box and label it your 2 box;
You put 8 whimmies in the next box and label it your 3 box.
Continuing the process, you put 16, 32, 64, 128, 256, 512 whimmies into further boxes labeled 4, 5, 6, 7, 8, and 9, respectively.

The total now in the boxes is:

$$1 + 2 + 4 + 8 + 16 + 32 + 64 + 128 + 256 + 512 = 1023$$

You have 10 boxes filled and 77 whimmybobs left. So you box them by the same process as before, putting in 1, 2, 4, 8, 16, 32, respectively, so that in 6 more boxes you have inserted 63 whimmybobs and have labeled them accordingly. You now have 14 whimmies left, so you wipe your brow, and begin again, packing 1, 2, 4 into 3 new boxes; you now have 7 whimmybobs left. Pack these into 3 more boxes 1, 2, 4, respectively), and—whew!—you have all of the whimmybobs stored.

Now sit back to examine what you have on your shelves:

Boxes with 2^0 whimmybobs:	[0]	[0]	[0]	[0]
2^1	[1]	[1]	[1]	[1]
2^2	[2]	[2]	[2]	[2]
2^3	[3]	[3]		
2^4	[4]	[4]		
2^5	[5]	[5]		
2^6	[6]			
2^7	[7]			
2^8	[8]			
2^9	[9]			

} 22 total boxes

The following week your customer requests x number of whimmybobs. You receive the order and quickly convert this decimal number x to binary. Now your system is ready to go to work. Examine your binary number and, for every 1 in position 2^k, hand your customer a box labeled k; if a 0 is in the 2^k position, don't hand over a box.

If the number x of whimmies requested is between 1 and 1023, you will pick your boxes from the first group of ten. If it's between 1024 and 1086, you will give all of the first ten boxes (a total of 1023 whimmybobs) plus some boxes from the second group. If the number requested is between 1087 and 1093, you will give all of the boxes in the first and second groups (a total of 1086 whimmybobs) and some boxes from the third group. And, if the number is more than 1093, you will give all of the boxes from the first three groups (a total of 1093) and some from the remaining group. Here are some examples:

(1) Your customer wants 425 whimmybobs. Convert to binary:

$$425$$
$$\underline{-256} \longrightarrow 2^8$$
$$169$$
$$\underline{-128} \longrightarrow 2^7$$
$$41$$
$$\underline{-\ 32} \longrightarrow 2^5$$
$$9$$
$$\underline{-\ 8} \longrightarrow 2^3$$
$$1$$
$$\underline{-\ 1} \longrightarrow 2^0$$
$$0$$

The binary representation is 110101001_2. From the first group of boxes you hand over the boxes numbered 8, 7, 5, 3, and 0. They will collectively contain exactly 425 whimmies.

(2) Your customer orders 1051 whimmies. You supply the entire first group of boxes (ten boxes containing 1023 whimmies), then compute that $1051 - 1023 = 28$ more whimmybobs are needed. Convert 28 to binary.

$$28$$
$$\underline{-16} \longrightarrow 2^4$$
$$12$$
$$\underline{-\ 8} \longrightarrow 2^3$$
$$4$$
$$\underline{-\ 4} \longrightarrow 2^2$$
$$0$$

The binary representation is 11100_2. From the second group you give the boxes numbered 4, 3, and 2. In them will be $2^4 + 2^3 + 2^2 = 28$ whimmies.

Octal and Hexadecimal Representation

PROBLEM 2-16   Convert the octal number 4647132_8 to decimal.

Solution

$$4647132_8 = 4(8^6) + 6(8^5) + 4(8^4) + 7(8^3) + 1(8^2) + 3(8^1) + 2(8^0)$$
$$= 4(262\,144) + 6(32\,768) + 4(4096) + 7(512) + 1(64) + 3(8) + 2(1)$$
$$= 1\,265\,242_{10}$$

PROBLEM 2-17   Convert 214.078125_{10} to base 8.

Solution   Work in two parts:

Integer part   *Fractional part*

$$\begin{array}{r} 26 \\ 8\overline{)214}, \quad R = 6 \\ 3 \\ 8\overline{)26}, \quad R = 2 \\ 0 \\ 8\overline{)3}, \quad R = 3 \end{array}$$

write   write

$$\underbrace{326}_{} \quad + \quad \underbrace{.05}_{}$$

$$0.078125$$
$$\underline{\qquad\qquad 8}$$
INT $= 0 \longleftarrow 0.625000$
$$\underline{\qquad\qquad 8}$$
INT $= 5 \longleftarrow 5.000000$

Thus $214.078125_{10} = 326.05_8$.

PROBLEM 2-18 Convert the binary number 100110100111001011010 to (a) octal and (b) hexadecimal representation.

Solution (a) For the binary–octal conversion, start from the right and group the digits by threes to get:

$$\underbrace{100}\ \underbrace{110}\ \underbrace{100}\ \underbrace{111}\ \underbrace{001}\ \underbrace{011}\ \underbrace{010} = 4647132_8$$
$$\quad 4\quad\ \ 6\quad\ \ 4\quad\ \ 7\quad\ \ 1\quad\ \ 3\quad\ \ 2$$

(b) For the binary–hexadecimal conversion, starting from the right and group the digits by fours:

$$\underbrace{0001}\ \underbrace{0011}\ \underbrace{0100}\ \underbrace{1110}\ \underbrace{0101}\ \underbrace{1010} = 134E5A_{16}$$
$$\quad 1\qquad 3\qquad 4\qquad E\qquad 5\qquad A_{16}$$

PROBLEM 2-19 Convert the octal number 4613_8 to binary.

Solution Just replace each octal digit by its own three-bit binary equivalent.

$$\underbrace{4}\quad \underbrace{6}\quad \underbrace{1}\quad \underbrace{3}$$
$$4\ \ 6\ \ 1\ \ 3_8 = 100\quad 110\quad 001\quad 011_2$$

PROBLEM 2-20 Convert the hexadecimal value $134E5A_{16}$ to decimal.

Solution

$$134E5A_{16} = 1(16^5) + 3(16^4) + 4(16^3) + E(16^2) + 5(16^1) + A(16^0) \qquad \textit{recall:}\ \text{E is 14 and A is 10.}$$
$$= 1(1\,048\,576) + 3(65\,536) + 4(4096) + 14(256) + 5(16) + 10(1)$$
$$= 1\,265\,242_{10}$$

PROBLEM 2-21 Convert the hexadecimal number $8D2B_{16}$ to octal representation.

Solution This is done most easily by passing through binary: hexadecimal → binary → octal.

$$8D2B_{16} = 1000\quad 1101\quad 0010\quad 1011 = 1000110100101011_2$$

Now regroup this into groups of three (starting from the right)

$$1000110100101011_2 = 001\quad 000\quad 110\quad 100\quad 101\quad 011$$
$$= 1\qquad 0\qquad 6\qquad 4\qquad 5\qquad 3_8$$

Complements

PROBLEM 2-22 Find the complement of the following in the given base: (a) 372_{10}, (b) 65.432_{10}, (c) 1011001_2, (d) 1101011_2.

Solution

(a) $r = 10$ and $p = 3$ so

$$A^{**}\ \text{of}\ 372_{10} = 10^3 - 372 = 1000 - 372 = 628_{10}$$

Working a second way, we find the nine's-complement as $A^* = 999 - 372 = 627$. Adding 1 gives $A^{**} = 628_{10}$.

(b) $r = 10$ and $p = 2$ so

$$A^{**}\ \text{of}\ 65.432_{10} = 10^2 - 65.432 = 100 - 65.432 = 34.568_{10}$$

Or find the nine's-complement: $A^* = 99.999 - 65.342 = 34.567$.
The ten's-complement will be $A^{**} = 34.568$.

note: You always add the 1 to the *rightmost* column.

(c) $r = 2$ and $p = 7$ so

$$A^{**} \text{ of } 1011001_2 = 2^7 - 1011001 = \begin{array}{r} 10000000 \\ - 1011001 \\ \hline 100111_2 \end{array}$$

The easier way is to find the one's-complement, $A^* = 100110_2$, and then add 1 to find the two's-complement: $A^{**} = 100111_2$.

(d) For 1101011_2, the one's-complement is $A^* = 10100_2$ and two's-complement is $A^{**} = 10101_2$.

PROBLEM 2-23   Use complements to perform the operations. Check the results in decimal notation.
(a) $11011_2 - 1101_2$ **(b)** $10011_2 - 11010_2$

Solution

(a) This result will be positive, so:

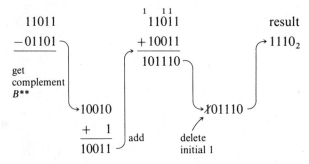

Check: $11011_2 - 1101_2 = 27_{10} - 13_{10} = 14_{10} = 1110_2$

(b) This result will be negative so:

$$\begin{array}{r} 10011 \\ -11010 \\ \hline \end{array} \quad \begin{array}{l} \text{get} \\ \text{complement} \\ B^{**} \end{array}$$

$$\begin{array}{r} \overset{1}{00101} \\ + 1 \\ \hline 110 \end{array} \Big] \text{Add}$$

$$\begin{array}{r} \overset{11}{10011} \\ + 110 \\ \hline 11001 \end{array} \text{— take the complement of this}$$

$$\longrightarrow \begin{array}{r} 00110 \\ + 1 \\ \hline 111 \end{array} \Big] \text{add minus sign}$$

$$\text{Result} - 111 \leftarrow$$

Check: $10011_2 - 11010_2 = 19_{10} - 26_{10} = -7_{10} = -111_2$

Modular Arithmetic

PROBLEM 2-24   Find the following values: **(a)** In modulo 3, find 12, $1 + 2$, 22; **(b)** In modulo 8, find 173, -42, $5 + 6$, $7 + 3$.

Solution

(a) $12 = 4 \cdot 3 + 0 = 0 \bmod 3$
$ 1 + 2 = 3 = 1 \cdot 3 + 0 = 0 \bmod 3$
$ 22 = 7 \cdot 3 + 1 = 1 \bmod 3$

(b) $173 = 21 \cdot 8 + 5 = 5 \bmod 8$
$ -42 = -6 \cdot 8 + 6 = 6 \bmod 8$
$ 5 + 6 = 11 = 1 \cdot 8 + 3 = 3 \bmod 8$
$ 7 + 3 = 10 = 1 \cdot 8 + 2 = 2 \bmod 8$

PROBLEM 2-25   Write an algorithm in pseudocode that will identify the modular value b of any number a in any modular base n.

Solution   Recall that if $a = \bmod\ n$, then $a = kn + b$.

Begin
1. Input a
2. Input n
3. $k \leftarrow \text{INT}(a/n)$
4. $b \leftarrow a - k * n$
5. Output b
End

Supplementary Exercises

PROBLEM 2-26   Convert these rational numbers to a terminating or repeating decimal: (a) $\frac{5}{2}$, (b) $\frac{1}{12}$, (c) $\frac{3}{11}$, (d) $\frac{4}{7}$.

PROBLEM 2-27   Which of the following are rational numbers? (a) $\sqrt{2}(1 + \sqrt{2})$, (b) $\sqrt{\frac{18}{32}}$, (c) $(4 - 2\sqrt{2}) + (1 + 2\sqrt{2})$, (d) $(5\sqrt{2})\sqrt{2}$, (e) $0.13113111311113\ldots$

PROBLEM 2-28   (a) Write $0.\overline{25} + 0.\overline{7}$ as a repeating decimal. (b) Write $\frac{2}{9} + \frac{7}{11}$ as a repeating decimal.

PROBLEM 2-29   Write each number in its expanded notation: (a) 123.456_{10}, (b) 76.54_8, (c) 1100.1001_2, (d) $C6F.2A_{16}$.

PROBLEM 2-30   Convert these binary numbers to decimal representation: (a) 10111.01_2, (b) 111111.1_2, (c) 100000000_2.

PROBLEM 2-31   Convert these decimal numbers to binary representation: (a) 46_{10}, (b) 1000_{10}, (c) 0.9375_{10}, (d) $\frac{17}{32}$.

PROBLEM 2-32   Convert these decimal numbers to binary, looking for repeating patterns: (a) 3.6_{10}, (b) 0.55_{10}, (c) 4.4_{10}.

PROBLEM 2-33   Convert each of these octal values to binary: (a) 47_8, (b) 526_8, (c) 3105_8.

PROBLEM 2-34   Convert these binary values to octal: (a) 1001101_2, (b) $10\hbar101_2$, (c) 100100101011_2.

PROBLEM 2-35   Convert each hexadecimal number to binary; then convert the result to decimal: (a) AB_{16}, (b) $E4D_{16}$, (c) $3F6_{16}$.

PROBLEM 2-36   Perform the following operations in binary; check your result by converting everything to decimal: (a) $1010_2 + 0110_2$, (b) $11010_2 + 1101_2$, (c) $11010_2 - 1101_2$.

PROBLEM 2-37   Write the representation for a number in base 5 (using digits 0, 1, 2, 3, 4). Now convert the following numbers to their equivalents in base 5: (a) 135_{10}, (b) 81.32_{10}, (c) 10110_2.

PROBLEM 2-38   Using Problem 2-15 as your basis, calculate the number of boxes needed to supply your customer with the following number of whimmybobs: (a) 505, (b) 823, (c) 1079, (d) 1089.

PROBLEM 2-39   Find the complement A^{**} of each of the following: (a) 111_{10}, (b) 111_2, (c) 81.65_{10}, (d) 1110110_2.

PROBLEM 2-40 Using complements, perform the operations: (a) $516_{10} - 274_{10}$, (b) $1100111_2 - 1010100_2$, (c) $100011_2 - 110010_2$.

PROBLEM 2-41 (a) What is 721 in mod 6? (b) What is -186 in mod 3? (c) What is $(5 + 8 + 3 + 6)$ in mod 9?

PROBLEM 2-42 Create the modular arithmetic tables for (a) addition and (b) multiplication for a modulo 3 system.

PROBLEM 2-43 Prove that $\sqrt{3}$ is irrational; then, using your calculator, find its value as far as you can.

Answers to Supplementary Exercises

2-26 (a) $\frac{5}{2} = 2.5$ (b) $\frac{1}{12} = 0.8\overline{3}$ (c) $\frac{3}{11} = 0.\overline{27}$

(d) $\frac{4}{7} = 0.\overline{571428}$

2-27 (b), (c), and (d)

2-28 (a) $1.\overline{03}$ (b) $0.\overline{85}$

2-29 (a) $1(10^2) + 2(10^1) + 3(10^0) + 4(10^{-1}) + 5(10^{-2}) + 6(10^{-3})$
(b) $7(8^1) + 6(8^0) + 5(8^{-1}) + 4(8^{-2})$
(c) $1(2^3) + 1(2^2) + 0(2^1) + 0(2^0) + 1(2^{-1}) + 0(2^{-2}) + 0(2^{-3}) + 1(2^{-4})$
(d) $C(16^2) + 6(16^1) + F(16^0) + 2(16^{-1}) + A(16^{-2})$

2-30 (a) 23.25_{10} (b) 63.5_{10} (c) 256_{10}

2-31 (a) 101110_2 (b) 1111101000_2 (c) 0.1111_2
(d) 0.10001_2

2-32 (a) $11.\overline{1001}_2$ (b) $0.\overline{10001}_2$ (c) $100.\overline{0110}_2$

2-33 (a) 100111_2 (b) 101010110_2
(c) 11001000101_2

2-34 (a) 115_8 (b) 5.5_8 (c) 4453_8

2-35 (a) 171_{10} (b) 3661_{10} (c) 1014_{10}

2-36 (a) 10000_2 (b) 100111_2 (c) 1101_2

2-37 generic representation: $vwxyz.ab_5 = v \cdot 5^4 + w \cdot 5^3 + x \cdot 5^2 + y \cdot 5^1 + z \cdot 5^0 + a \cdot 5^{-1} + b \cdot 5^{-2}$
(a) 1020_5 (b) 311.13_5 (c) 42_5

2-38 (a) 7 boxes (b) 7 boxes (c) 12 boxes
(d) 17 boxes

2-39 (a) 889 (b) 001 (c) 18.35 (d) 1010

2-40 (a) 242 (b) 10011 (c) 1111

2-41 (a) 1 mod 6 (b) 0 mod 3 (c) 4 mod 9

2-42 (a)

Addition mod 3	0 1 2
0	0 1 2
1	1 2 0
2	2 0 1

(b)

Multiplication mod 3	0 1 2
0	0 0 0
1	0 1 2
2	0 2 1

3 ARITHMETIC OPERATIONS WITH COMPUTERS

THIS CHAPTER IS ABOUT

☑ **Order of Operations**
☑ **Number Notation for Computers**
☑ **Floating Point Representation**
☑ **Floating Point Arithmetic**
☑ **Error Analysis in Computation**

3-1. Order of Operations

A. Writing arithmetic expressions for computers

In working a variety of mathematical problems, you will encounter many types of arithmetic expressions to evaluate. For example, the following are all types of expressions found in algebra or trigonometry:

$$(u^2 + v^2)^{1/2} \qquad \frac{-b \pm \sqrt{b^2 - 4ac}}{2a} \qquad r\cos(t) \qquad \sqrt{x^2 + y^2} \qquad \textbf{(3-1)}$$

When you want to evaluate such expressions using a computer, you will need special notation for some operations. The symbols used in most computer languages are $+$, $-$, $*$, $/$ for the familiar arithmetic operations, $\uparrow$ for exponentiation, and ( ) or [ ] for grouping.

EXAMPLE 3-1 Write the four expressions from (3-1) using computer notation.

Solution

$$(U \uparrow 2 + V \uparrow 2) \uparrow 0.5$$
$$(-B + SQR(B \uparrow 2 - 4 * A * C))/(2 * A)$$
$$(-B - SQR(B \uparrow 2 - 4 * A * C))/(2 * A)$$
$$R * COS(T)$$
$$SQR(X \uparrow 2 + Y \uparrow 2)$$

note: The SQR and COS are *built-in* (*library*) *functions* and will be explained in detail later (Chapter 13). You've already met SQR in Chapter 1.

The use of parentheses in Example 3-1 is very important to clarify which operations are to be done first. In the quadratic expression, for example, it was mandatory to use parentheses on $(2 * A)$ in the denominator. If these were omitted, the numerator would be divided by 2 and then the entire expression would be multiplied by A. That's not what we want! By placing the parentheses, we direct the computer to perform the operations in a specific order. This is because most programming languages work on what's called a **precedence of operations**. That is, they perform arithmetic operations in a strictly prescribed order, as described in the next section.

B. Precedence of operations

1. Basic arithmetic operations

In elementary mathematics, you learned that $5 - 4 + 3$ really means $(5 - 4) + 3 = 4$ and not $5 - (4 + 3) = -2$. That is, you perform the operations from left to right as you encounter them in the expression. To override this left-to-right convention, you can use parentheses, as in $5 - (4 + 3)$, because operations inside parentheses always take precedence over those on the outside.

If you have an expression mixing addition and multiplication, as in $3 + 2 * 4$, the multiplication takes precedence. Here we mean $3 + (2 * 4) = 11$ and not $(3 + 2) * 4 = 20$, though the parentheses may again be used to override the convention. Multiplication always takes precedence over addition and subtraction, and, since division is multiplication by inverses, division assumes the same precedence level as multiplication. We therefore have the following rules:

(1) Parentheses take precedence over all orders of operations.
(2) Multiplication and division take precedence over addition and subtraction.
(3) When operating at a fixed level of precedence (e.g., at all levels of addition and subtraction or multiplication and division), you perform those operations from left to right.

Put another way, you solve an expression in the following manner:

(1) Perform all operations inside parentheses.
(2) Perform multiplication and division from left to right.
(3) Perform addition and subtraction from left to right.

EXAMPLE 3-2   Find the value of the following expressions:

(a) $4 + 5 - 6 + 7$
(b) $4 * (5 - (6 + 7))$
(c) $4 + 5 * 6 - 7$
(d) $4 * 5/6 - 7$
(e) $4/5 * 6 + 7$
(f) $(4 - 5)/6 + 7$

Solution

(a) Since all the operations are at the same level (i.e., are all addition and subtraction), we merely go from left to right:

$$4 + 5 - 6 + 7 = 9 - 6 + 7 = 3 + 7 = 10$$

(b) The inner pair of parentheses takes precedence, then the outer pair:

$$4 * (5 - (6 + 7)) = 4 * (5 - 13) = 4 * (-8) = -32$$

(c) In this "mixture," do the multiplication first, then the other operations, from left to right:

$$4 + 5 * 6 - 7 = 4 + 30 - 7 = 34 - 7 = 27$$

(d) Do the second-level operations first, from left to right:

$$4 * 5/6 - 7 = 20/6 - 7 = 3\tfrac{1}{3} - 7 = -3\tfrac{2}{3}$$

(e) Same order as **(d)**:

$$4/5 * 6 + 7 = \tfrac{4}{5} * 6 + 7 = \tfrac{24}{5} + 7 = 4\tfrac{4}{5} + 7 = 11\tfrac{4}{5}$$

(f) First perform the operation inside the parentheses, then the division, then the addition:

$$(4 - 5)/6 + 7 = (-1)/6 + 7 = -\tfrac{1}{6} + 7 = 6\tfrac{5}{6}$$

---

2. Exponentiation

If an exponentiation operator is mixed in with arithmetic operations in an expression, it takes precedence. For example, the term $4 * 3 \uparrow 2$ means $4 * (3 \uparrow 2) = 4 * 9 = 36$, not $(4 * 3) \uparrow 2 = 12 \uparrow 2 = 144$. Blocking off with parentheses, however, will override the exponentiation precedence. In fact, using parentheses in the evaluation of something like $(3^2 + 4^2)^{1/2}$ is mandatory, for to write this as $3\uparrow 2 + 4\uparrow 2\uparrow 0.5$ would produce $9 + (16)^{1/2} = 9 + 4 = 13$ rather than the intended $(9 + 16)^{1/2} = 25^{1/2} = 5$.

note: If you're unsure about how the computer will perform the operations, use parentheses to establish the order exactly as you want it.

With multiple exponentiations you again work from left to right.

EXAMPLE 3-3 Find the value of the following expressions: **(a)** $2 + 5\uparrow 3$, **(b)** $3\uparrow 3 - 4\uparrow 2$, **(c)** $6 - 5\uparrow 2 * 3$.

Solution

(a) $2 + 5\uparrow 3 = 2 + (5\uparrow 3) = 2 + (5 * 5 * 5) = 2 + 125 = 127$
(b) $3\uparrow 3 - 4\uparrow 2 = (3\uparrow 3) - (4\uparrow 2) = (3 * 3 * 3) - (4 * 4) = 27 - 16 = 11$
(c) $6 - 5\uparrow 2 * 3 = 6 - (5\uparrow 2) * 3 = 6 - 25 * 3 = 6 - 75 = -69$

---

3. Unary and binary operations

In addition and subtraction operations such as $2 - 3$ and $x + 7$, you use two *operands*—variables or constants—to provide the data for each $+$ and $-$ operation. Such operations are called **binary operations** because the $+$ or $-$ affects two operands. But you may also use the $+$ and $-$ to affect only one operand, such as in $2\uparrow - 3$ or $x\uparrow + 7$. This type of operation is called a **unary operation** because the $+$ or $-$ influences only one operand (makes it positive or negative, respectively). Note that you must inspect the adjacent operands to determine if the symbol $+$ or $-$ is unary or binary. Unary operations take precedence over any other operation unless parentheses are present to override them.

EXAMPLE 3-4 Find the value of **(a)** $4 * 2\uparrow - 3 + 5$, **(b)** $2\uparrow + 3/4 * 5$, **(c)** $-2 * 4/3\uparrow - 1$.

Solution

(a) Observe first that the -3 exponent is a unary operation. Then do the exponentiation and follow standard operations:

$$4 * 2\uparrow - 3 + 5 = 4 * \frac{1}{2^3} + 5 = 4 * \frac{1}{8} + 5 = \frac{4}{8} + 5 = 5\tfrac{1}{2}$$

(b) You have unary $+3$ for the exponent. Follow standard rules:

$$2\uparrow + 3/4 * 5 = 2^3/4 * 5 = 8/4 * 5 = 2 * 5 = 10$$

(c) Do unary, then exponentiation, then multiplication and division:

$$-2 * 4/3\uparrow - 1 = -2 * 4/\tfrac{1}{3} = -8/\tfrac{1}{3} = -24$$

---

4. Summary table of precedence

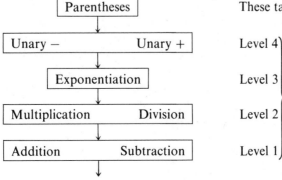

Parentheses	These take precedence at any level.
Unary $-$ Unary $+$	Level 4
Exponentiation	Level 3
Multiplication Division	Level 2
Addition Subtraction	Level 1

all done *LEFT TO RIGHT*

C. Evaluating arithmetic expressions

Any time you evaluate an arithmetic expression you progressively reduce it to a simpler form. This process of evaluation may be expressed as an algorithm:

1. Do operations inside parentheses.
2. Do unary operations.
3. Do exponentiation, left to right.
4. Do multiplication and division, left to right.
5. Do addition and subtraction, left to right.

EXAMPLE 3-5 Evaluate the following expressions:

(a) $2 + (3 - 4 * 2 \uparrow - 1)/5 * 3 \uparrow - 2 + 6 - (3 * 2)$
(b) $4 * (-3 \uparrow 2)/(2 * 5 \uparrow - 1) + 6 * (2 * 4 \uparrow - 2)$

Solution

(a) First work inside any parentheses, looking for the highest-level operations, then continue to follow the algorithm.

$$2 + (3 - 4 * 2\uparrow - 1)/5 * 3\uparrow - 2 + 6 - (3 * 2) = 2 + (3 - 4 * \tfrac{1}{2})/5 * 3\uparrow - 2 + 6 - (6)$$
$$= 2 + (3 - 2)/5 * 3\uparrow - 2 + 6 - 6$$
$$= 2 + (1)/5 * 3\uparrow - 2 + 6 - 6$$
$$= 2 + 1/5 * \frac{1}{3^2} + 6 - 6$$
$$= 2 + 1/5 * \tfrac{1}{9} + 6 - 6$$
$$= 2 + \tfrac{1}{45} + 6 - 6$$
$$= 2\tfrac{1}{45}$$

(b) Follow the algorithm.

$$4 * (-3\uparrow 2)/(2 * 5\uparrow - 1) + 6 * (2 * 4\uparrow - 2) = 4 * ((-3)^2)/(2 * \tfrac{1}{5}) + 6 * (2 * \tfrac{1}{16})$$
$$= 4 * 9/\tfrac{2}{5} + 6 * \tfrac{1}{8}$$
$$= 36/\tfrac{2}{5} + \tfrac{6}{8}$$
$$= 36 * \tfrac{5}{2} + \tfrac{3}{4}$$
$$= 18 * 5 + \tfrac{3}{4}$$
$$= 90\tfrac{3}{4}$$

3-2. Number Notation for Computers

A. The normalized exponential number system

A computer's memory is generally partitioned into "words" of equal length; these words are capable of storing numbers in a specific base—whether decimal, binary, or whatever—that is dictated by the computer's hardware. The storage system must fulfill two requirements. First, it must be able to store numbers in such a way that it doesn't consume unnecessary storage space with lots of place-holder zeros (as in very large or very small numbers such as 93,000,000 or 0.0000000005). Second, it must be able to have a **radix point** (decimal point, binary point, etc., depending on the base the computer uses) that has a fixed position for every number, so the machine can perform arithmetic.

To achieve these two goals, a computer usually converts each number to a specific system of notation called the **normalized exponential number system**. In this system, each number x has the form

$$x = M \times b^e$$

where b is the **base** of the number x $(b > 1)$, M is the **mantissa** and has value $0.1 \leq |M| < 1$, and e is the **exponent**, which is an integer. In this form, the mantissa M of a number x always has a nonzero digit to the immediate right of the radix point and a zero to the point's immediate left. The mantissa has at most t *significant digits* $(t > 0)$ and may be positive or negative, as may the exponent. For example, -147 written in the normalized exponential system is:

$$-147 = -0.147 \times 10^3$$

where $b = 10$ (decimal base), $M = -0.147$, and $e = 3$. The value of t, which supplies the **precision** of the number, is $t = 3$.

note: To identify what the exponent should be, take the number x and move its radix point to the right or left, as necessary, until the radix is to the immediate left of the leftmost nonzero digit in x. The total number of moves will be the value of the exponent. The *direction* you move the point will determine if e is positive or negative: if you move it to the left, e is positive, but if you move it to the right, e is negative:

$$147_8 = 0.147 \times 8^3 \qquad 59.4_{10} = 0.594 \times 10^2 \qquad 0.0000709_{10} = 0.709 \times 10^{-4}$$

$$2037.2_{10} = 0.20372 \times 10^4$$

EXAMPLE 3-6   Find the base 10 normalized exponential form for the following real numbers (go to four significant digits): **(a)** 427.3, **(b)** -5.28, **(c)** -0.0046, **(d)** 16.413, **(e)** 5280.1.

Solution   Using $b = 10$ and $t = 4$, we have:

	Real number	Normalized exponential form	Mantissa	Base	Exponent
(a)	427.3	0.4273×10^3	0.4273	10	3
(b)	-5.28	-0.5280×10^1	-0.5280	10	1
(c)	-0.0046	-0.4600×10^{-2}	-0.4600	10	-2
(d)	16.413	0.1641×10^2	0.1641	10	2
(e)	5280.1	0.5280×10^4	0.5280	10	4

note: To meet the requirement that $t = 4$ significant digits, we had to add zeros to the mantissa in **(b)** and **(c)**, and we had to cut off digits from it in **(d)** and **(e)**.

EXAMPLE 3-7   Find the base 2 normalized exponential form for the following real numbers for six significant digits: **(a)** 11.25, **(b)** -5, **(c)** 0.40625, **(d)** -0.75, **(e)** 27.5.

Solution   We have $b = 2$ and $t = 6$. After converting each real number to its equivalent binary expansion (see Section 2.3), we have:

	Real number	Binary form	Normalized exponential form	Mantissa	Base	Exponent
(a)	11.25	1011.01	0.101101×2^4	0.101101	2	4
(b)	$-5.$	$-101.$	-0.101000×2^3	-0.101000	2	3
(c)	0.40625	0.01101	0.110100×2^{-1}	0.110100	2	-1
(d)	-0.75	-0.11	-0.110000×2^0	-0.110000	2	0
(e)	27.5	11011.1	0.110111×2^5	0.110111	2	5

B. Scientific notation

Another exponential form computers commonly use is **scientific notation**. Used solely for decimal numbers, it also writes a given number x as $x = M \times b^e$, only here b is always 10, and the mantissa M is in the range $1 \leq M < 10$. In this notation, therefore, you set the radix point so that it is to the immediate right of the leftmost nonzero digit. In scientific notation, no precision value is used; all digits in the mantissa are significant.

> *note:* As with normalized exponentiation, you move the radix point in x to the right or to the left to calculate the exponent e. If you must move it to the left into the correct position, e is positive. If you must move it to the right, e is negative. Count the number of places you move it, and that number becomes e:

$$176 = 1.76 \times 10^2 \qquad 1492.5 = 1.4925 \times 10^3 \qquad 0.000209 = 2.09 \times 10^{-4}$$

EXAMPLE 3-8   Express the following real numbers in scientific notation: **(a)** 93000000, **(b)** 14.689, **(c)** 0.003164, **(d)** -11.83764.

Solution

	Real number	Scientific notation	Mantissa	Base	Exponent
(a)	93000000.	9.3×10^7	9.3	10	7
(b)	14.689	1.4689×10^1	1.4689	10	1
(c)	0.003164	3.164×10^{-3}	3.164	10	-3
(d)	-11.83764	-1.183764×10^1	-1.183764	10	1

3-3. Floating Point Representation

When a number is recorded in normalized exponential form, it is ready to be stored as a computer word. The computer converts it to internal computer representation using something called the **floating point system**. In this system, the computer identifies and individually stores the sign, exponent, and t-most significant digits of the normalized exponential form of a number.

A computer word consists of a collection of N "letters"—we'll call them bits—and is divided into three blocks of bits called **fields**. The leftmost field, consisting of one bit, designates the sign of the number: a 0 entry indicates a positive number, and a 1 entry indicates a negative number. The second field stores the exponent in s bits (actually, the exponent is first converted to another number called the *characteristic*, but more on that later). The third field stores the mantissa in the remaining $m = N - s - 1$ bits. For a machine that has N bits per word for storage, a word would look like this:

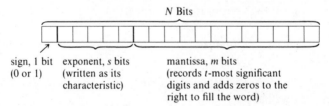

If we increase s, we get a wider range of exponents and thus a wider range of numbers that can be represented and used by the computer.

A. Binary representation

A computer that works in binary represents a decimal number in binary form and generally allots 32 bits for each word (number) in storage; it is therefore called a 32-bit machine. Such a machine typically has $s = 7$ bits for the exponent and $m = 32 - 7 - 1 = 24$ bits for the mantissa. Its internal computer representation looks like this:

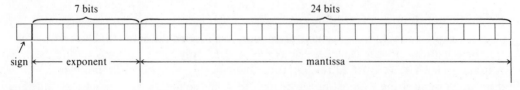

Now for that information on characteristics we promised. In the actual storage space of most machines, the exponent is represented by a nonnegative value called its *characteristic*, c, which is recorded in the computer's base and is limited to s bits. We calculate c for a binary machine as follows: From the given value of s, find $2^{s-1} = S$; then

**BINARY
CHARACTERISTIC**

$$c = S + e$$

For a value $s = 7$, then, $c = 2^6 + e = 64 + e$. Applying this equation to a given exponent e, we find that

- When $e = 4$, its characteristic $c = 64 + 4 = 68$; written in binary, c becomes 1000010_2, the stored description of $e = 4$.
- When $e = -20$, $c = 64 - 20 = 44$; the stored binary representation of $e = -20$ is therefore $c = 44 = 0011100_2$.

With $c = 64 + e$ for a 32-bit machine, then, we have a range of exponents extending from -64 ($c = 64 - 64 = 0 = 0000000_2$) to $+63$ ($c = 64 + 63 = 127 = 1111111_2$).

EXAMPLE 3-9 Given the real number $x = -639.421875$, create the internal computer representation in binary normalized exponential form using $s = 7$ and $m = 24$.

Solution First convert the real number x to binary notation:

$$639_{10} = 512 + 64 + 32 + 16 + 8 + 4 + 2 + 1$$
$$= 2^9 + 2^6 + 2^5 + 2^4 + 2^3 + 2^2 + 2^1 + 2^0$$
$$= 1\,001\,111\,111_2$$

$$0.421875_{10} = 0.25 + 0.125 + 0.03125 + 0.015625$$
$$= 2^{-2} + 2^{-3} + 2^{-5} + 2^{-6}$$
$$= 0.011011_2$$

Thus we have $x = -1001111111.011011_2$. Converting this to normalized exponential form gives us $x = -0.1001111111.011011 \times 2^{10}$. Since $e = 10$ and $s = 7$, $c = 2^6 + 10 = 64 + 10 = 74 = 64 + 8 + 2$, so in binary, the characteristic $c = 1001010_2$. The internal representation of x is therefore

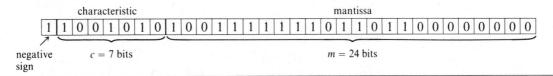

characteristic   mantissa

negative sign $c = 7$ bits $m = 24$ bits

EXAMPLE 3-10 Form the 32-bit floating point format for the following decimal numbers: (a) 293.75, (b) -0.578125, (c) -21.0234375.

Solution Convert the number to binary notation; then write it in normalized exponential form.

(a) $293.75_{10} = 256 + 32 + 4 + 1 + \frac{1}{2} + \frac{1}{4}$
$$= 2^8 + 2^5 + 2^2 + 2^0 + 2^{-1} + 2^{-2}$$
$$= 100100101.11_2$$
$$= 0.10010010111 \times 2^9$$

Calculate $c = 64 + 9 = 73_{10} = 64 + 8 + 1 = 1\,001\,001_2$

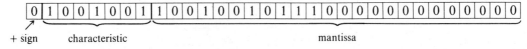

+ sign   characteristic   mantissa

(b) $-0.578125_{10} = -(0.5 + 0.0625 + 0.015625)$
$$= -(2^{-1} + 2^{-4} + 2^{-6})$$
$$= -0.100101_2$$
$$= -0.100101 \times 2^6$$

Characteristic $c = 64 - 0 = 64_{10} = 1\,000\,000_2$

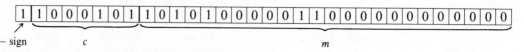

(c) $-21.0234375_{10} = -(16 + 4 + 1 + 0.015625 + 0.0078125)$

$$= -\left(2^4 + 2^2 + 2^0 + \frac{1}{2^6} + \frac{1}{2^7}\right)$$

$$= -10101.0000011_2$$

$$= -0.101010000011 \times 2^5$$

Now $e = 5$, so $c = 69_{10} = 64 + 4 + 1 = 1\,000\,101_2$

| 1 | 1 | 0 | 0 | 0 | 1 | 0 | 1 | 1 | 0 | 1 | 0 | 1 | 0 | 0 | 0 | 0 | 0 | 1 | 1 | 0 | 0 | 0 | 0 | 0 | 0 | 0 | 0 | 0 | 0 | 0 | 0 |

↗
− sign c m

B. Decimal representation

Examples 3-9 and 3-10 showed you how to store a computer word in a 32-bit binary machine. If the computer happens to work in decimal format, however, the memory word is usually represented by eight digits (instead of by 32 bits). As before, the word is divided into three fields: the leftmost field (1 digit) carries either a 0 or a 1 to identify the sign of the number; the second field (two digits) is for the exponent's characteristic; and the third field (five digits) is for the mantissa. We say that a computer memory constructed in this way has a precision of five significant digits. Everything is written in decimal, instead of in binary:

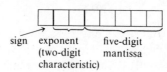

sign   exponent   five-digit
       (two-digit   mantissa
       characteristic)

The value for the two-digit characteristic lies between 00 and 99. You obtained it by putting the real number into normalized exponential form and then adding the exponent to the number 50; that is, $c = 50 + e$ for the eight-digit decimal computer. Thus the actual exponent e can be as large as 49 ($c = 99$) or as small as -50 ($c = 00$). The mantissa must fall within the range ± 0.99999. Thus the magnitude of the real numbers represented in decimal floating point form can be as large as 0.99999×10^{49} or as small as 0.00001×10^{-50}.

EXAMPLE 3-11 Form the eight-digit floating point format for the following numbers: **(a)** 400,000,000, **(b)** 0.0000123, **(c)** -0.01536.

Solution

(a) In normalized exponential form, this number is 0.4×10^9. Since the exponent $e = 9$, the characteristic is $c = 50 + 9$ and the mantissa is 0.40000. (Remember: We have a built-in precision of five digits.) The floating point format is

| 0 | 5 | 9 | 4 | 0 | 0 | 0 | 0 |

↗
+ sign c m

(b) This number is 0.123×10^{-4}. Thus $c = 50 - 4$.

| 0 | 4 | 6 | 1 | 2 | 3 | 0 | 0 |

↗
+ sign c m

(c) This number is -0.1536×10^{-1}. Thus $c = 50 - 1$.

$$\boxed{1}\;\boxed{4}\;\boxed{9}\;\boxed{1}\;\boxed{5}\;\boxed{3}\;\boxed{6}\;\boxed{0}$$

$$\underset{-\text{ sign}}{\nearrow} \quad \underbrace{}_{c} \quad \underbrace{}_{m}$$

3-4. Floating Point Arithmetic

Now that you know how to express a number in floating point representation, let's do some computer arithmetic.

agreement: For this section, all numbers will already be expressed in normalized exponential form.

To add and subtract in normalized exponential form, you must be sure the exponents on the numbers are the same. If they are different, you adjust the number that has the smaller exponent by adding zeros to the front of it. The computer usually does this automatically:

$$0.473 \times 10^4 \quad \text{different exponents;}$$
$$\pm 0.936 \times 10^6 \quad \text{adjust the number with the smaller exponent}$$

$$\to 0.00473 \times 10^6 \quad \text{same exponents;}$$
$$\pm 0.936 \quad \times 10^6 \quad \text{you can now perform the operation}$$

note: You may move the radix on any normalized number back and forth as long as you make a corresponding adjustment in the exponent so that the value remains the same. This movement, called **floating**, gives floating point arithmetic its name.

warning: In floating point arithmetic all mantissas are *truncated* to t digits—that is, extra digits beyond the required magnitude of precision are lopped off, without rounding up or down. This truncation, of course, could cause error in subsequent computations; such matters will be discussed in the next section.

A. Real addition

If the numbers being added have the same exponent, add the mantissas and use the same exponent. If the numbers have different exponents, however, you must adjust the number that has the smaller exponent to make its exponent correspond with the remaining one, as explained above.

EXAMPLE 3-12   Perform the following operations (assume $t = 4$):

(a) $0.1423 \times 10^3 + 0.4165 \times 10^3$ **(b)** $0.6112 \times 10^4 + 0.1826 \times 10^5$ **(c)** $0.8243 \times 10^1 + 0.7436 \times 10^1$

Solution

(a)
$$0.1432 \times 10^3$$
$$+0.4165 \times 10^3$$
$$\overline{0.5597 \times 10^3} \text{ Sum}$$

(b) $0.6112 \times 10^4 \longrightarrow$
$$0.0611 \times 10^5 \longleftarrow \text{ the value } 0.06112 \times 10^5 \text{ truncated}$$
$$0.1826 \times 10^5 \qquad +0.1826 \times 10^5 \qquad \text{to } t = 4$$
$$\overline{0.2437 \times 10^5} \text{ Truncated sum}$$

(c)
$$0.8243 \times 10^1$$
$$+0.7436 \times 10^1$$
$$\overline{1.5679 \times 10^1} \longrightarrow 0.1567 \times 10^2 \text{ Truncated normalized form}$$

note: Always truncate your values *before* you perform addition or subtraction because you're assuming the computer can only manipulate $t = 4$ digits; therefore it must truncate before it performs the operation. In this way, for $t = 4$ we would have

$$0.73247 \qquad \text{change} \qquad 0.7324$$
$$+0.13328 \qquad \text{to} \qquad +0.1332$$

so that the correct answer would be 0.8656, not 0.8657.

B. Real subtraction

The same rules apply for subtraction as for addition.

EXAMPLE 3-13 Perform the following operations (assume $t = 4$):

(a) $0.5729 \times 10^2 - 0.1483 \times 10^2$ **(b)** $0.8846 \times 10^4 - 0.8523 \times 10^4$ **(c)** $0.4444 \times 10^5 - 0.2222 \times 10^4$

Solution

(a) $\quad 0.5729 \times 10^2$
$\quad -0.1483 \times 10^2$
$\quad \overline{0.4246 \times 10^2}$ Difference

(b) $\quad 0.8846 \times 10^4$
$\quad -0.8523 \times 10^4$
$\quad \overline{0.0323 \times 10^4} \longrightarrow 0.3230 \times 10^3$ Difference

(c) $\quad 0.4444 \times 10^5 \qquad\qquad 0.4444 \times 10^5$
$\quad -0.2222 \times 10^4 \longrightarrow -0.0222 \times 10^5 \longleftarrow$ the value 0.02222×10^5 truncated to $t = 4$
$\qquad\qquad\qquad\qquad\quad \overline{0.4222 \times 10^5}$ Truncated difference

---

C. Real multiplication and division

To multiply two numbers in normalized exponential form, multiply their mantissas and add their exponents. To divide, divide their mantissas and subtract their exponents. Remember to normalize the results.

EXAMPLE 3-14 Perform the following operations (assume $t = 4$):

(a) $(0.2510) \times 10^2) \times (0.1418 \times 10^3)$ **(c)** $(0.5560 \times 10^4) \div (0.7515 \times 10^2)$
(b) $(0.3333 \times 10^{-3}) \times (0.1212) \times 10^4)$ **(d)** $(0.9288 \times 10^2) \div (0.3855 \times 10^5)$

Solution

(a) $(0.2510 \times 10^2) \times 0.1418 \times 10^3 = (0.2510 \times 0.1418) \times 10^{2+3} = 0.03559 \times 10^5 = 0.3559 \times 10^4$
(b) $(0.3333 \times 10^{-3}) \times (0.1212 \times 10^4) = (0.3333 \times 0.1212) \times 10^{-3+4} = 0.04039 \times 10^1 = 0.4039 \times 10^0$
(c) $(0.5560 \times 10^4) \div (0.7515 \times 10^2) = (0.5560 \div 0.7515) \times 10^{4-2} = 0.7398 \times 10^2$
(d) $(0.9288 \times 10^2) \div (0.3855 \times 10^5) = (0.9288 \div 0.3855) \times 10^{2-5} = 2.4093 \times 10^{-3} = 0.2409 \times 10^{-2}$

note: For multiplication and division, wait to truncate the answer until after you have put it into normalized exponential notation.

3-5. Error Analysis in Computation

Modern computers, in spite of all precautions, do commit errors when performing operations on real numbers. The very fact that they are finite machines precludes 100% accuracy, for they must store real numbers in memory with a fixed number of significant digits. Arithmetic operations, therefore, are bound to produce some amount of error. When we analyze errors, then, we don't try to eliminate them, but we do try to control them and discern their magnitude.

A. Types of error

There are two types of mathematical errors you should concern yourself with right now: *absolute errors* and *relative errors*:

- **absolute error**, e_a: The absolute value of the difference between the true value of a number, $\bar{a}$, and the approximate value, a:

$$e_a = |\bar{a} - a| = |a - \bar{a}| \tag{3-2}$$

note: Because we always think about error in positive terms (for example, we have a 10% error, not a -10% error) you can subtract either the approximate value from the true value or the true value from the approximate value, as in eq. (3-2).

- **relative error**, r: The absolute value of the absolute error divided by the true value; it measures how close the approximation a is to the true value $\bar{a}$ relative to the size of $\bar{a}$:

$$r = \left| \frac{e_a}{\bar{a}} \right|$$

EXAMPLE 3-15 For the addition problem of Example 3-12c, calculate the **(a)** absolute error and **(b)** relative error.

Solution The problem is $0.8243 \times 10^1 + 0.7436 \times 10^1$. The true value of the sum is 0.15679×10^2, but because of the truncation requirement of $t = 4$, the sum is given as 0.1567×10^2.

(a) The absolute error is

$$e_a = |0.1567 \times 10^2 - 0.15679 \times 10^2| = |-0.00009 \times 10^2| = 0.009$$

(b) The relative error is

$$r = \frac{0.009}{0.15679 \times 10^2} = \frac{0.009}{15.679} \approx 0.0057$$

or about 0.06%. Although this is very small, if the addition process is used many times in a given algorithm, the accumulated error may be substantial, that is, more than a scientist would allow for the total process. If a 3¢ error is made by the bank on one of our car payments, that is "small." But if the bank does this on 72 successive payments, that is "substantial."

B. Sources of error

There are basically three sources of error:

- **initial error**—an error that is present in the numbers used in a computation. If the data are obtained from measurements, for example, the device used may automatically produce some initial error. To estimate this error we need to know the accuracy of the sensor and the efficiency of the data transmission.

- **rounding error**—an error made when real numbers are stored in a format that has finite precision. In certain problems we may perform many calculations with numbers that are only approximate representations of the actual numbers, and considerable compounded errors could result.

- **truncation error**—an error made when terms are dropped (truncated) from an infinite process. For example, when using the infinite series expansion of

$$\cos x = 1 - \frac{x^2}{2!} + \frac{x^4}{4!} - \frac{x^6}{6!} + \frac{x^8}{8!} - \cdots$$

if we retain only the first four terms, then the truncation error in the value of $\cos x$ consists of the value of the terms neglected, that is,

$$\frac{x^8}{8!} - \frac{x^{10}}{10!} + \cdots$$

EXAMPLE 3-16 Large rounding errors can occur when two large numbers of almost the same size are subtracted. Find the absolute error when doing

$$\frac{243}{715} - \frac{197}{585}$$

Solution   The correct answer is $4/1287 = 0.31080 \times 10^{-2}$. But when the numbers are represented in the computer in normalized exponential form, we have $(0.33986 \times 10^0) - (0.33675 \times 10^0) = 0.3110 \times 10^{-2}$. The absolute error generated by this internal rounding is $|0.3110 \times 10^{-2} - 0.31080 \times 10^{-2}| = 0.0002 \times 10^{-2}$. If many such operations are compounded, a significant error could be produced.

C. Propagation of error

All three types of error can build up larger errors through progressive computations. If, for example, the error in approximate quantity a is e_a and the true value is $\bar{a}$, and the error in approximate quantity b is e_b and the true value is $\bar{b}$, then in computing $a + b = c$, we will produce yet a third error, e_c:

$$c = a + b = \bar{a} + e_a + \bar{b} + e_b = \bar{c} + e_c \tag{3-3}$$

Hence the error $e_c = e_a + e_b$ and the relative error is (by mathematical manipulation of eqs. 3-2 and 3-3):

$$r = \frac{e_c}{\bar{c}} = \left(\frac{a}{a+b}\right)\left(\frac{e_a}{\bar{a}}\right) + \left(\frac{b}{a+b}\right)\left(\frac{e_b}{\bar{b}}\right)$$

Thus we multiply the relative error in a by $a/(a+b)$, and the relative error in b by $b/(a+b)$, and then add them together to obtain the relative error in measuring the sum c of a and b. If a and b are subtracted, then the relative error in their difference d is:

$$r = \frac{e_d}{\bar{d}} = \left(\frac{a}{a-b}\right)\left(\frac{e_a}{\bar{a}}\right) - \left(\frac{b}{a-b}\right)\left(\frac{e_b}{\bar{b}}\right)$$

For multiplication, just the relative errors add together to produce the relative error of the product m:

$$r = \frac{e_m}{\bar{m}} = \frac{e_a}{\bar{a}} + \frac{e_b}{\bar{b}}$$

And, finally, for division, just the relative errors subtract to give us the relative error of the quotient q:

$$r = \frac{e_q}{\bar{q}} = \frac{e_a}{\bar{a}} - \frac{e_b}{\bar{b}}$$

EXAMPLE 3-17   Find the relative error in the answer to each of the following operations (use $t = 4$ precision): (a) $\bar{a} - \bar{b} = 162.78 - 162.31$, (b) $\bar{a} \times \bar{b} = 34.612 \times 12.783$.

Solution

(a) $\bar{a} = 162.78$ $a = 0.1627 \times 10^3$ $e_a = 0.08 = 0.8000 \times 10^{-1}$
$ \bar{b} = 162.31$ $b = 0.1623 \times 10^3$ $e_b = 0.01 = 0.1000 \times 10^{-1}$

$$\overline{} \ a - b = 0.0004 \times 10^3$$
$$= 0.4000 \times 10^0$$

$$\frac{e_d}{\bar{d}} = \left(\frac{0.1627 \times 10^3}{0.4000 \times 10^0}\right)\left(\frac{0.8000 \times 10^{-1}}{0.16278 \times 10^3}\right) - \left(\frac{0.1623 \times 10^3}{0.4000 \times 10^0}\right)\left(\frac{0.1000 \times 10^{-1}}{0.16231 \times 10^3}\right)$$

$$= 0.1999 - 0.0249 = 0.1750 = 17.5\% \text{ relative error.}$$

(b) $\bar{a} = 34.612$ $a = 0.3461 \times 10^2$ $e_a = 0.002 = 0.2000 \times 10^{-2}$
$ \bar{b} = 12.783$ $b = 0.1278 \times 10^2$ $e_b = 0.003 = 0.3000 \times 10^{-2}$

$$\frac{e_m}{\bar{m}} = \frac{0.2000 \times 10^{-2}}{0.34612 \times 10^2} + \frac{0.3000 \times 10^{-2}}{0.12783 \times 10^2}$$

$$= 0.5778 \times 10^{-4} + 0.2346 \times 10^{-3} = 2.923 \times 10^{-4} = 0.0002923 = 0.029\% \text{ relative error}$$

By comparing the relative errors of **(a)** and **(b)** of Example 3-17, you can see that it makes a great deal of difference which operation is being done and what magnitude of numbers we are using.

SUMMARY

1. When evaluating an expression, the order in which you perform the operations is very important.
2. Parentheses take precedence over all arithmetic operations. Unary operations go first, then exponentiation, then the multiplication and division operations, from left to right. Then do the addition and subtraction operations, from left to right. You can remember this order using PEMDAS—Parentheses, Exponentiation, Multiplication, Division, Addition, Subtraction.
3. In order for computers to store numbers effectively, each value is translated to a normalized exponential form $M \times b^e$, where $0.1 \leq M < 1$, $b > 1$, and e is an integer. A precision number t dictates the number of significant digits in the mantissa M.
4. A computer word that stores a number usually has three fields: one for the sign, one for the exponent, one for the mantissa. The hardware construction determines the base value and the precision value.
5. A computer representing numbers in decimal form would likely use 8-digit words; a computer using binary form would likely use 32-bit words (1 bit = 1 binary digit).
6. When computers process real numbers in arithmetic operations, errors are inevitable. Usually round-off in the form of truncating numbers is the chief cause.
7. The common sources of error in computation are initial error—when the incoming data is faulty; rounding error—when numbers inside the machine are chopped to fit the precision length; and truncation error—when an infinitely defined function is terminated after a finite number of terms.
8. Although the error in any individual computation may be quite small, the accumulation or propagation of small errors could produce a sizeable error when many successive operations are perfomed.

RAISE YOUR GRADES

Can you...?

☑ explain the difference between unary and binary operations
☑ recite the precedence order of all the arithmetic operations
☑ change any real number into either scientific notation or normalized exponential form
☑ explain the use of fields in representing a number in computer memory
☑ convert a real number into a string of digits that the computer memory can store efficiently
☑ perform the operations in floating point arithmetic
☑ explain absolute error and relative error
☑ give the three main sources of error
☑ calculate the relative error for each of the simple arithmetic operations

SOLVED PROBLEMS

Order of Operations

PROBLEM 3-1   Find the value of the following:

(a) $8 + 7 - 6 - 5$ (c) $8 - 7 * 6 + 5$
(b) $8 * (7 - 6)/5$ (d) $8/(7 - 6) * 5$

Solution   Always use the PEMDAS order:

(a) $8 + 7 - 6 - 5 = (8 + 7) - 6 - 5 = 15 - 6 - 5 = (15 - 6) - 5 = 9 - 5 = 4$
(b) $8 * (7 - 6)/5 = 8 * (1)/5 = (8 * 1)/5 = 8/5$

(c) $8 - 7 * 6 + 5 = 8 - 42 + 5 = -34 + 5 = -29$
(d) $8/(7 - 6) * 5 = 8/1 * 5 = (8/1) * 5 = 8 * 5 = 40$

PROBLEM 3-2   Find the value of the following:

(a) $4 * 3 \uparrow 2 - 1$ **(d)** $2 \uparrow (4 - 3 * 1)$
(b) $4/(3 - 1) \uparrow 2$ **(e)** $1 \uparrow (2 \uparrow (3 - 4)$
(c) $3 * (4 \uparrow 2 * 1)$

Solution   Use PEMDAS; work parentheses first, then exponentiation, then ordinary arithmetic.

(a) $4 * 3 \uparrow 2 - 1 = 4 * 9 - 1 = 36 - 1 = 35$
(b) $4/(3 - 1) \uparrow 2 = 4/(2) \uparrow 2 = 4/4 = 1$
(c) $3 * (4 \uparrow 2 * 1) = 3 * (16 * 1) = 3 * 16 = 48$
(d) $2 \uparrow (4 - 3 * 1) = 2 \uparrow (4 - 3) = 2 \uparrow 1 = 2$

(e) $1 \uparrow (2 \uparrow (3 - 4)) = 1 \uparrow (2 \uparrow (-1)) = 1 \uparrow \left(\dfrac{1}{2^1}\right) = 1 \uparrow \left(\dfrac{1}{2}\right) = \sqrt{1} = 1$

PROBLEM 3-3   Find the value of

(a) $2 \uparrow -3 * 8/4$ **(c)** $5 - ((4 \uparrow -1) * 8) + 2$
(b) $-(2 * 3 \uparrow 3)/9$ **(d)** $7/(-2 \uparrow -3) * 4/5$

Solution   Note the unary operations first. Then proceed with PEMDAS.

(a) $2 \uparrow (-3) * 8/4 = \dfrac{1}{2^3} * 8/4 = \left(\dfrac{1}{8}\right) * 8/4 = 1/4 = \dfrac{1}{4}$

(b) $-(2 * 3 \uparrow 3)/9 = -(2 * 27)/9 = -(54)/9 = -6$

(c) $5 - ((4 \uparrow (-1)) * 8) + 2 = 5 - \left(\dfrac{1}{4} * 8\right) + 2 = 5 - 2 + 2 = 5$

(d) $7/((-2) \uparrow (-3)) * 4/5 = 7 \Big/ \left(\dfrac{1}{(-2)^3}\right) * 4/5$

$$= 7 \Big/ \left(-\dfrac{1}{8}\right) * 4/5$$

$$= -56 * 4/5 = -224/5 = -44\tfrac{4}{5}$$

PROBLEM 3-4   Evaluate the "monster expression"

$$2 \uparrow 0 * 16 * (3 * (4 \uparrow -2)/4 \uparrow -1) \uparrow 2/9 - 1$$

Solution   Perform first all unary operations, then parentheses, then all exponentiation.

$$2 \uparrow 0 * 16 * \left(3 * \left(\dfrac{1}{4^2}\right) \Big/ \left(\dfrac{1}{4}\right)\right) \uparrow 2/9 - 1$$

$$2 \uparrow 0 * 16 * \left(\dfrac{3}{16} \Big/ \dfrac{1}{4}\right) \uparrow 2/9 - 1$$

$$2 \uparrow 0 * 16 * \left(\dfrac{3}{4}\right) \uparrow 2/9 - 1$$

$$1 * 16 * \left(\dfrac{9}{16}\right) \Big/ 9 - 1$$

$$16 * \dfrac{9}{16} \Big/ 9 - 1$$

$$9/9 - 1$$

$$1 - 1$$

$$0$$

Number Notation for Computers

PROBLEM 3-5 Convert each real number to normalized exponential form (base 10), using precision value $t = 4$: (a) 5, (b) 25.6, (c) 0.005168, (d) 6 billion, (e) -426.18, (f) 1.00038.

Solution The form to match is $M \times 10^e$ where $0.1 \le M < 1$ and e is an integer exponent.

(a) $5 = 0.5000 \times 10^1$
(b) $25.6 = 0.2560 \times 10^2$
(c) $0.005168 = 0.5168 \times 10^{-2}$
(d) $6{,}000{,}000{,}000 = 0.6000 \times 10^{10}$
(e) $-426.18 = -0.4261 \times 10^3$
(f) $1.00038 = 0.1000 \times 10^1$ (Note the large rounding error here.)

PROBLEM 3-6 Convert each real number to normalized exponential form (base 2), using precision value $t = 8$: (a) 16, (b) 3.625, (c) -0.01171875.

Solution First convert the real number to binary form; then to the nomalized form.

(a) $16_{10} = 10\,000_2 = 0.10000000 \times 2^5$
(b) $3.625_{10} = 2 + 1 + \frac{1}{2} + \frac{1}{8} = 11.101_2 = 0.11101000 \times 2^2$
(c) $-0.01171875 = -(\frac{1}{128} + \frac{1}{256}) = -(2^{-7} + 2^{-8}) = -0.00000011_2$
$\qquad = -0.11000000 \times 2^{-6}$

PROBLEM 3-7 Convert each real number to scientific notation: (a) $\dfrac{1}{5\,000\,000}$, (b) 123456.78, (c) -523.746.

Solution

(a) $\dfrac{1}{5\,000\,000} = \dfrac{1}{5} \cdot \dfrac{1}{1\,000\,000} = 0.2 \times 10^{-6} = 2.0 \times 10^{-7}$ *Note:* Exponent $-6 + -1$ (one move to the right) = exponent -7.
(b) $123456.78 = 1.2345678 \times 10^5$
(c) $-523.746 = -5.23746 \times 10^2$

Floating Point Representation

PROBLEM 3-8 Create the binary internal computer representation using a field of 7 for the characteristic and a field of 24 for the mantissa for (a) 1326.75_{10}, (b) -3998753_{10}.

Solution Convert first to binary form, then convert to normalized exponential form, then calculate the characteristic and convert it to binary.

(a) $1326.75 = 1024 + 256 + 32 + 8 + 4 + 2 + \frac{1}{2} + \frac{1}{4}$
$\qquad = 10100101110.11_2$
$\qquad = 0.1010010111011 \times 2^{11}$
Since $e = 11$ and $s = 7$, $c = 2^{s-1} + e = 2^6 + 11 = 75 = 64 + 8 + 2 + 1 = 1\,001\,011_2$. The internal representation is

$+$ sign $c = 7$ bits $m = 24$ bits

(b) $-3\,998\,753 = -(2\,097\,152 + 1\,048\,576 + 524\,288 + 262\,144 + 65\,536 + 1024 + 32 + 1)$
(Find these by deducting from x the largest power of 2 possible.)
$\qquad = -(2^{21} + 2^{20} + 2^{19} + 2^{18} + 2^{16} + 2^{10} + 2^5 + 2^0)$
$\qquad = -1\,111\,010\,000\,010\,000\,100\,001_2$
$\qquad = -0.111101000001000100001 \times 2^{22}$

Since $e = 22$, $s = 7$, then $c = 2^6 + 22 = 64 + 16 + 4 + 2 = 1\,010\,110_2$. The internal representation is

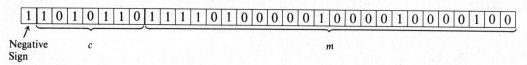

Negative c m
Sign

Note that we are close to losing some accuracy in the mantissa with this large number.

Floating Point Arithmetic

PROBLEM 3-9   Perform the floating point addition using a precision number $t = 4$ for

(a) $0.9126 \times 10^2 + 0.3287 \times 10^2$ (b) $0.2598 \times 10^3 + 0.1749 \times 10^4$

Solution

(a) 0.9126×10^2
    $+0.3287 \times 10^2$
    $\overline{1.2413 \times 10^2} \longrightarrow 0.1241 \times 10^3$ Truncated normalized form

(b) $0.2598 \times 10^3 \longrightarrow 0.0259 \times 10^4$
    $0.1749 \times 10^4 \qquad +0.1749 \times 10^4$
    $\qquad\qquad\qquad \overline{0.2008 \times 10^4}$ Truncated sum

PROBLEM 3-10   Perform the floating point operations with precision $t = 4$ for

(a) $(0.4727 \times 10^5) - (0.1948 \times 10^4)$ (b) $(0.6113 \times 10^2) \times (0.1286 \times 10^4)$
(c) $(0.8927 \times 10^1) \div (0.4196 \times 10^3)$

Solution

(a) $0.4727 \times 10^5 \qquad\qquad 0.4727 \times 10^5$
    $-0.1948 \times 10^4 \longrightarrow -0.0194 \times 10^5$
    $\qquad\qquad\qquad\qquad \overline{0.4533 \times 10^5}$ Truncated difference

(b) $(0.6113 \times 10^2) \times (0.1286 \times 10^4) = (0.6113 \times 0.1286) \times 10^{2+4} = 0.07861 \times 10^6 = 0.7861 \times 10^5$

(c) $(0.8927 \times 10^1) \div (0.4196 \times 10^3) = (0.8927 \div 0.4196) \times 10^{1-3} = 2.1275 \times 10^{-2} = 0.2127 \times 10^{-1}$

Error Analysis in Computation

PROBLEM 3-11   Calculate the absolute and relative error in the computation $0.6283 \times 10^3 + 0.7496 \times 10^3$.

Solution   The true value of the sum is 1.3779×10^3 but the truncation in normalized exponential form gives the result 0.1377×10^4. The absolute error then is

$$e_a = |a - \bar{a}| = 0.00009 \times 10^3 = 0.9000 \times 10^{-1} \qquad \text{The exponent is } 3 + -4 \text{ (4 moves to the right)} = -1.$$

and the relative error is

$$r = \frac{0.9000 \times 10^{-1}}{1.3779 \times 10^3} = 0.6531 \times 10^{-4} \approx 0.00006$$

This is a very small relative error at 0.006%.

PROBLEM 3-12   The formula for computing e^x for any x is

$$e^x = 1 + x + \frac{x^2}{2!} + \frac{x^3}{3!} + \frac{x^4}{4!} + \cdots$$

In computing $e^{1.47}$ we will use only the first five terms. What is the error made in truncating this series?

Solution By the formula with five terms:

$$e^{1.47} \approx 1 + (1.47) + \frac{(1.47)^2}{2} + \frac{(1.47)^3}{6} + \frac{(1.47)^4}{24}$$

$$\approx 1 + (1.47) + 1.08045 + 0.52942 + 0.19456$$

$$\approx 4.2744$$

By a calculator computation $e^{1.47} = 4.3492$. So the truncation error is $4.3492 - 4.2744 = 0.0748$.

   If we had used one additional term in the approximate computation, namely $(1.47)^5/120$, whose value is 0.0572, we could have cut the error considerably. And one more term beyond that one would have cut the error yet another 0.0163. Good judgment is important when deciding where to perform the truncation.

Propagation of Error

PROBLEM 3-13 ·Given that a computer uses a precision value $t = 4$ (i.e., the mantissa in the normalized form retains four digits), find the relative error in the arithmetic operation of adding $x = \frac{2}{7}$ to itself seven times.

Solution The value x will be stored as 0.2857. The relative error in this first computation is

$$r = \frac{\frac{2}{7} - 0.2857}{\frac{2}{7}} \approx 0.00005$$

Now we add x to itself.

$$2x = 0.2857 + 0.2857 = 0.5714 \times 10^0$$

$$3x = 0.5714 + 0.2857 = 0.8571 \times 10^0$$

$$4x = 0.8571 + 0.2857 = 1.1428 \times 10^0 = 0.1142 \times 10^1$$

$$5x = (0.1142 \times 10^1) + (0.2857 \times 10^0)$$
$$= (0.1142 \times 10^1) + (0.0285 \times 10^1)$$
$$= 0.1427 \times 10^1$$

$$6x = (0.1427 \times 10^1) + (0.2857 \times 10^0)$$
$$= (0.1427 \times 10^1) + (0.0285 \times 10^1)$$
$$= 0.1712 \times 10^1$$

$$7x = (0.1712 \times 10^1) + (0.2857 \times 10^0)$$
$$= (0.1712 \times 10^1) + (0.0285 \times 10^1)$$
$$= 0.1997 \times 10^1$$

So $7x = 7(\frac{2}{7}) \approx 1.997$.
The absolute error is $2.000 - 1.997 = 0.003$ and the relative error is $0.003/2 = 0.0015$, which is *30 times* the original relative error. It is obvious that the error has propagated through the calculation.

Supplementary Exercises

PROBLEM 3-14   Find the values of the expressions

(a) $(9 \uparrow -2 * 16) \uparrow 0.5 * (9/2 \uparrow 2)$ **(b)** $(-3 * -4) * -5/2 \uparrow 3 * 4 \uparrow 0.05/5 - 3$ **(c)** $(4 \uparrow 3) \uparrow (2/3 \uparrow 0)$

PROBLEM 3-15   Find the value of the expressions

(a) $(3.5 * 4/7) \uparrow 6 - 50$ **(b)** $8/7/6 * 21/4 - 3 \uparrow 0$ **(c)** $-3 \uparrow -2 - 2/4 \uparrow +2$

PROBLEM 3-16   Let $A = 2$, $B = 4$, $C = -3$, $D = 5$, $E = 1$. Find the value of the following:

(a) $D \uparrow A - B + C * E/B$ **(b)** $A - E/D * A \uparrow C$ **(c)** $C * -B + C \uparrow A/C * -A$

PROBLEM 3-17   Find the base 10 normalized exponential form using precision value $t = 5$ for **(a)** 456.7, **(b)** 0.0010203, **(c)** 14 trillion.

PROBLEM 3-18   Find the base 2 normalized exponential form using precision value $t = 8$ for **(a)** 438.25, **(b)** 9.9375, **(c)** 14.90625.

PROBLEM 3-19   A positive binary number in 32-bit format has a characteristic 1 001 101 and a mantissa 100 011 011 011 000 000 000 000. What is the number in decimal format?

PROBLEM 3-20   Write the following real numbers in scientific notation: **(a)** 0.0372, **(b)** -10.0085, **(c)** 4 003 056, **(d)** $-1/400\,000$.

PROBLEM 3-21   To input large or small numbers most machines require exponential notation. Often the letter E is used to designate an exponential base 10. Thus the number 5280 may be written as $5.28E + 3$. Write the following numbers in exponential notation: **(a)** 98765, **(b)** 0.0042, **(c)** 11 million, **(d)** $1/800\,000$.

PROBLEM 3-22   Change the following from exponential notation to real number form: **(a)** $0.486E - 3$, **(b)** $5.762E + 5$, **(c)** $0.5638E + 6$, **(d)** $93.74E - 5$.

PROBLEM 3-23   Suppose you have a 32-bit machine. Create the internal computer representation for the following numbers given that the characteristic field is seven bits long: **(a)** 469.53125, **(b)** -27843, **(c)** 255.96875.

PROBLEM 3-24   Using precision value $t = 4$, perform the following floating point addition operations:

(a) $(0.7288 \times 10^2) + (0.4963 \times 10^2)$ **(b)** $(0.5361 \times 10^{-3}) + (0.2792 \times 10^{-2})$

PROBLEM 3-25   Using precision value $t = 4$, perform the following floating point operations:

(a) $(0.5912 \times 10^2) - (0.4498 \times 10^1)$ **(b)** $(0.4847 \times 10^{-1}) \times (0.5119 \times 10^2)$
(c) $(0.2468 \times 10^3) \div (0.0481 \times 10^{-4})$

PROBLEM 3-26   Obtain the results of the following operations assuming precision $t = 4$:

(a) $0.8134 \times 10^{-2} - 0.8127 \times 10^{-2}$ **(b)** $0.3145 \times 10^3 \div 0.6621 \times 10^{-5}$
(c) $0.9999 \times 10^{-2} \times 0.7777 \times 10^4$

PROBLEM 3-27   Calculate the absolute and relative errors in the floating point computations for

(a) $0.1593 \times 10 + 0.9296 \times 10$ **(b)** $0.4444 \times 10 + 0.5599 \times 10$.

PROBLEM 3-28   The infinite series formula for finding $\sin x$ is:

$$\sin x = x - \frac{x^3}{3!} + \frac{x^5}{5!} - \frac{x^7}{7!} + \frac{x^9}{9!} - \cdots$$

If we are finding sin(0.95), how large is the truncation error if we use **(a)** 5 terms? **(b)** 4 terms? **(c)** 3 terms?

PROBLEM 3-29   Using precision value $t = 4$, find the relative error in the arithmetic operation of adding $\frac{2}{3}$ to itself nine times.

PROBLEM 3-30   If the actual distance from the earth to the sun on January 1 is 91,637,145 miles and it is approximated as 92,000,000, what are the absolute and relative errors made?

PROBLEM 3-31 The actual distance from the center of the earth to the equator at sea level is $6.36816E + 6$ meters. If this value is roughly estimated as $6.4E + 6$ meters, what are the absolute and relative errors made?

PROBLEM 3-32 We want to find the value of $x - y$ where $x = 333.66$ and $y = 333.22$. If our computer uses precision value $t = 4$, find the relative error in the computation.

Answers to Supplementary Exercises

3-14 (a) 1 (b) −6 (c) 4096

3-15 (a) 14 (b) 0 (c) $-\frac{1}{72}$

3-16 (a) $20\frac{1}{4}$ (b) $\frac{79}{40}$ (c) 18

3-17 (a) 0.45670×10^3 (b) 0.10203×10^{-2}
(c) 0.14000×10^{14}

3-18 (a) 0.11011011×2^9 (b) 0.10011111×2^4
(c) 0.11101110×2^4

3-19 4534

3-20 (a) 3.72×10^{-2} (b) -1.00085×10^1
(c) 4.003056×10^6 (d) -2.5×10^{-6}

3-21 (a) $9.8765E + 4$ (b) $4.2E - 3$ (c) $1.1E + 7$
(d) $1.25E - 6$

3-22 (a) 0.000486 (b) 576200 (c) 563800
(d) 0.0009374

3-23 (a) 01 001 001 111 010 101 100 010 000 000 000
(b) 11 001 111 110 110 011 000 011 000 000 000
(c) 01 001 000 111 111 111 111 100 000 000 000

3-24 (a) 0.1225×10^3 (b) 0.3328×10^{-2}

3-25 (a) 0.5463×10^2 (b) 0.2481×10^1
(c) 0.5130×10^8

3-26 (a) 0.7000×10^{-5} (b) 0.4750×10^8
(c) 0.7776×10^2

3-27 (a) 0.006 and 0.023%
(b) 0.000004 and 0.066%

3-28 (a) 0.000000018 (b) 0.0000017 (c) 0.00014

3-29 0.0083 or 0.083%

3-30 362,855. and 0.396%

3-31 21 840 and 0.342%

3-32 9.09%

# 4 ALGEBRAIC OPERATIONS WITH COMPUTERS

THIS CHAPTER IS ABOUT

☑ **Solving Equations**
☑ **Obtaining Approximate Solutions—The Half-Interval Method**
☑ **Evaluating Polynomials—Horner's Method**
☑ **Linear Interpolation**
☑ **Polynomial Interpolation**
☑ **Congruences**
☑ **Random Number Generation**

4-1. Solving Equations

recall: An equation is a symbolic statement of equality between two quantities. To **solve an equation** is to find a set of numbers which make the equation true (that is, that change it into an identity). When solving an equation, remember to follow the commonly accepted **axioms of equality** to reduce the equation to a simpler form:

AXIOMS OF EQUALITY

(1) If $x = y$, then $x + a = y + a$
(2) If $x = y$, then $x - a = y - a$
(3) If $x = y$ and $m \neq 0$, then $xm = ym$
(4) If $x = y$ and $n \neq 0$, then $x/n = y/n$

A good way to sum up these axioms is: To retain equality, always do to one side of the equation what you do to the other side.

A. Linear equations in one unknown

An equation that can be written in the form $ax + b = 0$ is a **linear equation in one unknown**. To solve a linear equation in one unknown, you simply apply the axioms of equality.

EXAMPLE 4-1 Use the axioms of equality to solve the following equations:

(a) $4(x - 2) = 3 + 5(x - 3)$
(b) $5y - 4(y + 1) + 7(2y - 3) = 8$
(c) $4z - 3 + 5(z - 6) = 9z - 33$
(d) $2(7t - 1) - 6(2t + 4) = 2(t + 5)$

Solution

(a)
$$4(x - 2) = 3 + 5(x - 3)$$
$$4x - 8 = 3 + 5x - 15$$
$$4x - 8 = 5x - 12$$
$$4x - 8 - 5x = 5x - 12 - 5x \qquad \text{Axiom 2}$$
$$-x - 8 = -12$$
$$-x - 8 + 8 = -12 + 8 \qquad \text{Axiom 1}$$
$$-x = -4$$
$$x = 4 \qquad \text{Axiom 4}$$

Check: $4(4 - 2) \overset{?}{=} 3 + 5(4 - 3)$
$4(2) \overset{?}{=} 3 + 5(1)$
$8 \overset{\checkmark}{=} 8$

78

(b) $5y - 4(y + 1) + 7(2y - 3) = 8$

$\qquad 5y - 4y - 4 + 14y - 21 = 8$

$\qquad\qquad\qquad 15y - 25 = 8$

$\qquad\qquad 15y - 25 + 25 = 8 + 25 \qquad$ Axiom 1

$\qquad\qquad\qquad\qquad 15y = 33$

$\qquad\qquad\qquad\qquad\quad y = \frac{33}{15} \qquad$ Axiom 4

(c) $4z - 3 + 5(z - 6) = 9z - 33$

$\qquad 4z - 3 + 5z - 30 = 9z - 33$

$\qquad\qquad\quad 9z - 33 = 9z - 33$

The equation has become an identity, thus every value of z is a solution.

(d) $2(7t - 1) - 6(2t + 4) = 2(t + 5)$

$\qquad 14t - 2 - 12t - 24 = 2t + 10$

$\qquad\qquad\qquad 2t - 26 = 2t + 10$

$\qquad\quad 2t - 26 - 2t = 2t + 10 - 2t$

$\qquad\qquad\qquad\quad -26 = 10$

This equation has become an impossible case, so no value of t is a solution.

Now to apply the axioms of equality to a word problem.

EXAMPLE 4-2   Suppose you need to cut a piece of carpet into a rectangle whose area is 162 ft^2 and whose length is twice its width. What will be the dimensions of the carpet?

Solution   In cutting the carpet, set width $= x$. Then length is $2x$. But the area is

$$A = (\text{length})(\text{width}) = (2x)(x) = 162$$

Setting this up as an equation, we have

$$2x^2 = 162$$
$$x^2 = 81$$

Thus $x = \pm 9$. Using -9 for width, however, is impossible. So width is 9 feet and the length is twice that, or 18 feet. Checking shows that Area $= (18 \text{ ft})(9 \text{ ft}) = 162 \text{ ft}^2$.

B. Higher-degree equations in one unknown

Simple rules apply to solving linear equations. But if the variable is raised to a positive integer power n greater than 1, we have a **higher-degree equation in one unknown**, and the solution process now is more complicated.

recall: If the variable is raised to a positive integer power n, we say that the equation has **degree n**.

The standard form of the second-degree equation, also called the **quadratic equation**, is

$$ax^2 + bx + c = 0$$

This may always be solved by the **quadratic formula**:

$$x = \frac{-b \pm \sqrt{b^2 - 4ac}}{2a}$$

which produces one of the following solutions:

- If the **discriminant** $b^2 - 4ac$ is positive, we get two real distinct roots.
- If the discriminant is zero, we have one real repeated root.
- If the discriminant is negative, we have *complex conjugate roots*.

  note: In real arithmetic we cannot have a negative quantity under the square root sign. But the number $i = \sqrt{-1}$ has been invented to allow us to write these kinds of things in a simpler form. So if $b^2 - 4ac$ is negative, then

$$\sqrt{b^2 - 4ac} = \sqrt{-1(4ac - b^2)} = i\sqrt{4ac - b^2}$$

is used in the quadratic formula. The roots then become

$$x = \frac{-b \pm i\sqrt{4ac - b^2}}{2a} = \frac{-b}{2a} \pm i\left(\frac{\sqrt{4ac - b^2}}{2a}\right)$$

These are complex numbers (note their form as $A \pm iB$) which are complex conjugates of each other.

We can create an algorithm in pseudocode to solve the quadratic equation by writing the formula in computer notation (Section 3-1) as:

$$R1 = (-B + (B\uparrow 2 - 4 * A * C)\uparrow 0.5)/(2 * A)$$

$$R2 = (-B - (B\uparrow 2 - 4 * A * C)\uparrow 0.5)/(2 * A)$$

where $R1$ and $R2$ are the respective roots. To save computer work, however, we'll use the fact that the product of the roots is $R1 \cdot R2 = C/A$ and solve for the second root in terms of the first. We then have $R2 = C/(A * R1)$. An algorithm for solving the quadratic equation would therefore be:

ALGORITHM FOR THE QUADRATIC FORMULA

Begin
1. Input A, B, C
2. $D \leftarrow B\uparrow 2 - 4 * A * C$ D is the discriminant.
3. IF $D < 0$ THEN
   a. Output "COMPLEX ROOTS OF FORM $X + iY$
      AND $X - iY$ WHERE $X = $"; $-B/(2 * A)$;
      "AND $Y = $"; $\mathrm{SQR}(-D)/(2 * A)$
4. ELSE
   a. $R1 \leftarrow (-B + \mathrm{SQR}(D)/(2 * A)$
   b. $R2 \leftarrow C/(A * R1)$
5. Output $R1, R2$
End

EXAMPLE 4-3 Use the quadratic formula algorithm to find the solutions to

(a) $2x^2 + 5x - 3 = 0$ **(b)** $x^2 - 10x + 25 = 0$ **(c)** $x^2 + 2x + 5 = 0$

Solution By the quadratic formula algorithm we get

(a) $D = 25 - 4(2)(-3) = 49$ and

$$R1 = \frac{-5 + \sqrt{49}}{4} = \frac{-5 + 7}{4} = \frac{1}{2} \quad \text{and} \quad R2 = \frac{-3}{(2 \cdot \frac{1}{2})} = -3$$

(b) Here $D = 100 - 4(1)(25) = 0$ so

$$R1 = \frac{10 \pm 0}{2} = 5 \quad \text{and} \quad R2 = \frac{25}{(1 \cdot 5)} = 5$$

(c) Here $D = 4 - 4(1)(5) = -16$ so $-B/(2 * A) = -2/2 = -1$ and $\mathrm{SQR}(-D)/(2 * A) = 4/2 = 2$. The roots are $R1 = -1 + 2i$ and $R2 = -1 - 2i$.

Explicit solution formulas do exist for third-degree and fourth-degree equations but their forms are very complicated. The general fifth-degree equation does not have an explicit solution.

C. Linear equations in two unknowns

An equation of the form $ax + by = c$ where a, b, and c are real constants is a **linear equation in two unknowns**. If both a and b are nonzero, then a solution is a pair of real numbers (u, v) that makes the equation an identity. A general, natural language algorithm for finding a solution pair for a linear equation in two variables would be:

ALGORITHM FOR SOLVING A LINEAR EQUATION IN TWO UNKNOWNS

1. Assign a value to one variable.
2. Solve for the second variable using the axioms of equality.
3. Output the x and y values as an ordered pair.

There will usually be an infinite number of solution pairs for a given linear equation in two unknowns.

EXAMPLE 4-4 Find three solution pairs (u, v) that solve the linear equation $4x - 3y = 6$.

Solution Let x be the variable to which we assign values. Follow the algorithm.

Solution Pairs

For $x = 1$, we have
$$4 - 3y = 6$$
$$-3y = 6 - 4 = 2$$
$$y = -\frac{2}{3}$$
$\left(1, -\frac{2}{3}\right)$

For $x = 2$, we have
$$8 - 3y = 6$$
$$-3y = 6 - 8 = -2$$
$$y = \frac{2}{3}$$
$\left(2, \frac{2}{3}\right)$

For $x = -3$, we have
$$-12 - 3y = 6$$
$$-3y = 6 + 12 = 18$$
$$y = -6$$
$(-3, -6)$

Plugging in more values of x in Example 4-4 will produce additional number pairs, all of which have a common geometric property: Each pair, in the Euclidian plane, lies on the same straight line. The line is called the **graph** of the linear equation (Figure 4-1).

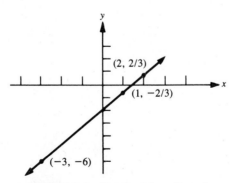

FIGURE 4-1 Graph of $4x - 3y = 6$.

D. Systems of two linear equations

A **system of equations** is simply a collection of two or more equations that share the same variables and that are somehow related to each other. For example,

$$\begin{Bmatrix} x + y = -11 \\ xy = 14 \end{Bmatrix} \quad \text{and} \quad \begin{Bmatrix} 2w + y = 1 \\ -3w = 12 \end{Bmatrix}$$

are both systems of two equations in two unknowns. A *solution* of a system in two variables is an ordered pair (x, y) that solves all equations in the system. To find the ordered pair(s) (x, y) of a system is known as *solving* the system. The general form of a system of two linear equations in two unknowns looks like this:

SYSTEM OF TWO LINEAR EQUATIONS IN TWO UNKNOWNS
$$\begin{cases} a_1 x + b_1 y = c_1 \\ a_2 x + b_2 y = c_2 \end{cases}$$

where $a_1, a_2, b_1, b_2, c_1, c_2$ are real constants.

Although there are more advanced techniques for solving a system of two linear equations, we do have a general algorithm available that will serve us nicely:

ALGORITHM FOR SOLVING A SYSTEM OF TWO LINEAR EQUATIONS IN TWO UNKNOWNS

1. Divide the first equation by a_1, the coefficient of x.
2. Solve the first equation for x in terms of y.
3. Insert this expression for x in place of every x in the second equation.
4. Solve the resulting single-variable equation for y.
5. Plug the value of y into the equation for x formed in step 2.
6. Output the values of x and y as an ordered pair.

EXAMPLE 4-5   Solve $\begin{cases} 4x + 8y = 20 \\ 3x + 4y = 11 \end{cases}$.

Solution   Follow the algorithm for solving a system of two linear equations in two unknowns.

1. Divide the first equation by 4:
$$\begin{cases} 4x + 8y = 20 \\ 3x + 4y = 11 \end{cases} = \begin{cases} x + 2y = 5 \\ 3x + 4y = 11 \end{cases}$$

2. Solve the first equation for x:
$$x = 5 - 2y$$

3. Insert this into the second equation:
$$3(5 - 2y) + 4y = 11$$

4. Solve this single-variable equation:
$$15 - 6y + 4y = 11$$
$$15 - 2y = 11$$
$$-2y = -4$$
$$y = 2$$

5. Plug this y value into the equation for x:
$$x = 5 - 2y = 5 - 2(2) = 5 - 4 = 1$$

6. The solution is the pair $(1, 2)$.

Sometimes you have to create a system of equations yourself to solve a problem, as in the next example.

EXAMPLE 4-6   Two boys working together can rake a large yard in $3\frac{3}{4}$ hours. If Mark works alone for 3 hours and then Dan finishes the job by working alone for another 5 hours, how long would it take each boy to do the entire yard alone?

Solution   Let x be the number of hours it would take Mark to do the job alone and let y be the number of hours it would take Dan to do the job alone. Then in one hour, Mark can do $1/x$ of the job and Dan can do $1/y$ of the job. We know that Mark, when working alone, toils for 3 hours, so he does $3/x$ of the job. Likewise Dan, when working alone, completes in 5 hours the remaining $5/y$ of the job. Count the entire job as 1 unit. Hence

$$\frac{3}{x} + \frac{5}{y} = 1 \quad \longleftarrow \quad \text{First equation for the system}$$

Since the original condition of the boys' working together gets the job done in $3\frac{3}{4}$ hours, we can see that Mark's $3\frac{3}{4}$ hours plus Dan's $3\frac{3}{4}$ hours $= 1$ job. We therefore write

$$\frac{3\frac{3}{4}}{x} + \frac{3\frac{3}{4}}{y} = 1 \longleftarrow \text{ Second equation for the system}$$

We now have a system of equations to solve:

$$\left\{ \begin{array}{l} \dfrac{3}{x} + \dfrac{5}{y} = 1 \\[2mm] \dfrac{3\frac{3}{4}}{x} + \dfrac{3\frac{3}{4}}{y} = 1 \end{array} \right\} \tag{4-1}$$

We can run this through the algorithm now, or we can make it easier on ourselves and simplify the system by setting $1/x = U$ and $1/y = V$. Then the system is

$$\left\{ \begin{array}{l} 3U + 5V = 1 \\ (3\frac{3}{4})U + (3\frac{3}{4})V = 1 \end{array} \right\} \tag{4-2}$$

We're now ready to apply the algorithm.

1. Divide the second equation by $3\frac{3}{4}$ (i.e., by $\frac{15}{4}$): $\left\{ \begin{array}{l} 3U + 5V = 1 \\ U + V = \frac{4}{15} \end{array} \right\}$

   note: Starting with the second equation keeps the whole process tidier; we must be sure, however, to make the substitution into the *first* equation when we get to step 5.

2. Solve the second equation for U: $U = \frac{4}{15} - V$

3. Insert this into the first equation: $3(\frac{4}{15} - V) + 5V = 1$

4. Solve this for V:
$$\frac{12}{15} - 3V + 5V = 1$$
$$2V = \frac{1}{5}$$
$$V = \frac{1}{10}$$

5. Put this back into the U equation:
$$U = \frac{4}{15} - V = \frac{4}{15} - \frac{1}{10}$$
$$= \frac{8}{30} - \frac{3}{30} = \frac{5}{30} = \frac{1}{6}$$

6. Therefore, the solution set is $\{U = \frac{1}{6}, V = \frac{1}{10}\}$

We have now solved the system in eq. (4-2), but we have yet to reach the answer we seek from the system in eq. (4-1). Remember that we substituted $U = 1/x$ and $V = 1/y$, so we have:

$$U = \frac{1}{6} = \frac{1}{x} \qquad V = \frac{1}{10} = \frac{1}{y}$$

$$x = 6 \qquad\qquad y = 10$$

The original system therefore has the solution $\{x = 6, y = 10\}$. That is, Mark and Dan would have to work 6 and 10 hours, respectively, to rake the lawn if working alone.

A system of two linear equations has one of three possible solution sets. It will have one solution pair, an infinite number of solution pairs, or no solution pairs. There is a handy way to quickly check what type of solution a system of two equations will produce. Given any system

$$\left\{ \begin{array}{l} a_1 x + b_1 y = c_1 \\ a_2 x + b_2 y = c_2 \end{array} \right\}$$

examine the various **ratios of the coefficients**. One of three conditions will exist:

- If $a_1 b_2 - a_2 b_1 \neq 0$, then the solution is a *single ordered pair*. The graphs of the two equations— which are two straight lines—intersect at exactly one point, the solution point.

- If $a_1b_2 - a_2b_1 = 0$, we have $a_1b_2 = a_2b_1$, or $a_1/a_2 = b_1/b_2$; that is, the coefficients of x and y have the same ratios. Now, if $a_1/a_2 = b_1/b_2 = c_1/c_2$ (that is, if *all* constants share an identical ratio), then the equations are actually the same equation and there are an *infinite number of solution pairs*. The graphs of the two equations are straight lines that lie one upon the other and intersect at every point on them.

- If $a_1b_2 - a_2b_1 = 0$ and $a_1/a_2 = b_1/b_2 \neq c_1/c_2$ (that is, if the ratio of the constants c_1 and c_2 does *not* equal that of the a's and b's), then there is *no solution* to the system. The graphs are two parallel lines, which never intersect and therefore have no solution points.

Having followed through the algorithm for solving a system of two linear equations, we can see that any system of n linear equations in n unknowns,

$$\begin{cases} a_{11}x_1 + a_{12}x_2 + \cdots + a_{1n}x_n = c_1 \\ a_{21}x_1 + a_{22}x_2 + \cdots + a_{2n}x_n = c_2 \\ \vdots \qquad\qquad\qquad\quad \vdots \\ a_{m1}x_1 + a_{m2}x_2 + \cdots + a_{mn}x_n = c_n \end{cases}$$

where $x_1, x_2, \ldots, x_n$ are each a different variable, can be solved in basically the same way. This could get complicated by umpteen substitutions, though, so mathematicians have worked out three **elementary operations** that you may do to a system of linear equations to make it easier to work with or solve. Each operation on a system produces an *equivalent system*, i.e., a system that has the same solution as the original system. You may:

(1) Interchange the order in which two equations occur.
(2) Multiply any equation by a nonzero number.
(3) Add a multiple of one equation to another equation of the system.

There's a special way to use these elementary operations to reduce a system so much that it actually solves itself. Called the **Gauss-Jordan elimination process**, it manipulates a system so that we eventually get each variable alone into its own equation. Since the process is easier to understand once you see how it's done, let's look at another example.

EXAMPLE 4-7   Use the elementary operations to solve the system

$$\begin{cases} x - 2y + z = -1 \\ 2x - 3y + 4z = -5 \\ 3x - 4y + 2z = 1 \end{cases}$$

Solution   By applying the elementary operations, create equivalent systems until each equation has only one (different) variable in it.

$$\begin{cases} x - 2y + z = -1 \\ 2x - 3y + 4z = -5 \\ 3x - 4y + 2z = 1 \end{cases} \quad \begin{matrix} \text{I} \\ \text{II} \\ \text{III} \end{matrix}$$

note: For easier reference, we've numbered the system's equations as I, II, and III.

Look at the first equation and notice that both x and z have a coefficient of 1. Either variable would be equally easy to work with, but we'll choose x as the variable to remain in this equation. Now we want to *eliminate* the variable x from the other two equations because we want only eq. I to have an x. To do this, we can apply elementary operation 3 and add (-2) times eq. I to eq. II to get rid of the x in eq. II. Then we can add (-3) times eq. I to eq. III to get rid of its x:

eq. I: $(-2)(x - 2y + z = -1) \longrightarrow -2x + 4y - 2z = 2$
$\underline{+ \quad 2x - 3y + 4z = -5} \longleftarrow$ eq. II
$0x + y + 2z = -3 \longleftarrow$ new eq. II

eq. I: $(-3)(x - 2y + z = -1) \longrightarrow -3x + 6y - 3z = 3$
$\underline{+ \quad 3x - 4y + 2z = 1} \longleftarrow$ eq. III
$0x + 2y - z = 4 \longleftarrow$ new eq. III

The new (equivalent) system is therefore:

$$\begin{cases} x - 2y + z = -1 \\ y + 2z = -3 \\ 2y - z = 4 \end{cases} \quad \begin{matrix} \text{I} \\ \text{II} \\ \text{III} \end{matrix}$$

Since we've allocated x to eq. I, we must allocate either y or z to eq. II. The variable y has 1 as its coefficient, so it would be the easiest to work with. We therefore must get rid of the y's in eqs. I and III so that only eq. II has a y. Again using elementary operation 3, we now add $(+2)$ times eq. II to eq. I and dispose of the $-2y$ in eq. I. We add (-2) times eq. II to eq. III and eliminate the $+2y$ in eq. III:

$$\text{eq. II:} \quad (2)(y + 2z = -3) \longrightarrow 2y + 4z = -6$$
$$\underline{+x - 2y + z = -1} \longleftarrow \text{eq. I}$$
$$x - 0y + 5z = -7 \longleftarrow \text{new eq. I}$$

$$\text{eq. II:} \quad (-2)(y + 2z = -3) \longrightarrow -2y - 4z = 6$$
$$\underline{+ \quad 2y - z = 4} \longleftarrow \text{eq. III}$$
$$0y - 5z = 10 \longleftarrow \text{new eq. III}$$

And the new system looks like:

$$\begin{cases} x + 5z = -7 \\ y + 2z = -3 \\ -5z = 10 \end{cases} \quad \begin{matrix} \text{I} \\ \text{II} \\ \text{III} \end{matrix}$$

We've now allocated x to eq. I and y to eq. II; therefore z must belong only in eq. III. Before we start to eliminate the z's from eqs. I and II, though, let's make the z in eq. III have a coefficient 1. Apply elementary operation 2 and multiply eq. III by $(-1/5)$ to get:

$$\begin{cases} x + 5z = -7 \\ y + 2z = -3 \\ z = -2 \end{cases} \quad \begin{matrix} \text{I} \\ \text{II} \\ \text{III} \end{matrix}$$

We're now ready to eliminate the z's in I and II. Add (-5) times eq. III to eq. I, and add (-2) times eq. III to eq. II:

$$\text{eq. III:} \quad (-5)(z = -2) \longrightarrow -5z = 10$$
$$\underline{+x + 5z = -7} \longleftarrow \text{eq. I}$$
$$x + 0z = 3 \longleftarrow \text{new eq. I}$$

$$\text{eq. III:} \quad (-2)(z = -2) \longrightarrow -2z = 4$$
$$\underline{+y + 2z = -3} \longleftarrow \text{eq. II}$$
$$y + 0z = 1 \longleftarrow \text{new eq. II}$$

The new and final system is

$$\begin{cases} x = 3 \\ y = 1 \\ z = -2 \end{cases}$$

which is its own solution to the original system of this example.

Congratulations. You have just succeeded with the Gauss-Jordan elimination process. The Supplementary Exercises will help determine if you can "do it by yourself"!

There is an algorithm for this process, and it's based on the following general form for a system of N equations in N unknowns:

$$\begin{cases} a_{11}x_1 + a_{12}x_2 + a_{13}x_3 + \cdots + a_{1n}x_n = c_1 \\ a_{21}x_1 + a_{22}x_2 + a_{23}x_3 + \cdots + a_{2n}x_n = c_2 \\ a_{31}x_1 + a_{32}x_2 + a_{33}x_3 + \cdots + a_{3n}x_n = c_3 \\ \vdots \qquad \vdots \qquad \vdots \qquad \qquad \vdots \qquad \vdots \\ a_{n1}x_1 + a_{n2}x_2 + a_{n3}x_3 + \cdots + a_{nn}x_n = c_n \end{cases}$$

in which each x_i stands for a different variable and in which we label a given coefficient as $a_{i,j}$, where i is the row and j is the column (so that if we're referring to a_{23} we mean the value of the coefficient that is in the second row (i.e., second equation) and the third column (i.e., the third term in the equation). To create the algorithm, we first need a few more identifiers to name things with. Let I be the equation we use to eliminate the i-th unknown from the remaining equations (*note:* We use capital letters to identify an equation and lowercase letters to identify a specific term). That is, we'll keep the first unknown in the first equation and eliminate it from all others, and keep the second unknown in the second equation and eliminate it from all others, and so on, just as we did in Example 4-7.

Now use J as the generic term to describe the equation we're eliminating the i-th unknown from as just described. Thus when $I = J$ we're talking about the same equation, but when $I = 2$ and $J = 3$, we're talking about the second and third equations, respectively. To allow consideration of all of the terms of each equation, rename the constants on the right side to $a_{1,n+1}, a_{2,n+2}, \ldots, a_{n,n+1}$. Thus rename c_p as $a_{p,n+1}$. Now for the algorithm.

GAUSS-JORDAN ELIMINATION PROCESS ALGORITHM

Begin
1. FOR $I = 1$ TO N DO
   a. FOR $J = 1$ TO N DO
      1. IF $J \neq I$ THEN
         When the I-th equation is not the same as the J-th equation, the algorithm will use the I-th equation to eliminate the i-th term from the other equations.

         a. $F \leftarrow a(j,i)/a(i,i)$
         This sets up the correct ratio of the i-th term in I and the i-th term in J so that the i-th term in J can eventually be subtracted out.

         b. FOR $K = 1$ TO $(N + 1)$ DO
            a. $a(j,k) \leftarrow a(j,k) - F * a(i,k)$
         K is a dummy variable to help the algorithm know which terms to compare. The value of K goes to $N + 1$ because we have to multiply both sides of equations I and J when we manipulate them, as in Example 4-7. $N + 1$ takes us to the final column (the constant on the right side of the equality symbol).

2. FOR $I = 1$ TO N DO
   a. $x(i) \leftarrow a(i, n + 1)/a(i,i)$
   b. Output $x(i)$
End

The nested loops in step 1 eventually produce a system of N equations, each equation with only one, unique variable in it. This step performs any final division in case a variable has a coefficient other than 1 (say if the final equation is $-5z = 10$). It outputs x_i ($i = 1, 2, 3, \ldots, n$), the solution values of the system.

A major flaw in this algorithm is the possibility that the coefficient a_{ii} is zero. If this happens, the easiest way to resolve the problem is to interchange this row with another row for which the problem does not exist. You should also always be aware of the possibility of round-off errors that could occur. In certain cases, the error could be sizeable.

4-2. Obtaining Approximate Solutions—The Half-Interval Method

If the equation you are trying to solve contains terms in which the variable is raised to a power that is not an integer, or if the equation contains terms that are not algebraic but instead are exponential or trigonometric, then the standard processes of solving don't apply. A few such equations might be:

$$x^{3/2} - 5x + 3 = 0 \qquad 2x + 1 + \sin 2x = 0 \qquad e^x - 4x = 1$$

To solve this type of equation we usually use an algorithm that generates successive approximations to a root, with each approximation closer than the previous one. The algorithm continues to run until the desired degree of accuracy has been obtained. One such algorithm uses the **half-interval method**, a procedure based on the divide-and-conquer principle outlined in Chapter 1. The half-interval method finds the root, or closest approximation of the root, of an exponential or trigonometric polynomial. Its general algorithm (with explanation) goes like this:

GENERAL ALGORITHM FOR THE HALF-INTERVAL METHOD

1. Using the axioms of equality, adjust the equation so that it has the form $f(x) = 0$.

2. Choose an arbitrary, not-too-big interval (x_1, x_2) so that $f(x_1)$ and $f(x_2)$ have opposite signs. That is, find an x_1 and x_2 such that

$$f(x_1) > 0 \quad \text{and} \quad f(x_2) < 0$$

or $\qquad f(x_1) < 0 \quad \text{and} \quad f(x_2) > 0$

This will mean that the value of x for which $f(x) = 0$ is somewhere in the open interval from x_1 and x_2.

3. Next you must find out where, between x_1 and x_2, the actual root resides, so divide the interval in half and record its midpoint:

$$x_{mid} = \frac{x_1 + x_2}{2}$$

This produces two new open intervals (x_1, x_{mid}) and (x_{mid}, x_2).

4. Plug x_{mid} into your $f(x)$ equation, just in case you stumbled upon the root by taking the midpoint. If $f(x_{mid}) = 0$, you did indeed find it, so stop. If $f(x_{mid}) \neq 0$, however, the value of x for which $f(x) = 0$ is in one of those two new intervals.

5. Now you must find which interval the root is in. To do this, test the endpoints of each interval:

   - Plug x_1 and x_{mid} into your equation. If $f(x_1)$ and $f(x_{mid})$ have opposite signs, then $f(x) = 0$ is in the left-hand interval (x_1, x_{mid}). In this case, rename x_{mid} as x_2 and continue the search by halving your new interval (x_1, x_2) as in step 3.

   - If you didn't find the root in the left-hand interval, plug x_{mid} and x_2 into your equation. If $f(x_{mid})$ and $f(x_2)$ have opposite signs, then $f(x) = 0$ is in the right-hand interval (x_{mid}, x_2). In this case, rename x_{mid} as x_1 and continue the search by halving this new interval (x_1, x_2) as in step 3.

6. Repeat steps 3-5 until x_{mid} meets the precision requirements.

note: By repeating the process, you get a series of progressively smaller and smaller intervals, all containing the root. Your approximation, therefore, gets closer and closer to the actual value of the root you seek. The algorithm will stop after it either (1) finds the actual root by stumbling upon it as x_{mid} or (2) approximates the root to the requested degree of accuracy.

EXAMPLE 4-8 Use the half-interval method to find a root of

$$f(x) = x^{3/2} - 5x = +3$$

Solution We already have the equation in the form $f(x) = 0$. Observe that if we choose $x_1 = 0$ and $x_2 = 1$, we have:

$$f(0) = 0 - 0 + 3 = 3 \quad \text{and} \quad f(1) = 1 - 5 + 3 = -1$$

so a root lies in our initially chosen interval $(0, 1)$. The midpoint is $x_{\text{mid}} = 0.5$, and we check to see if it's the root we seek:

$$f(0.5) = 0.353 - 2.5 + 3 = 0.853$$

It's not the root $[f(0.5) \neq 0]$, so we know a root exists in either $(x_1, x_{\text{mid}}) = (0, 0.5)$, or in $(x_{\text{mid}}, x_2) = (0.5, 1)$. Notice that $f(0)$ is positive, $f(0.5)$ is positive, and $f(1)$ is negative. Therefore, according to step 5 of the half-interval algorithm, the root lies in the interval $(0.5, 1)$. So we go through the process again. The midpoint of the new interval $(0.5, 1)$ is 0.75.

$$f(0.75) = 0.649 - 3.75 + 3 = -0.101$$

It's not a root, but since $f(0.5)$ is positive, $f(0.75)$ is negative, and $f(1)$ is negative, we know the root of $f(x) = 0$ lies in the interval $(0.5, 0.75)$. Back to step 3. The new midpoint is 0.625.

$$f(0.625) = 0.494 - 3.125 + 3 = 0.369$$

This is not a root, but since $f(0.5)$ is positive, $f(0.625)$ is positive, and $f(0.75)$ is negative, the root lies in $(0.625, 0.75)$. The new midpoint is 0.6875.

$$f(0.6875) = 0.570 - 3.4375 + 3 = 0.1325$$

Since $f(0.625)$ is positive, $f(0.6875)$ is positive, and $f(0.75)$ is negative, the root lies in $(0.6875, 0.75)$. The new midpoint is 0.71875.

$$f(0.71875) = 0.609 - 3.59375 + 3 = 0.01525$$

Since $f(0.6875)$ is positive, $f(0.71875)$ is positive, and $f(0.75)$ is negative, the root lies in $(0.71875, 0.75)$. As our final result, we will use the latest midpoint:

$$\text{root} = \frac{0.71875 + 0.75}{2} = 0.734375$$

The process in Example 4-8 does not converge very rapidly to an accurate result, and you'd probably run out of pencil or paper before you reached one by hand. But if we program a computer to do the computations, thousands of iterations are easily done and a very accurate result may be found. An algorithm in pseudocode for the half-interval method is as follows (minor modifications will cover the case when $f(x) > 0$ is on the left and $f(x) < 0$ is on the right):

HALF-INTERVAL ALGORITHM

Begin
1. Input $f(X)$ — $f(X)$ is the function in question.
2. Input $X\text{lft}$, $X\text{rgt}$, p — $X\text{lft}$ (x_{left}) is an arbitrary value that makes $f(x) < 0$ and $X\text{rgt}$ (x_{right}) is an arbitrary value that makes $f(x) > 0$; together they define your first interval. p is the predetermined error bound.
3. $F\text{lft} \leftarrow f(X\text{lft})$ — Put $X\text{lft}$ and $X\text{rgt}$ into your function $f(X)$; call their answers $F\text{lft}$ and $F\text{rgt}$, respectively. Note: $F\text{lft}$ will always be negative and $F\text{rgt}$ will always be positive.

4. $F\text{rgt} \leftarrow f(X\text{rgt})$
5. REPEAT
   a. $X\text{mid} \leftarrow 0.5 * (X\text{lft} + X\text{rgt})$ $X\text{mid}$ (x_{midpoint}) is the midpoint computed from the two end values of your interval.

   b. $F\text{mid} \leftarrow f(X\text{mid})$ Put $X\text{mid}$ into the function; call its answer $F\text{mid}$.

   c. IF $F\text{mid} = 0$ THEN   If the answer to $f(X\text{mid}) = 0$, then
      1. Output $X\text{mid}$ $X\text{mid}$ is the root.
      2. Stop
   d. IF $F\text{mid} < 0$ THEN   If $F\text{mid}$ is negative, your root will be
      1. $X\text{lft} \leftarrow X\text{mid}$ between $X\text{mid}$ and $X\text{rgt}$ (remember
      2. $F\text{lft} \leftarrow F\text{mid}$ that $F\text{rgt}$ is always positive). Rename $X\text{mid}$ as the left side of a new, smaller interval $X\text{lft}$ and $X\text{rgt}$, and rename the function value for use in the next trip through the loop.

   e. ELSE   If $F\text{mid}$ is positive, your root will be
      1. $X\text{rgt} \leftarrow X\text{mid}$ between $X\text{lft}$ and $X\text{mid}$ (remember
      2. $F\text{rgt} \leftarrow F\text{mid}$ that $F\text{lft}$ always is negative). Rename to make $F\text{mid}$ the right-hand side of a new, smaller interval $X\text{lft}$ and $X\text{rgt}$, and rename the function value for use in the next trip through the loop.

6. UNTIL $|F\text{mid}| < p$
7. Output $X\text{mid}$
End

4-3. Evaluating Polynomials—Horner's Method

Scientific and engineering applications often require you to evaluate various mathematical expressions.

note: To **evaluate an expression** means to plug in a value and solve for the answer.

Some expressions may be well-known functions for which value tables exist, but others are complicated enough to require special treatment to solve. One such special type of expression is the polynomial

$$p(x) = a_0 + a_1 x + a_2 x^2 + \cdots + a_n x^n \tag{4-3}$$

It is particularly important that we have a way to evaluate eq. (4-3) with a high degree of accuracy and a minimum number of calculations. (Think about it: The fewer calculations the computer must perform, the less time and storage it will take, and the less chance it has of accumulating errors.)

A typical way to evaluate a polynomial is, of course, to multiply everything out. Unfortunately, this is a very inefficient way of doing things, for it requires the highest number of multiplications. For example, to evaluate a polynomial of degree n by the "normal" evaluation process, we have:

$p(x) = a_0 + a_1 x^1 \longrightarrow$ requires 1 multiplication

$p(x) = a_0 + a_1 x^1 + a_2 x^2$
$ = a_0 + a_1 \cdot x + a_2 \cdot x \cdot x \longrightarrow$ requires 3 multiplications

$p(x) = a_0 + a_1 x^1 + a_2 x^2 + a_3 x^3$
$ = a_0 + a_1 \cdot x + a_2 \cdot x \cdot x$
$ + a_3 \cdot x \cdot x \cdot x \longrightarrow$ requires 6 multiplications

$\vdots$

$p(x) = a_0 + a_1 x^1 + a_2 x^2 + a_3 x^3$

$ + \cdots + a_n x^n \longrightarrow$ requires $1 + 2 + 3 + \cdots + n = \dfrac{n(n + 1)}{2}$ multiplications

That means that for a polynomial of degree 8 you must perform a horrendous 36 separate multiplications. Extend it to a higher-degree polynomial—say, 100—and you have a good reason to buy stock in a paper mill. By a simple method of reorganizing, however, you can dramatically cut the number of multiplications you—or the computer—must perform. In calculating, say, x^4, if you use the previously determined x^3 value and multiply it by x, you save effort: instead of doing $x \cdot x \cdot x \cdot x$ (3 multiplications), you just do $x \cdot x^3$ (1 multiplication). In this way, you can evaluate any polynomial of degree n with only $2n - 1$ multiplications. Let's look at how this works.

EXAMPLE 4-9   Prove that only $2n - 1$ multiplications are needed to evaluate the nth-degree polynomial.

Solution   Write

$$p(x) = a_0 + \underbrace{a_1 \cdot x + a_2 \cdot x_2 + a_3 \cdot x_3 + \cdots + a_n \cdot x_n}_{n \text{ multiplications}}$$

where
$$\left. \begin{array}{l} x_2 = x \cdot x \\ x_3 = x \cdot x_2 \\ x_4 = x \cdot x_3 \\ \quad \vdots \\ x_n = x \cdot x_{n-1} \end{array} \right\} n - 1 \text{ multiplications}$$

Total of $2n - 1$ multiplications

note: The proof in Example 4-9 demonstrates the possibility of using iteration for a simple algorithm to evaluate the nth-degree polynomial.

Though we've got the procedure down to only $2n - 1$ multiplications, mathematicians aren't satisfied with this. We can make yet one more reduction in the number of multiplications by using the **Horner method**. To see how this works, inspect the polynomial

$$p(x) = a_0 + a_1x^1 + a_2x^2 + a_3x^3 + a_4x^4 \tag{4-4}$$

Rewrite it as

$$p(x) = a_0 + x \cdot (a_1 + x(a_2 + x(a_3 + x \cdot a_4))) \tag{4-5}$$

Notice that the number of multiplications for the fourth-degree polynomial in eq. (4-5) is only 4, which happens to also equal n. (Notice also that it takes no more additions for eq. 4-5 than for the original form in eq. 4-4). Thankfully, we can do this reduction process on any polynomial, so we now have a way to evaluate an nth-degree polynomial with only n multiplications. Horner's method is one of the fastest, most efficient, and most economical ways to evaluate a polynomial. Table 4-1 compares the three methods we've discussed.

TABLE 4-1   Number of Multiplications Needed

Degree of polynomial	Ordinary method $\dfrac{n(n+1)}{2}$	Improved method $(2n - 1)$	Horner method n
1	1	1	1
2	3	3	2
3	6	5	3
4	10	7	4
5	15	9	5
6	21	11	6
$\vdots$	$\vdots$	$\vdots$	$\vdots$
100	5050	199	100
200	20 100	399	200
500	125 250	999	500
1000	500 500	1999	1000

The general form for Horner's method is

HORNER'S METHOD

$$p(x) = a_0 + a_1 x^1 + a_2 x^2 + \cdots + a_{n-2} x^{n-2} + a_{n-1} x^{n-1} + a_n x^n \tag{4-6}$$

$$= a^0 + x^1(a_1 + x^2(a_2 + \cdots + x^{n-2}(a_{n-2} + x^{n-1}(a_{n-1} + x^n \cdot a_n)))) \tag{4-7}$$

note: The form of the Horner-reduced polynomial in eq. (4-7) is a set of nested parentheses, which will allow us to use an iteration process when evaluating the polynomial with a computer.

You can write the algorithm for Horner's method by first designating the polynomial as a series of subscripted variable coefficients (as in eq. 4-6), and then converting that to computer notation:

$$a(0) + a(1) * x + a(2) * x \uparrow 2 + \cdots + a(n-2) * x \uparrow (n-2) + a(n-1) * x \uparrow (n-1) + a(n) * x \uparrow n$$

HORNER'S ALGORITHM

Begin
1. Input x x is the value you want to plug into the equation.
2. $p \leftarrow a(n)$
3. For $I = n - 1$ TO 0 in steps of -1 DO
   a. $p \leftarrow p * x + a(I)$
4. Output p
End

EXAMPLE 4-10   Use Horner's method to find the value of $p(x) = 2.30 + 1.60x + 4.15x^2 - 2.73x^3$ at $x = 1.73$.

Solution   First rewrite $p(x)$ in reduced form, as in eq. (4.7):

$$p(x) = 2.3 + x(1.6 + x(4.15 + x(-2.73)))$$

Next plug in $x = 1.73$, starting with the innermost parentheses and working from the right:

$$p(x) = 2.3 + x(1.6 + x(4.15 + x(-2.73)))$$

$$= \qquad\qquad\qquad\qquad -4.7229$$
$$= \qquad\qquad\qquad -0.5729$$
$$= \qquad\qquad\quad 0.991117$$
$$= \qquad\qquad 0.608883$$
$$= \qquad\quad 1.05336759$$
$$= \qquad 3.35336759$$

(Notice that this algorithm follows the rules for the order of operations: it performs all inner parenthetical work first, doing multiplication then addition within the parentheses.) The original coefficients were significant to only two decimal places, so the final result is $f(1.73) = 3.35$.

---

note: Because only three multiplications and three additions were required in Example 4-10, Horner's method minimizes the amount of round-off error that could creep into the computations.

4-4. Linear Interpolation

Suppose we are given a function that is defined only by a table of values (x_i, y_i). Then to approximate the value of the function for values of x that are not tabulated, we must *interpolate* the values, that is, make an educated estimate.

If we approximate the function between any pair of consecutive points as the straight line that exists between those points, we have created a **linear approximation** of the function between the two points. This process is called **linear interpolation** (see Figure 4-2). By connecting the given consecutive points across the graph from left to right with line segments, we obtain the **polygonal approximation** to the function (see Figure 4-3).

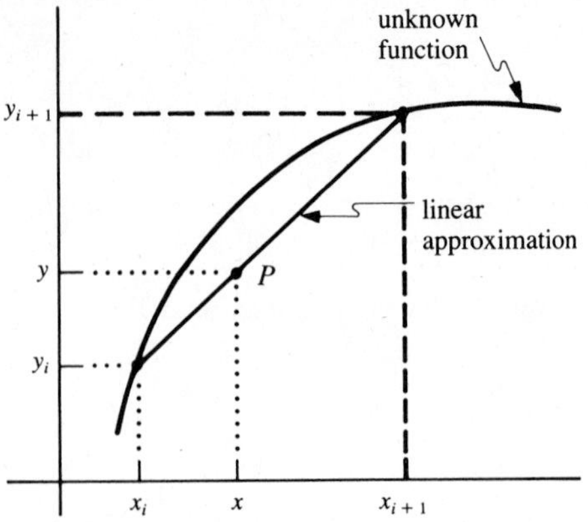

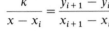

FIGURE 4-2 Linear approximation.

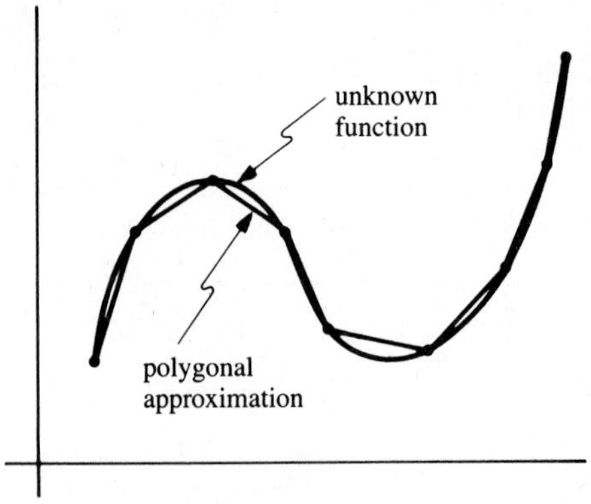

FIGURE 4-3 Polygonal approximation.

To approximate the value of a function at a point x not in the table, then, we first locate two consecutive points x_i and x_{i+1} that enclose the point x. Then we find the x-value of the point P on the linear approximation segment. Geometrically, it looks like Figure 4-4.

Fortunately, this collection of lines does have mathematical meaning, for it eventually produces a useful formula. Using similar triangles APB and ADC we get the relationship.

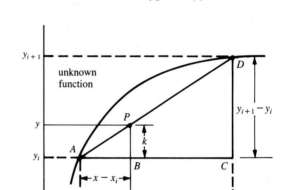

FIGURE 4-4

$$\frac{k}{x - x_i} = \frac{y_{i+1} - y_i}{x_{i+1} - x_i}$$

or

$$k = \left(\frac{y_{i+1} - y_i}{x_{i+1} - x_i}\right) \cdot (x - x_i) \qquad \textbf{(4-8)}$$

Now examine Figure 4-4 and notice that for any x in $x_i < x < x_{i+1}$, the linear approximation to y is $y_i + k$ or, substituting from eq. (4-8).

LINEAR APPROXIMATION FORMULA
$$y = y_i + \left(\frac{y_{i+1} - y_i}{x_{i+1} - x_i}\right)(x - x_i) \qquad \textbf{(4-9)}$$

From eq. (4-9) we can prepare an algorithm to compute the linear approximation of any x when we are given a table of values $(x_1, y_1), (x_2, y_2), (x_3, y_3), \ldots, (x_n, y_n)$ under the condition that the given points x_i ($i = 1, 2, 3, \ldots, n$) are *equally spaced* along the horizontal axis. Call the common distance between any pair of points h units; that is,

$$x_{i+1} - x_i = h \qquad \textbf{(4-10)}$$

We therefore have divided a portion of the x-axis into a series of intervals of length h, with each interval beginning and ending with two consecutive x values from our given table of values (see Figure 4-5a).

The first task, given any arbitrary value of x for which the y-value is desired, is to locate the interval in which x lies.

If we move h units to the left of x_1 (our first given x-value) and put a value there and denote it our *basepoint* x_0 (see Figure 4-5b), then any of the predetermined, given x_i values can be designated as

$$x_i = x_0 + h \cdot i$$

This may also be rearranged as

$$\frac{x_i - x_0}{h} = i \qquad \textbf{(4-11)}$$

which will be useful in all sorts of manipulations later.

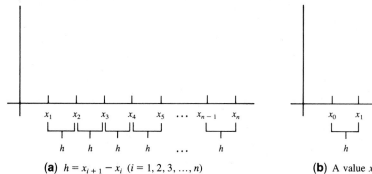

(a) $h = x_{i+1} - x_i \ (i = 1, 2, 3, \ldots, n)$

(b) A value x_0 is added as a basepoint; the length of interval x_0 to x_1 is h.

FIGURE 4-5

Okay. Now that we've set everything up on the axis, it's time to do the general calculations for an arbitrary value of x so that we can get to the algorithm. To discover what interval x falls into, consider that the quantity $x - x_0$ tells us how far x is to the right of our basepoint x_0 (refer to Figure 4-5b, pretending x is between x_3 and x_4). Dividing this quantity by h gives us a real number R. This number R will have an integer part and a decimal fraction part. It represents the number of h-steps (and h-partial-steps) needed to move from x_0 to x. The integer part of R, call it J, is the number of whole h-steps we need to take to get to the *left end* of the interval in which x resides. Thus we can say that x resides in the interval (x_J, x_{J+1}). Now refer back to eqs. (4-9) and (4-10) and notice that since $h = x_{i+1} - x_i$ (eq. 4-10), we may now write eq. (4-9) for the linear approximation of y as

$$y = y_i + \left(\frac{y_{i+1} - y_i}{h}\right)(x - x_i)$$

For our arbitrary x in the interval (x_J, x_{J+1}), this now becomes

$$y = y_J + \left(\frac{y_{J+1} - y_J}{h}\right)(x - x_J) \tag{4-12}$$

In the second term of eq. (4-12), the quotient

$$\frac{x - x_J}{h} = \frac{x - x_0}{h} - \frac{x_J - x_0}{h} \tag{4-13}$$

(Just expand it algebraically on the right and see the cancelation.) But notice that in the text following eq. (4-11) we designated $(x - x_0)/h$ as R and that from eq. (4-11) we can say $(x_J - x_0)/h = J$. Thus by eq. (4-13) we have $(x - x_J)/h = R - J$. Now we can finally simplify eq. (4-12) to read:

$$y = y_J + (y_{J+1} - y_J)(R - J) \tag{4-14}$$

which is the shortened equivalent of eq. (4-9). Whew!

Examining eq. (4-14) we find that the algorithm for linear interpolation will only need to use the variables y_J, y_{J+1}, R, and J. Our given table of values will supply the numbers for y_J and y_{J+1} and R and J are easily computed as $(x - x_0)/h$ and $(x_J - x_0)/h$, respectively.

The algorithm in pseudocode is:

LINEAR INTERPOLATION ALGORITHM

Begin
1. Input $(x_1, y_1), (x_2, y_2), (x_3, y_3), \ldots, (x_n, y_n)$ The given collection of number pairs (x_i, y_i), $i = 1, 2, \ldots, n$.

2. Input x x is in the interval (x_1, x_n) and is different from all x_i. It is $y = f(x)$ we are trying to calculate.

3. $h \leftarrow x_2 - x_1$
4. $x_0 \leftarrow x_1 - h$
5. $R \leftarrow (x - x_0)/h$ R equals the number of h-steps and h-partial-steps to get from x_0 to x.

6. $J \leftarrow \text{INT}(R)$
7. $y \leftarrow y_J + (y_{J+1} - y_J) * (R - J)$ Perform the function in eq. (4-14).
8. Output y
End

note: If the given table values along the horizontal axis are not equally spaced, then we need to search the list of x_i-values for the first x_i smaller than the given x. This will locate the interval such that $x_J \le x < x_{J+1}$, and the y values may be obtained in the usual way.

EXAMPLE 4-11   Given the points of a function $(1,3), (1.5, 2.4), (2,2), (2.5, 1.1), (3, 0.3)$, find the approximate y-value for **(a)** $x = 1.8$ and **(b)** $x = 2.7$.

Solution   The given table x-values are equally spaced so we can plug directly into the algorithm. Following the algorithm steps:

(a) $h = x_2 - x_1 = 1.5 - 1.0 = 0.5$
$x_0 = x_1 - h = 1.0 - 0.5 = 0.5$

$R = \dfrac{x - x_0}{h} = \dfrac{1.8 - 0.5}{0.5} = \dfrac{1.3}{0.5} = 2.60$

J = integer part of $R = 2$
$Y = y_2 + (y_3 - y_2)(2.60 - 2)$
   $= 2.4 + (2 - 2.4)(0.60)$
   $= 2.4 - 0.24 = 2.16$

(b) $h = x_2 - x_1 = 1.5 - 1.0 = 0.5$
$x_0 = x_1 - h = 1.0 - 0.5 = 0.5$

$R = \dfrac{x - x_0}{h} = \dfrac{2.7 - 0.5}{0.5} = \dfrac{2.2}{0.5} = 4.4$

J = integer part $R = 4$
$Y = y_4 + (y_5 - y_4)(4.4 - 4)$
   $= 1.1 + (.3 - 1.1)(0.4)$
   $= 1.1 - (0.8)(0.4)$
   $= 1.1 - 0.32 = 0.78$

4-5. Polynomial Interpolation

In the previous section we explored how to approximate the value of a particular point on a function that is described only by a table of values. In this section, you will learn how to approximate the actual *equation* of a function given only a handful of points. But first,

recall: You obtain a *polygonal approximation* of a function by "connecting the dots": that is, you plot your given points and then connect them consecutively across the graph, from left to right, with straight line segments (refer to Figure 4-3).

If the points that determine a polygonal approximation of a function are spaced far apart, and if the actual curve oscillates (bends rapidly between the points), the polygonal approximation will not be a very close portrayal of the function. Fortunately, there is a more accurate way than polygonal approximation to fit a polynomial curve to a prescribed set of points. It is known as the **Lagrange interpolation formula**, and is based on the premise that,

- for two points in a plane, the unique polynomial containing them is a straight line, a first-degree polynomial.
- for three points in a plane, the unique polynomial containing them is a parabola, a second-degree polynomial.
- for four points, the unique polynomial containing them is a third-degree polynomial.
- for n points, the unique polynomial containing them is a $(n - 1)$-degree polynomial.

Suppose $(x_1, y_1), (x_2, y_2), (x_3, y_3), \ldots, (x_n, y_n)$ are the given points of an unknown function. The unique polynomial containing these points is

LAGRANGE INTERPOLATION FORMULA

$$p(x) = \left[\frac{(x - x_2)(x - x_3)\cdots(x - x_n)}{(x_1 - x_2)(x_1 - x_3)\cdots(x_1 - x_n)} \right] y_1 + \left[\frac{(x - x_1)(x - x_3)\cdots(x - x_n)}{(x_2 - x_1)(x_2 - x_3)\cdots(x_2 - x_n)} \right] y_2$$
$$+ \cdots + \left[\frac{(x - x_1)(x - x_2)\cdots(x - x_{n-1})}{(x_n - x_1)(x_n - x_2)\cdots(x_n - x_{n-1})} \right] y_n$$

EXAMPLE 4-12   Find the unique polynomial that contains the points $(1, 3)$, $(3, -2)$, and $(4, 2)$.

Solution   The Lagrange interpolation formula gives

$$p(x) = \left[\frac{(x-3)(x-4)}{(1-3)(1-4)}\right] \cdot 3 + \left[\frac{(x-1)(x-4)}{(3-1)(3-4)}\right] \cdot (-2) + \left[\frac{(x-1)(x-3)}{(4-1)(4-3)}\right] \cdot 2$$

$$= \frac{x^2 - 7x + 12}{6} \cdot 3 + \frac{x^2 - 5x + 4}{-2} \cdot (-2) + \frac{x^2 - 4x + 3}{3} \cdot 2$$

$$= \frac{1}{2}x^2 - \frac{7}{2}x + 6 + x^2 - 5x + 4 + \frac{2}{3}x^2 - \frac{8}{3}x + 2$$

$$= \frac{13}{6}x^2 - \frac{67}{6}x + 12$$

Check: $p(1) = \frac{13}{6} - \frac{67}{6} + \frac{72}{6} = \frac{18}{6} = 3$

$p(3) = \frac{117}{6} - \frac{201}{6} + \frac{72}{6} = -\frac{12}{6} = -2$

$p(4) = \frac{208}{6} - \frac{268}{6} + \frac{72}{6} = \frac{12}{6} = 2$

Figure 4-6 is a graph of the function.

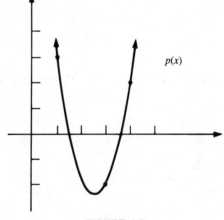

FIGURE 4-6

EXAMPLE 4-13   Find the third-degree polynomial that passes through the points $(-3, -9)$, $(-1, 1)$, $(0, -3)$, and $(2, 1)$.

Solution   The Lagrange interpolation formula gives

$$p(x) = \left[\frac{(x+1)(x-0)(x-2)}{(-3+1)(-3)(-3-2)}\right](-9) + \left[\frac{(x+3)(x-0)(x-2)}{(-1+3)(-1)(-1-2)}\right](1)$$

$$+ \left[\frac{(x+3)(x+1)(x-2)}{(0+3)(0+1)(0-2)}\right](-3) + \left[\frac{(x+3)(x+1)(x-0)}{(2+3)(2+1)(2-0)}\right](1)$$

$$= \frac{x(x^2 - x - 2)}{-30}(-9) + \frac{x(x^2 + x - 6)}{6} + \frac{(x+3)(x^2 - x - 2)}{-6}(-3) + \frac{x(x^2 + 4x + 3)}{30}$$

$$= \frac{3}{10}x^3 - \frac{3}{10}x^2 - \frac{6}{10}x + \frac{1}{6}x^3 + \frac{1}{6}x^2 - x + \frac{1}{2}(x^3 + 2x^2 - 5x - 6) + \frac{1}{30}x^3 + \frac{2}{15}x^2 + \frac{1}{10}x$$

$$= \left(\frac{3}{10} + \frac{1}{6} + \frac{1}{2} + \frac{1}{30}\right)x^3 + \left(-\frac{3}{10} + \frac{1}{6} + 1 + \frac{2}{15}\right)x^2 + \left(-\frac{6}{10} - 1 - \frac{5}{2} + \frac{1}{10}\right)x - 3$$

$$p(x) = 1x^3 + 1x^2 - 4x - 3$$

4-6. Congruences

In Section 2-6 we introduced the concept of modular arithmetic. On a practical level, modular arithmetic is useful in computers because computers are finite machines with finite storage space. With modular, computer scientists can save space as they study groups and symmetries, create secret codes, and even generate lists of random numbers.

recall: In *modulo n* we can write any number of integers in only n digits; if $a = b$ mod n, then $a = k \cdot n + b$.

Let p be a fixed integer greater than 1. Consider other integers m and n. We say that **m is congruent to n modulo p** provided $m - n$ is divisible by p. We write this as $m \equiv n(\text{mod } p)$ where the $\equiv$ is read, "is congruent to." If we write $m = q_1 \cdot p + r_1$ and $n = q_2 \cdot p + r_2$, where q_1 and q_2 are integers and r_1 and r_2 are in the set of nonnegative integers less than p, it is easy to see that if $r_1 = r_2$ then

$$
\begin{aligned}
m - n &= (q_1 \cdot p + r_1) - (q_2 \cdot p + r_2) \\
&= (q_1 \cdot p - q_2 \cdot p) + (r_1 - r_2) \\
&= (q_1 - q_2) \cdot p + 0
\end{aligned}
$$

so $m - n$ is divisible by p. This gives the property:

$$
m \equiv n(\text{mod } p) \quad \text{if and only if} \quad r_1 = r_2
$$

That is, $m \equiv n(\text{mod } p)$ if and only if m and n leave the same remainders when divided by p.

note: The congruence formula $m \equiv n(\text{mod } p)$ is the same as saying $m = n \bmod p$, the format that you learned in Section 2-6, only with different variables. To prove this, consider:

if $m = n \bmod p$, then $m = k \cdot p + n$ (by definition of modulo, eq. 2-11)   **(4-15)**

if $m \equiv n(\text{mod } p)$, then $(m - n)/p = k$ where k is an integer (by definition of congruence)

But $(m - n)/p = k$ can be rewritten as $m - n = k \cdot p$ or $m = k \cdot p + n$, which is the same as eq. (4-15). Therefore we can say that $m \equiv n(\text{mod } p)$ is the same thing as saying $m = n \bmod p$. The congruence notation is simply more specific.

If we say that two integers are **congruent modulo n**, we mean that the modulo n value of the first integer equals—that is, is congruent to—the modulo n value of the other integer. The difference between any pair of integers that are congruent modulo n will be n or a multiple of n. For example, 7 and 15 are congruent modulo 4 because $7 = 3 \bmod 4$ and $15 = 3 \bmod 4$; the difference between 7 and 15 is 8, a multiple of 4.

EXAMPLE 4-14 (a) Find a set of numbers such that each element is congruent modulo 2 to all other elements in the set. (b) Find a set of numbers such that each element is congruent modulo 7 to all other elements in the set.

Solution

(a) The entire set of even integers has this property. The difference between any pair of them is a multiple of 2, so they are congruent modulo 2. The entire set of odd integers also has this property. The union of these two sets consists of all of the integers.

(b) One set satisfying the requirement is the collection of all multiples of 7, $\{\ldots, -21, -14, -7, 0, 7, 14, 21, 28, 35, \ldots\}$. The difference between any pair is a multiple of 7. There are six other sets also having this property: just add the same integer (from 1 to 6) to every number above. The union of these seven sets consists of all of the integers. We say that these seven lists *partition the integers*.

EXAMPLE 4-15 Your bicycle odometer has a three-digit display. When you bought it the reading was 000. Now the reading is 426. How many miles have you cycled?

Solution We don't know exactly how many miles you've cycled because the actual value could be 426, or 1426, or 2426, or 3426, or .... you get the point. But we do know that the number of miles is in the set $\{m\}$ where $m \equiv 426 \bmod 1000$.

4-7. Random Number Generation

There are many situations that require a "random" choice from a collection or a space: Projected voting preferences (Gallup polls) depend on a random sampling of voters; quality control depends on a random sampling of objects from production lines; calls for jury duty depend on randomly chosen birthdates and driver's license numbers. To assure impartiality, randomness is an essential ingredient in sampling.

In many sampling situations, the basis for choosing the objects or values is the *random number generator*. This is an algorithm which produces a sequence of numbers that satisfy the statistical concept of randomness. This concept requires that every member of the set being sampled has an equal chance of being chosen.

Computers are often used to generate tables of random digits. For single-digit numbers, each digit should have a probability of 1/10 of being chosen. Thus a table of 10 000 random single-digit numbers should theoretically contain 1000 of each of the digits 0 through 9. If a computer generates such a table, we could check the efficiency and correctness of the random generation process by counting the digits and comparing the actual totals against the theoretical totals of 1000 per digit. A statistician would then decide on the efficiency of the process using well-founded mathematical principles.

A variety of algorithms exist for generating random numbers. One of the most efficient and successful ones is the **linear congruential generator**, which uses modular arithmetic. It is defined by

LINEAR CONGRUENTIAL GENERATOR
$$x_{n+1} \equiv (ax_n + c)(\bmod m) \tag{4-16}$$

where x_{n+1} is the next number in the random number sequence, a is the multiplier, x_n is the preceding random number, c is the increment, and m is the chosen modulo. To use the formula, we provide the computer with values for x_0 (the initial values), a, c, and m, and then the computer generates subsequent random values by performing the congruence operation in eq. (4-16). Because it is limited by our choice of x_0, a, c, and m, the computer produces random numbers in a *periodic sequence*; that is, it creates a finite number of random values before the pattern of values begins to repeat itself.

EXAMPLE 4-16   Let $x_0 = 5$, $a = 5$, $c = 7$, and $m = 11$. Generate a random sequence.

Solution   The formula is $x_{n+1} \equiv (5x_n + 7) \,(\bmod 11)$, so we have:

$$x_0 \equiv 5$$
$$x_1 \equiv (5 \cdot 5 + 7) \,(\bmod 11) \equiv 10$$
$$x_2 \equiv (5 \cdot 10 + 7) \,(\bmod 11) \equiv 2$$
$$x_3 \equiv (5 \cdot 2 + 7) \,(\bmod 11) \equiv 6 \qquad \text{A periodic sequence of five random numbers}$$
$$x_4 \equiv (5 \cdot 6 + 7) \,(\bmod 11) \equiv 4$$
$$x_5 \equiv (5 \cdot 4 + 7) \,(\bmod 11) \equiv 5$$
$$x_6 \equiv (5 \cdot 5 + 7) \,(\bmod 11) \equiv 10$$
$$x_7 \equiv (5 \cdot 10 + 7) \,(\bmod 11) \equiv 2 \qquad \text{A repeat of the above sequence}$$
$$\vdots \qquad\qquad \vdots$$

Some combinations of x_0, a, c, and m will generate longer sequences. The choice $x_0 = 7$, $a = 7$, $c = 3$, and $m = 11$, for example, produces a sequence of period 10.

When a and c are chosen correctly, the maximum number of random numbers that can be generated without repetition by eq. (4-16) is equal to our chosen modulo m. To generate very long sequences of random numbers, then, we want the largest m possible. But the size of m is limited by a computer's hardware: if a machine has word length N (see Section 3-3) and uses base b arithmetic (see Section 3-4), the largest value of m that we can choose is given by:

$$m = b^N + i \quad \text{where } i = -1, 0, \text{ or } 1$$

That is, the maximum modulo we can insert into eq. (4-16)—and therefore the greatest number of random, nonrepeating numbers we can produce—is either

$$m = b^N - 1, \qquad m = b^N + 0, \qquad \text{or} \qquad m = b^N + 1$$

Remember, of course, that to reach a periodic sequence of length m we must also choose the parameters a and c "correctly." The values of a and c must be in accordance with the following rules:

(1) If m is a multiple of 4, then $a - 1$ is a multiple of 4.
(2) For each prime factor, p, of m, the value of $a - 1$ is a multiple of p.
(3) c and m are *relatively prime*; that is, the only positive integer that divides both c and m is 1.
   The choice of x_0 is immaterial.

EXAMPLE 4-17 **(a)** In a machine with 8-bit words and using binary arithmetic, how long can a periodic sequence of random numbers be? **(b)** If the machine has 32-bit words in binary, how long can the sequence of random numbers be?

Solution

(a) If $N = 8$, $b = 2$, then $m = 2^8 - 1 = 255$ or $m = 2^8 = 256$ or $m = 2^8 + 1 = 257$ random numbers may be generated before the sequence repeats. This of course, assumes that a and c are appropriately chosen to get this sequence.

(b) If $N = 32$ and $b = 2$, then $m = 2^{32} - 1 = 4,294,967,295$ or $m = 2^{32} = 4,294,967,296$ or $m = 2^{32} + 1 = 4,294,967,297$ random numbers may be generated before the sequence repeats.

SUMMARY

1. The axioms of equality are the basic operations used to solve linear equations. The equation may have exactly one solution, no solutions, or an infinite number of solutions.
2. The value of the discriminant of a quadratic equation determines the nature of the solutions.
3. A linear equation in two unknowns has an infinite number of solution pairs; the graph of this infinite set is a straight line.
4. A system of linear equations may have a unique solution, no solution, or an infinite number of solutions; you can analyze the ratios of the coefficients to make this determination.
5. A system of linear equations may be solved by producing equivalent systems using the elementary operations:

   (1) Interchange the order in which two equations occur.
   (2) Multiply any equation by a nonzero number.
   (3) Add a multiple of one equation to another equation of the system.

6. The roots of some equations can only be approximated, and there are many methods available. An easy one is the half-interval method.
7. Horner's method is an efficient and effective way to evaluate complicated polynomials for a given value.
8. If a function is given only as a table of values, then you can approximate new table values by the linear interpolation method.
9. The Lagrange interpolation formula creates a polynomial function from a predetermined collection of points.
10. Computers sometimes simplify their internal number storage and manipulation using the modular system of congruences.
11. To create sequences of random numbers for scientific use, a linear congruential generator is used. Periodic sequences of any desired length are usually possible.

RAISE YOUR GRADES

Can you...?

- ☑ recite the axioms of equality and use them to solve a linear equation
- ☑ use the quadratic formula algorithm to solve a quadratic equation with a computer
- ☑ solve a linear equation and draw its graph
- ☑ solve a system of two linear equations
- ☑ examine the ratios of the coefficients to decide the form of a solution to a system of two linear equations
- ☑ recite the elementary operations for a system of linear equations
- ☑ use the Gauss-Jordan elimination process to solve a system of more than two equations
- ☑ apply the half-interval method to approximate the root of a polynomial
- ☑ use Horner's algorithm to efficiently evaluate a polynomial
- ☑ explain the linear interpolation process
- ☑ use the Lagrange interpolation formula to find a polynomial from a given set of points of the function
- ☑ define a congruence and its properties
- ☑ explain how a random number generator operates and how random integer sequences are formed

SOLVED PROBLEMS

Solving Equations

PROBLEM 4-1 Use the axioms of equality to solve

(a) $\frac{3}{x} - \frac{5}{x} + \frac{1}{2x} = 8$ (b) $4(y - 1) - 6(2 - y) = 5(2y + 3)$ (c) $5(1 - z) + 4z - 3(z + 2) = -(4z + 1)$

Solution

(a) Temporarily replace $1/x$ by t to simplify the problem. Then
$$3t - 5t + t/2 = 8$$
$$-3t/2 = 8 \longrightarrow t = -16/3$$
Thus
$$x = \frac{1}{t} = \frac{1}{-16/3} = -\frac{3}{16}$$

(b) $4(y - 1) - 6(2 - y) = 5(2y + 3)$
$4y - 4 - 12 + 6y = 10y + 15$
$10y - 16 = 10y + 15$
$-16 = 15$ Impossible!
No value of y will solve this equation.

(c) $5(1 - z) + 4z - 3(z + 2) = -(4z + 1)$
$5 - 5z + 4z - 3z - 6 = -4z - 1$
$(5 - 6) + (-5z + 4z - 3z) = -4z - 1$
$-1 + (-4z) = -4z - 1$ Identity!
Every value of z will solve this equation.

PROBLEM 4-2   A rectangular plot of land has a perimeter of 210 ft and its length is twice its width. What are the dimensions?

Solution   Let x = width. Then $2x$ = length. The perimeter is

$$x + 2x + x + 2x = 210$$
$$6x = 210$$
$$x = 35$$

So the width is 35 ft and the length is 70 ft.

Higher-Degree Equations in One Unknown

PROBLEM 4-3   Find one or more roots to the following equations:

(a) $(x + 2)^3 = 0$ **(b)** $x^2 - 9x + 3 = 0$ **(c)** $x^2 - 8x + 25 = 0$

Solution

(a) Since $(x + 2)^3 = 0$ then $x + 2 = 0$ or $x = -2$.
(b) Use the quadratic formula; $a = 1, b = -9, c = 3$, so

$$x = \frac{9 \pm \sqrt{81 - 4(1)(3)}}{2} = \frac{9 \pm \sqrt{69}}{2}$$

The solutions are $x = \dfrac{9 + \sqrt{69}}{2}$ and $x = \dfrac{9 - \sqrt{69}}{2}$

(c) Use the quadratic formula; $a = 1, b = -8, c = 25$, so

$$x = \frac{8 \pm \sqrt{64 - 4(1)(25)}}{2} = \frac{8 \pm \sqrt{64 - 100}}{2} = \frac{8 \pm 6i}{2} = 4 \pm 3i$$

The solutions are $x = 4 + 3i$ and $x = 4 - 3i$.

Linear Equations in Two Unknowns

PROBLEM 4-4   Find a set of solution pairs that solve the linear equation $5x - 2y = 12$, then draw the graph.

Solution   Let y be the variable to which we assign values. To make it easier to find x when y is assigned, we may rearrange the equation to

$$5x = 2y + 12$$

$$x = \frac{2y + 12}{5}$$

For $y = 0, x = \frac{12}{5}$
For $y = 1, x = \frac{14}{5}$
For $y = 2, x = \frac{16}{5}$
For $y = -2, x = \frac{8}{5}$
For $y = -6, x = 0$

The graph is as shown in Figure 4-7.

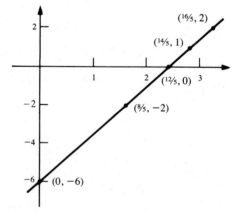

FIGURE 4-7

Systems of Two Linear Equations

PROBLEM 4-5   Solve the system $\begin{Bmatrix} 3x - 4y = -2 \\ 6x + 7y = 3 \end{Bmatrix}$.

Solution   Following the algorithm:

1. Divide first equation by 3:

$$x - \tfrac{4}{3}y = -\tfrac{2}{3}$$

2. Solve it for x:

$$x = \tfrac{4}{3}y - \tfrac{2}{3}$$

3. Insert this into second equation:

$$6(\tfrac{4}{3}y - \tfrac{2}{3}) + 7y = 3$$

4. Solve this for y:

$$8y - 4 + 7y = 3$$
$$15y = 7$$
$$y = \tfrac{7}{15}$$

5. Put this back into the x-equation:

$$x = \tfrac{4}{3}y - \tfrac{2}{3}$$
$$= \tfrac{4}{3}(\tfrac{7}{15}) - \tfrac{2}{3} = \tfrac{28}{45} - \tfrac{30}{45}$$
$$= -\tfrac{2}{45}$$

The unique solution pair is $(-\tfrac{2}{45}, \tfrac{7}{15})$.

PROBLEM 4-6   Find two numbers whose sum is 51 and whose difference is 15.

Solution   Let the unknown numbers be x and y. Then we have the system

$$\begin{cases} x + y = 51 \\ x - y = 15 \end{cases}$$

Following the algorithm, we have from the first equation $x = 51 - y$. Plug that into the second equation and solve for y:

$$51 - y - y = 15$$
$$-2y = 15 - 51$$
$$y = 18$$

Now stick that back into the first equation, and we get

$$x + 18 = 51$$
$$x = 33$$

So, $x = 33$ and $y = 18$.

PROBLEM 4-7   Johnny needs to work 100 math problems, so his mom comes up with an incentive! He gets 50¢ for each problem solved correctly but has 30¢ deducted for each incorrect one. He ends up with a net gain of $22.00. How many did he do correctly? (Don't you wish your mom did that?)

Solution   Let $x =$ number correct and $y =$ number incorrect. We have the following relationships:

$$\begin{cases} x + y = 100 \text{ (problems in all)} \\ (0.50x) - (0.30)y = 22.00 \text{(dollars earned)} \end{cases}$$

Following the algorithm but solving for y first (just to be arbitrary and to prove you can do it that way), $y = 100 - x$ so we have

$$(0.50)x - (0.30)(100 - x) = 22.00$$
$$0.50x - 30.00 + 0.30x = 22.00$$
$$0.80x = 52.00$$
$$x = \frac{520}{8} = 65 \qquad \text{correct answers}$$
$$y = 100 - 65 = 35 \qquad \text{incorrect answers}$$

Check: $(65)(0.50) - (35)(0.30) \overset{?}{=} 22.00$
$$32.50 - 10.50 \overset{\checkmark}{=} 22.00$$

PROBLEM 4-8 A theater has a seating capacity of 300. Tickets are $4 for adults and $1.50 for children. If a sold-out performance brings in $760.00, how many of each kind of ticket were sold?

Solution Let x = number of adults and y = number of children. Then our two-equation system is

$$\begin{cases} x + y = 300 \\ 4x + 1.5y = 760 \end{cases}$$

Solve the first equation for x: $\qquad\qquad\qquad\qquad\qquad x = 300 - y$

Substitute this into the second equation: $\qquad 4(300 - y) + 1.5y = 760$

Solve for y: $\qquad\qquad\qquad\qquad\qquad\qquad 1200 - 4y + 1.5y = 760$

$$1200 - 2.5y = 760$$

$$-2.5y = 760 - 1200 = -440$$

$$y = \frac{-440}{-2.5} = 176$$

Then $x = 300 - y = 124$

The solution is $(124, 176)$

Check: $\qquad (124)(\$4) + (176)(\$1.50) \overset{?}{=} \$760$

$$\$496 \quad + \quad \$264 \quad \overset{\checkmark}{=} \$760$$

PROBLEM 4-9 By checking the ratios of coefficients determine if the given system has one solution, an infinite number of solutions, or no solutions.

$$\text{(a)} \begin{cases} 3x - 5y = 4 \\ -2x + 6y = 3 \end{cases} \qquad \text{(b)} \begin{cases} 4x - 2y = 3 \\ -8x + 4y = 11 \end{cases} \qquad \text{(c)} \begin{cases} -x + 2y = 6 \\ -3x + 6y = 18 \end{cases}$$

Solution In each case we form the ratios $\dfrac{a_1}{a_2}, \dfrac{b_1}{b_2}, \dfrac{c_1}{c_2}$ and compare them.

(a) $\dfrac{a_1}{a_2} = \dfrac{3}{-2}, \dfrac{b_1}{b_2} = -\dfrac{5}{6}, \dfrac{c_1}{c_2} = \dfrac{4}{3}$ $\qquad$ Since $\dfrac{a_1}{a_2} \neq \dfrac{b_1}{b_2}$ there is a unique solution.

(b) $\dfrac{a_1}{a_2} = \dfrac{4}{-8}, \dfrac{b_1}{b_2} = -\dfrac{2}{4}, \dfrac{c_1}{c_2} = \dfrac{3}{11}$ $\qquad$ Since $\dfrac{a_1}{a_2} = \dfrac{b_1}{b_2} \neq \dfrac{c_1}{c_2}$ we have no solution.

(c) $\dfrac{a_1}{a_2} = \dfrac{-1}{-3}, \dfrac{b_1}{b_2} = \dfrac{2}{6}, \dfrac{c_1}{c_2} = \dfrac{6}{18}$ $\qquad$ Since $\dfrac{a_1}{a_2} = \dfrac{b_1}{b_2} = \dfrac{c_1}{c_2}$ we have an infinite number of solutions.

PROBLEM 4-10 Use the elementary operations to solve the system

$$\begin{cases} x + 2y - 4z = -4 & \quad \text{I} \\ 4x - y + 2z = 11 & \quad \text{II} \\ -2x + 3y - z = -3 & \quad \text{III} \end{cases}$$

Solution Label the equations I, II, III at each stage, as above. Multiply I by (-4) and add to II, then multiply I by (2) and add to III. The new system will be:

$$\begin{cases} x + 2y - 4z = -4 & \quad \text{I} \\ 0x - 9y + 18z = 27 & \quad \text{II} \\ 0x + 7y - 9z = -11 & \quad \text{III} \end{cases}$$

Divide II by (-9):

$$\begin{cases} x + 2y - 4z = -4 & \quad \text{I} \\ y - 2z = -3 & \quad \text{II} \\ 7y - 9z = -11 & \quad \text{III} \end{cases}$$

Multiply II by (-2) and add to I, then multiply II by (-7) and add to III:

$$\begin{cases} x + 0y + 0z = 2 & \quad \text{I} \\ y - 2z = -3 & \quad \text{II} \\ 0y + 5z = 10 & \quad \text{III} \end{cases}$$

Divide III by 5:

$$\begin{cases} x & = 2 \\ y - 2z = -3 \\ z = 2 \end{cases}$$

Then multiply III by $(+2)$ and add to II:

$$\begin{cases} x & = 2 \\ y + 0z = 1 \\ z = 2 \end{cases}$$

The solution is now obvious.

Obtaining Approximate Solutions

PROBLEM 4-11   Use the half-interval method to find a root of the equation $e^x - 4x = 1$.

Solution   First rewrite the problem as a function $f(x) = 0$:

$$f(x) = e^x - 4x - 1 = 0$$

A sharp observer will note that $x = 0$ is a zero since $f(0) = e^0 - 4(0) - 1 = 1 - 0 - 1 = 0$
But let's find another zero of $f(x)$ using the half-interval method. First locate an interval in which $f(x) = 0$ will lie (it's easiest to choose two consecutive integers):

$$\left.\begin{array}{l} f(2) = e^2 - 8 - 1 = 7.389 - 9 = -1.61 \\ f(3) = e^3 - 12 - 1 = 20.08 - 13 = 7.08 \end{array}\right\}$$ The zero lies in $(2, 3)$ because $f(2)$ and $f(3)$ have opposite signs.

The midpoint is 2.5, so evaluate function $f(x)$ there:

$$\left.f(2.5) = e^{2.5} - 10 - 1 = 12.18 - 11 = 1.28\right\}$$ $f(2.5)$ is positive and $f(2)$ is negative, so the zero lies in $(2, 2.5)$.

New midpoint is 2.25.

$$\left.f(2.25) = e^{2.25} - 9 - 1 = 9.487 - 10 = -0.513\right\}$$ The zero lies in $(2.25, 2.50)$ because $f(2.25)$ is negative and $f(2.50)$ is positive.

New midpoint is 2.375.

$$\left.f(2.375) = e^{2.375} - 9.5 - 1 = 10.751 - 10.5 = 0.251\right\}$$ $f(2.375)$ is positive so now the zero lies in $(2.25, 2.375)$.

New midpoint is 2.3125.

$$\left.f(2.3125) = e^{2.3125} - 9.25 - 1 = 10.099 - 10.25 = -0.15\right\}$$ $f(2.3125)$ is negative, so the zero lies in $(2.3125, 2.375)$.

If we use the next midpoint as the approximate value for the zero of the function our final solution is 2.34375.

This answer, of course, could be made much more accurate if we used a computer to iterate the process 100 times rather than perform it five times by hand computation.

PROBLEM 4-12   As a modification of the half-interval method, we could use an interpolation method in which the new approximation is the x-intercept of the line joining the two endpoints $(x_{\text{left}}, f(x_{\text{left}}))$ and $(x_{\text{right}}, f(x_{\text{right}}))$. Derive the formula for this process.

Solution Use the following notation: the interval points given are (XL, YL) and (XR, YR). Suppose the left point lies below the horizontal axis, i.e., $YL < 0$, and that $YR > 0$ so the x-intercept (the point at which $f(x) = 0$) is between XL and XR (see Figure 4-8).

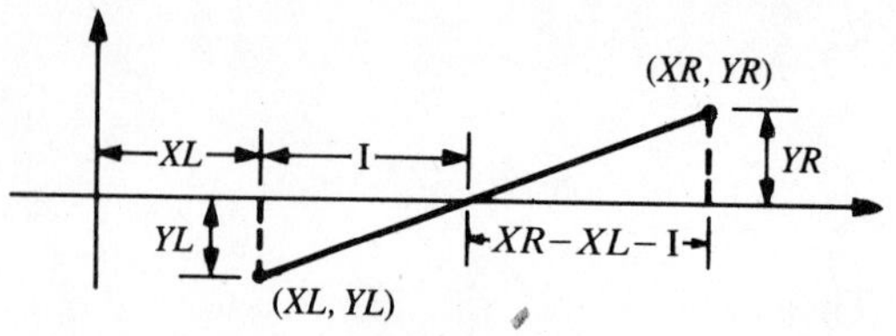

FIGURE 4-8

Observing that the triangles are similar, we set up the ratios

$$\frac{I}{YL} = \frac{XR - XL - I}{YR}$$

Cross multiply and solve for I:

$$I \cdot YR = YL \cdot (XR - XL) - I \cdot YL$$

$$I \cdot (YR - YL) = YL \cdot (XR - XL)$$

$$I = \left(\frac{XR - XL}{YR - YL}\right) \cdot YL$$

Thus the new approximation for the solution is

$$x = XL + \left(\frac{XR - XL}{YR - YL}\right) \cdot YL \qquad \textbf{(4-17)}$$

We've already said that the x-intercept is in the interval (XL, XR), and we're given that YL is negative and YR is positive. If $f(x)$ is negative, then the x-intercept is in the interval (x, XR), so rename x as the new XL and insert its value $f(x)$ in place of the former YL. If $f(x)$ is positive, however, then the x-intercept is in the interval (XL, x), so rename x as the new XR and insert its value $f(x)$ in place of the former YR. Now perform the process again, getting a new value of x from eq. (4-17).

Evaluating Polynomials

PROBLEM 4-13   Use Horner's method to find the value at $x = 3.15$ of the polynomial

$$p(x) = -25.18 + 0.41x - 2.66x^2 - 1.43x^3 + 0.12x^4$$

Solution   First rewrite $p(x)$ in reduced form:

$$p(x) = -25.18 + x(0.41 - x(2.66 + x(1.43 - x(0.12))))$$

Then let $x = 3.15$ and begin evaluating in the innermost parentheses.

$$
\begin{aligned}
p(x) &= -25.18 + x(0.41 - x(2.66 + x(1.43 - x(0.12)))) \\
&= 1.43 - (3.15)(0.12) \\
&= 2.66 + (3.15)(1.052) \\
&= 0.41 + (3.15)(5.9738) \\
&= -25.18 + (3.15)(-18.40747) \\
&= -83.163531
\end{aligned}
$$

The digits are significant to two decimal places, so the result is $p(3.15) = -83.16$. (Note the nesting of the braces from the right to left. This is the recursive evaluation.)

Linear Interpolation

PROBLEM 4-14   If an unknown function contains the points $(-1, 2)$, $(0, -1)$, $(2, -3)$, $(4, 1)$, $(5, -1)$, construct the polygonal approximation.

Solution   Graph the five points and connect them with line segments to form the (open) polygon of Figure 4-9.

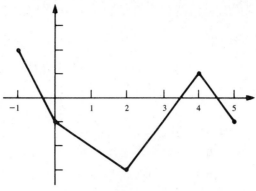

FIGURE 4-9

PROBLEM 4-15 Marine engineers are measuring water depths in a shipping channel. Rough water conditions allow only four readings to be taken (in addition to depth of zero at each edge). Refer to Figure 4-10.

A portable computer is on-site and programmed with an algorithm to use linear approximation. The number pairs $(0.2, 16.2)$, $(0.4, 25.4)$, $(0.6, 21.5)$, $(0.8, 19.7)$ are input. What do the engineers discover are the depths at **(a)** $x = 0.3$ and **(b)** $x = 0.5$?

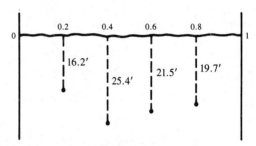

FIGURE 4-10

Solution The intervals are evenly spaced, so follow the algorithm steps.

(a) First solve for n then continue with the algorithm:

$$h = x_2 - x_1 = 0.4 - 0.2 = 0.2$$

$$x_0 = x_1 - h = 0$$

$$R = \frac{x - x_0}{h} = \frac{0.3 - 0}{0.2} = 1.5$$

$$J = \text{integer part of } R = 1$$

$$Y = y_1 + (y_2 - y_1)(R - J)$$
$$= 16.2 + (25.4 - 16.2)(1.5 - 1)$$
$$= 16.2 + (9.2)(0.5) = 16.2 + 4.6 = 20.8 \text{ ft}$$

(b) We still have $h = 0.2$ and $x_0 = 0$

$$R = \frac{0.5 - 0}{0.2} = 2.5$$

$$J = 2$$

$$Y = y_2 + (y_3 - y_2)(R - J)$$
$$= 25.4 + (21.5 - 25.4)(2.5 - 2)$$
$$= 25.4 + (-3.9)(0.5) = 25.4 - 1.95 = 23.45 \text{ ft}$$

PROBLEM 4-16 Given three points in the plane (x_1, y_1), (x_2, y_2), (x_3, y_3), use the Lagrange interpolation formula to derive the parabola passing through these points.

Solution The Lagrange formula gives

$$p(x) = \left[\frac{(x - x_2)(x - x_3)}{(x_1 - x_2)(x_1 - x_3)}\right] y_1 + \left[\frac{(x - x_1)(x - x_3)}{(x_2 - x_1)(x_2 - x_3)}\right] y_2 + \left[\frac{(x - x_1)(x - x_2)}{(x_3 - x_1)(x_3 - x_2)}\right] y_3$$

Since $x_1, x_2, x_3, y_1, y_2, y_3$ are all known values, we can make the following substitutions to simplify the notation:

$$\frac{y_1}{(x_1 - x_2)(x_1 - x_3)} = A \qquad \frac{y_2}{(x_2 - x_1)(x_2 - x_3)} = B \qquad \frac{y_3}{(x_3 - x_1)(x_3 - x_2)} = C$$

Then

$$p(x) = A[x^2 - (x_2 + x_3)x + x_2 \cdot x_3] + B[x^2 - (x_1 + x_3)x + x_1 \cdot x_3]$$
$$+ C[x^2 - (x_1 + x_2)x + x_1 \cdot x_2]$$
$$= (A + B + C)x^2 - [A(x_2 + x_3) + B(x_1 + x_3) + C(x_1 + x_2)]x$$
$$+ [Ax_2x_3 + Bx_1x_3 + Cx_1x_2]$$

Since the coefficients are all known values, this can be further simplified to:

$$Qx^2 - Rx + S$$

which is the equation of a parabola.

PROBLEM 4-17 Use the Lagrange interpolation formula to find the curve passing through $(-3, 1)$ and $(2, 7)$.

Solution Since we have only two points, this "curve" is a straight line. The Lagrange formula gives:

$$p(x) = \left[\frac{x - x_2}{x_1 - x_2}\right]y_1 + \left[\frac{x - x_1}{x_2 - x_1}\right]y_2$$
$$= \frac{xy_1 - x_2y_1}{x_1 - x_2} + \frac{xy_2 - x_1y_2}{x_2 - x_1}$$
$$= \frac{x(y_1 - y_2) + (x_1y_2 - x_2y_1)}{x_1 - x_2} = \left[\frac{y_1 - y_2}{x_1 - x_2}\right]x + \left[\frac{x_1y_2 - x_2y_1}{x_1 - x_2}\right]$$

as the formula for the equation of the "curve" between two given points. Plugging in our given points, we find the equation of our "curve" is:

$$\left[\frac{y_1 - y_2}{x_1 - x_2}\right]x + \left[\frac{x_1y_2 - x_2y_1}{x_1 - x_2}\right] = \left(\frac{+1 - 7}{-3 - 2}\right)x + \left(\frac{-3 \cdot 7 - 2 \cdot 1}{-3 \cdot 2}\right)$$
$$= \left(\frac{-6}{-5}\right)x + \left(\frac{-23}{-5}\right)$$
$$= \frac{6}{5}x + \frac{23}{5}$$

note: The "two-point" Lagrange interpolation formula (section 4.5) is in rough form when first constructed since the variable is in both terms. By algebraically revising it to put it in a polished form $(Ax + B)$, we make it easier to insert the numbers and to obtain the desired result right away. You can see that if we need to determine the equations of a hundred pairs of numbers, it's much better to do one algebraic simplification first and then 100 numbers crunches rather than use the original version every time.

Congruences

PROBLEM 4-18 Suppose that $m_1 \equiv n_1 \pmod{p}$ and that $m_2 \equiv n_2 \pmod{p}$. Prove that $m_1 + m_2 \equiv (n_1 + n_2) \pmod{p}$.

Solution Since $m_1 \equiv n_1 \pmod{p}$, $m_1 = n_1 + a \cdot p$ for some integer a. Since $m_2 \equiv n_2 \pmod{p}$, $m_2 = n_2 + b \cdot p$ for some integer b. Then

$$m_1 + m_2 = n_1 + n_2 + ap + bp$$
$$= n_1 + n_2 + (a + b)p$$

Because a and b are integers, $a + b$ is an integer, so $(m_1 + m_2) = (n_1 + n_2) + (\text{an integer}) \cdot p$. Thus $m_1 + m_2 \equiv (n_1 + n_2) \,(\text{mod } p)$

PROBLEM 4-19 Which positive integers less than 100 are congruent to zero modulo 12?

Solution If a number x is congruent to zero modulo 12, then $x - 0 = a \cdot 12$; i.e., $x = 12a$ for some a. The muitiples of 12 form this set. Those less than 100 are $\{12, 24, 36, 48, 60, 72, 84, 96\}$.

Random Number Generation

PROBLEM 4-20 Use the linear congruential generator to develop a periodic sequence of random numbers of length at least 6.

Solution The generator is $x_{n+1} \equiv (ax_n + c)\,(\text{mod } m)$ and we will pick $x_0 = 3, a = 5, c = 4, m = 7$. The formula now is $x_{n+1} \equiv (5x_n + 4)\,(\text{mod } 7)$.

$$x_0 \equiv 3$$
$$\left. \begin{array}{l} x_1 = (5 \cdot 3 + 4)\,(\text{mod } 7) \equiv 5 \\ x_2 = (5 \cdot 5 + 4)\,(\text{mod } 7) \equiv 1 \\ x_3 = (5 \cdot 1 + 4)\,(\text{mod } 7) \equiv 2 \\ x_4 = (5 \cdot 2 + 4)\,(\text{mod } 7) \equiv 0 \\ x_5 = (5 \cdot 0 + 4)\,(\text{mod } 7) \equiv 4 \\ x_6 = (5 \cdot 4 + 4)\,(\text{mod } 7) \equiv 3 \end{array} \right\} \quad \text{Periodic sequence of length 6}$$

Supplementary Exercises

PROBLEM 4-21 The sum of a number and five times its reciprocal is 6. What is the number?

PROBLEM 4-22 Carlos can paint a house in 6 days and Juan can paint the same house in 5 days. How many days would it take to paint the house if they worked together?

PROBLEM 4-23 A coin-operated vending machine was emptied and found to contain \$30 in quarters and dimes. If there were 150 coins in the machine, how many were quarters and how many were dimes?

PROBLEM 4-24 The sum of three consecutive integers is 114. What are the numbers?

PROBLEM 4-25 The length of a rectangle is 8 inches greater than its width and the area is 128 square inches. Find the length and width.

PROBLEM 4-26 A ball is thrown into the air with an initial velocity v_0 of 128 feet per second. The height h of the ball after t seconds is given by the formula $h = v_0 \cdot t - 16t^2$. **(a)** How long does it take the ball to come back to the ground? (*Hint*: at $h = 0$) **(b)** How high does the ball go?

PROBLEM 4-27 Ted is 4 years older than Wanda. In 6 years the sum of their ages will be 46. What are their ages now?

PROBLEM 4-28 The square of a number is 6 less than five times the number. What is the number?

PROBLEM 4-29 You are required to construct an open box whose base is a square and whose height is 4 inches shorter than twice its length. **(a)** Find the polynomial that expresses the total surface area of the box if the square base has sides of z inches. **(b)** What is the total surface area of the box if the base is $10'' + 10''$? **(c)** What would the dimensions be if the total surface area of the box is 1540 in^2?

PROBLEM 4-30   Mr. MacDonald had a piece of property bounded on one side by a river. When he enclosed a rectangular subportion of the property adjacent to the river, he used 500 feet of fencing for three sides of the pasture and let the river be the fourth side. What should the dimensions of the enclosure be if the area is 30,000 ft^2?

PROBLEM 4-31   Solve the system $\begin{Bmatrix} -2x + y = -4 \\ 6x - 3y = 12 \end{Bmatrix}$.

PROBLEM 4-32   Solve the system $\begin{Bmatrix} 4x - 3y = 5 \\ x + 5y = 7 \end{Bmatrix}$.

PROBLEM 4-33   Solve the system $\begin{Bmatrix} x + y - 4z = -7 \\ 2x - 3y + 7z = 26 \\ 9x - y - 2z = 29 \end{Bmatrix}$.

PROBLEM 4-34   Use the half-interval method to find a positive root of $e^{2x} - \frac{1}{2}x = 2$.

PROBLEM 4-35   Use Horner's method to evaluate the given polynomial at the indicated point:

(a) $f(x) = 1.3 + 0.45x - 1.21x^2$ at $x = 0.95$
(b) $f(x) = 2.614 - 1.93x + 2.68x^2 - 1.321x^3$ at $x = 1.27$

PROBLEM 4-36   Use the modified linear interpolation method of problem 4-12 to find a positive root of (a) $\cos x - x = 0$, (b) $e^{-x} = x$.

PROBLEM 4-37   Given the points of a function $(2, 9)$, $(4, 3)$, $(6, -5)$, $(8, -1)$, and $(10, 3)$, find the linear approximate y value for (a) $x = 3$ and (b) $x = 7.6$.

PROBLEM 4-38   Find the quadratic polynomial that contains the points $(1, 6)$, $(3, 16)$, $(-1, 12)$.

PROBLEM 4-39   Find the cubic polynomial that contains the points $(0, -1)$, $(1, 4)$, $(2, 13)$, $(3, 32)$.

PROBLEM 4-40   Find a set of numbers such that each element is congruent modulo 10 to all other elements in the set.

PROBLEM 4-41   Suppose that $m_1 \equiv n_1$ (mod p) and $m_2 \equiv n_2$ (mod p). (a) Prove that $-m_1 \equiv -n_1$ (mod p). (b) Prove that $m_1 \cdot m_2 \equiv n_2$ (mod p).

PROBLEM 4-42   Given $p \geq 2$, prove that for integers a and b (a) $a \equiv a$ (mod p), (b) $a \equiv b$ (mod p) implies $b \equiv a$ (mod p).

PROBLEM 4-43   Given a binary computer with word length 3, determine the maximum sequence length for random number generation.

PROBLEM 4-44   Generate the random number sequences for

(a) $x_{n+1} = (x_n + 2) \bmod 7$ (b) $x_{n+1} = (5x_n + 3) \bmod 8$ (c) $x_{n+1} = (4x_n + 2) \bmod 9$

Answers to Supplementary Exercises

4-21   5 or 1

4-22   2.72 days

4-23   100 quarters, 50 dimes

4-24   37, 38, 39

4-25   8 by 16

4-26   (a) 8 seconds   (b) 256 feet

4-27   Ted is 19, Wanda is 15

4-28 $x = 2$ or $x = 3$

4-29 (a) $9z^2 - 16z$ (b) 740 in^2
(c) $14'' \times 14'' \times 24''$

4-30 $150' \times 200'$ or $100' \times 300'$

4-31 $(a, 2a - 4)$ for any a

4-32 $(2, 1)$

4-33 $(4, 1, 3)$

4-34 0.392578

4-35 (a) 0.635 (b) 1.779

4-36 (a) 0.739085 (b) 0.56714

4-37 (a) 6 (b) -1.8

4-38 $p(x) = 2x^2 - 3x + 7$

4-39 $p(x) = x^3 - x^2 + 5x - 1$

4-40 All multiples of 10

4-43 m will be 7, 8, or 9

4-44 (a) 3, 5, 0, 2, 4, 6, 1
(b) 5, 4, 7, 6, 1, 0, 3, 2
(c) 6, 8, 7, 3, 5, 4, 0, 2, 1

5 MATRICES

THIS CHAPTER IS ABOUT

- ☑ **Matrices Defined**
- ☑ **Matrix Operations**
- ☑ **Special Types of Matrices**
- ☑ **Properties of Matrices**
- ☑ **Solving Systems with Matrices**
- ☑ **Arrays**

5-1. Matrices Defined

A **matrix** is any rectangular array of numbers that is made up of one or more rows and one or more columns (something like an ice cube tray or a checkers board). The **dimension**, or shape, of a matrix specifies its number of rows and columns. If the matrix has m rows and n columns, it is called an $m \times n$ (read "m-by-n") matrix. Its dimension is m rows by n columns. For example,

$$\mathbf{A} = \begin{bmatrix} 1 & 2 \\ 3 & 4 \end{bmatrix} \qquad \text{is a } 2 \times 2 \text{ matrix}$$

$$\mathbf{B} = \begin{bmatrix} 5 & \frac{1}{2} & 3 & 1 \\ 2 & 4 & 8 & \frac{6}{5} \end{bmatrix} \qquad \text{is a } 2 \times 4 \text{ matrix}$$

$$\mathbf{C} = \begin{bmatrix} 5 & 3 & 8 \\ 7 & 4 & -2 \\ 1 & 9 & 6 \\ 0 & -8 & -\frac{1}{2} \end{bmatrix} \qquad \text{is a } 4 \times 3 \text{ matrix}$$

note: As you will see in this chapter, a matrix is a valuable mathematical tool for displaying data systematically and compactly. It has applications in describing and solving equations, analyzing graphs, and creating schedules, among other important processes. A matrix also allows a single letter or symbol to represent a large body of information so that we may manipulate the information more easily.

A $1 \times n$ matrix is called a **row vector**; an $m \times 1$ matrix is called a **column vector**.

$$\mathbf{D} = \begin{bmatrix} 1 & 5 & -\frac{3}{4} & 6 \end{bmatrix} \qquad \text{is a row vector}$$

$$\mathbf{E} = \begin{bmatrix} 4 \\ 2 \\ 4 \end{bmatrix} \qquad \text{is a column vector}$$

The numbers appearing in a matrix are called the **entries** or **elements** of the matrix. You determine the **location**, or **address**, of any element in a matrix by specifying the row and the column in which the element lies. The element that lies in row i and column j is usually designed as a_{ij} or $a(i, j)$ or $\mathbf{A}(i, j)$. The first notation is usually used in mathematical writing, while the latter is used in programming situations.

A general $m \times n$ matrix $\mathbf{A}$ is written as:

GENERAL
$m \times n$
MATRIX

$$\mathbf{A} = \begin{bmatrix} a_{11} & a_{12} & a_{13} & \cdots & a_{1j} & \cdots & a_{1n} \\ a_{21} & a_{22} & a_{23} & \cdots & a_{2j} & \cdots & a_{2n} \\ a_{31} & a_{32} & a_{33} & \cdots & a_{3j} & \cdots & a_{3n} \\ \vdots & & & & & & \vdots \\ a_{i1} & a_{i2} & a_{i3} & \cdots & a_{ij} & \cdots & a_{in} \\ \vdots & & & & & & \vdots \\ a_{m1} & a_{m2} & a_{m3} & \cdots & a_{mj} & \cdots & a_{mn} \end{bmatrix} \tag{5-1}$$

Note that a_{32} is the address of the element in the $i = 3$ (or third) row and the $j = 2$ (or second) column; a_{51} is the address of the element in the fifth row and the first column.

The i-th row of $\mathbf{A}$ is $[a_{i1} \quad a_{i2} \quad \cdots \quad a_{ij} \quad \cdots \quad a_{in}]$, where i may take any value in $1 \leq i \leq m$, and the j-th column of $\mathbf{A}$ is

$$\begin{bmatrix} a_{1j} \\ a_{2j} \\ \vdots \\ a_{ij} \\ \vdots \\ a_{mj} \end{bmatrix}$$

where j may take any value in $1 \leq j \leq n$. If we know the dimension of a matrix $\mathbf{A}$, we can make the notation for the matrix more compact by writing it as

$$\mathbf{A} = [a_{ij}] \tag{5-2}$$

in which a_{ij} denotes the various entries of $\mathbf{A}$; the brackets show that we're talking about the *whole matrix*, not just a particular element a_{ij}.

note: Eq. (5-2) is simply a compact version of the $m \times n$ matrix in eq. (5-1).

EXAMPLE 5-1   Given $\mathbf{M} = \begin{bmatrix} 1 & 3 & \frac{1}{2} \\ 2 & -5 & 4 \end{bmatrix}$, find $a_{12}, a_{21},$ and a_{23}.

Solution

a_{12} is in the first row and second column: $a_{12} = 3$

a_{21} is in the second row and first column: $a_{21} = 2$

a_{23} is in the second row and third column: $a_{23} = 4$

Two matrices $\mathbf{A}$ and $\mathbf{B}$ are **equal**, written $\mathbf{A} = \mathbf{B}$, if they have precisely the same dimension and if their corresponding entries are equal, that is, if $a_{ij} = b_{ij}$ for all i, j.

EXAMPLE 5-2

The matrices $\mathbf{A} = \begin{bmatrix} 4 & 6 \\ 2 & -1 \end{bmatrix}$ and $\mathbf{B} = \begin{bmatrix} 5-1 & 2+4 \\ 7-5 & 3-4 \end{bmatrix}$ are equal.

For what values are $\mathbf{P} = \begin{bmatrix} m & n \\ s+1 & t+3 \end{bmatrix}$ and $\mathbf{Q} = \begin{bmatrix} 8 & 2 \\ -4 & 1 \end{bmatrix}$ equal?

Solution   For $\mathbf{P}$ to equal $\mathbf{Q}$, we must have $m = 8$, $n = 2$, $s + 1 = 4$, and $t + 3 = 1$. The values for m and n are obvious, and we solve for s and t: $s = -5$ and $t = -2$.

5-2. Matrix Operations

A. Sum and difference

If $A = [a_{ij}]$ and $B = [b_{ij}]$ are $m \times n$ matrices, then the *sum* of A and B, written $A + B$, is the matrix $C = [c_{ij}]$, where for every element c_{ij},

$$c_{ij} = a_{ij} + b_{ij} \qquad 1 \le i \le m, 1 \le j \le n$$

Thus you obtain the sum matrix C by adding the corresponding elements of A and B together.

To perform subtraction, you must first consider a new form. If matrix $A = [a_{ij}]$, then the *negative* of A is $-A = [-a_{ij}]$ (that is, you change the sign on every element of A to its opposite sign). Thus *subtraction* of matrices is defined as $B - A = B + (-A)$.

EXAMPLE 5-3   Given $A = \begin{bmatrix} 4 & 1 & 6 \\ 2 & -5 & 3 \end{bmatrix}$ and $B = \begin{bmatrix} -1 & 9 & 7 \\ 5 & 6 & -2 \end{bmatrix}$, find

(a) $A + B$ (b) $A - B$ (c) $B - A$

Solution

(a) Add the corresponding entries together:

$$A + B = \begin{bmatrix} 4 + (-1) & 1 + 9 & 6 + 7 \\ 2 + 5 & -5 + 6 & 3 + (-2) \end{bmatrix} = \begin{bmatrix} 3 & 10 & 13 \\ 7 & 1 & -1 \end{bmatrix}$$

(b) To subtract B from A, take the negative of B,

$$-B = \begin{bmatrix} 1 & -9 & -7 \\ -5 & -6 & 2 \end{bmatrix}$$

and add it to A:

$$A - B = A + (-B) = \begin{bmatrix} 4 + 1 & 1 + (-9) & 6 + (-7) \\ 2 + (-5) & -5 + (-6) & 3 + 2 \end{bmatrix} = \begin{bmatrix} 5 & -8 & -1 \\ -3 & -11 & 5 \end{bmatrix}$$

(c) Since $B - A = -(A - B)$, we can get this easily by using $(A - B)$ from part (b):

$$B - A = \begin{bmatrix} -5 & 8 & 1 \\ 3 & 11 & -5 \end{bmatrix}$$

important: Two matrices must have exactly the same dimension for you to be able to add or subtract them. If they do not, their sum and difference are not defined.

B. Scalar multiplication

Besides adding and subtracting matrices, you can also multiply a matrix by a real number called a **scalar**. If k is a scalar and A is an $m \times n$ matrix, then kA is the $m \times n$ matrix obtained by multiplying each element of A by the scalar k; that is,

SCALAR MULTIPLICATION $\qquad\qquad kA = [ka_{ij}]$

EXAMPLE 5-4   Find $3M$ and $-2M$ if

$$M = \begin{bmatrix} 2 & 4 \\ -3 & 1 \end{bmatrix}$$

Solution

$$3M = 3 \begin{bmatrix} 2 & 4 \\ -3 & 1 \end{bmatrix} = \begin{bmatrix} 6 & 12 \\ -9 & 3 \end{bmatrix} \quad \text{and} \quad -2M = -2 \begin{bmatrix} 2 & 4 \\ -3 & 1 \end{bmatrix} = \begin{bmatrix} -4 & -8 \\ 6 & -2 \end{bmatrix}$$

C. Matrix multiplication

If $\mathbf{A} = [a_{ij}]$ is an $m \times p$ matrix and $\mathbf{B} = [b_{ij}]$ is a $p \times n$ matrix, then the product of $\mathbf{A}$ and $\mathbf{B}$, denoted $\mathbf{AB}$, is the $m \times n$ matrix $\mathbf{C} = [c_{ij}]$ defined by

$$c_{ij} = a_{i1}b_{1j} + a_{i2}b_{2j} + \cdots + a_{ip}b_{pj} \qquad 1 \le i \le m, 1 \le j \le n \tag{5-3}$$

In expanded notation this is

$$
\begin{bmatrix}
a_{11} & a_{12} & \cdots & a_{1p} \\
a_{21} & a_{22} & \cdots & a_{2p} \\
\vdots & & & \\
a_{i1} & a_{i2} & \cdots & a_{ip} \\
\vdots & & & \\
a_{m1} & a_{m2} & \cdots & a_{mp}
\end{bmatrix}
\begin{bmatrix}
b_{11} & b_{12} & \cdots & b_{1j} & \cdots & b_{1n} \\
b_{21} & b_{22} & & b_{2j} & & b_{2n} \\
\vdots & \vdots & & \vdots & & \vdots \\
b_{p1} & b_{p2} & & b_{pj} & & b_{pn}
\end{bmatrix}
=
\begin{bmatrix}
c_{11} & c_{12} & \cdots & c_{1n} \\
c_{21} & c_{22} & & c_{2n} \\
\vdots & \vdots & c_{ij} & \vdots \\
c_{m1} & c_{m2} & & c_{mn}
\end{bmatrix}
$$

important: For the product $\mathbf{AB}$ to be defined, the number of columns of $\mathbf{A}$ *must* equal the number of rows of $\mathbf{B}$. The resulting product has dimension $m \times n$, in which m equals the number of rows of $\mathbf{A}$ and n equals the number of columns of $\mathbf{B}$. For example, if you multiply a 3×4 matrix by a 4×2 matrix, your answer will be a 3×2 matrix.

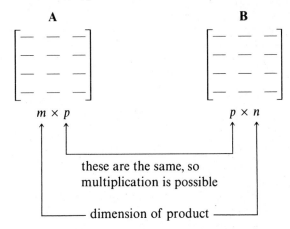

EXAMPLE 5-5   Let

$$\mathbf{A} = \begin{bmatrix} 1 & 2 \\ -1 & 3 \\ 3 & -4 \end{bmatrix} \qquad \mathbf{B} = \begin{bmatrix} 0 & 1 & 3 \\ 5 & 2 & 1 \end{bmatrix}$$

Is the product $\mathbf{AB}$ defined? If so, find it.

Solution   Since $\mathbf{A}$ is 3×2 and $\mathbf{B}$ is 2×3, the number of columns of $\mathbf{A}$ (which is 2) is equal to the number of rows of $\mathbf{B}$ (which is 2). Thus the product is defined. The resulting matrix will be 3×3. Now we're ready for the manipulations. Substituting from eq. (5-3) we have the following entries for c_{11}, c_{12}, and c_{13}, in which you multiply the first row of $\mathbf{A}$ by the first, second, and third columns of $\mathbf{B}$, respectively:

$$\begin{bmatrix} 1 & 2 \\ -1 & 3 \\ 3 & -4 \end{bmatrix}\begin{bmatrix} 0 & 1 & 3 \\ 5 & 2 & 1 \end{bmatrix} \qquad \begin{bmatrix} 1 & 2 \\ -1 & 3 \\ 3 & -4 \end{bmatrix}\begin{bmatrix} 0 & 1 & 3 \\ 5 & 2 & 1 \end{bmatrix} \qquad \begin{bmatrix} 1 & 2 \\ -1 & 3 \\ 3 & -4 \end{bmatrix}\begin{bmatrix} 0 & 1 & 3 \\ 5 & 2 & 1 \end{bmatrix}$$

$$c_{11} = 1 \cdot 0 + 2 \cdot 5 = 10 \qquad c_{12} = 1 \cdot 1 + 2 \cdot 2 = 5 \qquad c_{13} = 1 \cdot 3 + 2 \cdot 1 = 5$$

$$\begin{bmatrix} 10 & - & - \\ - & - & - \\ - & - & - \end{bmatrix} \qquad \begin{bmatrix} 10 & 5 & - \\ - & - & - \\ - & - & - \end{bmatrix} \qquad \begin{bmatrix} 10 & 5 & 5 \\ - & - & - \\ - & - & - \end{bmatrix}$$

(It's more complicated than difficult because you always have to keep track of where you are.) When solving for c_{21}, c_{22}, and c_{23}, you simply take the second row of **A** and multiply it by the first, second, and third columns of **B**, respectively; follow the same process for entries c_{31}, c_{32}, and c_{33} only use the third row of **A**. The new matrix then becomes

$$\mathbf{AB} = \begin{bmatrix} 1 & 2 \\ -1 & 3 \\ 3 & -4 \end{bmatrix} \begin{bmatrix} 0 & 1 & 3 \\ 5 & 2 & 1 \end{bmatrix} = \begin{bmatrix} 1\cdot0+2\cdot5 & 1\cdot1+2\cdot2 & 1\cdot3+2\cdot1 \\ -1\cdot0+3\cdot5 & -1\cdot1+3\cdot2 & -1\cdot3+3\cdot1 \\ 3\cdot0-4\cdot5 & 3\cdot1-4\cdot2 & 3\cdot3-4\cdot1 \end{bmatrix} = \begin{bmatrix} 10 & 5 & 5 \\ 15 & 5 & 0 \\ -20 & -5 & 5 \end{bmatrix}$$

EXAMPLE 5-6   Find **CD** and **DC** if

$$\mathbf{C} = \begin{bmatrix} 1 & -2 & 0 \\ 3 & 1 & 3 \end{bmatrix} \quad \text{and} \quad \mathbf{D} = \begin{bmatrix} 4 \\ 1 \\ -6 \end{bmatrix}$$

Solution

$$\mathbf{CD} = \begin{bmatrix} 1 & -2 & 0 \\ 3 & 1 & 3 \end{bmatrix} \begin{bmatrix} 4 \\ 1 \\ -6 \end{bmatrix} = \begin{bmatrix} 1\cdot4-2\cdot1+0\cdot(-6) \\ 3\cdot4+1\cdot1+3\cdot(-6) \end{bmatrix} = \begin{bmatrix} 2 \\ -5 \end{bmatrix}$$

DC is not defined since the number of columns of **D** (which is 1) does not equal the number of rows of **C** (which is 2).

The process of matrix multiplication can be readily put into algorithm form. Given $m \times p$ matrix **A** with elements $\mathbf{A}(i, j)$ and $p \times n$ matrix **B** with entries $\mathbf{B}(i, j)$:

Begin
  1. FOR $i = 1$ TO m DO — This takes each row of matrix **A**.
    a. FOR $j = 1$ TO n DO — This takes each column of matrix **B**.
      1. $\mathbf{C}(i, j) \leftarrow 0$ — This initializes the entry to zero, avoiding carryover from previous computations.

      2. FOR $k = 1$ TO p DO — This multiplies the correct row entry
        a. $\mathbf{C}(i, j) \leftarrow \mathbf{C}(i, j) + (\mathbf{A}(i, k) * \mathbf{B}(k, j))$ — of **A** by the correct column entry of
        b. Output $\mathbf{C}(i, j)$ — **B** and sums the resulting products to produce the new entry c_{ij} in the product matrix.

End

EXAMPLE 5-7   Find $\mathbf{A}^2$ if

$$\mathbf{A} = \begin{bmatrix} 3 & 0 & 0 \\ 0 & 2 & 0 \\ 0 & 0 & 1 \end{bmatrix}$$

Solution

$$\mathbf{A}^2 = \mathbf{A} \cdot \mathbf{A} = \begin{bmatrix} 3 & 0 & 0 \\ 0 & 2 & 0 \\ 0 & 0 & 1 \end{bmatrix} \begin{bmatrix} 3 & 0 & 0 \\ 0 & 2 & 0 \\ 0 & 0 & 1 \end{bmatrix}$$

$$= \begin{bmatrix} 3\cdot3+0\cdot0+0\cdot0 & 3\cdot0+0\cdot2+0\cdot0 & 3\cdot0+0\cdot0+0\cdot1 \\ 0\cdot3+2\cdot0+0\cdot0 & 0\cdot0+2\cdot2+0\cdot0 & 0\cdot0+2\cdot0+0\cdot1 \\ 0\cdot3+0\cdot0+1\cdot0 & 0\cdot0+0\cdot2+1\cdot0 & 0\cdot0+0\cdot0+1\cdot1 \end{bmatrix}$$

$$= \begin{bmatrix} 9 & 0 & 0 \\ 0 & 4 & 0 \\ 0 & 0 & 1 \end{bmatrix}$$

5-3. Special Types of Matrices

Some matrices have patterns or forms that make them special. Before we explore how to use matrices in applications, let's take a look at some of these special forms.

- If a matrix has an equal number of rows and columns, (that is, if $m = n$), then this $n \times n$ matrix is called a **square matrix**.

$$\begin{bmatrix} 4 & 7 \\ 3 & 9 \end{bmatrix} \quad \text{and} \quad \begin{bmatrix} 2 & 1 & 3 \\ 3 & 3 & 1 \\ 4 & 2 & -7 \end{bmatrix} \quad \text{are both square matrices}$$

If A is square, then $A^2, A^3, \ldots, A^n$ are always defined because the number of rows in one multiplier will always equal the number of columns in the other. If matrices A and B are both square matrices of the same dimension, then both AB and BA are defined but, in general, $AB \neq BA$.

EXAMPLE 5-8 Find AB and BA if

$$A = \begin{bmatrix} 4 & 1 \\ 2 & 3 \end{bmatrix} \quad \text{and} \quad B = \begin{bmatrix} 0 & -1 \\ 5 & 2 \end{bmatrix}$$

Solution

$$AB = \begin{bmatrix} 4 & 1 \\ 2 & 3 \end{bmatrix}\begin{bmatrix} 0 & -1 \\ 5 & 2 \end{bmatrix} = \begin{bmatrix} 4\cdot 0 + 1\cdot 5 & 4\cdot(-1) + 1\cdot 2 \\ 2\cdot 0 + 3\cdot 5 & 2\cdot(-1) + 3\cdot 2 \end{bmatrix} = \begin{bmatrix} 5 & -2 \\ 15 & 4 \end{bmatrix}$$

$$BA = \begin{bmatrix} 0 & -1 \\ 5 & 2 \end{bmatrix}\begin{bmatrix} 4 & 1 \\ 2 & 3 \end{bmatrix} = \begin{bmatrix} 0\cdot 4 - 1\cdot 2 & 0\cdot 1 - 1\cdot 3 \\ 5\cdot 4 + 2\cdot 2 & 5\cdot 1 + 2\cdot 3 \end{bmatrix} = \begin{bmatrix} -2 & -3 \\ 24 & 11 \end{bmatrix}$$

- If a square matrix $A = [a_{ij}]$ has the property that $a_{ij} = 0$ when $i \neq j$, then A is a **diagonal matrix**. A diagonal matrix has the form

$$A = \begin{bmatrix} a_{11} & 0 & \cdot & \cdot & \cdot & 0 \\ 0 & a_{22} & 0 & \cdot & & \cdot \\ \cdot & & 0 & & \cdot & 0 \\ 0 & \cdot & & \cdot & \cdot & a_{nn} \end{bmatrix}$$

The elements $a_{11}, a_{22}, \ldots, a_{nn}$ are on the *main diagonal* of the matrix.

$$\begin{bmatrix} 9 & 0 & 0 \\ 0 & 4 & 0 \\ 0 & 0 & 1 \end{bmatrix}, \quad \begin{bmatrix} 10 & 0 & 0 \\ 0 & 0 & 0 \\ 0 & 0 & 1 \end{bmatrix}, \quad \text{and} \quad \begin{bmatrix} 14 & 0 & 0 & 0 \\ 0 & 7 & 0 & 0 \\ 0 & 0 & 3 & 0 \\ 0 & 0 & 0 & 2 \end{bmatrix} \quad \text{are all diagonal matrices.}$$

- If all the main diagonal elements of a diagonal matrix are 1, the matrix is called an **identity matrix**. Thus the definition of an $n \times n$ identity matrix may be written

$$A = [a_{ij}] \quad \text{where} \quad a_{ij} = \begin{cases} 0 & \text{if} \quad i \neq j \\ 1 & \text{if} \quad i = j \end{cases}$$

and such a matrix is denoted by I_n.

$$I_3 = \begin{bmatrix} 1 & 0 & 0 \\ 0 & 1 & 0 \\ 0 & 0 & 1 \end{bmatrix} \quad \text{and} \quad I_4 = \begin{bmatrix} 1 & 0 & 0 & 0 \\ 0 & 1 & 0 & 0 \\ 0 & 0 & 1 & 0 \\ 0 & 0 & 0 & 1 \end{bmatrix} \quad \text{are identity matrices.}$$

- If *all* of the elements of an $m \times n$ matrix are zero, it is called a **zero matrix** and is denoted $\mathbf{0}_{m \times n}$ or just $\mathbf{0}$ if the dimensions are understood.

$$\begin{bmatrix} 0 & 0 & 0 \\ 0 & 0 & 0 \end{bmatrix} \quad \text{and} \quad \begin{bmatrix} 0 & 0 \\ 0 & 0 \end{bmatrix} \quad \text{are zero matrices.}$$

- If $\mathbf{A} = [a_{ij}]$ is an $m \times n$ matrix, then the $n \times m$ matrix $\mathbf{A}^T = [a_{ij}^T]$, where $a_{ij}^T = a_{ji}$, is called the **transpose of A**. You obtain the transpose by interchanging the rows and columns of the matrix: make the first row of $\mathbf{A}$ the first column of $\mathbf{A}^T$; make the second row of $\mathbf{A}$ the second column of $\mathbf{A}^T$, etc.

EXAMPLE 5-9   Find $\mathbf{A}^T$ and $\mathbf{B}^T$, given

$$\mathbf{A} = \begin{bmatrix} 2 & 1 & 3 \\ 0 & 4 & -1 \end{bmatrix} \quad \text{and} \quad \mathbf{B} = \begin{bmatrix} 5 & 1 \\ 3 & 0 \end{bmatrix}$$

Solution   Write the rows of the original matrix as the columns of the transpose:

$$\mathbf{A}^T = \begin{bmatrix} 2 & 0 \\ 1 & 4 \\ 3 & -1 \end{bmatrix} \quad \mathbf{B}^T = \begin{bmatrix} 5 & 3 \\ 1 & 0 \end{bmatrix}$$

A simple algorithm will find the transpose of a square matrix.

Begin
1. Input elements of an $n \times n$ matrix $\mathbf{A}$ in the form $\mathbf{A}(i, j)$
   2. FOR $i = 1$ TO N DO
      a. FOR $j = 1$ TO N DO
         1. $\mathbf{B}(j, i) \leftarrow \mathbf{A}(i, j)$
         2. Output $\mathbf{B}(j, i)$ $\mathbf{B}(j, i)$ are the elements of matrix
End $\mathbf{B} = \mathbf{A}^T$.

- If $\mathbf{A}$ is a square matrix and $\mathbf{A}^T = \mathbf{A}$, then $\mathbf{A}$ is a **symmetric matrix**. The elements are then related by $a_{ij} = a_{ji}$. The elements not on the main diagonal are mirror images across the main diagonal.

$$\mathbf{A} = \begin{bmatrix} 4 & 2 \\ 2 & 6 \end{bmatrix} \quad \text{is a } 2 \times 2 \text{ symmetric matrix.}$$

$$\mathbf{B} = \begin{bmatrix} 1 & 5 & -1 \\ 5 & 2 & 6 \\ -1 & 6 & 3 \end{bmatrix} \quad \text{is a } 3 \times 3 \text{ symmetric matrix.}$$

5-4. Properties of Matrices

A. Theorems on matrix addition

If $\mathbf{A} = [a_{ij}]$ is an $m \times n$ matrix, $\mathbf{B} = [b_{ij}]$ is an $m \times n$ matrix, $\mathbf{C} = [c_{ij}]$ is an $m \times n$ matrix, and $\mathbf{0}_{m \times n}$ is a zero matrix, then

    A1. $\mathbf{A} + \mathbf{B} = \mathbf{B} + \mathbf{A}$ Commutative Law
    A2. $(\mathbf{A} + \mathbf{B}) + \mathbf{C} = \mathbf{A} + (\mathbf{B} + \mathbf{C})$ Associative Law
    A3. $\mathbf{A} + \mathbf{0} = \mathbf{0} + \mathbf{A} = \mathbf{A}$ Additive Identity
    A4. $\mathbf{A} + (-\mathbf{A}) = (-\mathbf{A}) + \mathbf{A} = \mathbf{0}$ Additive Inverse

These properties of matrix addition are easy to check using the definitions.

B. Theorems on scalar multiplication

If s and t are real numbers and **A** and **B** are matrices of the same dimension, then

B1. $(s + t)\mathbf{A} = s\mathbf{A} + t\mathbf{A}$
B2. $s(t\mathbf{A}) = (st)\mathbf{A}$
B3. $s(\mathbf{A} + \mathbf{B}) = s\mathbf{A} + s\mathbf{B}$
B4. $-s\mathbf{A} = (-s)\mathbf{A} = s(-\mathbf{A})$
B5. $(s\mathbf{A}) = s\mathbf{A}$
B6. $0\mathbf{A} = \mathbf{0}$ (that is, zero times a matrix produces a zero matrix.)

As with the addition theorems, scalar multiplication theorems for matrices are relatively easy to prove, so we will explore them in the Solved Problems section.

C. Theorems on matrix multiplication

Let **A**, **B**, and **C** be $m \times n$, $n \times p$, and $n \times p$ matrices, respectively. Then

C1. $\mathbf{A}(\mathbf{B} + \mathbf{C}) = \mathbf{AB} + \mathbf{AC}$ Distributivity

Let **A**, **B**, and **C** be $p \times q$, $n \times p$, and $n \times p$ matrices, respectively. Then

C2. $(\mathbf{B} + \mathbf{C})\mathbf{A} = \mathbf{BA} + \mathbf{CA}$ Distributivity

Let **A**, **B**, and **C** be $m \times n$, $n \times p$, and $p \times q$ matrices, respectively. Then

C3. $\mathbf{A}(\mathbf{BC}) = (\mathbf{AB})\mathbf{C}$ Associativity

The proofs of these theorems are considerably more difficult; we will only demonstrate their validity.

EXAMPLE 5-10   Let

$$\mathbf{A} = \begin{bmatrix} 2 & 1 \\ 0 & 3 \end{bmatrix} \qquad \mathbf{B} = \begin{bmatrix} -1 & 0 \\ 4 & 2 \end{bmatrix} \quad \text{and} \quad \mathbf{C} = \begin{bmatrix} 0 & -3 \\ 2 & 5 \end{bmatrix}$$

Demonstrate the validity of Theorems C1, C2, and C3.

Solution

C1. $\mathbf{A}(\mathbf{B} + \mathbf{C}) = \begin{bmatrix} 2 & 1 \\ 0 & 3 \end{bmatrix}\begin{bmatrix} -1 & -3 \\ 6 & 7 \end{bmatrix} = \begin{bmatrix} 4 & 1 \\ 18 & 21 \end{bmatrix}$

$\mathbf{AB} = \begin{bmatrix} 2 & 1 \\ 0 & 3 \end{bmatrix}\begin{bmatrix} -1 & 0 \\ 4 & 2 \end{bmatrix} = \begin{bmatrix} 2 & 2 \\ 12 & 6 \end{bmatrix}$

$\mathbf{AC} = \begin{bmatrix} 2 & 1 \\ 0 & 3 \end{bmatrix}\begin{bmatrix} 0 & -3 \\ 2 & 5 \end{bmatrix} = \begin{bmatrix} 2 & -1 \\ 6 & 15 \end{bmatrix}$ same

$\mathbf{AB} + \mathbf{AC} = \begin{bmatrix} 4 & 1 \\ 18 & 21 \end{bmatrix}$

note: Especially when multiplying matrices, the order in which you put them is very important because quite frequently $\mathbf{AB} \neq \mathbf{BA}$.

C2. $(\mathbf{B} + \mathbf{C})\mathbf{A} = \begin{bmatrix} -1 & -3 \\ 6 & 7 \end{bmatrix}\begin{bmatrix} 2 & 1 \\ 0 & 3 \end{bmatrix} = \begin{bmatrix} -2 & -10 \\ 12 & 27 \end{bmatrix}$

$\mathbf{BA} = \begin{bmatrix} -1 & 0 \\ 4 & 2 \end{bmatrix}\begin{bmatrix} 2 & 1 \\ 0 & 3 \end{bmatrix} = \begin{bmatrix} -2 & -1 \\ 8 & 10 \end{bmatrix}$

$\mathbf{CA} = \begin{bmatrix} 0 & -3 \\ 2 & 5 \end{bmatrix}\begin{bmatrix} 2 & 1 \\ 0 & 3 \end{bmatrix} = \begin{bmatrix} 0 & -9 \\ 4 & 17 \end{bmatrix}$ same

$\mathbf{BA} + \mathbf{CA} = \begin{bmatrix} -2 & -10 \\ 12 & 27 \end{bmatrix}$

C3. $\mathbf{A(BC)} = \begin{bmatrix} 2 & 1 \\ 0 & 3 \end{bmatrix} \left(\begin{bmatrix} -1 & 0 \\ 4 & 2 \end{bmatrix} \begin{bmatrix} 0 & -3 \\ 2 & 5 \end{bmatrix} \right)$

$\qquad\quad = \begin{bmatrix} 2 & 1 \\ 0 & 3 \end{bmatrix} \begin{bmatrix} 0 & 3 \\ 4 & -2 \end{bmatrix} = \begin{bmatrix} 4 & 4 \\ 12 & -6 \end{bmatrix}$

$\quad\mathbf{(AB)C} = \left(\begin{bmatrix} 2 & 1 \\ 0 & 3 \end{bmatrix} \begin{bmatrix} -1 & 0 \\ 4 & 2 \end{bmatrix} \right) \begin{bmatrix} 0 & -3 \\ 2 & 5 \end{bmatrix}$ same

$\qquad\quad = \begin{bmatrix} 2 & 2 \\ 12 & 6 \end{bmatrix} \begin{bmatrix} 0 & -3 \\ 2 & 5 \end{bmatrix} = \begin{bmatrix} 4 & 4 \\ 12 & -6 \end{bmatrix}$

D. Theorems on transposes

If r is a real number and $\mathbf{A}$ and $\mathbf{B}$ are matrices whose dimensions allow addition and multiplication, then

D1. $(\mathbf{A}^T)^T = \mathbf{A}$
D2. $(\mathbf{A} + \mathbf{B})^T = \mathbf{A}^T + \mathbf{B}^T$
D3. $(r\mathbf{A})^T = r\mathbf{A}^T$
D4. $(\mathbf{AB})^T = \mathbf{B}^T \cdot \mathbf{A}^T$

Proofs of the transpose theorems will also be explored in the Solved Problems.

E. Determinants

Every square matrix $\mathbf{A}$ has a certain scalar number, called the **determinant of** $\mathbf{A}$ and denoted det $\mathbf{A}$, associated with it. As you will see later, the determinant is useful in solving systems of equations. The value of the determinant for an $n \times n$ matrix involves a rather complicated formula using $n!$ terms, so we will only define the cases $n = 1$, $n = 2$, and $n = 3$. Determinants for larger matrices require additional properties and a computer algorithm to determine their value.

- For a 1×1 matrix $\mathbf{A} = [a_{11}]$, the value of det $\mathbf{A}$ is a_{11}.

- For a 2×2 matrix $\mathbf{B} = \begin{bmatrix} a_{11} & a_{12} \\ a_{21} & a_{22} \end{bmatrix}$, the value of det $\mathbf{B}$ is $a_{11}a_{22} - a_{12}a_{21}$.

- For a 3×3 matrix $\mathbf{C} = \begin{bmatrix} a_{11} & a_{12} & a_{13} \\ a_{21} & a_{22} & a_{23} \\ a_{31} & a_{32} & a_{33} \end{bmatrix}$, the value of det $\mathbf{C}$ is

$$a_{11}a_{22}a_{33} + a_{12}a_{23}a_{31} + a_{13}a_{21}a_{32} - a_{13}a_{22}a_{31} - a_{11}a_{23}a_{32} - a_{12}a_{21}a_{33}$$

A pictorial representation makes it much easier to remember these values (we follow mathematical convention and indicate the determinant of a matrix by enclosing the elements of the matrix in long vertical bars):

$$\det \mathbf{A} = \det[a_{11}] = |a_{11}| = a_{11}$$

$$\det \mathbf{B} = \det \begin{bmatrix} a_{11} & a_{12} \\ a_{21} & a_{22} \end{bmatrix} = \begin{vmatrix} a_{11} & a_{12} \\ a_{21} & a_{22} \end{vmatrix} = a_{11}a_{22} - a_{12}a_{21}$$

second   first
term   term

$$a_{11}a_{22}a_{33} \quad + \quad a_{12}a_{23}a_{31} \quad + \quad a_{13}a_{21}a_{32}$$

$$\det \mathbf{C} = \det \begin{bmatrix} a_{11} & a_{12} & a_{13} \\ a_{21} & a_{22} & a_{23} \\ a_{31} & a_{32} & a_{33} \end{bmatrix} = \begin{vmatrix} a_{11} & a_{12} & a_{13} \\ a_{21} & a_{22} & a_{23} \\ a_{31} & a_{32} & a_{33} \end{vmatrix} \quad \begin{vmatrix} a_{11} & a_{12} & a_{13} \\ a_{21} & a_{22} & a_{23} \\ a_{31} & a_{32} & a_{33} \end{vmatrix} \quad \begin{vmatrix} a_{11} & a_{12} & a_{13} \\ a_{21} & a_{22} & a_{23} \\ a_{31} & a_{32} & a_{33} \end{vmatrix}$$

first term   second term   third term

$$- a_{13}a_{22}a_{31} \quad - \quad a_{11}a_{23}a_{32} \quad - \quad a_{12}a_{21}a_{33}$$

$$\begin{vmatrix} a_{11} & a_{12} & a_{13} \\ a_{21} & a_{22} & a_{23} \\ a_{31} & a_{32} & a_{33} \end{vmatrix} \quad \begin{vmatrix} a_{11} & a_{12} & a_{13} \\ a_{21} & a_{22} & a_{23} \\ a_{31} & a_{32} & a_{33} \end{vmatrix} \quad \begin{vmatrix} a_{11} & a_{12} & a_{13} \\ a_{21} & a_{22} & a_{23} \\ a_{31} & a_{32} & a_{33} \end{vmatrix}$$

fourth term   fifth term   sixth term

(5-4)

Eq. (5-4) shows exactly which entries you multiply within the matrix to get each term of the 3×3 determinant value equation. A shorter, quicker way to write it is to attach a copy of the first two columns of the matrix to the right side of the matrix and then to draw the arrows and positive and negative signs as follows:

$$\det \mathbf{C} = \det \begin{bmatrix} a_{11} & a_{12} & a_{13} \\ a_{21} & a_{22} & a_{23} \\ a_{31} & a_{32} & a_{33} \end{bmatrix} = \begin{vmatrix} a_{11} & a_{12} & a_{13} \\ a_{21} & a_{22} & a_{23} \\ a_{31} & a_{32} & a_{33} \end{vmatrix} \begin{matrix} a_{11} & a_{12} \\ a_{21} & a_{22} \\ a_{31} & a_{32} \end{matrix}$$

fourth   fifth   sixth   first   second   third
term   term   term   term   term   term

$$= a_{11}a_{22}a_{33} + a_{12}a_{23}a_{31} + a_{13}a_{21}a_{32} - a_{13}a_{22}a_{31} - a_{11}a_{23}a_{32} - a_{12}a_{21}a_{33} \quad \textbf{(5-5)}$$

The procedures of eqs. (5-4) and (5-5) produce the same answer, only you'll probably find eq. (5-5) a lot more convenient to use.

EXAMPLE 5-11   Find the determinants of the matrices

$$\mathbf{S} = \begin{bmatrix} 5 & -2 \\ 6 & -1 \end{bmatrix}, \quad \mathbf{T} = \begin{bmatrix} 4 & -3 & 0 \\ 1 & 5 & 2 \\ -2 & 1 & 6 \end{bmatrix}, \quad \text{and} \quad \mathbf{I}_3$$

Solution

$$\det \mathbf{S} = \begin{vmatrix} 5 & -2 \\ 6 & -1 \end{vmatrix} = 5 \cdot (-1) - (6) \cdot (-2) = -5 + 12 = 7$$

$$\det \mathbf{T} = \begin{vmatrix} 4 & -3 & 0 \\ 1 & 5 & 2 \\ -2 & 1 & 6 \end{vmatrix} \begin{matrix} 4 & -3 \\ 1 & 5 \\ -2 & 1 \end{matrix} = 4 \cdot 5 \cdot 6 + (-3) \cdot 2 \cdot (-2) + 0 \cdot 1 \cdot 1$$
$$- 0 \cdot 5 \cdot (-2) - 4 \cdot 2 \cdot 1 - (-3) \cdot 1 \cdot 6$$
$$= 120 + 12 + 0 - 0 - 8 - (-18)$$
$$= 142$$

$$\det \mathbf{I}_3 = \begin{vmatrix} 1 & 0 & 0 \\ 0 & 1 & 0 \\ 0 & 0 & 1 \end{vmatrix} \begin{matrix} 1 & 0 \\ 0 & 1 \\ 0 & 0 \end{matrix} = 1 + 0 + 0 - 0 - 0 - 0 = 1$$

recall: I_3 is the identity matrix of a 3×3 matrix.

note: The determinant of an identity matrix will always equal 1.

One other useful property of determinants is

$$\det(\mathbf{AB}) = (\det \mathbf{A}) \cdot (\det \mathbf{B}) \text{ for all square matrices } \mathbf{A} \text{ and } \mathbf{B} \qquad (5\text{-}6)$$

Evaluating the determinant of a matrix that has a dimension higher than 3×3 is difficult when you use the above process. For example, you would need to make $4! = 24$ product evaluations, each involving 4 numbers, to get the value of a 4×4 determinant, and $5! = 120$ product evaluations, each involving 5 numbers, for a 5×5 determinant, etc. Continuing this calculation is impractical even when using a computer, since $n!$, even for moderately sized n, gets *very* large very quickly. To evaluate a large-sized determinant, then, you'd use a computer algorithm that relies on some of the properties of determinants (such as creating the value 0 in some key positions) and special procedures. For the full story on this, consult any linear algebra or matrix theory book. Once you master the material in this chapter, those advanced procedures should be easier to understand.

F. Inverses of matrices

Given a square $n \times n$ matrix $\mathbf{A}$, another square $n \times n$ matrix $\mathbf{B}$ is the **inverse of A** if *both* of the products $\mathbf{AB}$ and $\mathbf{BA}$ are equal to the identity matrix $\mathbf{I}_n$. The inverse of a matrix $\mathbf{A}$, denoted by $\mathbf{A}^{-1}$, is unique. Thus, if $\mathbf{A}$ has an inverse (and not all matrices do!), then $\mathbf{AA}^{-1} = \mathbf{A}^{-1}\mathbf{A} = \mathbf{I}_n$.

note: This notation for the inverse of a matrix is *not* a reciprocal: $\mathbf{A}^{-1} \neq \dfrac{1}{\mathbf{A}}$.

Each matrix that has an inverse is said to be **invertible**. To test for invertibility you use determinants: an $n \times n$ matrix $\mathbf{A}$ is invertible if and only if $\det(\mathbf{A}) \neq 0$.

EXAMPLE 5-12   Verify the test for invertibility by showing that the following matrix does not have an inverse:

$$\mathbf{A} = \begin{bmatrix} 1 & 3 \\ 2 & 6 \end{bmatrix}$$

Solution   Suppose $\mathbf{A}$ has an inverse and designate it as

$$\mathbf{A}^{-1} = \begin{bmatrix} a & b \\ c & d \end{bmatrix}$$

Then

$$\mathbf{I}_2 = \mathbf{AA}^{-1} = \begin{bmatrix} 1 & 3 \\ 2 & 6 \end{bmatrix}\begin{bmatrix} a & b \\ c & d \end{bmatrix} = \begin{bmatrix} a + 3c & b + 3d \\ 2a + 6c & 2b + 6d \end{bmatrix} = \begin{bmatrix} 1 & 0 \\ 0 & 1 \end{bmatrix} \qquad (5\text{-}7)$$

For eq. (5-7) to be true we must have

$$a + 3c = 1 \qquad\qquad\qquad b + 3d = 0$$
$$\underline{2a + 6c = 0} \qquad\qquad\qquad \underline{2b + 6d = 1}$$

Contradiction, multiplying the first equation by 2 gives us $2a + 6c = 2$, which does not agree with the second equation $2a + 6c = 0$.

Also a contradiction.

Therefore eq. (5-7) is false and $\mathbf{A}^{-1}$ does not exist. Now to confirm that we do *not* have $\det(\mathbf{A}) \neq 0$:

$$\det(\mathbf{A}) = \begin{vmatrix} 1 & 3 \\ 2 & 6 \end{vmatrix} = 1 \cdot 6 - 3 \cdot 2 = 6 - 6 = 0$$

In the next example, we'll explain how to find the inverse of an invertible matrix.

EXAMPLE 5-13 Show that matrix $\mathbf{B} = \begin{bmatrix} 4 & 2 \\ 1 & 1 \end{bmatrix}$ is invertible and find its inverse.

Solution Since $\det(\mathbf{B}) = \begin{vmatrix} 4 & 2 \\ 1 & 1 \end{vmatrix} = 4 \cdot 1 - 2 \cdot 1 = 2 \neq 0$, matrix $\mathbf{B}$ is invertible. Designate the inverse as

$\mathbf{B}^{-1} = \begin{bmatrix} a & b \\ c & d \end{bmatrix}$. Then

$$\mathbf{I}_2 = \mathbf{B} \cdot \mathbf{B}^{-1} \begin{bmatrix} 1 & 0 \\ 0 & 1 \end{bmatrix} = \begin{bmatrix} 4 & 2 \\ 1 & 1 \end{bmatrix} \begin{bmatrix} a & b \\ c & d \end{bmatrix} = \begin{bmatrix} 4a + 2c & 4b + 2d \\ a + c & b + d \end{bmatrix}$$

Thus

$$\begin{cases} 4a + 2c = 1 \\ a + c = 0 \end{cases} \quad \text{and} \quad \begin{cases} 4b + 2d = 0 \\ b + d = 1 \end{cases}$$

which gives us two systems of linear equations. We can now solve for a, b, c, and d by solving these two systems using the elementary operations (see Section 4-1):

$$\begin{cases} 4a + 2c = 1 \\ a + c = 0 \end{cases} \qquad \begin{cases} 4b + 2d + 0 \\ b + d = 1 \end{cases}$$
$$\Downarrow \qquad\qquad\qquad \Downarrow$$
$$\begin{cases} 4a + 2c = 1 \\ \underline{2a + -2c = 0} \end{cases} \qquad \begin{cases} 4b + 2d = 0 \\ \underline{-2b + 2d = 2} \end{cases}$$
$$2a = 1 \qquad\qquad 2b = -2$$
$$a = 1/2 \qquad\qquad b = -1$$
$$c = -1/2 \qquad\qquad d = 2$$

so the inverse of $\mathbf{B}$ is

$$\mathbf{B}^{-1} = \begin{bmatrix} a & b \\ c & d \end{bmatrix} = \begin{bmatrix} \frac{1}{2} & -1 \\ -\frac{1}{2} & 2 \end{bmatrix}$$

Check: We should also have $\mathbf{B}^{-1} \cdot \mathbf{B} = \mathbf{I}_2$:

$$\mathbf{B}^{-1} \cdot \mathbf{B} = \begin{bmatrix} \frac{1}{2} & -1 \\ -\frac{1}{2} & 2 \end{bmatrix} \begin{bmatrix} 4 & 2 \\ 1 & 1 \end{bmatrix} = \begin{bmatrix} 1 & 0 \\ 0 & 1 \end{bmatrix}$$

---

note: In Example 5-13, it is more convenient to "factor out" the value $\frac{1}{2}$ from $\mathbf{B}^{-1}$ when writing the matrix (the process is the reverse of scalar multiplication—see Theorem B2 of Section 5-4):

$$\mathbf{B}^{-1} = \frac{1}{2} \begin{bmatrix} 1 & -2 \\ -1 & 4 \end{bmatrix}$$

As you will soon see, this simplified version is easier to work with.

Example 5-13 leads us to the general definition of the inverse of a 2×2 matrix: If $\mathbf{A} = \begin{bmatrix} a & b \\ c & d \end{bmatrix}$

with $\det(\mathbf{A}) = ad - bc \neq 0$ then

INVERSE OF A 2 × 2 MATRIX $\qquad \mathbf{A}^{-1} = \dfrac{1}{\det(\mathbf{A})} \begin{bmatrix} d & -b \\ -c & a \end{bmatrix}$

Sometimes it is easier to remember this as a natural language algorithm. To find the inverse of a 2 × 2 matrix,

---

ALGORITHM FOR FINDING THE INVERSE OF A 2 × 2 MATRIX

1. Interchange the elements on the main diagonal.
2. Change the sign on the off-diagonal elements (the *off diagonal* is the one that goes from the upper right corner to the lower left corner).
3. Multiply the entire matrix by scalar 1/det(**A**) (be sure to put it in the front).

In the next section, you will learn how to find the inverse of a matrix that has a dimension higher than 2 × 2.

5-5. Solving Systems with Matrices

Now that we've hashed over the various types and qualities of matrices, it's time to put them to work for us.

A. Notation

Remember in Chapter 4 how you slaved over manipulating linear equations in a system to solve the system (see Section 4-1)? Well, now you don't have to work so hard, because with matrices and determinants you can virtually ignore all of the variables and work only with the coefficients and constants of your equations. You do so by representing your system in matrix notation **AX** = **C**, in which **A** is a **coefficient matrix** (a matrix made up of the coefficients of your system) and **X** and **C** are each a column vector.

EXAMPLE 5-14   Represent the following linear system in matrix notation:

$$\begin{cases} 4x + 2y = 7 \\ x + y = -2 \end{cases}$$ **(5-8)**

Solution   We will form a coefficient matrix **A** consisting of the coefficients on the left side of the equations. (Be sure to align the *x*-terms and the *y*-terms beneath one another first.) For the above system the coefficient matrix is:

$$\mathbf{A} = \begin{bmatrix} 4 & 2 \\ 1 & 1 \end{bmatrix}$$

Now, also form two column vectors (recall: a column vector is a matrix with only one column) for the system—one showing the variables, and one created from the constants on the right side of the equations. We then have:

$$\mathbf{X} = \begin{bmatrix} x \\ y \end{bmatrix} \quad \text{and} \quad \mathbf{C} = \begin{bmatrix} 7 \\ -2 \end{bmatrix}$$

All parts of our system are now in matrix notation, and if we write **AX** = **C**, we obtain

$$\begin{bmatrix} 4 & 2 \\ 1 & 1 \end{bmatrix} \begin{bmatrix} x \\ y \end{bmatrix} = \begin{bmatrix} 7 \\ -2 \end{bmatrix}$$

as the matrix representation of eq. (5-8). If you multiply out the matrix product on the left and equate its elements to the corresponding elements on the right, the original two linear equations of eq. (5-8) reappear.

Let's try another one.

EXAMPLE 5-15   Represent the following system in matrix notation:

$$\begin{cases} -2x + 5y = 3 \\ x - 4y = 8 \end{cases}$$

Solution   The coefficient matrix is $\mathbf{A} = \begin{bmatrix} -2 & 5 \\ 1 & -4 \end{bmatrix}$; the variable column matrix is $\mathbf{X} = \begin{bmatrix} x \\ y \end{bmatrix}$; and the constant column matrix is $\mathbf{C} = \begin{bmatrix} 3 \\ 8 \end{bmatrix}$. Thus $\mathbf{AX} = \mathbf{C}$ is

$$\begin{bmatrix} -2 & 5 \\ 1 & -4 \end{bmatrix} \begin{bmatrix} x \\ y \end{bmatrix} = \begin{bmatrix} 3 \\ 8 \end{bmatrix}$$

This representation is identical to the original system.

B. Solution methods

Knowing how to set up a system in matrix notation doesn't do you much good until you know how to use it to solve the system, so here we go. If a linear system is represented as $\mathbf{AX} = \mathbf{C}$, then two matrix operations will be required to obtain its solution:

ALGORITHM FOR SOLVING $AX = C$

1. Find $\mathbf{A}^{-1}$
2. Multiply on the *left* by $\mathbf{A}^{-1}$ on *both* sides of the equation to produce $\mathbf{A}^{-1} \cdot (\mathbf{AX}) = \mathbf{A}^{-1} \cdot \mathbf{C}$. Equivalently, this is

$$(\mathbf{A}^{-1} \cdot \mathbf{A})\mathbf{X} = \mathbf{A}^{-1} \cdot \mathbf{C}$$

$$\mathbf{IX} = \mathbf{A}^{-1} \cdot \mathbf{C}$$

$$\mathbf{X} = \mathbf{A}^{-1} \cdot \mathbf{C} \qquad \textbf{(5-9)}$$

3. Output the results from this matrix equality (you will end up with a straightforward equality of the variable column on the left and a constant column on the right).

EXAMPLE 5-16   Solve the system from Example 5-14 using the algorithm for solving $\mathbf{AX} = \mathbf{C}$.

Solution   The system $\begin{cases} 4x + 2y = 7 \\ x + y = -2 \end{cases}$ is written as

$$\begin{bmatrix} 4 & 2 \\ 1 & 1 \end{bmatrix} \begin{bmatrix} x \\ y \end{bmatrix} = \begin{bmatrix} 7 \\ -2 \end{bmatrix}$$

First we want to find the inverse of $\mathbf{A} = \begin{bmatrix} 4 & 2 \\ 1 & 1 \end{bmatrix}$ which we did in Example 5-13. This result is

$$\mathbf{A}^{-1} = \frac{1}{2} \begin{bmatrix} 1 & -2 \\ -1 & 4 \end{bmatrix}$$

(note: We use the factored version for simplicity.) Hence we form the new matrix equation

$$\mathbf{A}^{-1} \cdot \mathbf{AX} = \mathbf{A}^{-1} \cdot \mathbf{C}$$

$$\underbrace{\frac{1}{2} \begin{bmatrix} 1 & -2 \\ -1 & 4 \end{bmatrix} \begin{bmatrix} 4 & 2 \\ 1 & 1 \end{bmatrix}}_{\mathbf{A}^{-1} \cdot \mathbf{A} = \mathbf{I}_2} \begin{bmatrix} x \\ y \end{bmatrix} = \frac{1}{2} \begin{bmatrix} 1 & -2 \\ -1 & 4 \end{bmatrix} \begin{bmatrix} 7 \\ -2 \end{bmatrix}$$

$$\begin{bmatrix} 1 & 0 \\ 0 & 1 \end{bmatrix}\begin{bmatrix} x \\ y \end{bmatrix} = \frac{1}{2}\begin{bmatrix} 1\cdot 7 + (-2)\cdot(-2) \\ -1\cdot 7 + 4\cdot(-2) \end{bmatrix} = \frac{1}{2}\begin{bmatrix} 11 \\ -15 \end{bmatrix}$$

or

$$\begin{bmatrix} x \\ y \end{bmatrix} = \begin{bmatrix} \frac{11}{2} \\ -\frac{15}{2} \end{bmatrix}$$

The solution is easy to read: $x = \frac{11}{2}$, $y = -\frac{15}{2}$

Check: $\begin{cases} 4x + 2y = 7 \\ x + y = -2 \end{cases}$ $\longrightarrow$ $\begin{aligned} 4(\frac{11}{2}) + 2(-\frac{15}{2}) = \frac{44}{2} - \frac{30}{2} = \frac{14}{2} = 7 \quad \checkmark \\ \frac{11}{2} + (-\frac{15}{2}) = -\frac{4}{2} = -2 \quad \checkmark \end{aligned}$

EXAMPLE 5-17 Solve the system from Example 5-15 using matrix methods:

$$\begin{cases} -2x + 5y = 3 \\ x - 4y = 8 \end{cases}$$

Solution The system in matrix notation (**AX = C**) is written as

$$\begin{bmatrix} -2 & 5 \\ 1 & -4 \end{bmatrix}\begin{bmatrix} x \\ y \end{bmatrix} = \begin{bmatrix} 3 \\ 8 \end{bmatrix}$$

To get $\mathbf{A}^{-1}$ we follow the algorithm for finding the inverse of a 2×2 matrix, described earlier (Section 5-4):

$$\begin{bmatrix} -2 & 5 \\ 1 & -4 \end{bmatrix} \xrightarrow[\substack{\text{interchange elements} \\ \text{in the main diagonal} \\ \text{and change the signs} \\ \text{of the off-diagonal} \\ \text{elements}}]{} \begin{bmatrix} -4 & -5 \\ -1 & -2 \end{bmatrix} \xrightarrow[\substack{\text{multiply new matrix} \\ \text{by } \frac{1}{\det(\mathbf{A})}}]{} \frac{1}{3}\begin{bmatrix} -4 & -5 \\ -1 & -2 \end{bmatrix}$$

get det(**A**)

$$\det(\mathbf{A}) = \begin{vmatrix} -2 & 5 \\ 1 & -4 \end{vmatrix} = 8 - 5 = 3$$

Thus we have $\mathbf{A}^{-1}\cdot \mathbf{AX} = \mathbf{A}^{-1}\cdot \mathbf{C}$:

$$\underbrace{\frac{1}{3}\begin{bmatrix} -4 & -5 \\ -1 & -2 \end{bmatrix}\begin{bmatrix} -2 & 5 \\ 1 & -4 \end{bmatrix}}_{I_2}\begin{bmatrix} x \\ y \end{bmatrix} = \frac{1}{3}\begin{bmatrix} -4 & -5 \\ -1 & -2 \end{bmatrix}\begin{bmatrix} 3 \\ 8 \end{bmatrix}$$

$$\begin{bmatrix} 1 & 0 \\ 0 & 1 \end{bmatrix}\begin{bmatrix} x \\ y \end{bmatrix} = \frac{1}{3}\begin{bmatrix} -12 - 40 \\ -3 - 16 \end{bmatrix} = \frac{1}{3}\begin{bmatrix} -52 \\ -19 \end{bmatrix} = \begin{bmatrix} -\frac{52}{3} \\ -\frac{19}{3} \end{bmatrix}$$

The solution is $x = -\frac{52}{3}$, $y = -\frac{19}{3}$

Check:

$$\begin{cases} -2x + 5y = 3 \\ x - 4y = 8 \end{cases} \longrightarrow \begin{aligned} -2(-\frac{52}{3}) + 5(-\frac{19}{3}) = \frac{104}{3} - \frac{95}{3} = \frac{9}{3} = 3 \\ -\frac{52}{3} - 4(-\frac{19}{3}) = -\frac{52}{3} + \frac{76}{3} = \frac{24}{3} = 8 \end{aligned}$$

C. Higher-order systems

You are not limited to solving only systems with two linear equations using matrices. For a system of three linear equations, it is equally simple to represent the system in matrix notation. In this case the coefficient matrix **A** would be 3×3, and the column vectors for the variables **X** and the column vector for the constants **C** would each be 3×1.

EXAMPLE 5-18 Represent the following system of equations in matrix notation:

$$\begin{cases} 2x - y + z = 5 \\ x + 3y - z = 5 \\ y + 2z = 8 \end{cases}$$

Solution We represent the system as $\mathbf{AX} = \mathbf{C}$ or

$$\begin{bmatrix} 2 & -1 & 1 \\ 1 & 3 & -1 \\ 0 & 1 & 2 \end{bmatrix} \begin{bmatrix} x \\ y \\ z \end{bmatrix} = \begin{bmatrix} 5 \\ 5 \\ 8 \end{bmatrix}$$

You solve a matrix equation like that of Example 5-18 by the same general method you use for a two-equation system, by first computing the inverse of the 3×3 matrix $\mathbf{A}$. Though finding the inverse of a 3×3 matrix is more complicated than for a 2×2 matrix, an easy-to-follow algorithm can lead you through the steps needed to find $\mathbf{A}^{-1}$:

ALGORITHM FOR FINDING INVERSE OF 3 × 3 MATRIX

1. Construct an **augmented matrix** by appending to the right side of matrix $\mathbf{A}$ the identity matrix $\mathbf{I}_3$:

$$\begin{bmatrix} a_{11} & a_{12} & a_{13} & 1 & 0 & 0 \\ a_{21} & a_{22} & a_{23} & 0 & 1 & 0 \\ a_{31} & a_{32} & a_{33} & 0 & 0 & 1 \end{bmatrix}$$

note: Use the dashed line as shown to separate the main matrix from the stuck-on $\mathbf{I}_3$ matrix.

2. Operate on this new matrix by using the elementary row operations (see Section 4-1):
   (1) Interchange any two rows.
   (2) Multiply any row by a nonzero constant.
   (3) Add to any row a constant multiple of any other row.

   Perform these operations until the main matrix to the left of the dashed line has been changed into the identity matrix. The 3×3 matrix to the right of the dashed line will be the inverse, $\mathbf{A}^{-1}$, of $\mathbf{A}$.

note: A "row" of the augmented matrix includes all *six* numbers across, on both sides of the dashed line.

EXAMPLE 5-19 Find $\mathbf{A}^{-1}$ if

$$\mathbf{A} = \begin{bmatrix} 2 & -1 & 1 \\ 1 & 3 & -1 \\ 0 & 1 & 2 \end{bmatrix}$$

Solution

$$\begin{bmatrix} 2 & -1 & 1 & 1 & 0 & 0 \\ 1 & 3 & -1 & 0 & 1 & 0 \\ 0 & 1 & 2 & 0 & 0 & 1 \end{bmatrix}$$

Multiply row (1) by $\frac{1}{2}$ to make $a_{11} = 1$
$\longrightarrow$

$$\begin{bmatrix} 1 & -\frac{1}{2} & \frac{1}{2} & \frac{1}{2} & 0 & 0 \\ 1 & 3 & -1 & 0 & 1 & 0 \\ 0 & 1 & 2 & 0 & 0 & 1 \end{bmatrix}$$

Multiply row (1) by -1 and add it to row (2) to make $a_{21} = 0$
$\longrightarrow$

$$\begin{bmatrix} 1 & -\frac{1}{2} & \frac{1}{2} & \frac{1}{2} & 0 & 0 \\ 0 & \frac{7}{2} & -\frac{3}{2} & -\frac{1}{2} & 1 & 0 \\ 0 & 1 & 2 & 0 & 0 & 1 \end{bmatrix}$$

a_{31} already equals 0, so multiply row (3) by $\frac{1}{2}$ and add it to row (1) to make $a_{12} = 0$
$\longrightarrow$

$$\begin{bmatrix} 1 & 0 & \frac{3}{2} & \frac{1}{2} & 0 & \frac{1}{2} \\ 0 & \frac{7}{2} & -\frac{3}{2} & -\frac{1}{2} & 1 & 0 \\ 0 & 1 & 2 & 0 & 0 & 1 \end{bmatrix}$$

Multiply row (2) by $\frac{2}{7}$ to make $a_{22} = 1$
$\longrightarrow$

$$\begin{bmatrix} 1 & 0 & \frac{3}{2} & \vdots & \frac{1}{2} & 0 & \frac{1}{2} \\ 0 & 1 & -\frac{3}{7} & \vdots & -\frac{1}{7} & \frac{2}{7} & 0 \\ 0 & 1 & 2 & \vdots & 0 & 0 & 1 \end{bmatrix}$$

Multiply row (2) by -1 and add it to row (3) to make $a_{32} = 0$

$$\begin{bmatrix} 1 & 0 & \frac{3}{2} & \vdots & \frac{1}{2} & 0 & \frac{1}{2} \\ 0 & 1 & -\frac{3}{7} & \vdots & -\frac{1}{7} & \frac{2}{7} & 0 \\ 0 & 0 & \frac{17}{7} & \vdots & \frac{1}{7} & -\frac{2}{7} & 1 \end{bmatrix}$$

Multiply row (3) by $\frac{7}{17}$ to make $a_{33} = 1$

$$\begin{bmatrix} 1 & 0 & \frac{3}{2} & \vdots & \frac{1}{2} & 0 & \frac{1}{2} \\ 0 & 1 & -\frac{3}{7} & \vdots & -\frac{1}{7} & \frac{2}{7} & 0 \\ 0 & 0 & 1 & \vdots & \frac{1}{17} & -\frac{2}{17} & \frac{7}{17} \end{bmatrix}$$

Multiply row (3) by $-\frac{3}{2}$ and add it to row (1) to make $a_{13} = 0$

$$\begin{bmatrix} 1 & 0 & 0 & \vdots & \frac{7}{17} & \frac{3}{17} & -\frac{2}{17} \\ 0 & 1 & -\frac{3}{7} & \vdots & -\frac{1}{7} & \frac{2}{7} & 0 \\ 0 & 0 & 1 & \vdots & \frac{1}{17} & -\frac{2}{17} & \frac{7}{17} \end{bmatrix}$$

And, finally, multiply row (3) by $\frac{3}{7}$ and add it to row (2) to make $a_{23} = 0$

$$\begin{bmatrix} 1 & 0 & 0 & \vdots & \frac{7}{17} & \frac{3}{17} & -\frac{2}{17} \\ 0 & 1 & 0 & \vdots & -\frac{2}{17} & \frac{4}{17} & \frac{3}{17} \\ 0 & 0 & 1 & \vdots & \frac{1}{17} & -\frac{2}{17} & \frac{7}{17} \end{bmatrix}$$

This is **I**   This is **A**$^{-1}$

A simpler way to write $\mathbf{A}^{-1}$ is

$$\frac{1}{17}\begin{bmatrix} 7 & 3 & -2 \\ -2 & 4 & 3 \\ 1 & -2 & 7 \end{bmatrix}$$

note: The manipulations in Example 5-19 are exactly what you did in Chapter 4 when you solved a three-equation system, only here you don't have to carry along all the variables.

Now that you know how to get the inverse of a 3×3 matrix, you can use the same algorithm you applied to a two-equation system to solve a three-equation system.

EXAMPLE 5-20   Find the solution to the system of Example 5-18.

Solution   Since we have just computed the inverse of the coefficient matrix in Example 5-19, we obtain the solution from eq. (5-9):

$$\mathbf{X} = \mathbf{A}^{-1} \cdot \mathbf{C}$$

Here

$$\mathbf{X} = \frac{1}{17}\begin{bmatrix} 7 & 3 & -2 \\ -2 & 4 & 3 \\ 1 & -2 & 7 \end{bmatrix}\begin{bmatrix} 5 \\ 5 \\ 8 \end{bmatrix}$$

or

$$\begin{bmatrix} x \\ y \\ z \end{bmatrix} = \frac{1}{17}\begin{bmatrix} 7 \cdot 5 + 3 \cdot 5 - 2 \cdot 8 \\ -2 \cdot 5 + 4 \cdot 5 + 3 \cdot 8 \\ 1 \cdot 5 - 2 \cdot 5 + 7 \cdot 8 \end{bmatrix} = \frac{1}{17}\begin{bmatrix} 34 \\ 34 \\ 51 \end{bmatrix} = \begin{bmatrix} 2 \\ 2 \\ 3 \end{bmatrix}$$

The solution is $x = 2$, $y = 2$, $z = 3$.

When solving a matrix equation by the method of Example 5-20, you are guaranteed a solution if **A** has an inverse. But if **A** does not have an inverse (i.e., if det **A** = 0), there may or may not be a solution. Other methods are then required. One such method is an algorithm that follows almost exactly the same process as the one for finding the inverse, except it is a faster method of solving the original system. Here it is:

ALGORITHM FOR SOLVING ANY SYSTEM WITH MATRICES

1. Augment the column matrix of constants onto the right side of the original coefficient matrix to obtain $[\mathbf{A} \mid \mathbf{C}]$.
2. Perform row operations on this augmented matrix until the portion originally occupied by $\mathbf{A}$ is changed into the identity matrix:
   a. Call the element in column J of row J (that is, a_{ij} where $i = j$) the pivotal element of row J. Using row operation (2), create a 1 in the pivotal position of the first row.
   b. Using row operation (3), create a 0 in the non-pivotal positions of the first column.
   c. Using row operation (2), create a 1 in the pivotal position of the second row.
   d. Using row operation (3), create a 0 in the non-pivotal positions of the second column.
   e. Using row operation (2), create a 1 in the pivotal position of the third row.
   f. Using row operation (3), create a 0 in the non-pivotal positions of the third column.

The identity is automatically produced in the left portion of the augmented matrix, and the portion originally occupied by $\mathbf{C}$ will now be a column vector whose entries are the solution values. (Tricky, huh?)

EXAMPLE 5-21   Solve the following linear system without finding its inverse:

$$\begin{cases} 3x - y + 3z = -6 \\ -x + y + 2z = -2 \\ 4x - 2y - z = 0 \end{cases}$$

Solution   First form the augmented matrix

$$[\mathbf{A} \mid \mathbf{C}] = \begin{bmatrix} 3 & -1 & 3 & \mid & -6 \\ -1 & 1 & 2 & \mid & -2 \\ 4 & -2 & -1 & \mid & 0 \end{bmatrix}$$

note: As we perform row operations on this matrix, we are actually doing the operations on the equations themselves. This matrix method is thus a shorthand version of our earlier algebraic method (Section 4-1).

step 2a.

Multiply row (1) by $\frac{1}{3}$ to get the pivotal element ready:

$$\begin{bmatrix} 1 & -\frac{1}{3} & 1 & \mid & -2 \\ -1 & 1 & 2 & \mid & -2 \\ 4 & -2 & -1 & \mid & 0 \end{bmatrix}$$

$\longrightarrow$

step 2b.

Add row (1) to row (2). Then multiply row (1) by (-4) and add it to row (3):

$$\begin{bmatrix} 1 & -\frac{1}{3} & 1 & \mid & -2 \\ 0 & \frac{2}{3} & 3 & \mid & -4 \\ 0 & -\frac{2}{3} & -5 & \mid & 8 \end{bmatrix}$$

step 2c.

Multiply row (2) by $\frac{3}{2}$:

$$\begin{bmatrix} 1 & -\frac{1}{3} & 1 & \vdots & -2 \\ 0 & 1 & \frac{9}{2} & \vdots & -6 \\ 0 & -\frac{2}{3} & -5 & \vdots & 8 \end{bmatrix}$$

step 2d.

Multiply row (2) by $\frac{1}{3}$ and add to row (1). Then, multiply row (2) by $\frac{2}{3}$ and add to row (3):

$$\longrightarrow \begin{bmatrix} 1 & 0 & \frac{5}{2} & \vdots & -4 \\ 0 & 1 & \frac{9}{2} & \vdots & -6 \\ 0 & 0 & -2 & \vdots & 4 \end{bmatrix}$$

step 2e.

Multiply row (3) by $(-\frac{1}{2})$:

$$\begin{bmatrix} 1 & 0 & \frac{5}{2} & \vdots & -4 \\ 0 & 1 & \frac{9}{2} & \vdots & -6 \\ 0 & 0 & 1 & \vdots & -2 \end{bmatrix}$$

step 2f.

Multiply row (3) by $(-\frac{5}{2})$ and add to row (1). Then, multiply row (3) by $(-\frac{9}{2})$ and add to row (2):

$$\longrightarrow \begin{bmatrix} 1 & 0 & 0 & \vdots & 1 \\ 0 & 1 & 0 & \vdots & 3 \\ 0 & 0 & 1 & \vdots & -2 \end{bmatrix}$$

The solution is easily read from the matrix: $x = 1$, $y = 3$, $z = -2$.

5-6. Arrays

Different algorithms work differently. Some algorithms take a few input data items, process them, output the results, then go back for more input items, process these, output the results, etc. Such algorithms can deal with a large volume of data in only a small amount of computer storage.

Often, however, large lists or tables of data items are needed, and sometimes these must be referred to more than once, either as a whole or for a particular item in the list or table. These lists or tables of data are called **arrays**.

A **one-dimensional array** is a *list* of data items. You give the list a name, such as *LIST,* (or *FRED,* if you're feeling imaginative that day), and refer to its members through a *subscripted variable name,* using the parentheses notation: *LIST*(1), *LIST*(2),..., *LIST*(N). (This is the computer programmer's equivalent to the items $list_1$, $list_2$,..., $list_n$.) A few of the algorithms you studied in Chapter 1 used one-dimensional arrays, such as the Binary Search algorithm. Other examples would include algorithms that average elements in an array, find the largest and smallest element in an array, or sort arrays (reorganize their elements).

Not all arrays are one-dimensional. For many purposes, data are better organized in a two-dimensional planar surface (like squares of a checkerboard) or in three dimensions (like a Rubic's cube). (For now, we'll not deal with the 3-D version.)

A **two-dimensional array** is a *table* of data items. We give the table a name, such as *TABLE* (original, huh?), and refer to its members by *double-subscripted variables* in parentheses notation:

$$\begin{array}{cccc} TABLE(1,1) & TABLE(1,2) & \cdots & TABLE(1,N) \\ TABLE(2,1) & TABLE(2,2) & \cdots & TABLE(2,N) \\ \vdots & \vdots & & \vdots \\ TABLE(M,1) & TABLE(M,2) & \cdots & TABLE(M,N) \end{array}$$

(You might already have noticed that a two-dimensional array is exactly the same as a matrix, in which $A(1,1)$, $A(1,2)$,..., $A(i,j)$ would be the computer equivalents of the a_{11}, a_{12},..., a_{ij} notation you use with the matrix.) In algorithms that manipulate arrays, then, you need the array concept to get data in and out of the computer. You handle input by writing:

1. FOR $i = 1$ TO M Do
   a. FOR $j = 1$ TO N Do
      1. Input $TABLE(i,j)$

Outputting a table is handled similarly:

1. FOR $i = 1$ TO M Do
   a. FOR $j = 1$ TO N Do
      1. Output $TABLE(i,j)$

The array entries are referred to by their double-subscripted variables, and after you've allowed for input and output, you use separate algorithms for their processing.

EXAMPLE 5-22 Suppose your class contains M students and each has taken N exams. The scores could be recorded in a two-dimensional array as follows:

$$SCORE(1,1) \quad SCORE(1,2) \quad \cdots \quad SCORE(1,N)$$
$$SCORE(2,1) \quad SCORE(2,2) \quad \cdots \quad SCORE(2,N)$$
$$\vdots \qquad\qquad \vdots \qquad\qquad \vdots$$
$$SCORE(M,1) \quad SCORE(M,2) \quad \cdots \quad SCORE(M,N)$$

Now suppose that the exams each carry a "weight," that is, the score of exam k is multiplied by weight W before all the scores for each student are totaled. Create the algorithm to find each student's total score.

Solution Let $W(k)$ be the one-dimensional array holding the weight values. Assume you've already input the scores and weights. The processing algorithm will be:

Begin
1. FOR $i = 1$ TO M Do
   a. $SUM(i) \leftarrow 0$ This sets each student's total score at zero to begin.

   b. FOR $j = 1$ TO N Do   This step adds the scores of
      1. $SUM(i) \leftarrow SUM(i) + SCORE(i,j) * W(j)$ successive tests after multiplying each
      2. Output $SUM(i)$ score by the respective weight $W(j)$.
End

Notice that this outputs a list instead of a table.

SUMMARY

1. A matrix is a rectangular array of numbers, arranged into rows and columns, that is used to compactly and systematically display data.
2. The elements of a matrix are indexed (located) by the double subscripts, which specify the row and column position for the element.
3. Two matrices are equal only if their corresponding elements are equal in every position.
4. The arithmetic operations of negation, addition, subtraction, and multiplication may be performed with matrices.
5. Many of the matrix operations can be readily performed by computers using rather simple algorithms.
6. The determinant of a square matrix is a unique number. This value is used when solving a system of equations using matrix theory.
7. If a square matrix **A** has a nonzero determinant, then it has an inverse $\mathbf{A}^{-1}$ with the property that $\mathbf{A}\mathbf{A}^{-1} = \mathbf{A}^{-1}\mathbf{A} = \mathbf{I}$ (the identity matrix). Computing the inverse of a matrix **A** is part of solving systems of linear equations.
8. Arrays are lists or tables of data items used in computer data processing.

RAISE YOUR GRADES

Can you...?

☑ define the dimension of a matrix
☑ describe a row vector and a column vector
☑ use subscripts to define the location of a matrix element

☑ describe how two matrices can be equal
☑ add and subtract matrices
☑ describe the size qualifications for matrix multiplication
☑ find the product of correctly sized matrices
☑ define a diagonal matrix, an identity matrix, a zero matrix, and a symmetric matrix
☑ find the transpose of a given matrix
☑ describe how to find the value of determinants of square matrices
☑ define the inverse of a matrix
☑ find the inverse of a matrix
☑ represent a linear system in matrix notation
☑ use the inverse matrix to solve a linear system
☑ use the augmented matrix to find the inverse of a matrix
☑ use the augmented matrix to solve a linear system
☑ define an array and describe how it may be used in data processing

SOLVED PROBLEMS

Matrices Defined

PROBLEM 5-1   Give the dimension of the following matrices:

$$A = \begin{bmatrix} 2 & 1 & 3 \\ 4 & 5 & 9 \end{bmatrix} \qquad B = \begin{bmatrix} 5 \\ 8 \\ 2 \end{bmatrix} \qquad C = \begin{bmatrix} 5 & 4 & 2 \\ 9 & 8 & 1 \\ 6 & 3 & 0 \end{bmatrix} \qquad D = [7 \quad 3 \quad 6 \quad 1]$$

Solution   Matrix **A** has 2 rows and 3 columns. It is 2×3. Matrix **B** is a column vector and is 3×1. Matrix **C** is 3×3. Matrix **D** is a row vector and is 1×4.

PROBLEM 5-2   For the matrices defined in Problem 5-1, find the value of the elements a_{13}, b_{21}, c_{33}, d_{14}.

Solution a_{13} (in matrix **A**) is the entry in the first row and third column $= 3$; b_{21} (in matrix **B**) is the element in the second row and first (only) column $= 8$; $c_{33} = 0$; $d_{14} = 1$.

PROBLEM 5-3   Under what conditions are the following matrices equal?

$$A = \begin{bmatrix} p+s & s-r \\ 4 & 3 \end{bmatrix} \quad \text{and} \quad B = \begin{bmatrix} 7 & 2 \\ p-q & r-q \end{bmatrix}$$

Solution   For **A** and **B** to be equal, their corresponding elements must be equal. Thus we get four equations to solve:

$$\begin{cases} p + s = 7 \\ p - q = 4 \end{cases} \quad \begin{array}{l}\text{subtract the second equation} \\ \text{from the first to get rid} \\ \text{of the } p\end{array} \longrightarrow s + q = 3 \searrow \begin{array}{l}\text{add these} \\ \text{together to}\end{array}$$

$$\begin{cases} s - r = 2 \\ r - q = 3 \end{cases} \quad \begin{array}{l}\text{add the second equation} \\ \text{to the first to get rid of the } r\end{array} \longrightarrow s - q = s \nearrow \begin{array}{l}\text{get rid of} \\ \text{the } q\end{array}$$

$$2s = 8$$

Now just solve for *s* and do some back-and-forth substituting to obtain $s = 4, q = -1, p = 3$, and $r = 2$.

Matrix Operations

PROBLEM 5-4   Let

$$A = \begin{bmatrix} 1 & 2 & 5 \\ 0 & 4 & -1 \end{bmatrix} \quad \text{and} \quad B = \begin{bmatrix} 2 & 3 & -1 \\ -1 & 2 & 0 \end{bmatrix}$$

Find   **(a) A + B**   **(b)** −2**B**   **(c) B − 2A**   **(d)** 3**A** + 5**B**.

Solution

(a) $A + B = \begin{bmatrix} 1+2 & 2+3 & 5-1 \\ 0-1 & 4+2 & -1+0 \end{bmatrix} = \begin{bmatrix} 3 & 5 & 4 \\ -1 & 6 & -1 \end{bmatrix}$

(b) $-2B = -2\begin{bmatrix} 2 & 3 & -1 \\ -1 & 2 & 0 \end{bmatrix} = \begin{bmatrix} -4 & -6 & 2 \\ 2 & -4 & 0 \end{bmatrix}$

(c) $B - 2A = \begin{bmatrix} 2 & 3 & -1 \\ -1 & 2 & 0 \end{bmatrix} - \begin{bmatrix} 2 & 4 & 10 \\ 0 & 8 & -2 \end{bmatrix} = \begin{bmatrix} 0 & -1 & -11 \\ -1 & -6 & 2 \end{bmatrix}$

(d) $3A + 5B = \begin{bmatrix} 3 & 6 & 15 \\ 0 & 12 & -3 \end{bmatrix} + \begin{bmatrix} 10 & 15 & -5 \\ -5 & 10 & 0 \end{bmatrix} = \begin{bmatrix} 13 & 21 & 10 \\ -5 & 22 & -3 \end{bmatrix}$

PROBLEM 5-5   Find the products.

(a) $AB = \begin{bmatrix} 2 & -1 & 0 \\ -3 & 4 & 5 \\ 0 & 1 & 6 \end{bmatrix}\begin{bmatrix} 3 & 1 \\ -4 & 0 \\ 2 & 5 \end{bmatrix}$ **(b)** $CD = \begin{bmatrix} 4 \\ 2 \\ -6 \\ 3 \end{bmatrix}[8 \quad 2]$

Solution

(a) This is a 3 × 3 times a 3 × 2, so it will produce a 3 × 2:

$$AB = \begin{bmatrix} 2 \cdot 3 + (-1) \cdot (-4) + 0 \cdot 2 & 2 \cdot 1 + (-1) \cdot 0 + 0.5 \\ (-3) \cdot 3 + 4 \cdot (-4) + 5 \cdot 2 & (-3) \cdot 1 + 4 \cdot 0 + 5 \cdot 5 \\ 0 \cdot 3 + 1 \cdot (-4) + 6 \cdot 2 & 0 \cdot 1 + 1 \cdot 0 + 6 \cdot 5 \end{bmatrix} = \begin{bmatrix} 10 & 2 \\ -15 & 22 \\ 8 & 30 \end{bmatrix}$$

(b) This is a 4 × 1 times a 1 × 2, giving, a 4 × 2:

$$CD = \begin{bmatrix} 4 \cdot 8 & 4 \cdot 2 \\ 2 \cdot 8 & 2 \cdot 2 \\ (-6) \cdot 8 & (-6) \cdot 2 \\ 3 \cdot 8 & 3 \cdot 2 \end{bmatrix} = \begin{bmatrix} 32 & 8 \\ 16 & 4 \\ -48 & -12 \\ 24 & 6 \end{bmatrix}$$

PROBLEM 5-6   Find A^3 where

$$A = \begin{bmatrix} 1 & 0 \\ 2 & 4 \end{bmatrix}$$

Solution   Remember that $A^3 = A \cdot A \cdot A = (A \cdot A) \cdot A = A^2 \cdot A$. So we must first solve for A^2:

$$A^2 = A \cdot A = \begin{bmatrix} 1 & 0 \\ 2 & 4 \end{bmatrix}\begin{bmatrix} 1 & 0 \\ 2 & 4 \end{bmatrix} = \begin{bmatrix} 1 \cdot 1 + 0 \cdot 2 & 1 \cdot 0 + 0 \cdot 4 \\ 2 \cdot 1 + 4 \cdot 2 & 2 \cdot 0 + 4 \cdot 4 \end{bmatrix} = \begin{bmatrix} 1 & 0 \\ 10 & 16 \end{bmatrix}$$

and multiply A^2 by A gives us A^3:

$$A^3 = A^2 \cdot A = \begin{bmatrix} 1 & 0 \\ 10 & 16 \end{bmatrix}\begin{bmatrix} 1 & 0 \\ 2 & 4 \end{bmatrix} = \begin{bmatrix} 1 & 0 \\ 42 & 64 \end{bmatrix}$$

PROBLEM 5-7   Let

$$A = \begin{bmatrix} 2 & -3 \\ 0 & 1 \end{bmatrix} \quad B = \begin{bmatrix} 4 & 0 \\ 2 & 1 \end{bmatrix} \quad C = \begin{bmatrix} 2 & 0 \\ 0 & 5 \end{bmatrix} \quad D = \begin{bmatrix} 0 & -3 \\ 1 & 0 \end{bmatrix}$$

Find   **(a) AB** and **BA**   **(b) CD**   **(c) A^T, B^T, and C^T**   **(d) $(A + C)^T$**   **(e) det B.**

Solution

(a) $AB = \begin{bmatrix} 2 & -3 \\ 0 & 1 \end{bmatrix}\begin{bmatrix} 4 & 0 \\ 2 & 1 \end{bmatrix} = \begin{bmatrix} 2 & -3 \\ 2 & 1 \end{bmatrix}$

$\quad BA = \begin{bmatrix} 4 & 0 \\ 2 & 1 \end{bmatrix}\begin{bmatrix} 2 & -3 \\ 0 & 1 \end{bmatrix} = \begin{bmatrix} 8 & -12 \\ 4 & -5 \end{bmatrix}$ *note:* **AB ≠ BA**

(b) $CD = \begin{bmatrix} 2 & 0 \\ 0 & 5 \end{bmatrix}\begin{bmatrix} 0 & -3 \\ 1 & 0 \end{bmatrix} = \begin{bmatrix} 0 & -6 \\ 5 & 0 \end{bmatrix}$

(c) To get the transpose, write the rows of the original as the columns of the transpose:

$$A = \begin{bmatrix} 2 & -3 \\ 0 & 1 \end{bmatrix} \quad \text{so} \quad A^T = \begin{bmatrix} 2 & 0 \\ -3 & 1 \end{bmatrix}$$

$$B^T = \begin{bmatrix} 4 & 2 \\ 0 & 1 \end{bmatrix} \quad C^T = \begin{bmatrix} 2 & 0 \\ 0 & 5 \end{bmatrix} \quad \text{*note:* } C = C^T \text{ here.}$$

(d) $A + C = \begin{bmatrix} 4 & -3 \\ 0 & 6 \end{bmatrix} \Rightarrow (A + C)^T = \begin{bmatrix} 4 & 0 \\ -3 & 6 \end{bmatrix}$

(e) $\det B = \det \begin{bmatrix} 4 & 0 \\ 2 & 1 \end{bmatrix} = 4 \cdot 1 - 0 \cdot 2 = 4$

Properties of Matrices

PROBLEM 5-8   Prove Theorem A1, that for similarly dimensioned matrices **A** and **B**,

$$A + B = B + A$$

Solution   If **A** is represented as $[a_{ij}]$ and **B** is represented as $[b_{ij}]$, then the general element in row i and column j of $A + B$ is $a_{ij} + b_{ij}$. The general element in row i and column j of $B + A$ is $b_{ij} + a_{ij}$. Since the entries of the matrices are real numbers and real number addition is commutative (i.e., $a_{ij} + b_{ij} = b_{ij} + a_{ij}$), then $A + B$ and $B + A$ have the same numbers in every position. They are thus equal.

PROBLEM 5-9   Given that matrix **A** is $m \times n$ and s is a real number, prove that **(a)** $-sA = (-s)A = s(-A)$ (Theorem B4) and **(b)** $0A = 0$ (Theorem B6).

Solution   Let

$$A = \begin{bmatrix} a_{11} & a_{12} & \cdots & a_{1n} \\ a_{21} & a_{22} & \cdots & a_{2n} \\ \vdots & & & \vdots \\ a_{m1} & a_{m2} & \cdots & a_{mn} \end{bmatrix}$$

Then

(a)
$$(-s)A = \begin{bmatrix} (-s)a_{11} & (-s)a_{12} & \cdots & (-s)a_{1n} \\ (-s)a_{21} & (-s)a_{22} & \cdots & (-s)a_{2n} \\ \vdots & & & \vdots \\ (-s)a_{m1} & (-s)a_{m2} & \cdots & (-s)a_{mn} \end{bmatrix} = \begin{bmatrix} -sa_{11} & -sa_{12} & \cdots & -sa_{1n} \\ -sa_{21} & -sa_{22} & \cdots & -sa_{2n} \\ \vdots & & & \vdots \\ -sa_{m1} & -sa_{m2} & \cdots & -sa_{mn} \end{bmatrix} = -sA$$

$$= s\begin{bmatrix} -a_{11} & -a_{12} & \cdots & -a_{1n} \\ -a_{21} & -a_{22} & \cdots & -a_{2n} \\ \vdots & & & \vdots \\ -a_{m1} & -a_{m2} & \cdots & -a_{mn} \end{bmatrix} = s(-A)$$

(b)

$$0\mathbf{A} = 0 \begin{bmatrix} a_{11} & a_{12} & \cdots & a_{1n} \\ a_{21} & a_{22} & \cdots & a_{2n} \\ \vdots & & & \vdots \\ a_{m1} & a_{m2} & \cdots & a_{mn} \end{bmatrix} = \begin{bmatrix} 0 \cdot a_{11} & 0 \cdot a_{12} & \cdots & 0 \cdot a_{1n} \\ 0 \cdot a_{21} & 0 \cdot a_{22} & \cdots & 0 \cdot a_{2n} \\ \vdots & & & \vdots \\ 0 \cdot a_{m1} & 0 \cdot a_{m2} & \cdots & 0 \cdot a_{mn} \end{bmatrix}$$

$$= \begin{bmatrix} 0 & 0 & \cdots & 0 \\ 0 & 0 & \cdots & 0 \\ \vdots & & & \vdots \\ 0 & 0 & \cdots & 0 \end{bmatrix} = \mathbf{0}$$

PROBLEM 5-10   Let

$$\mathbf{A} = \begin{bmatrix} 3 & -1 & 0 \\ -2 & 4 & 1 \end{bmatrix}, \qquad \mathbf{B} = \begin{bmatrix} 6 \\ 1 \\ 3 \end{bmatrix}, \qquad \mathbf{C} = [1 \quad 0 \quad 5]$$

Verify that $\mathbf{A}(\mathbf{BC}) = (\mathbf{AB})\mathbf{C}$ (Theorem C3).

Solution

$$\mathbf{BC} = \begin{bmatrix} 6 \\ 1 \\ 3 \end{bmatrix} [1 \quad 0 \quad 5] = \begin{bmatrix} 6 & 0 & 30 \\ 1 & 0 & 5 \\ 3 & 0 & 15 \end{bmatrix}$$

$$\mathbf{A}(\mathbf{BC}) = \begin{bmatrix} 3 & -1 & 0 \\ -2 & 4 & 1 \end{bmatrix} \begin{bmatrix} 6 & 0 & 30 \\ 1 & 0 & 5 \\ 3 & 0 & 15 \end{bmatrix} = \begin{bmatrix} 17 & 0 & 85 \\ -5 & 0 & -25 \end{bmatrix}$$

$$\mathbf{AB} = \begin{bmatrix} 3 & -1 & 0 \\ -2 & 4 & 1 \end{bmatrix} \begin{bmatrix} 6 \\ 1 \\ 3 \end{bmatrix} = \begin{bmatrix} 17 \\ -5 \end{bmatrix} \qquad \text{same}$$

$$(\mathbf{AB})\mathbf{C} = \begin{bmatrix} 17 \\ -5 \end{bmatrix} [1 \quad 0 \quad 5] = \begin{bmatrix} 17 & 0 & 85 \\ -5 & 0 & -25 \end{bmatrix}$$

PROBLEM 5-11   Show that it is possible to have $\mathbf{AB} = 0$ with $\mathbf{A} \neq 0$, and $\mathbf{B} \neq 0$.

Solution   Let

$$\mathbf{A} = \begin{bmatrix} 0 & 0 & 0 \\ 1 & -3 & 7 \\ -2 & 6 & -5 \end{bmatrix} \quad \text{and} \quad \mathbf{B} = \begin{bmatrix} 3 & -6 & 9 \\ 1 & -2 & 3 \\ 0 & 0 & 0 \end{bmatrix}$$

Then

$$\mathbf{AB} = \begin{bmatrix} 0 & 0 & 0 \\ 1 & -3 & 7 \\ -2 & 6 & -5 \end{bmatrix} \begin{bmatrix} 3 & -6 & 9 \\ 1 & -2 & 3 \\ 0 & 0 & 0 \end{bmatrix} = \begin{bmatrix} 0 & 0 & 0 \\ 0 & 0 & 0 \\ 0 & 0 & 0 \end{bmatrix}$$

PROBLEM 5-12   Let

$$\mathbf{A} = \begin{bmatrix} 3 & 4 \\ 1 & -2 \end{bmatrix} \quad \text{and} \quad \mathbf{B} = \begin{bmatrix} -1 & 2 \\ 3 & 4 \end{bmatrix}$$

Verify that $(\mathbf{AB})^T = \mathbf{B}^T \mathbf{A}^T$ (Theorem D4).

Solution

$$\mathbf{AB} = \begin{bmatrix} 3 & 4 \\ 1 & -2 \end{bmatrix} \begin{bmatrix} -1 & 2 \\ 3 & 4 \end{bmatrix} = \begin{bmatrix} 9 & 22 \\ -7 & -6 \end{bmatrix}$$

Thus

$$(\mathbf{AB})^T = \begin{bmatrix} 9 & -7 \\ 22 & -6 \end{bmatrix}$$

But

$$\mathbf{B}^T = \begin{bmatrix} -1 & 3 \\ 2 & 4 \end{bmatrix} \quad \text{and} \quad \mathbf{A}^T = \begin{bmatrix} 3 & 1 \\ 4 & -2 \end{bmatrix} \quad \text{same}$$

so $\quad \mathbf{B}^T \cdot \mathbf{A}^T = \begin{bmatrix} -1 & 3 \\ 2 & 4 \end{bmatrix}\begin{bmatrix} 3 & 1 \\ 4 & -2 \end{bmatrix} = \begin{bmatrix} 9 & -7 \\ 22 & -6 \end{bmatrix}$

PROBLEM 5-13   Prove that the inverse of an $n \times n$ matrix **A**, when it exists, is unique.

Solution   Assume there are two distinct matrices **B** and **C**, both of which are inverses of **A**.

Then $\mathbf{AC} = \mathbf{I}$	since **C** is an inverse of **A**
and $\mathbf{BA} = \mathbf{I}$	since **B** is an inverse of **A**
Thus $\mathbf{B} = \mathbf{BI}$	by the identity property
$= \mathbf{B(AC)}$	since $\mathbf{AC} = \mathbf{I}$
$= \mathbf{(BA)C}$	by the associativity property (Theorem C3)
$= \mathbf{IC}$	since $\mathbf{BA} = \mathbf{I}$
$= \mathbf{C}$	by the identity property

We conclude that $\mathbf{B} = \mathbf{C}$. Thus the inverse is unique.

PROBLEM 5-14   Let

$$\mathbf{A} = \begin{bmatrix} 5 & -2 \\ -3 & 3 \end{bmatrix} \quad \mathbf{B} = \begin{bmatrix} -1 & 3 \\ -2 & 7 \end{bmatrix}$$

Find det(**A**) and det(**B**). Then verify that det(**AB**) = (det **A**) $\cdot$ (det **B**) (see eq. 5-6).

Solution

$$\det(\mathbf{A}) = \det\begin{bmatrix} 5 & -2 \\ -3 & 3 \end{bmatrix} = \begin{vmatrix} 5 & -2 \\ -3 & 3 \end{vmatrix} = 5 \cdot 3 - (-2) \cdot (-3) = 15 - 6 = 9$$

$$\det(\mathbf{B}) = \det\begin{bmatrix} -1 & 3 \\ -2 & 7 \end{bmatrix} = \begin{vmatrix} -1 & 3 \\ -2 & 7 \end{vmatrix} = -1 \cdot 7 - (3) \cdot (-2) = -7 + 6 = -1$$

Then

$$\mathbf{AB} = \begin{bmatrix} 5 & -2 \\ -3 & 3 \end{bmatrix}\begin{bmatrix} -1 & 3 \\ -2 & 7 \end{bmatrix} = \begin{bmatrix} -1 & 1 \\ -3 & 12 \end{bmatrix}$$

Thus $\quad \det(\mathbf{AB}) = \begin{vmatrix} -1 & 1 \\ -3 & 12 \end{vmatrix} = -1 \cdot 12 - (1) \cdot (-3) = -12 + 3 = -9 = (\det \mathbf{A}) \cdot (\det \mathbf{B})$

PROBLEM 5-15   Find det(**M**) if

$$\mathbf{M} = \begin{bmatrix} 4 & 3 & -2 \\ -1 & 0 & 5 \\ 2 & 7 & 1 \end{bmatrix}$$

Solution   By the shortcut method, copy the first two columns to the right of the matrix, then multiply through according to the arrow/sign pattern:

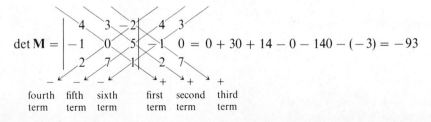

PROBLEM 5-16 Let

$$T = \begin{bmatrix} 4 & -1 \\ -3 & 2 \end{bmatrix}$$

Find **(a)** det **T**   **(b)** T^{-1} **(c)** product $T \cdot T^{-1}$ **(d)** $\det(T - \lambda I)$ for real number λ.

Solution

(a) $\det T = \begin{vmatrix} 4 & -1 \\ -3 & 2 \end{vmatrix} = 8 - 3 = 5$

(b) $T^{-1} = \dfrac{1}{\det T} \begin{bmatrix} 2 & 1 \\ 3 & 4 \end{bmatrix} = \dfrac{1}{5} \begin{bmatrix} 2 & 1 \\ 3 & 4 \end{bmatrix}$

(c) $T \cdot T^{-1} = \begin{bmatrix} 4 & -1 \\ -3 & 2 \end{bmatrix} \dfrac{1}{5} \begin{bmatrix} 2 & 1 \\ 3 & 4 \end{bmatrix} = \dfrac{1}{5} \begin{bmatrix} 8-3 & 4-4 \\ -6+6 & -3+8 \end{bmatrix} = \dfrac{1}{5} \begin{bmatrix} 5 & 0 \\ 0 & 5 \end{bmatrix} = \begin{bmatrix} 1 & 0 \\ 0 & 1 \end{bmatrix}$

(d) $\det(T - \lambda I) = \det\left(\begin{bmatrix} 4 & -1 \\ -3 & 2 \end{bmatrix} - \begin{bmatrix} \lambda & 0 \\ 0 & \lambda \end{bmatrix} \right)$

$$= \det\left(\begin{bmatrix} 4-\lambda & -1 \\ -3 & 2-\lambda \end{bmatrix} \right) = \begin{vmatrix} 4-\lambda & -1 \\ -3 & 2-\lambda \end{vmatrix}$$

$$= (4-\lambda) \cdot (2-\lambda) - (-1) \cdot (-3)$$
$$= 8 - 6\lambda + \lambda^2 - 3$$
$$= \lambda^2 - 6\lambda + 5$$

PROBLEM 5-17 Show that the following matrices are inverses of each other:

$$A = \begin{bmatrix} -1 & -2 & 3 \\ 0 & -1 & 2 \\ 3 & 1 & 0 \end{bmatrix} \quad \text{and} \quad B = \begin{bmatrix} 2 & -3 & 1 \\ -6 & 9 & -2 \\ -3 & 5 & -1 \end{bmatrix}$$

Solution We need show only that $AB = I_3$ (from $A \cdot A^{-1} = I_3$).

$$AB = \begin{bmatrix} -1 & -2 & 3 \\ 0 & -1 & 2 \\ 3 & 1 & 0 \end{bmatrix} \begin{bmatrix} 2 & -3 & 1 \\ -6 & 9 & -2 \\ -3 & 5 & -1 \end{bmatrix} = \begin{bmatrix} -2+12-9 & 3-18+15 & -1+4-3 \\ 0+6-6 & 0-9+10 & 0+2-2 \\ 6-6+0 & -9+9+0 & 3-2+0 \end{bmatrix} = \begin{bmatrix} 1 & 0 & 0 \\ 0 & 1 & 0 \\ 0 & 0 & 1 \end{bmatrix}$$

PROBLEM 5-18 An $n \times n$ matrix **A** is called a **permutation matrix** if **A** has exactly one 1 in each row and in each column and if all the other entries are 0. The product of two permutation matrices is itself a permutation matrix. Verify this with

$$A = \begin{bmatrix} 0 & 0 & 1 \\ 1 & 0 & 0 \\ 0 & 1 & 0 \end{bmatrix} \quad \text{and} \quad B = \begin{bmatrix} 0 & 1 & 0 \\ 0 & 0 & 1 \\ 1 & 0 & 0 \end{bmatrix}$$

Solution

$$A \cdot B = \begin{bmatrix} 0 & 0 & 1 \\ 1 & 0 & 0 \\ 0 & 1 & 0 \end{bmatrix} \begin{bmatrix} 0 & 1 & 0 \\ 0 & 0 & 1 \\ 1 & 0 & 0 \end{bmatrix} = \begin{bmatrix} 1 & 0 & 0 \\ 0 & 1 & 0 \\ 0 & 0 & 1 \end{bmatrix} = I_3$$

which is itself a permutation matrix.

PROBLEM 5-19 Show that if **A** is an invertible matrix then

$$\det(A^{-1}) = \frac{1}{\det A}$$

Solution Since **A** is invertible, then A^{-1} exists and $A^{-1} \cdot A = I$. Therefore

$$\det(A^{-1} \cdot A) = \det(I)$$

or, applying eq. (5-6),

$$\det(A^{-1}) \cdot \det(A) = \det(I) = 1 \qquad \textbf{(5-10)}$$

Since **A** is invertible, then $\det(A)$ is not zero, so from eq. (5-10) $\det(A^{-1}) = 1/\det A$.

Solving Systems with Matrices

PROBLEM 5-20 Put the following system of equations into matrix form and then solve it by first finding its inverse:

$$\begin{cases} -x - 2y + 3z = 5 \\ - y + 2z = -2 \\ 3x + y = 3 \end{cases}$$

Solution The matrix equation is **AX = C** or

$$\begin{bmatrix} -1 & -2 & 3 \\ 0 & -1 & 2 \\ 3 & 1 & 0 \end{bmatrix} \begin{bmatrix} x \\ y \\ z \end{bmatrix} = \begin{bmatrix} 5 \\ -2 \\ 3 \end{bmatrix}$$

This coefficient matrix is the matrix **A** of Problem 5-17, and its inverse is matrix **B** of the same problem. So if **AX = C**, then

$$A^{-1} \cdot AX = A^{-1} \cdot C \quad \text{or} \quad X = A^{-1} \cdot C$$

(which, you might recall, is eq. 5-9). Thus,

$$X = \begin{bmatrix} 2 & -3 & 1 \\ -6 & 9 & -2 \\ -3 & 5 & -1 \end{bmatrix} \begin{bmatrix} 5 \\ -2 \\ 3 \end{bmatrix} = \begin{bmatrix} 19 \\ -54 \\ -28 \end{bmatrix}$$

The solution is $x = 19, y = -54, z = -28$.

PROBLEM 5-21 Using matrix methods, solve the following system without finding its inverse:

$$\begin{cases} x - y + 2z = 7 \\ 3x + 2y - z = -10 \\ -x + 3y + z = -2 \end{cases}$$

Solution Form the augmented matrix

$$\left[\begin{array}{ccc|c} 1 & -1 & 2 & 7 \\ 3 & 2 & -1 & -10 \\ -1 & 3 & 1 & -2 \end{array} \right]$$

and perform the elementary row operations to change the left portion to I_3:

$$\left[\begin{array}{ccc|c} 1 & -1 & 2 & 7 \\ 3 & 2 & -1 & -10 \\ -1 & 3 & 1 & -2 \end{array} \right] \xrightarrow[\substack{\text{Multiply row (1)} \\ \text{by } -3 \text{ and add} \\ \text{to row (2).} \\ \text{Then add row (1)} \\ \text{to row (3).}}]{} \left[\begin{array}{ccc|c} 1 & -1 & 2 & 7 \\ 0 & 5 & -7 & -31 \\ 0 & 2 & 3 & 5 \end{array} \right] \xrightarrow[\substack{\text{Multiply row (2)} \\ \text{by } \frac{1}{5}}]{}$$

$$\left[\begin{array}{ccc|c} 1 & -1 & 2 & 7 \\ 0 & 1 & -\frac{7}{5} & -\frac{31}{5} \\ 0 & 2 & 3 & 5 \end{array} \right] \xrightarrow[\substack{\text{Add row (2)} \\ \text{to row (1).} \\ \text{Then multiply row (2)} \\ \text{by } (-2) \text{ and add} \\ \text{to row (3).}}]{} \left[\begin{array}{ccc|c} 1 & 0 & \frac{3}{5} & \frac{4}{5} \\ 0 & 1 & -\frac{7}{5} & -\frac{31}{5} \\ 0 & 0 & \frac{29}{5} & \frac{87}{5} \end{array} \right] \xrightarrow[\substack{\text{Multiply row (3)} \\ \text{by } \frac{5}{29}}]{}$$

$$\left[\begin{array}{ccc|c} 1 & 0 & \frac{3}{5} & \frac{4}{5} \\ 0 & 1 & -\frac{7}{5} & -\frac{31}{5} \\ 0 & 0 & 1 & 3 \end{array} \right] \xrightarrow[\substack{\text{Multiply row (3) by} \\ \frac{7}{5} \text{ and add to} \\ \text{row (2).} \\ \text{Then multiply} \\ \text{row (3) by } (-\frac{3}{5}) \\ \text{and add to row (1).}}]{} \left[\begin{array}{ccc|c} 1 & 0 & 0 & -\frac{5}{5} \\ 0 & 1 & 0 & -\frac{10}{5} \\ 0 & 0 & 1 & 3 \end{array} \right]$$

The solution is $x = -1, y = -2, z = 3$.

PROBLEM 5-22   Given a one-dimensional array *PARTS(I)* with 200 numerical entries, construct an algorithm to (**a**) find the sum of the original entries; (**b**) find the average of the original entries; (**c**) increase each entry value by 5.

Solution   Assume the data elements *PARTS(I)* to *PARTS*(200) are in memory. Then

(**a**) Begin
   1. *sum* ← 0
   2. FOR *I* = 1 TO 200 DO
      a. *sum* ← *sum* + *PARTS(I)*
   3. Output *sum*
   End

(**b**) Begin
   1. *sum* ← 0
   2. FOR *I* = 1 TO 200 DO
      a. *sum* ← *sum* + *PARTS(I)*
      b. *ave* ← *sum*/200
   3. Output *ave*
   End

(**c**) Begin
   1. FOR *J* = 1 TO 200 DO
      a. *PARTS(J)* ← *PARTS(J)* + 5
   End

Supplementary Exercises

PROBLEM 5-23   Let

$$A = \begin{bmatrix} 4 & -2 \\ 1 & 5 \\ -3 & 6 \end{bmatrix} \quad \text{and} \quad B = \begin{bmatrix} -2 & 3 & 4 \\ -5 & 0 & 1 \end{bmatrix}$$

If the product is denoted $P = A \cdot B$ then what is the value of (**a**) p_{12}, (**b**) p_{32}, (**c**) p_{21}?

PROBLEM 5-24   Given

$$A = \begin{bmatrix} 4 & -5 \\ -1 & 3 \end{bmatrix} \quad B = \begin{bmatrix} -2 & 6 \\ 3 & -8 \end{bmatrix} \quad C = \begin{bmatrix} 0 & 5 \\ 2 & 0 \end{bmatrix}$$

find the value of (**a**) det(A), (**b**) det(B), (**c**) AB, (**d**) 3B − 4C, (**e**) C^T.

PROBLEM 5-25   Using the matrices of Problem 5-24, find (**a**) $(BC)^2$, (**b**) C^3, (**c**) ABC.

PROBLEM 5-26   Using the matrices of Problem 5-24, determine whether or not the following statements are true in this case: (**a**) $(A - C)B = AB - CB$ and (**b**) $(B + C)^2 = B^2 + BC + CB + C^2$.

PROBLEM 5-27   Convert the matrix equation to a system of linear equations

$$\begin{bmatrix} 3 & 0 & 1 \\ -4 & 2 & 2 \\ 1 & 5 & 6 \end{bmatrix} \begin{bmatrix} x \\ y \\ z \end{bmatrix} = \begin{bmatrix} 1 \\ 3 \\ 3 \end{bmatrix}$$

PROBLEM 5-28   Let

$$A = \begin{bmatrix} 1 & 4 & 7 \\ 2 & 5 & 8 \\ 3 & 6 & 9 \end{bmatrix}$$

Find the indicated products (**a**) $A \begin{bmatrix} 2 \\ 0 \\ 3 \end{bmatrix}$, (**b**) $\begin{bmatrix} 1 & 0 & 1 \end{bmatrix} A$, (**c**) $A \begin{bmatrix} -1 & -2 \\ 3 & 4 \\ -5 & -6 \end{bmatrix}$.

PROBLEM 5-29   Let

$$A = \begin{bmatrix} 0 & 1 \\ 0 & 0 \end{bmatrix} \qquad B = \begin{bmatrix} 0 & 1 & 1 \\ 0 & 0 & 1 \\ 0 & 0 & 0 \end{bmatrix} \qquad C = \begin{bmatrix} 1 & 0 \\ 0 & 0 \end{bmatrix}$$

Show that (a) $A^2 = 0$, (b) $B^3 = 0$ when $B \neq 0$, (c) $C^2 = C$.

PROBLEM 5-30   Prove that the sum and product of diagonal matrices are diagonal matrices.

PROBLEM 5-31   (a) If $A = \begin{bmatrix} 0 & 1 \\ 1 & 0 \end{bmatrix}$ show that $A^2 = I_2$. (b) If $B = \begin{bmatrix} 1 & 0 \\ 0 & -1 \end{bmatrix}$ show that $B^2 = I_2$.

PROBLEM 5-32   Show that $A + A^T$ is symmetric.

PROBLEM 5-33   If A and B are symmetric, show that $A + B$ is symmetric.

PROBLEM 5-34   Let

$$A = \begin{bmatrix} 2 & 1 \\ 5 & 3 \end{bmatrix} \qquad B = \begin{bmatrix} 4 & 3 \\ 3 & 2 \end{bmatrix}$$

Find (a) A^{-1}, (b) B^{-1}, (c) $(AB)^{-1}$, (d) $B^{-1} \cdot A^{-1}$.

PROBLEM 5-35   Show that the inverse of $\begin{bmatrix} a & 0 & 0 \\ 0 & b & 0 \\ 0 & 0 & c \end{bmatrix}$ is $\begin{bmatrix} \frac{1}{a} & 0 & 0 \\ 0 & \frac{1}{b} & 0 \\ 0 & 0 & \frac{1}{c} \end{bmatrix}$.

PROBLEM 5-36   Show that the following matrix equation has no solution: $\begin{bmatrix} 1 & -4 \\ -2 & 8 \end{bmatrix}\begin{bmatrix} x \\ y \end{bmatrix} = \begin{bmatrix} 2 \\ 2 \end{bmatrix}$.

PROBLEM 5-37   Find det A if $A = \begin{bmatrix} 1 & 0 & 0 \\ 1 & 1 & 1 \\ 1 & 2 & 4 \end{bmatrix}$.

PROBLEM 5-38   Find det B if $B = \begin{bmatrix} 1 & 0 & 5 \\ -1 & 3 & 2 \\ 0 & 2 & 2 \end{bmatrix}$.

PROBLEM 5-39   Find det C if $C = \begin{bmatrix} 4 & 3 & -6 \\ -2 & 0 & 5 \\ 7 & 1 & -8 \end{bmatrix}$.

PROBLEM 5-40   Solve the following matrix equations:

(a) $\begin{bmatrix} 3 & -1 \\ 4 & -2 \end{bmatrix}\begin{bmatrix} x \\ y \end{bmatrix} = \begin{bmatrix} 8 \\ 10 \end{bmatrix}$ (b) $\begin{bmatrix} -5 & 0 \\ 1 & 2 \end{bmatrix}\begin{bmatrix} x \\ y \end{bmatrix} = \begin{bmatrix} 10 \\ 2 \end{bmatrix}$

PROBLEM 5-41   Solve the following matrix equations:

a $\begin{bmatrix} 1 & 3 & 2 \\ 0 & -2 & 1 \\ 4 & 0 & -1 \end{bmatrix}\begin{bmatrix} x \\ y \\ z \end{bmatrix} = \begin{bmatrix} 6 \\ -1 \\ 3 \end{bmatrix}$ (b) $\begin{bmatrix} -3 & 0 & 4 \\ 1 & 1 & 2 \\ 0 & -2 & 1 \end{bmatrix}\begin{bmatrix} x \\ y \\ z \end{bmatrix} = \begin{bmatrix} -2 \\ 2 \\ 5 \end{bmatrix}$

PROBLEM 5-42   Solve the linear system $\begin{cases} x + 2y - 4z = -3 \\ -3x + 5y - z = 18 \\ 2x + z = -5 \end{cases}$

PROBLEM 5-43 Show that $M_1 = \begin{bmatrix} 2 & \frac{1}{3} \\ 3 & 2 \end{bmatrix}$ is not invertible but that $M_2 = \begin{bmatrix} 2 & .3333 \\ 3 & 2 \end{bmatrix}$ is invertible. Find M_2^{-1} and verify its correctness.

PROBLEM 5-44 A matrix is **upper triangular** if all of its elements below the main diagonal are zero.

(a) Show that an upper triangular matrix $T = \begin{bmatrix} 1 & r & s \\ 0 & 1 & t \\ 0 & 0 & 1 \end{bmatrix}$ is always invertible. **(b)** Show that

$T^{-1} = \begin{bmatrix} 1 & -r & rt - s \\ 0 & 1 & -t \\ 0 & 0 & 1 \end{bmatrix}$.

PROBLEM 5-45 Solve the system of equations $\begin{cases} \dfrac{3}{x} + \dfrac{4}{y} = 8 \\ \dfrac{2}{x} - \dfrac{6}{y} = 1 \end{cases}$.

PROBLEM 5-46 Use matrix methods to solve the system of equations $\begin{cases} 0.41x - 1.73y = -0.91 \\ 2.13x + 0.61y = \quad 4.87 \end{cases}$.

PROBLEM 5-47 Use matrix methods to solve the system of equations $\begin{cases} x_1 - x_2 - x_3 = 0 \\ -2x_1 + x_2 - 2x_3 = 0 \\ 3x_1 - 2x_2 + x_3 = 0 \end{cases}$.

Answers to Supplementary Exercises

5-23 (a) 12 (b) -9 (c) -27

5-24 (a) 7 (b) -2 (c) $\begin{bmatrix} -23 & 64 \\ 11 & -30 \end{bmatrix}$

(d) $\begin{bmatrix} -6 & -2 \\ 1 & -24 \end{bmatrix}$ (e) $\begin{bmatrix} 0 & 2 \\ 5 & 0 \end{bmatrix}$

5-25 (a) $\begin{bmatrix} 304 & -270 \\ -432 & 385 \end{bmatrix}$ (b) $\begin{bmatrix} 0 & 50 \\ 20 & 0 \end{bmatrix}$

(c) $\begin{bmatrix} 128 & -115 \\ -60 & 55 \end{bmatrix}$

5-26 (a) True in this case (b) True in this case

5.27 $\begin{cases} 3x \quad + z = 1 \\ -4x + 2y + 2z = 3 \\ x + 5y + 6z = 3 \end{cases}$

5-28 (a) $\begin{bmatrix} 23 \\ 28 \\ 33 \end{bmatrix}$ (b) $\begin{bmatrix} 4 & 10 & 16 \end{bmatrix}$ (c) $\begin{bmatrix} -24 & -28 \\ -27 & -32 \\ -30 & -36 \end{bmatrix}$

5-34 (a) $\begin{bmatrix} 3 & -1 \\ -5 & 2 \end{bmatrix}$ (b) $\begin{bmatrix} -2 & 3 \\ 3 & -4 \end{bmatrix}$

(c) $\begin{bmatrix} -21 & 8 \\ 29 & -11 \end{bmatrix}$ (d) same as part (c)

5-37 2

5-38 -8

5-39 49

5-40 (a) $(3, 1)$ (b) $(-2, 2)$

5-41 (a) $(1, 1, 1)$ (b) $(2, -2, 1)$

5-42 $(-3, 2, 1)$

5-45 $(\frac{1}{2}, 2)$

5-46 $(2, 1)$

5-47 $(3, 4, -1)$

EXAM 1 (Chapters 1–5)

1. Make the following number conversions:

   (a) 11011.01_2 to base 10   (c) 11011010110_2 to base 8   (e) 1101100111011011_2 to base 16

   (b) 152_{10} to base 2   (d) 3415_8 to base 10

2. Represent the decimal numbers 14, 45, and -19 in (a) modulo 6 and (b) modulo 11.

3. Find the value of the expressions (a) $2 * 4 - 6 + 8$, (b) $8/4 + 2 - 6$, and (c) $2 - 3 + 4 * 5/6 + 7 * 8$.

4. Evaluate the expression $2 * (1 - 3 \uparrow 2) + (5 - 6 * 2 \uparrow -1)/2 * (4 \uparrow 2 - 3 \uparrow 2)$.

5. (a) Express the real number $16\,540\,000$ in normalized exponential form and then in eight-digit floating point form.

   (b) Form the 32-bit floating point representation of the number 217.25.

6. Solve the system of equations

$$\begin{cases} -2x + 5y = 1 \\ 3x - 4y = 2 \end{cases}$$

7. Find the polynomial of degree 3 that passes through the points $(-1, -1)$, $(0, 3)$, $(1, 3)$, and $(2, 5)$.

8. When Ken and Deb wash the windows at their home together, they get the job done in 5 hours. If Ken does it alone, working at his usual efficient rate, it takes him 9 hours. How long would it take Deb, working alone to complete the job?

9. Let matrix $\mathbf{A} = \begin{bmatrix} 4 & 1 \\ -2 & 2 \end{bmatrix}$ and $\mathbf{B} = \begin{bmatrix} 3 & 5 \\ -1 & 1 \end{bmatrix}$. Find (a) det $\mathbf{A}$, (b) det $\mathbf{B}$, (c) $\mathbf{A} + \mathbf{B}$, (d) $2A - 3\mathbf{B}$, and (e) $\mathbf{AB}$.

10. Find the determinant of the matrix $\mathbf{C} = \begin{bmatrix} 2 & -1 & 3 \\ 4 & 1 & 5 \\ -3 & 6 & 2 \end{bmatrix}$.

11. Find the inverse of the 2×2 matrix $\mathbf{A} = \begin{bmatrix} -2 & 5 \\ 3 & -7 \end{bmatrix}$. Check your answer by solving for $\mathbf{A}^{-1}\mathbf{A}$.

12. Using the augmented matrix with row operations, find the solution to the linear system

$$\begin{cases} x + 2y - z = 4 \\ -3x \phantom{{}+2y} + 2z = -1 \\ \phantom{-3x+{}} y - 4z = -2 \end{cases}$$

Solutions to Exam 1

1. (a)
$$11011.01_2 = \begin{matrix} 1 & 1 & 0 & 1 & 1 & .0 & 1 \\ \downarrow & \downarrow & & \downarrow & \downarrow & & \downarrow \\ 2^4 & 2^3 & & 2^1 & 2^0 & & 2^{-2} \end{matrix}$$
$$16 + 8 + 2 + 1 + .25 = 27.25_{10}$$

(b) Using the algorithm method, we have:

$$152 \div 2 = 76, \quad R = 0$$
$$76 \div 2 = 38, \quad R = 0$$
$$38 \div 2 = 19, \quad R = 0$$
$$19 \div 2 = 9, \quad R = 1 \quad 10011000_2$$
$$9 \div 2 = 4, \quad R = 1$$
$$4 \div 2 = 2, \quad R = 0$$
$$2 \div 2 = 1, \quad R = 0$$
$$1 \div 2 = 0, \quad R = 1$$

(c) Group the binary number into binary triples and convert them to octal form using Table 2-1:

$$11011010110_2 = \underbrace{11}_{3} \ \underbrace{011}_{3} \ \underbrace{010}_{2} \ \underbrace{110}_{6} = 3326_8$$

(d) Expand the octal number to be the sum of the powers of eight:

$$3415_8 = 3(8^3) + 4(8^2) + 1(8^1) + 5(8^0)$$
$$= 1536 + 256 + 8 + 5$$
$$= 1805_{10}$$

(e) Group the binary number into 4-bit groups and convert them to hexadecimal form using Table 2-2.

$$1101100111011011_2 = \underbrace{1101}_{D} \ \underbrace{1001}_{9} \ \underbrace{1101}_{D} \ \underbrace{1011}_{B} = D9DB_{16}$$

2. (a) Extract all multiples of 6 and record the remainder in modulo base 6:

$$14 = 2 \cdot 6 + 2 = 2 \bmod 6$$
$$45 = 7 \cdot 6 + 3 = 3 \bmod 6$$
$$-19 = -4 \cdot 6 + 5 = 5 \bmod 6$$

(b) Extract all multiples of 11 and record the remainder in modulo base 11:

$$14 = 1 \cdot 11 + 3 = 3 \bmod 11$$
$$45 = 4 \cdot 11 + 1 = 1 \bmod 11$$
$$-19 = -2 \cdot 11 + 3 = 3 \bmod 11$$

3. Working from left to right, follow the order of operations:

(a) $2 * 4 - 6 + 8 = 8 - 6 + 8 = 2 + 8 = 10$
(b) $8/4 + 2 - 6 = 2 + 2 - 6 = 4 - 6 = -2$
(c) $2 - 3 + 4 * 5/6 + 7 * 8 = 2 - 3 + (20)/6 + 56 = 2 - 3 + \frac{10}{3} + 56 = \frac{175}{3}$

4. First work inside any parentheses, looking for the highest-level operations:

$$2 * (1 - 3 \uparrow 2) + (5 - 6 * 2 \uparrow -1)/2 * (4 \uparrow 2 - 3 \uparrow 2) = 2 * (1 - 9) + (5 - 6 * \tfrac{1}{2})/2 * (16 - 9)$$

$$= 2 * (-8) + (5 - 3)/2 * 7$$
$$= 2 * (-8) + 2/2 * 7$$
$$= -16 + 2/14$$
$$= -15\tfrac{6}{7}$$

5. (a) In the normalized exponential number system, each number x in base b has the form $x = M \times b^e$. In normalized exponential form, our number is

$$16\,540\,000_{10} = 0.1654 \times 10^8$$

In eight-digit floating point form, we have one bit for the sign of the number, two bits for the exponent (characteristic), and five bits for the mantissa. Since the character $c = 50 + e$ for an eight-digit computer, we have

$$16\,540\,000 = 0.1654 \times 10^8 \rightarrow c = 50 + 8 = 58 \rightarrow \boxed{0 \mid 5 \mid 8 \mid 1 \mid 6 \mid 5 \mid 4 \mid 0}$$

(b) First convert the number to binary and then to normalized exponential form:

$$217.25 = 128 + 64 + 32 + 8 + 4 + 1 + \tfrac{1}{4}$$
$$= 2^7 + 2^6 + 2^5 + 2^3 + 2^2 + 2^0 + 2^{-2}$$
$$= 11101101.01_2$$
$$= 0.1110110101 \times 2^8$$

In 32-bit floating point form, we have one bit for the sign, seven bits to represent the exponent (characteristic), and 24 bits to represent the mantissa. Since $c = 64 + e$ for a 32-bit computer, we have

$$c = 64 + 8 = 72_{10} = 64 + 8 = 1001000_2$$

and the 32-bit floating point representation is:

0	1	0	0	1	0	0	0	1	1	1	0	1	1	0	1	0	1	0	0	0	0	0	0	0	0	0	0	0	0	0	0

6. Follow the algorithm for solving a system of two linear equations in two unknowns. Divide the first equation by the coefficient of x and solve the first equation for x in terms of y:

$$\begin{cases} x - \tfrac{5}{2}y = -\tfrac{1}{2} \\ 3x - 4y = 2 \end{cases} \rightarrow \begin{cases} x = \tfrac{5}{2}y - \tfrac{1}{2} \\ 4y = 3x - 2 \end{cases}$$

Insert the expression for x in place of every x in the second equation and solve for y:

$$4y = 3(\tfrac{5}{2}y - \tfrac{1}{2}) - 2$$
$$4y = \tfrac{15}{2}y - \tfrac{3}{2} - 2$$
$$\tfrac{7}{2} = \tfrac{7}{2}y$$
$$y = 1$$

Finally, plug the value of y back into the equation you formed in the first step and solve for x:

$$x = \tfrac{5}{2}y - \tfrac{1}{2} = \tfrac{5}{2} - \tfrac{1}{2} = 2$$

The solution is the ordered pair $(2, 1)$.

7. Use the Lagrange Interpolation formula:

$$\mathscr{P}(x) = \left[\frac{(x-0)(x-1)(x-2)}{(-1)(-2)(-3)}\right](-1) + \left[\frac{(x+1)(x-1)(x-2)}{(1)(-1)(-2)}\right](3) + \left[\frac{(x+1)(x-0)(x-2)}{(2)(1)(-1)}\right](3)$$

$$+ \left[\frac{(x+1)(x-0)(x-1)}{(3)(2)(1)}\right](5)$$

$$= \frac{x^3 - 3x^2 + 2x}{-6}(-1) + \frac{x^3 - 2x^2 - x + 2}{2}(3) + \frac{x^3 - x^2 - 2x}{-2}(3) + \frac{x^3 - x}{6}(5)$$

$$= \left(\frac{1}{6} + \frac{3}{2} - \frac{3}{2} + \frac{5}{6}\right)x^3 + \left(-\frac{1}{2} - \frac{6}{2} + \frac{3}{2}\right)x^2 + \left(\frac{1}{3} - \frac{3}{2} + \frac{6}{2} - \frac{5}{6}\right)x + 3$$

$$= 1x^3 - 2x^2 + 1x + 3$$

8. The basic formula is

$$\frac{1}{K} + \frac{1}{D} = \frac{1}{T}$$

where K is Ken's time, D is Deb's time and T is the together time. Plugging in the given values, we solve for Deb's time:

$$\frac{1}{9} + \frac{1}{D} = \frac{1}{5}$$

$$\frac{1}{D} = \frac{1}{5} - \frac{1}{9} = \frac{9-5}{45} = \frac{4}{45}$$

$$D = \frac{45}{4}$$

9. (a) $\det \mathbf{A} = \begin{vmatrix} 4 & 1 \\ -2 & 2 \end{vmatrix} = 4 \cdot 2 - (1)(-2) = 10$

(b) $\det \mathbf{B} = \begin{vmatrix} 3 & 5 \\ -1 & 1 \end{vmatrix} = 3 - (-5) = 8$

(c) $\mathbf{A} + \mathbf{B} = \begin{bmatrix} 7 & 6 \\ -3 & 3 \end{bmatrix}$

(d) $2\mathbf{A} - 3\mathbf{B} = \begin{bmatrix} 8 & 2 \\ -4 & 4 \end{bmatrix} - \begin{bmatrix} 9 & 15 \\ -3 & 3 \end{bmatrix} = \begin{bmatrix} -1 & -13 \\ -1 & 1 \end{bmatrix}$

(e) $\mathbf{AB} = \begin{bmatrix} 4 & 1 \\ -2 & 2 \end{bmatrix}\begin{bmatrix} 3 & 5 \\ -1 & 1 \end{bmatrix} = \begin{bmatrix} 11 & 21 \\ -8 & -8 \end{bmatrix}$

10. $\det \mathbf{C} = \begin{vmatrix} 2 & -1 & 3 \\ 4 & 1 & 5 \\ -3 & 6 & 2 \end{vmatrix} \begin{matrix} 2 & -1 \\ 4 & 1 \\ -3 & 6 \end{matrix} = 4 + 15 + 72 - (-9 + 60 - 8) = 91 - (43) = 48$

11. Using the algorithm for finding the inverse of a 2×2 matrix, we get

$$\mathbf{A}^{-1} = \frac{1}{\det \mathbf{A}}\begin{bmatrix} -7 & -5 \\ -3 & -2 \end{bmatrix} = \begin{bmatrix} 7 & 5 \\ 3 & 2 \end{bmatrix}$$

Check: $\begin{bmatrix} 7 & 5 \\ 3 & 2 \end{bmatrix}\begin{bmatrix} -2 & 5 \\ 3 & -7 \end{bmatrix} = \begin{bmatrix} 1 & 0 \\ 0 & 1 \end{bmatrix}$

12. First write the linear system as an augmented matrix,

$$\begin{bmatrix} 1 & 2 & -1 & \vdots & 4 \\ -3 & 0 & 2 & \vdots & -1 \\ 0 & 1 & -4 & \vdots & -2 \end{bmatrix}$$

then follow the algorithm for solving any system with matrices, using the elementary row operations:

$$\begin{bmatrix} 1 & 2 & -1 & \vdots & 4 \\ -3 & 0 & 2 & \vdots & -1 \\ 0 & 1 & -4 & \vdots & -2 \end{bmatrix} \rightarrow \begin{bmatrix} 1 & 2 & -1 & \vdots & 4 \\ 0 & 6 & -1 & \vdots & 11 \\ 0 & 1 & -4 & \vdots & -2 \end{bmatrix}$$

(Add a multiple of row 1 to row 2 to get a 0 in the nonpivotal positions in column 1.)

$$\rightarrow \begin{bmatrix} 1 & 2 & -1 & \vdots & 4 \\ 0 & 1 & -4 & \vdots & -2 \\ 0 & 6 & -1 & \vdots & 11 \end{bmatrix}$$

(Interchange rows 2 and 3—this automatically puts a 1 in the pivotal position in row 2.)

$$\rightarrow \begin{bmatrix} 1 & 0 & 7 & \vdots & 8 \\ 0 & 1 & -4 & \vdots & -2 \\ 0 & 0 & 23 & \vdots & 23 \end{bmatrix}$$

(Add a multiple of row 2 to each of rows 1 and 3 to get a 0 in the nonpivotal positions in column 2.)

$$\rightarrow \begin{bmatrix} 1 & 0 & 7 & \vdots & 8 \\ 0 & 1 & -4 & \vdots & -2 \\ 0 & 0 & 1 & \vdots & 1 \end{bmatrix}$$

(Multiply row 3 by a nonzero constant to get a 1 in th pivotal position.)

$$\rightarrow \begin{bmatrix} 1 & 0 & 0 & \vdots & 1 \\ 0 & 1 & 0 & \vdots & 2 \\ 0 & 0 & 1 & \vdots & 1 \end{bmatrix}$$

$$\Rightarrow \begin{matrix} x = 1 \\ y = 2 \\ z = 1 \end{matrix}$$

(Add a multiple of row 3 to each of rows 1 and 2 to get a 0 in the nonpivotal positions in column 3. The portion originally occupied by the constants becomes a column vector whose entries are the solution values.)

6 LOGIC

THIS CHAPTER IS ABOUT

☑ **Connectives and Truth Tables**
☑ **Application to Programming**

The use of logic is very important in any formal science discipline. Computer science in particular applies logic directly to a number of concepts, such as to the creation and verification of algorithms, to the design of circuits, and to the problem-solving techniques of programming. Indeed, in our everyday lives we constantly use logical reasoning (well, at least we *think* we do!), perhaps in persuasion, arguments, or decision-making. But how do we know whether or not our arguments are logically sound? In this chapter we will explore the concepts of logic and establish some rules to follow, some principles to adhere to, and some ways to apply these principles.

6-1. Connectives and Truth Tables

A. Propositions and statements

Because there is no room for opinion, question, or half-truths in the world of logic, we define a **proposition** as a statement that is either true or false. For a proposition or statement to be **valid**, you must be able to assign a **truth value** to it; that is, you must be able to say it's true (truth value = T) or it's false (truth value = F) but not both. (Notice how nicely this fits into the yes/no, on/off world of computers.) There is no in-between.

EXAMPLE 6-1   Which of the following are valid propositions?

(a) Abrahim Lincoln was a U.S. President.
(b) Elephants have exactly two feet.
(c) The moon is not made of green cheese.
(d) In the set of natural numbers, 32 is greater than 25.
(e) I like ice cream.

(f) Is it raining today?
(g) $5 \cdot x = 12$.
(h) This sentence is false.
(i) The state of California will fall into the ocean tomorrow.

Solution

(a) This is a proposition. The statement has a truth value T.
(b) This is a proposition whose truth value is F.
(c) This is a proposition with truth value T.
(d) This is a proposition with truth value T.
(e) This is not a proposition. It is a comment or opinion. Depending on who "I" is, it is neither true nor false all of the time.
(f) This is not a proposition. It is a question.
(g) This equation is not a proposition since x is not specified and we cannot assign a truth value to it. But $5 \cdot 3 = 12$ is a proposition that is false; $5 \cdot \frac{12}{5} = 12$ is a proposition that is true.
(h) This is not a proposition. It is a **paradox** since for the sentence to be true it must have a truth value F, and for the sentence to be false it must have a truth value T.

(i) This is a proposition. It has a definite truth value of either T or F, though we do not know which at this time.

Each of the valid propositions in Example 6-1 is a **simple statement**. Each stands alone on its own, unrelated to any other statement. When you combine one or more simple statements—that is, relate them to one another somehow—you create a **compound statement**. For example, you can combine statements (a) and (b) of Example 6-1 to form any of the following compound statements:

"Abraham Lincoln was a U.S. president *and* elephants have exactly two feet."   **(6-1)**

"Abraham Lincoln was a U.S. president *or* elephants have exactly two feet."   **(6-2)**

"*If* Abraham Lincoln was a U.S. president, *then* elephants have exactly two feet."   **(6-3)**

"Abraham Lincoln was a U.S. president *if and only if* elephants have exactly two feet."   **(6-4)**

note: Statements (6-1) through (6-4) are each valid propositions because they are composed of valid propositions and can, themselves, be assigned a specific truth value. As will soon be explained, the truth value of compound statements like those of (6-1) through (6-4) depends on the truth values of their components.

A shorter way to write compound statements (6-1) through (6-4) is to assign an individual variable to each simple statement and then link the variables by special symbols. It is common to use the variables $p, q, r, \ldots$—as opposed to the algebraic variables $x, y, z, \ldots$—to represent simple statements. Thus, by assigning p to the statement "Abraham Lincoln was a U.S. president" and q to the statement "Elephants have exactly two feet," statements (6-1) through (6-4) become, respectively:

statement (6-1): "p and q"

statement (6-2): "p or q"

statement (6-3): "if p, then q"

statement (6-4): "p if and only if q"

EXAMPLE 6-2 Let the letter p denote "It is past 8:00 a.m." and let q denote the statement "I am out of bed." Combine these into compound statements using **(a)** "p and q" **(b)** "p or q" **(c)** "if p, then q" and **(d)** "p if and only if q."

Solution

(a) p and q: "It is past 8:00 a.m. and I am out of bed."
(b) p or q: "It is past 8:00 a.m. or I am out of bed."
(c) if p, then q: "If it is past 8:00 a.m., then I am out of bed."
(d) p if and only if q: "It is past 8:00 a.m. if and only if I am out of bed." (Don't you wish you could control time that way?)

B. Symbols, simple connectives, and truth tables

As in algebra, where you must use variables $x, y, z, \ldots$ and combine them with operators $+, -, *, /, \ldots$, in logic you must use special operators to connect propositions. These operators are called (surprisingly enough) **connectives**. The three main connectives are *and* (**conjunction**), *or* (**disjunction**), and *not* (**negation**). Each connective has a particular symbol that represents it, and you combine these symbols with variables to create compound logic statements. The simple connectives and symbols, along with their logic names, are as follows:

Connective	Symbol	Logic name	Interpretation
AND	$\wedge$	conjunction	"and"
OR	$\vee$	disjunction	"or"
NOT	$\sim$	negation	"not"

Going back to our Lincoln–elephants statements, we can now write statements (6-1) and (6-2) as:

$p \wedge q$ (read as "*p* and *q*"): "Abraham Lincoln was a U.S. president and elephants have exactly two feet." (This is a *conjunction*.)   **(6-5)**

$p \vee q$ (read as "*p* or *q*"): "Abraham Lincoln was a U.S. president or elephants have exactly two feet." (This is a *disjunction*.)   **(6.6)**

The use of the negative always applies to just one statement, in the same way that a negative sign applies only to one number: it makes it the opposite of what it was. (For example, -2 is the opposite of $+2$; $-(-2)$ is the opposite of -2.) To make the opposite of a statement, $\sim p$, you simply negate it, or make its truth value the opposite:

$\sim p$ (read as "not *p*"): "Abraham Lincoln was not a U.S. president." (This is a *negation*.)   **(6-7)**

$\sim q$ (read as "not *q*") "Elephants do not have exactly two feet." (This is a *negation*.)   **(6-8)**

note: A negation is a simple statement, not a compound one.

another note: $\sim(\sim p)$ means to take the negation of a negation. The negation of $(\sim p)$ would be: "Abraham Lincoln was not not a U.S. president," or "It is not the case that Abraham Lincoln was not a U.S. president," or "Abraham Lincoln was a U.S. president," which is equivalent to the original statement *p*. Therefore $\sim(\sim p)$ is equivalent to *p*, in much the same way that $-(-2)$ is equivalent to 2.

EXAMPLE 6-3   Form the negation (opposite) of the following propositions:

(a) *p*: It is raining.   (b) *q*: Leonardo da Vinci is a genius.

Solution

(a) $\sim p$ is the statement: "It is not the case that it is raining" or, more simply, "It is not raining."
(b) $\sim q$ is the statement: "It is not the case that Leonardo da Vinci is a genius" or, more simply, "Leonardo de Vinci is not a genius."

EXAMPLE 6-4   Form the conjuction of the following statements:

(a) *p*: The sun is shining;   *q*: I am hot.
(b) *p*: Lions eat meat;   *q*: Sheep are vegetarians.

Solution   To form the conjuction of *p* and *q*, denoted $p \wedge q$, write it exactly as "(statement *p*) and (statement *q*)."

(a) $p \wedge q$: "The sun is shining and I am hot."
(b) $p \wedge q$: "Lions eat meat and sheep are vegetarians."

EXAMPLE 6-5   Form the disjunction of the given statements:

(a) *p*: Ice is hard;   *q*: It is December.
(b) *p*: Ripe tomatoes are red;   *q*: Edison discovered America.

Solution   To form the disjunction of *p* and *q*, written $p \vee q$, simply write statement *p*, then the disjunction word "or" followed by the written statement *q*.

(a) $p \vee q$: "Ice is hard or it is December."
(b) $p \vee q$: "Ripe tomatoes are red or Edison discovered America."

Now to assign those truth values to compound statements as we promised. Recall that we commented that the truth value of any compound statement depends on the truth values of its components. This is always the case since each of the simple statements $p, q, r, \ldots$ of a compound statement may be either true or false. When you have two simple propositions *p* and *q*, you always have only four cases, or possibilities, to contend with; either:

(**1**) Both p and q have truth value T; or
(**2**) p has truth value T and q has truth value F; or
(**3**) p has truth value F and q has truth value T; or
(**4**) both p and q have truth value F.

The truth value of a compound statement (say, $p \wedge q$, or $p \vee q$) therefore depends on which truth value pair you have for p and q. Fortunately, by definition, logic is rigidly logical and there are well-established rules that compound statements always follow. For example, for any $p \wedge q$ ("p and q"), if both p and q are true, $p \wedge q$ is true; if p is true and q is false, $p \wedge q$ is false; if p is false and q is true, $p \wedge q$ is still false; and if both p and q are false, $p \wedge q$ is also false. Now trying to remember all that could become difficult, so logicians supply a shorthand version of this collection: a **truth table**. A truth table is simply a chart that identifies the truth values of a compound statement for individual cases (possible values of p and q). You write T's and F's and read the table across for particular values of p and q. The truth tables for conjunction, disjunction, and negation are as follows:

Truth Table for Conjunction ("p and q")

p	q	$p \wedge q$
T	T	T
T	F	F
F	T	F
F	F	F

Truth Table for Disjunction ("p or q")

p	q	$p \vee q$
T	T	T
T	F	T
F	T	T
F	F	F

Truth Table for Negation ("not p")

p	$\sim p$
T	F
F	T

EXAMPLE 6-6 Let p denote "Chicago has skyscrapers," and let q denote "Columbus was Japanese." Use the truth tables to find the truth values of the statements (**a**) $p \wedge q$ (**b**) $p \vee q$ (**c**) $\sim p$ (**d**) $\sim q$

Solution

(**a**) $p \wedge q$: "Chicago has skyscrapers and Columbus was Japanese." Statement p is true and q is false. Examining the truth table for conjunction, we find that when p is true and q is false, $p \wedge q$ is false.

(**b**) $p \vee q$: "Chicago has skyscrapers or Columbus was Japanese." We already know that statement p is true and statement q is false. Looking over the truth table for disjunction, we see that in these circumstances $p \vee q$ is true.

(**c**) $\sim p$: "Chicago does not have skyscrapers." According to the truth table for negation, when p is true (which it is), $\sim p$ is false.

(**d**) $\sim q$: "Columbus was not Japanese." The statement q is false so, according to the negation truth table, $\sim q$ is true.

important note: The only way for a disjunction statement to be false is when both p and q are false. This is because of the mathematician's use of "or." In our common English usage, the connective "or" may be used in two different ways. In the sentence "On my last roll in this bowling game I scored a strike or I left some pins standing," the word "or" is used in an *exclusive* sense because both occurrences could not have happened; the occurrence of one event excludes the possibility of the occurrence of the other. We can have one or the other but not both. But in the disjunction "Sara ate an ice cream cone or she stopped at the bank," we use the word "or" in an *inclusive* sense. *Either* action could have happened or *both* actions could have happened because the actions are not mutually exclusive. In mathematics and computer science, we agree to use the connective "or" in this inclusive sense. Thus for a "*p* or *q*" statement we're really saying "*p* and/or *q*": that is, $p \vee q$ is true if

(1) *p* is true; or if
(2) *q* is true; or if
(3) both are true.

C. Building additional connectives

Other connectives are built from the basic connectives of conjunction, disjunction, and negation. Like the basic connectives, these new ones also have special symbols to represent them. Other connectives are:

(1) implication: $p \rightarrow q$
(2) converse: $q \rightarrow p$
(3) equivalence: $p \leftrightarrow q$
(4) contrapositive: $\sim q \rightarrow \sim p$

We will examine them one by one.

1. Implication

Using statements *p* and *q*, the compound statement "if *p*, then *q*," denoted $p \rightarrow q$, is called an *implication*. Statement *p* is called the **antecedent** (that which comes before) and statement *q* is called the **consequent** (that which comes after).

EXAMPLE 6-7   Form an implication with the given statements.

(a) *p*: It is raining;   *q*: I will take my umbrella.
(b) *p*: It is Tuesday;   *q*: Tonight I will go golfing.
(c) *p*: You graduate from college;   *q*: You will be successful in your career.

Solution

(a) "If it is raining, then I will take my umbrella."
(b) "If it is Tuesday, then tonight I will go golfing."
(c) "If you graduate from college, then you will be successful in your career."

The statements *p* and *q* used in an implication do not need to be related by cause–effect. The implication merely asserts that we cannot simultaneously have the negation of *q* and the truth of *p*; it does not say that statement *p* *causes* statement *q*. (Take (c) of Example 6-7, for instance. Graduating from college may certainly be a contributing factor to being successful in your career, but it is hardly the sole cause.)

As with every other connective, we have four cases for which implication will occur, each of which affects the truth value of the entire statement. Take (c) of Example 6-7 again. The four possibilities are:

(1) If you graduate from college (*p* is true), and if you are successful in your career (*q* is true), then statement $p \rightarrow q$ is obviously true.
(2) If you graduate from college (*p* is true), but you aren't successful in your career (*q* is false), then the statement $p \rightarrow q$ is obviously false.
(3)–(4) If, however, you do not graduate from college (*p* is false), then statement $p \rightarrow q$ is *not false*, whether or not you are successful in your career (*q* is true or false). Therefore statement $p \rightarrow q$ (not necessarily the cause–effect behind it) is always true whenever the antecedent *p* is false. (Yes, this does mean that you can put anything in front of the consequent and still have the implication be true: "If the sky is plaid, then you will be successful in your career" is an implication with truth value T. Whether or not it makes sense or nonsense on a cause-and-effect level is of non consequence to logicians; they're only concerned with the truth values of the component statements as they are set up in the truth tables. Follow the truth tables and you can't go wrong.)

Examining the results of the four cases, we have the truth table for implication as follows:

Truth Table for Implication
("if p, then q")

p	q	$p \rightarrow q$
T	T	T
T	F	F
F	T	T
F	F	T

note: From the implication truth table it is clear that the only way for $p \rightarrow q$ to be false is if the consequent q is false. For example, when

$$p: \text{"3} \times 7 = 21\text{" and } q: \text{"2} \times 3 = 7\text{"}$$

the implication $p \rightarrow q$, "If $3 \times 7 = 21$, then $2 \times 3 = 7$," is false according to the truth table. However, if we have

$$p: \text{"2} \times 3 = 7\text{" and } q: \text{"3} \times 7 = 21\text{"}$$

the proposition $p \rightarrow q$ will read "If $2 \times 3 = 7$, then $3 \times 7 = 21$." Because p is false and q is true in this circumstance, $p \rightarrow q$ is true according to the truth table. It is obvious that the conclusion $3 \times 7 = 21$ is true even though the antecedent is nonsense.

Another way of expressing $p \rightarrow q$ is to write it as $(\sim p) \vee q$. The proof of this involves creating the truth table for the new compound statement and then comparing it with the implication table.

note: To *create a truth table* for a compound statement such as $(\sim p) \vee q$, you work within the parentheses first, then from left to right, much like you perform operations in algebra. First set up the four possible truth value cases for p and q, then find the truth values for $(\sim p)$ for the four cases by using the truth value of p and the truth table for negation; finally, use the truth values of $(\sim p)$ and q and the truth table for disjunction to find the truth values for $(\sim p) \vee q$.

Truth Table for $(\sim p) \vee q$

p	q	$\sim p$	$(\sim p) \vee q$
T	T	F	T
T	F	F	F
F	T	T	T
F	F	T	T

Notice that the truth values for $(\sim p) \vee q$ are indeed the same as for the implication $p \rightarrow q$ in the various cases. There are a number of ways to say the implication statement $p \rightarrow q$ in English. We could say:

(1) p implies q
(2) p only if q
(3) p is a sufficient condition for q
(4) q is a necessary condition for p

2. Converse

The reverse of the implication connective is known as the *converse* and is denoted $q \rightarrow p$. (You may be familiar with theorems that say something like "if $ab = ba$ then its converse, $ba = ab$, may also be true.") The truth table for converse is:

Truth Table
for Converse
("if *q*, then *p*")

p	*q*	*q* → *p*
T	T	T
T	F	T
F	T	F
F	F	T

EXAMPLE 6-8   Let *p* be "Christmas comes twice a year" and let *q* be "Squirrels eat peanuts." Find the converse *q* → *p* and its truth value.

Solution   Write the converse in the form "If (statement *q*), then (statement *p*)."

q → *p*: "If squirrels eat peanuts then Christmas comes twice a year." According to the truth table, this statement has truth value F because *p* is false and *q* is true.

3. Equivalence

A connective that expresses the compound statement "*p* if and only if *q*," denoted by *p* ↔ *q*, is called an *equivalence*. The truth values of the equivalence connective are:

Truth Table
for Equivalence
("*p* if and only if *q*")

p	*q*	*p* ↔ *q*
T	T	T
T	F	F
F	T	F
F	F	T

For example, "Herman Melville wrote *Moby Dick* if and only if mice have four legs" is a true equivalence because both statements *p* and *q* are true (though not at all related by rational cause and effect). "Clams have legs if and only if shrimp can fly" is also a true equivalence because both statements *p* and *q* are false.

An alternate way to describe the equivalence connective is with the statement "*p* is a necessary and sufficient condition for *q*." This is based on the concept of the converse connective. Using the converse, we can break the "*p* ↔ *q*" connective into two parts:

$$(p \leftrightarrow q) \leftrightarrow ((p \rightarrow q) \wedge (q \rightarrow p)) \tag{6-9}$$

Thus a theorem that is stated as an "if and only if" theorem must be proved in two parts: Show that *p* → *q* *and* that *q* → *p*.

EXAMPLE 6-9   Establish the truth of the statement $(p \leftrightarrow q) \leftrightarrow ((p \rightarrow q) \wedge (q \rightarrow p))$.

Solution   Show that the truth tables for both sides of the equivalence are identical for all cases of p and q.

left side

p	q	$q \leftrightarrow p$
T	T	T
T	F	F
F	T	F
F	F	T

right side

p	q	$p \rightarrow q$	$q \rightarrow p$	$(p \rightarrow q) \wedge (q \rightarrow p)$
T	T	T	T	T
T	F	F	T	F
F	T	T	F	F
F	F	T	T	T

↑ ————— same ————— ↑

note: Some people mistakenly confuse their own logic with mathematical logic by saying that if $p \rightarrow q$ then that automatically also means $q \rightarrow p$. This is faulty logic. For example, if we have

> p: "You are over 50 years old."
>
> q: "You are slowing down your pace."

$p \rightarrow q$ is usually true, but $q \rightarrow p$ is often false. Thus if you are over 50, you'll probably slow down, but if you slow down it is not necessarily the case that you are over 50. People slow down for lots of other reasons.

As a mathematical example, we can use p: "$x = 5$" and q: "x is an odd integer." In this case we can clearly see that $p \rightarrow q$ is true but $q \rightarrow p$ is false. Thus, we actually have

$$(p \rightarrow q) \not\rightarrow (q \rightarrow p) \qquad \text{or} \qquad (p \rightarrow q) \not\leftrightarrow (q \rightarrow p)$$

which can be verified by truth tables.

EXAMPLE 6-10   For p: "$3 + 4 = 7$" and q: "$5 + 9 = 12$" show that $p \rightarrow q$ doesn't necessarily mean $q \rightarrow p$.

Solution   According to the truth tables, if p is true and q is false (which they are), then the implication $p \rightarrow q$ is false. But notice that the converse, $q \rightarrow p$, is *true*, according to the truth tables, because q is false. Thus, $(p \rightarrow q) \not\rightarrow (q \rightarrow p)$ and (if we switched letters) $(q \rightarrow p) \not\leftrightarrow (p \rightarrow q)$.

4. Contrapositive

Another connective, called the *contrapositive* and designated by $\sim q \rightarrow \sim p$, is read as "if not q, then not p." Its truth table is

Truth Table
for Contrapositive
("if not q, then not p")

p	q	$\sim q \rightarrow \sim p$
T	T	T
T	F	F
F	T	T
F	F	T

which can be built from the negation and converse truth tables:

p	q	$\sim p$	$\sim q$	$\sim q \to \sim p$
T	T	F	F	T
T	F	F	T	F
F	T	T	F	T
F	F	T	T	T

Notice, though, that the results of this table are identical to the results of the implication table for each case. In expanded form we see that $p \to q$ has the same truth values as $\sim q \to \sim p$:

p	q	$p \to q$	$\sim p$	$\sim q$	$\sim q \to \sim p$
T	T	T	F	F	T
T	F	F	F	T	F
F	T	T	T	F	T
F	F	T	T	T	T

↑————— same —————↑

So we can say that

$$(p \to q) \leftrightarrow (\sim q \to \sim p) \tag{6-10}$$

that is, that $(p \to q)$ is equivalent to $(\sim q \to \sim p)$. In the next section, we'll explore this further.

D. Tautologies and contradictions

Take a look again at statement (6-10). The truth table of this statement, in expanded form, would read:

left side			right side			full statement
p	q	$p \to q$	$\sim p$	$\sim q$	$\sim q \to \sim p$	$(p \to q) \leftrightarrow (\sim q \to \sim p)$
T	T	T	F	F	T	T
T	F	F	F	T	F	T
F	T	T	T	F	T	T
F	F	T	T	T	T	T

That is, statement (6-10) is true for all possible truth values of p and q. This is an example of a **tautology**, a statement that is always true. Example 6-9 was also the statement of a tautology.

note: Understanding the entire set of connectives is important for eliminating faulty logic from a computer program. The contrapositive and the tautology are especially useful in restructuring a program that contains negative logic. For example, we can replace a negative proposal by a positive conclusion and a negative conclusion by a positive proposal and thus obtain positive logic. This is expressed in statement (6-10).

The opposite of a tautology would be a statement that is always false. This is referred to as a **contradiction** or **absurdity**.

EXAMPLE 6-11   Determine if each of the following is a tautology, a contradiction, or neither.

(a) $\sim \sim p \leftrightarrow p$ **(b)** $p \wedge \sim p$ **(c)** $(p \wedge q) \to q$

Solution

(a) This is a tautology. The negation of the negation of p is p. (Use the negation truth table.)

(b) This is a contradiction. We cannot simultaneously have the truth of statement p and its negation.

(c) This is a tautology that is also an implication.

EXAMPLE 6-12   Given the compound statement $(q \wedge p) \vee (q \wedge {\sim}p)$, determine if it is a tautology, a contradiction or neither.

Solution   Write its truth table:

p	q	$q \wedge p$	${\sim}p$	$q \wedge {\sim}p$	$(q \wedge p) \vee (q \wedge {\sim}p)$
T	T	T	F	F	T
T	F	F	F	F	F
F	T	F	T	T	T
F	F	F	T	F	F

The final values are neither all true nor all false, so the statement is neither a tautology nor a contradiction.

If you examine the truth table from Example 6-12, you can see that the resulting values of $(q \wedge p) \vee (q \wedge {\sim}p)$ are the same as those for q. This means that

$$(q \wedge p) \vee (q \wedge {\sim}p) \leftrightarrow q \tag{6-11}$$

which is a tautology. When you have a tautology formed by an equivalence statement, like this one is, the tautology becomes very helpful: you can use it to replace an expression in a given program by another, equivalent, expression. For example, if you had a program that utilizes the convoluted proposition on the left side of tautology (6-11), you could replace the entire mess in your program with the proposition on the right side of the statement, namely, q. This type of substitution is a key use of tautologies.

There are two special equivalence tautologies, called **idempotent tautologies**, that are particularly useful:

$$(p \wedge p) \leftrightarrow p \qquad \text{and} \qquad (p \vee p) \leftrightarrow p$$

EXAMPLE 6-13   Prove the idempotent tautologies.

Solution   As setting up the truth tables shows, they are true for all cases of p:

p	$p \wedge p$	$(p \wedge p) \leftrightarrow p$
T	T	T
F	F	T

p	$p \vee p$	$(p \vee p) \leftrightarrow p$
T	T	T
F	F	T

Knowing the idempotent properties allows you to substitute for $p \wedge p$ (or for $p \vee p$) the simpler expression p. Or anywhere p occurs, you are allowed to replace it with $p \wedge p$ (or with $p \vee p$), possibly allowing some other further simplification.

Some other tautologies that may be verified by truth tables include

(a) $\left.\begin{array}{l} p \wedge q \leftrightarrow q \wedge p \\ p \vee q \leftrightarrow q \vee p \end{array}\right\}$ Commutative property

(b) $\left.\begin{array}{l} (p \wedge q) \wedge r \leftrightarrow p \wedge (q \wedge r) \\ (p \vee q) \vee r \leftrightarrow p \vee (q \vee r) \end{array}\right\}$ Associative property

(c) $\left. \begin{array}{l} p \wedge (q \vee r) \leftrightarrow (p \wedge q) \vee (p \wedge r) \\ p \vee (q \wedge r) \leftrightarrow (p \vee q) \wedge (p \vee r) \end{array} \right\}$ Distributive property

(d) $\left. \begin{array}{l} \sim(p \wedge q) \leftrightarrow (\sim p) \vee (\sim q) \\ \sim(p \vee q) \leftrightarrow (\sim p) \wedge (\sim q) \end{array} \right\}$ DeMorgan's Laws

We've now introduced a third proposition variable *r* [in tautologies (b) and (c)]. Don't panic. As you'll see in the next example, you handle it just like you would any other proposition.

EXAMPLE 6-14   Use truth tables to establish the validity of the first distributive property.

Solution   Since this is a tautology that is based on an equivalence, you must only prove that the propositions on the left and right sides of the equivalence symbol produce the same truth values for all cases *p*, *q*, and *r*. Now that we have three variables (propositions), we can have *eight* possible cases, not just four, because the three individual propositions can be true or false in more combinations. The truth table will look like this:

				left side		right side	
p	*q*	*r*	$q \vee r$	$p \wedge (q \vee r)$	$p \wedge q$	$p \wedge r$	$(p \wedge q) \vee (p \wedge r)$
T	T	T	T	T	T	T	T
T	F	T	T	T	F	T	T
F	T	T	T	F	F	F	F
F	F	T	T	F	F	F	F
T	T	F	T	T	T	F	T
T	F	F	F	F	F	F	F
F	T	F	T	F	F	F	F
F	F	F	F	F	F	F	F

same, so the tautology is valid

EXAMPLE 6-15   Establish the validity of the first DeMorgan's Law:

$$\sim(p \wedge q) \leftrightarrow (\sim p) \vee (\sim q)$$

Solution

		right side			left side	
p	*q*	$\sim p$	$\sim q$	$(\sim p) \vee (\sim q)$	$p \wedge q$	$\sim(p \wedge q)$
T	T	F	F	F	T	F
T	F	F	T	T	F	T
F	T	T	F	T	F	T
F	F	T	T	T	F	T

same, so the tautology is valid

Some additional tautologies used in proofs and/or in restructuring program logic are:

(e) $(p \rightarrow q) \leftrightarrow ((\sim p) \vee q)$

(f) $\sim(p \rightarrow q) \leftrightarrow p \wedge \sim q$

(g) $\sim(p \leftrightarrow q) \leftrightarrow ((p \wedge \sim q) \vee (q \wedge \sim p))$

EXAMPLE 6-16　Establish the validity of tautology (g).

Solution

	left side							right side	
p	q	$p \leftrightarrow q$	$\sim(p \leftrightarrow q)$	$\sim q$	$p \wedge \sim q$	$\sim p$	$q \wedge \sim p$	$(p \wedge \sim q) \vee (q \wedge \sim p)$
T	T	T	F	F	F	F	F	F
T	F	F	T	T	T	F	F	T
F	T	F	T	F	F	T	T	T
F	F	T	F	T	F	T	F	F

↑ ——————— same ——————— ↑

There are also tautologies that are "one-way"; that is, they are also implications instead of equivalences. These are known as *implication tautologies*:

(h) $(p \wedge q) \rightarrow p$
　　　$(p \wedge q) \rightarrow q$

(i) $p \rightarrow (p \vee q)$
　　　$q \rightarrow (p \vee q)$

(j) $\sim p \rightarrow (p \rightarrow q)$
　　　$\sim(p \rightarrow q) \rightarrow p$

(k) $(p \wedge (p \rightarrow q)) \rightarrow q$
　　　$(\sim p \wedge (p \vee q)) \rightarrow q$

(l) $\sim q \wedge (p \rightarrow q) \rightarrow \sim p$

(m) $((p \rightarrow q) \wedge (q \rightarrow r)) \rightarrow (p \rightarrow r)$　Transitivity

Many of these one-way tautologies are very useful in restructuring program logic to allow for easier input or output of information. Also, you can make many mathematical proofs simpler when you use these statements in the steps of the proofs.

EXAMPLE 6-17　Establish the validity of the implication tautology $\sim p \rightarrow (p \rightarrow q)$. Then establish its counterpart, $\sim(p \rightarrow q) \rightarrow p$.

Solution　Use the negation and implication tables.

p	q	$\sim p$	$p \rightarrow q$	$\sim p \rightarrow (p \rightarrow q)$
T	T	F	T	T
T	F	F	F	T
F	T	T	T	T
F	F	T	T	T

Since we have already shown that $(A \rightarrow B) \leftrightarrow (\sim B \rightarrow \sim A)$ (the contrapositive), then the following statement is valid:

$$(\sim p \rightarrow (p \rightarrow q)) \leftrightarrow (\sim(p \rightarrow q) \rightarrow p)$$

and the counterpart of the original implication tautology is established as a tautology, too.

Tautologies can help you verify the validity of an argument. All you do is break the argument into its simple propositions, set it up as a symbolic statement, then test to see if the statement is a tautology.

EXAMPLE 6-18　Decide whether or not the following argument is valid: "If I ride my bicycle to school, then I will arrive tired. I am not tired when I arrive at school. Therefore, I do not ride my bicycle to school."

Solution   Assign symbols to the simple statements:

p: "I ride my bicycle to school."

q: "I arrive at school tired."

We now have the following symbolic statement $(p \rightarrow q) \wedge (\sim q) \rightarrow \sim p$ to represent the argument. If the symbolic statement is true for all cases of *p* and *q*, then the argument is valid.

p	q	$p \rightarrow q$	$\sim q$	$(p \rightarrow q) \wedge (\sim q)$	$\sim p$	$(p \rightarrow q) \wedge (\sim q) \rightarrow \sim p$
T	T	T	F	F	F	T
T	F	F	T	F	F	T
F	T	T	F	F	T	T
F	F	T	T	T	T	T

Now, consider the following set of statements:

p: "It is 10:00 in Detroit."

q: "It is 9:00 in Chicago."

r: "It is 7:00 in Los Angeles."

Using these simple propositions, we can create the following argument:

"If it is 10:00 in Detroit, then it is 9:00 in Chicago. If it is 9:00 in Chicago, then it is 7:00 in Los Angeles. Hence, if it is 10:00 in Detroit, then it is 7:00 in Los Angeles."

We have just created an example of the **transitive inference**. The transitive inference is valid if it is true for all cases *p*, *q*, and *r*. Written in symbolic form, the transitive inference becomes

TRANSITIVE INFERENCE
$$((p \rightarrow q) \wedge (q \rightarrow r)) \rightarrow (p \rightarrow r) \tag{6-12}$$

EXAMPLE 6-19   Establish the validity of the transitive inference.

Solution   Remember that there are eight possible cases with a three-variable argument. Use the conjunction and implication tables.

			left side			right side	full statement
p	q	r	$p \rightarrow q$	$q \rightarrow r$	$(p \rightarrow q) \wedge (q \rightarrow r)$	$p \rightarrow r$	$((p \rightarrow q) \wedge (q \rightarrow r)) \rightarrow (p \rightarrow r)$
T	T	T	T	T	T	T	T
T	F	T	F	T	F	T	T
F	T	T	T	T	T	T	T
F	F	T	T	T	T	T	T
T	T	F	T	F	F	F	T
T	F	F	F	T	F	F	T
F	T	F	T	F	F	T	T
F	F	F	T	T	T	T	T

Since the truth values are T for all possible cases, statement (6-12) is valid.

EXAMPLE 6-20   Decide whether or not the following argument is valid: "If I practice hard and I have talent, then I will play major league baseball. If I become a major league baseball player, then I will be happy. Thus, if I will not be happy, then I did not practice hard or I do not have talent."

Solution   Assign symbols to the statements

p: "I practice hard" r: "I will play major league baseball"

q: "I have talent" s: "I will be happy"

To represent the argument, we form the symbolic statement

$$(p \wedge q \rightarrow r) \wedge (r \rightarrow s) \rightarrow (\sim s \rightarrow (\sim p \vee \sim q))$$

Setting up a four-variable statement in truth table form can lead to a long, complicated table, so let's first see if we can simplify our symbolic statement. Careful examination will show that we can make the transitivity substitution on the left and the DeMorgan substitution on the right to begin cutting this critter down to size.

$$((p \wedge q) \rightarrow s) \rightarrow (\sim s \rightarrow \sim (p \wedge q))$$

Now make the contrapositive substitution on the right, and we have

$$((p \wedge q) \rightarrow s) \rightarrow (p \wedge q \rightarrow s)$$

This is obviously a taulology, so the argument is valid (and we never even had to touch a truth table!).

EXAMPLE 6-21   Decide whether or not the following argument is valid: "I will get a new car if and only if I have a job. I am happy or I don't have a job. If I get a new car, then I will be popular. Thus, I am happy or I am popular."

Solution   Assign symbols to the statements:

p: "I will get a new car." r: "I am happy."

q: "I have a job." s: "I am popular."

We represent the argument by the symbolic statement

$$(p \leftrightarrow q) \wedge (r \vee \sim q) \wedge (p \rightarrow s) \rightarrow (r \vee s)$$

If we can assign truth values so that all of the *premises* are true and the *conclusion* is false, we will have shown that the argument is not valid.

For the conclusion $(r \vee s)$ to be false, r must be false and s must be false (by the disjunction truth table). Can all the premises be true under these conditions?

Since r is false, the second premise is true if $\sim q$ is true; that is, q is false.

Since s is false, the third premise is true if p is false.

Hence, for the conclusion to be false and the second and third premises to be true, we must collectively have: p false, q false, r false, and s false. This assignment does make the first premise $(p \leftrightarrow q)$ true. Thus the assignment of values p false, q false, r false, and s false makes all three premises true while the conclusion is false, which is what we set out to establish. Thus the argument is invalid.

EXAMPLE 6-22   Derive a truth table for the logical expression $(p \rightarrow \sim q) \leftrightarrow (r \rightarrow (q \wedge \sim p))$. Is this a valid statement?

Solution   The truth value of the expression depends on the truth values of the individual parts. Build the truth table one-half of the expression at a time:

		left side				right side		
p	q	r	$\sim q$	$p \rightarrow \sim q$	$\sim p$	$(q \wedge \sim p)$	$r \rightarrow (q \wedge \sim p)$	expression
T	T	T	F	F	F	F	F	T
T	F	T	T	T	F	F	F	F
F	T	T	F	T	T	T	T	T
F	F	T	T	T	T	F	F	F
T	T	F	F	F	F	F	T	F
T	F	F	T	T	F	F	T	T
F	T	F	F	T	T	T	T	T
F	F	F	T	T	T	F	T	T

The expression is not a tautology; therefore it is not valid.

6-2. Application to Programming

When constructing logical programming statements, you may find that one of the hardest parts of deciphering a complicated statement is dealing with the negatives inherent in it. For example, consider the WHILE–DO program fragment:

$$\text{WHILE } (M > 0 \text{ AND NOT DONE}) \text{ OR } M = 10 \text{ DO}$$

If we assign symbols to the "statements"

$\quad p\colon M > 0 \qquad q\colon \text{DONE} \qquad r\colon M = 10$

we can now have

$\quad \sim p\colon M \le 0 \qquad \sim q\colon \text{NOT DONE} \qquad \sim r\colon M \ne 10$

and we can express the compound statement as:

$$(p \wedge \sim q) \vee r \tag{6-13}$$

Sometimes a REPEAT–UNTIL loop is better than a WHILE–DO loop because it may be easier to program, may run faster, or may allow more efficient connection to other parts of a program. By negating the WHILE–DO loop, we can then obtain a REPEAT–UNTIL loop:

$$\text{REPEAT}$$
$$\text{UNTIL NOT } ((M > 0 \text{ AND NOT DONE}) \text{ OR } M = 10) \tag{6-14}$$

The compound statement here is expressed symbolically as

$$\sim ((p \wedge \sim q) \vee r) \tag{6-15}$$

But this has a "nested not," a negation symbol nested in parentheses with another "not" controlling it. This is undesirable, so we will look for a logical equivalent and try to simplify the statement. First use DeMorgan's Law $[\sim(A \vee B) \leftrightarrow \sim A \wedge \sim B]$

$\quad \sim ((p \wedge \sim q) \vee r) \leftrightarrow \sim (p \wedge \sim q) \wedge \sim r \quad$ Use DeMorgan here.

$\qquad\qquad\qquad \leftrightarrow (\sim p \vee \sim \sim q) \wedge \sim r \quad$ Use the double negation property.

$\qquad\qquad\qquad \leftrightarrow (\sim p \vee q) \wedge \sim r$

On the right side we no longer have "nested nots," which usually makes the statement easier to understand than (6-15). So in the program fragment (6-14), we may now make the replacement

$$\text{REPEAT}$$
$$\text{UNTIL } (M \le 0 \text{ OR DONE}) \text{ AND } M \ne 10$$

Instead of checking the logical equivalence established above by way of the laws, etc., you could check it directly by using a truth table.

EXAMPLE 6-23 Show that the replacement we made in the program fragment above is a valid replacement. That is, show that

$$\sim [(p \wedge \sim q) \vee r] \leftrightarrow (\sim p \vee q) \wedge \sim r$$

p	q	r	$p \wedge \sim q$	$(p \wedge \sim q) \vee r$	$\sim[(p \wedge \sim q) \vee r]$	$(\sim p \vee q)$	$\sim r$	$(\sim p \vee q) \wedge \sim r$
T	T	T	F	T	F	T	F	F
T	F	T	T	T	F	F	F	F
F	T	T	F	T	F	T	F	F
F	F	T	F	T	F	T	F	F
T	T	F	F	F	T	T	T	T
T	F	F	T	T	F	F	T	F
F	T	F	F	F	T	T	T	T
F	F	F	F	F	T	T	T	T

$\qquad\qquad\qquad\qquad\qquad\qquad\qquad\llcorner\text{————— same —————}\lrcorner$

SUMMARY

1. A proposition is a statement that is either true or false but not both. A proposition is valid if we can assign a truth value to it.
2. Sometimes statements are combined to form compound statements. Special connective words are used to express such built-up statements.
3. The three main connectives are "and" (conjunction), "or" (disjunction), and "not" (negation); they use the special symbols $\wedge$, $\vee$, $\sim$, respectively.
4. A truth table is a chart that identifies the truth values of a compound statement for individual cases.
5. Additional important connectives are built from the basic three. These include implication ($p \rightarrow q$), converse ($q \rightarrow p$), equivalence ($p \leftrightarrow q$), and contrapositive ($\sim q \rightarrow \sim p$). Each connective has its own truth table.
6. A tautology is a statement that is always true. Its opposite is a contradiction or absurdity.
7. Some of the important tautologies include the commutative, associative, and distributive properties for logic and DeMorgan's Laws.
8. Tautologies are very useful in establishing the validity of an argument.
9. The properties of logical reasoning can often aid in restructuring programming statements to make the execution of the statement more efficient. Removing "nested not" statements is one particularly useful application.

RAISE YOUR GRADES

Can you ...?

☑ define what a proposition is
☑ decide if a given statement is a valid proposition
☑ define what a paradox is
☑ combine simple statements to form complex statements
☑ give the meanings of the basic connectives
☑ form the negation of a simple or complex statement
☑ form the conjunction or disjunction of simple statements
☑ write the truth tables for the basic connectives
☑ define and denote the additional connectives and form their truth tables
☑ define a tautology and illustrate it
☑ write down the idempotent tautologies
☑ write down DeMorgan's Laws and establish their validity via a truth table
☑ use tautologies to verify an argument as valid or invalid
☑ use logical replacement to simplify computer programming statements

SOLVED PROBLEMS

Connectives and Truth Tables

PROBLEM 6-1   Decide whether or not each of the following are valid propositions: (**a**) Fish swim. (**b**) The desert is ice-covered. (**c**) Snoopy is President of the USA. (**d**) This sentence is false. (**e**) Get lost! (**f**) $5 + 6 = 7$.

Solution   Remember that for a proposition to be valid, you must be able to assign a truth value to it.

(**a**) This is a proposition. The statement is true.
(**b**) This is a proposition; the statement is false.
(**c**) This is a proposition whose truth value is F.
(**d**) This is not a proposition. It is a paradox since for the sentence to be true it must have a truth value F and to be false it must have a truth value T.
(**e**) This cannot be true or false. It is not a proposition.
(**f**) This is a proposition that is false.

PROBLEM 6-2 Write the negation of the following compound statements: (**a**) Ken went skiing or he went to the mall. (**b**) Deb didn't call last night and she didn't pick up her papers.

Solution

(**a**) Label the individual propositions as:

u: Ken went skiing.

v: Ken went to the mall.

We have $u \vee v$. By DeMorgan's Law the negation of this statement is

$$\sim(u \vee v) \leftrightarrow \sim u \wedge \sim v$$

We write this result as

"Ken did not go skiing and he did not go to the mall."

(**b**) Let the propositions be labeled as:

w: Deb called last night.

x: Deb picked up her papers.

We have $\sim w \wedge \sim x$. By DeMorgan's Law the negation of this statement is

$$\sim(\sim w \wedge \sim x) \leftrightarrow \sim(\sim w) \vee \sim(\sim x)$$
$$\leftrightarrow \quad w \quad \vee \quad x$$

Thus we write the result as:

"Deb called last night or she picked up her papers."

PROBLEM 6-3 Write the following argument in symbol notation:

If you take the Tribune, then you are well informed. You take the Tribune. Therefore, you are well informed.

Solution Label the statements as follows:
p: You take the Tribune.
q: You are well informed.

We then have:

If p, then q; p; therefore q.

In symbol notation, we therefore have:

$$p \rightarrow q$$
$$\underline{p \qquad\quad}$$
$$\therefore q$$

recall: "Therefore" is written as $\therefore$ in mathematics.

PROBLEM 6-4 Form the negation of the following statements: (**a**) Wilt is tall. (**b**) Some cats meow. (**c**) All birds have feathers. (**d**) All primes are odd. (**e**) Kirsten is smart and Kim is thin.

Solution Negation means the opposite of the statement or, "It is not the case that...."

(a) "Wilt is not tall."
(b) "It is not the case that some cats meow"; that is, "No cats meow."
(c) "Some birds have no feathers."
(d) "Some primes are not odd."
(e) "Kirsten is dumb or Kim is fat." (Check this using DeMorgan's Law, as in Problem 6-2a.)

PROBLEM 6-5 For each of the following pairs of statements write the conjunction and the disjunction. Then decide the truth value of these new statements.

(a) *p*: Water is wet.   *q*: Five is an odd integer.
(b) *p*: Elephants have six legs.   *q*: Lions are furry.
(c) *p*: California is an eastern state.   *q*: A trumpet is a stringed instrument.

Solution

(a) $p \wedge q$: "Water is wet and five is an odd integer." Since both *p* and *q* are true, this is true.
   $p \vee q$: "Water is wet or five is an odd integer." Since both *p* and *q* are true, this is true.
(b) $p \wedge q$: "Elephants have six legs and lions are furry." Since *p* is false, this is false.
   $p \vee q$: "Elephants have six legs or lions are furry." Since *q* is true, this is true.
(c) $p \wedge q$: "California is an eastern state and a trumpet is a stringed instrument." This is false since both *p* and *q* are false.
   $p \vee q$: "California is an eastern state or a trumpet is a stringed instrument." This is false since both *p* and *q* are false.

PROBLEM 6-6 Given the set of statements

   p: I have a penny.

   q: I eat spinach.

   r: I am strong.

   s: Some apples are rotten.

   t: I am rich.

form the compound statements **(a)** $p \rightarrow t$ **(b)** $q \wedge r$ **(c)** $\sim q \wedge r$ **(d)** $\sim s$ **(e)** $p \vee \sim t$ **(f)** $r \leftrightarrow q$.

Solution Write out the statements, remembering that $p \rightarrow q$ means "if *p*, then *q*"; $p \wedge q$ means "*p* and *q*"; $p \vee q$ means "*p* or *q*"; $\sim p$ means "not *p*"; and $p \leftrightarrow q$ means "*p* if and only if *q*."

(a) $p \rightarrow t$: "If I have a penny, then I am rich."
(b) $q \wedge r$: "I eat spinach and I am strong."
(c) $\sim q \wedge r$: "I do not eat spinach and I am strong."
(d) $\sim s$: "No apples are rotten." (i.e., "It is not the case that some apples are rotten.")
(e) $p \vee \sim t$: "I have a penny or I am not rich."
(f) $r \leftrightarrow q$: "I am strong if and only if I eat spinach."

PROBLEM 6-7 Use the implication truth table to decide on the truth of the following statements:

(a) If the moon is solid gold, then Abe Lincoln was president.
(b) If $2 + 4 = 7$, then $8 - 5 = 1$.
(c) If Ohio is a state, then Lake Michigan is dry.

Solution The implication truth table is

p	q	$p \rightarrow q$
T	T	T
T	F	F
F	T	T
F	F	T

(a) p is F and q is T, so $p \rightarrow q$ is T.
(b) p is F and q is F, so $p \rightarrow q$ is T.
(c) p is T and q is F, so $p \rightarrow q$ is F.

PROBLEM 6-8 Form the converse and the contrapositive of the statements: (a) If I have been speeding, then I will pay a fine. (b) If I get paid, I will go out on Saturday.

Solution Recall that the converse is "if q, then p" and the contrapositive is "if not q, then not p."

(a) Converse: "If I pay a fine, then I've been speeding."
Contrapositive: "If I don't pay a fine, then I've not been speeding."
(b) Converse: "If I go out on Saturday, then I got paid."
Contrapositive: "If I don't go out on Saturday, then I didn't get paid."
Note that the contrapositive is valid when its original statement is valid, but that the converse is not necessarily valid when its original statement is valid.

PROBLEM 6-9 Let P be the proposition

> If it is snowing, then Mark is studying.

and let Q be the proposition

> If Dan passes, then the moon is green cheese.

Form the converse and contrapositive of both P and Q.

Solution If proposition P is written $a \rightarrow b$, then the converse is $b \rightarrow a$ and the contrapositive is $\sim b \rightarrow \sim a$. Therefore we have

$b \rightarrow a$: "If Mark is studying, then it is snowing."

$\sim b \rightarrow \sim a$: "If Mark is not studying, then it is not snowing."

If proposition Q is written $c \rightarrow d$, then

$d \rightarrow c$: "If the moon is green cheese, then Dan passes."

$\sim d \rightarrow \sim c$: "If the moon is not green cheese, then Dan does not pass."

PROBLEM 6-10 Show the equivalence of the statements

$$(p \vee q) \rightarrow r \quad \text{and} \quad (p \rightarrow r) \wedge (q \rightarrow r)$$

Solution Construct the truth table for each one, and check to see whether or not the truth values for all cases of p, q, and r are the same for both statements.

p	q	r	$p \vee q$	$(p \vee q) \rightarrow r$	$p \rightarrow r$	$q \rightarrow r$	$(p \rightarrow r) \wedge (q \rightarrow r)$
T	T	T	T	T	T	T	T
T	F	T	T	T	T	T	T
F	T	T	T	T	T	T	T
F	F	T	F	T	T	T	T
T	T	F	T	F	F	F	F
T	F	F	T	F	F	T	F
F	T	F	T	F	T	F	F
F	F	F	F	T	T	T	T

↑———— same ————↑

The truth values for all cases p, q, and r are the same. Thus the statements are equivalent.

PROBLEM 6-11 Find the truth value of the following statements when p is true and q is false:
(a) $(\sim p \vee q) \wedge \sim p$ (b) $\sim (p \vee q) \wedge (\sim p \vee (p \wedge q))$.

Solution   Use the truth tables.

(a) Since p is T, then $\sim p$ is F. We also have q is F. Thus $\sim p \vee q$ is F (according to the truth table for disjunction) so $(\sim p \vee q) \wedge \sim p$ is also F (according to the truth table for conjunction).

(b) Since p is T and q is F, we have $p \vee q$ is T and $p \wedge q$ is F. Thus $\sim(p \vee q)$ is F. Also $\sim p \vee (p \wedge q)$ is F. Since we have a false statement $\sim(p \vee q)$ in disjunction with another false statement $(\sim p \vee (p \wedge q))$, the entire expression is false, according to the disjunction truth table.

PROBLEM 6-12   Find the truth value when p is false, q is true, and r is false for the following expressions: **(a)** $\sim(p \vee \sim q) \wedge \sim r$ **(b)** $(p \vee q) \vee [(r \wedge \sim q) \vee (\sim r \wedge p)]$.

Solution   Use the truth tables and take one piece at a time.

(a) Since p is F and $\sim q$ is F, then $p \vee \sim q$ is F (disjunction table), so $\sim(p \vee \sim q)$ is T (negation table). But $\sim r$ is T so the whole expression is T, by the conjunction table.

(b) Under the conditions given, $(p \vee q)$ is T; $(r \wedge \sim q)$ is F; $(\sim r \wedge p)$ is F; $[(r \wedge \sim q) \vee (\sim r \wedge p)]$ is F. So the entire expression is T.

PROBLEM 6-13   Decide on the validity of the following argument:

> If Franklin D. Roosevelt was assassinated, then he is dead. Therefore, if he is dead, then he was assassinated.

Solution   The argument is not valid. If we label the statements

   p: F. D. R. was assassinated.

   q: He is dead.

the argument may be expressed as

> If p, then q. Therefore if q, then p.

Symbolically, this is

$$p \to q$$
$$\therefore q \to p$$

which is a violation of the converse property; the argument is invalid.

PROBLEM 6-14   Rewrite the following as one statement: **(a)** If a polygon is a triangle, then it has three sides. **(b)** If a polygon has three sides, then it is a triangle.

Solution   This is known as a *biconditional*, and we write

> "A polygon is a triangle if and only if it has three sides."

In symbols this is $p \leftrightarrow q$ (the equivalence connective) and we say that p is logically equivalent to q. The truth table for the biconditional (equivalence) is

p	q	$p \leftrightarrow q$
T	T	T
T	F	F
F	T	F
F	F	T

PROBLEM 6-15   The expression $[(p \to q) \wedge \sim q] \to \sim p$ is called *modus tollens*. Show that it is a tautology.

Solution   Create the truth table and see if the expression is true for all cases p and q.

p	q	$p \rightarrow q$	$\sim q$	$(p \rightarrow q) \wedge \sim q$	$\sim p$	$[(p \rightarrow q) \wedge \sim q] \rightarrow \sim p$
T	T	T	F	F	F	T
T	F	F	T	F	F	T
F	T	T	F	F	T	T
F	F	T	T	T	T	T

↑
True for all cases, so
it is a tautology.

Here's an example of a modus tollens statement: [("If it is raining, I will take my umbrella") and "I do not take my umbrella"] imply that ("it is not raining.")

PROBLEM 6-16   The sentence $[p \wedge (p \rightarrow q)] \rightarrow q$ is called *modus ponens*. Show that it is a tautology.

Solution   Build the truth table.

p	q	$p \rightarrow q$	$p \wedge (p \rightarrow q)$	q	$[p \wedge (p \rightarrow q)] \rightarrow q$
T	T	T	T	T	T
T	F	F	F	F	T
F	T	T	F	T	T
F	F	T	F	F	T

↑
It is a tautology.

Here's an example of a modus ponens statement: ["I got a B on the final exam" and ("If I get a B on the final exam, I will pass the course")] imply that ("I will pass the course.")

PROBLEM 6-17   Show that $\sim(\sim p \wedge q) \rightarrow (p \vee \sim q)$ is a tautology.

Solution   Create the truth table.

p	q	$\sim p$	$\sim p \wedge q$	$\sim(\sim p \wedge q)$	$\sim q$	$p \vee \sim q$	$\sim(\sim p \wedge q) \rightarrow (p \vee \sim q)$
T	T	F	F	T	F	T	T
T	F	F	F	T	T	T	T
F	T	T	T	F	F	F	T
F	F	T	F	T	T	T	T

↑
It is a tautology.

PROBLEM 6-18   Decide on the validity of the following arguments:

(a) If Myk takes the cake, then Gord will take the ice cream. Gord does not take the ice cream. Hence Myk does not take the cake.
(b) If you play backgammon, then you are intelligent. You play backgammon. Hence you are intelligent.

Solution

(a) The propositions may be represented in symbols as p: "Myk takes the cake"; q: "Gord takes the ice cream." Thus the argument, in symbols, is $[(p \rightarrow q) \wedge \sim q] \rightarrow \sim p$. But this is modus tollens, so it is a valid argument.
(b) These propositions may be represented in symbols as r: "You play backgammon"; s: "You are intelligent." In symbols, the argument becomes $[(r \rightarrow s) \wedge r] \rightarrow s$. But this is modus ponens and it is a valid argument.

PROBLEM 6-19   Test the validity of the argument:

If you have an engineering degree, you will make a lot of money. You do
not have an engineering degree. Thus you will not make a lot of money.

Solution   Use the symbols p: "You have an engineering degree"; q: "You will make a lot of money." The
argument is

$$p \rightarrow q$$
$$\frac{\sim p}{\therefore \sim q}$$

This is a false argument, as a table can verify:

p	q	$p \rightarrow q$	$\sim p$	$(p \rightarrow q) \wedge \sim p$	$\sim q$	$(p \rightarrow q) \wedge \sim p \rightarrow \sim q$
T	T	T	F	F	F	T
T	F	F	F	F	T	T
F	T	T	T	T	F	F
F	F	T	T	T	T	T

↑
Not true for all
cases, therefore we
do not have a
tautology.

The argument is false.

PROBLEM 6-20   Determine if the following argument is valid, assuming the Hawkeyes are playing
the Wolverines:

If the Iowa Hawkeyes lose the game, then the Michigan Wolverines win.
If Michigan wins, then they will go to the Rose Bowl. Therefore, if Iowa
loses, then Michigan goes to the Rose Bowl.

Solution   Use symbols to express the statements:

   p: "Iowa loses the game."
   q: "Michigan wins the game."
   r: "Michigan goes to the Rose Bowl."

The argument then is

$$p \rightarrow q$$
$$\frac{q \rightarrow r}{\therefore p \rightarrow r}$$

This is the transitivity property and hence is a valid argument.

PROBLEM 6-21   Determine if the following argument is valid:

If Susie uses Hotshot mouthwash, then she will have fewer cavities. Thus,
if Susie has fewer cavities, then she uses Hotshot mouthwash.

Solution   Use symbols to express the statements.

   p: "Susie uses Hotshot mouthwash."
   q: "Susie has fewer cavities."

The argument then is

$$\frac{p \to q}{\therefore q \to p}$$

This is a false argument. The converse of a given valid statement is not necessarily valid.

PROBLEM 6-22 Using the statements

    p: Tom passed the course.

    q: Susie received an A.

    s: Jimmy failed the course.

form the composite statements (**a**) $\sim p \wedge \sim s$ (**b**) $\sim (p \vee q)$ (**c**) $p \wedge (q \vee s)$.

Solution

(**a**) "Tom did not pass the course and Jimmy did not fail the course."

(**b**) We may use DeMorgan's Law to rewrite this as

$$\sim (p \vee q) \leftrightarrow \sim p \wedge \sim q$$

Hence we may phrase this in two ways:

        "It is not true that Tom passed the course or Susie received an A."

or

        "Tom did not pass the course and Susie did not receive an A."

(**c**) The distributive law allows us to rewrite this as:

$$p \wedge (q \vee s) \leftrightarrow (p \wedge q) \vee (p \wedge s)$$

Hence we may phrase this in two ways:

        "Tom passed the course and (Susie received an A or Jimmy failed the course)."

or

        "(Tom passed the course and Susie received an A) or (Tom passed the course and Jimmy failed the course)."

PROBLEM 6-23 Form a truth table to show that the following argument is invalid:

$$[(p \to q) \wedge (p \to r)] \to (q \to r)$$

Solution

p	q	r	$p \to q$	$(p \to r)$	$(p \to q) \wedge (p \to r)$	$q \to r$	$[(p \to q) \wedge (p \to r)] \to (q \to r)$
T	T	T	T	T	T	T	T
T	F	T	F	T	F	T	T
F	T	T	T	T	T	T	T
F	F	T	T	T	T	T	T
T	T	F	T	F	F	F	T
T	F	F	F	F	F	T	T
F	T	F	T	T	T	F	F
F	F	F	T	T	T	T	T

                                              ↑

The result is not always true, so the argument is invalid.

To illustrate this fallacy, consider the following:

> If we learn to swim, we will go to Hawaii. If we learn to swim, we will buy a boat. Therefore, if we go to Hawaii, then we will buy a boat.

This surely is not true for all visitors to our 50th state!

Application to Programming

PROBLEM 6-24   Suppose that in structuring a computer program we encounter the segment

1. IF $(p \wedge (p \vee q)) \wedge r$ THEN
   a. [sequence A of instructions]
2. ELSE
   a. [sequence B of instructions]

where p, q, and r are logical conditions. (For example, we could have

$$p: x > 5, q: y \leq 0, \text{ and } r: z = 10.)$$

If we can rewrite the expression in simpler form, the computer will do less work in making the check in the IF-THEN-ELSE expression.
Use a truth table to find a way to simplify the IF condition.

Solution   Form the table for the "mixed portion" $p \wedge (p \vee q)$:

p	q	$p \vee q$	$p \wedge (p \vee q)$
T	T	T	T
T	F	T	T
F	T	T	F
F	F	F	F

↑———— same ————↑

Note that the result matches p exactly. Thus we have an equivalence $p \wedge (p \vee q) \leftrightarrow p$, and we can replace $p \wedge (p \vee q)$ with p. We can now rewrite the original expression as

1. IF $p \wedge r$ THEN
   a. [sequence A of instructions]
2. ELSE
   a. [sequence B of instructions]

PROBLEM 6-25   Suppose a program contains the statement

$$((X \text{ NOT LESS THAN } 4) \quad \text{OR} \quad (Y \text{ NOT GREATER THAN } 6))$$
$$\text{AND} \quad (Z \text{ NOT LESS THAN } 9)$$

Find the truth value of the statements created by inserting the given values of the variables:

(a) $X = 0$, $Y = 13$, $Z = 0$
(b) $X = 5$, $Y = 5$, $Z = 5$
(c) $X = 10$, $Y = 5$, $Z = 10$

Solution   First give names to the statement components:

  p: X NOT LESS THAN 4 $[X \geq 4]$
  q: Y NOT GREATER THAN 6 $[Y \leq 6]$
  r: Z NOT LESS THAN 9 $[Z \geq 9]$

The program statement is $(p \lor q) \land r$. Form the truth table for this statement and identify the given cases:

p	q	r	$p \lor q$	$(p \lor q) \land r$	
T	T	T	T	T	⟵ case (c)
T	F	T	T	T	
F	T	T	T	T	
F	F	T	F	F	
T	T	F	T	F	⟵ case (b)
T	F	F	T	F	
F	T	F	T	F	
F	F	F	F	F	⟵ case (a)

(a) If $X = 0$, p is F; if $Y = 13$, q is F; if $Z = 0$, r is F. The statement is false.
(b) If $X = 5$, p is T; if $Y = 5$, q is T; if $Z = 5$, r is F. The statement is false.
(c) If $X = 10$, p is T; if $Y = 5$, q is T; if $Z = 10$, r is T. The statement is true.

PROBLEM 6-26   Suppose your computer program contains the statement

IF (((U GREATER THAN 4) OR (V NOT LESS THAN 8))

AND NOT ((U GREATER THAN 4) AND (V NOT LESS THAN 8)))

THEN NOT ((U GREATER THAN 4) OR (V NOT LESS THAN 8))

Show that this may be replaced by the simpler statement

IF ((U GREATER THAN 4) OR (V NOT LESS THAN 8))

THEN ((U GREATER THAN 4) AND (V NOT LESS THAN 8))

Solution   Give symbols to the sub-parts:

p: *U* GREATER THAN 4 q: *V* NOT LESS THAN 8

The first statement is

$$(p \lor q) \land \sim(p \land q) \to \sim(p \lor q)$$

Its truth table is

p	q	$p \lor q$	$p \land q$	$\sim(p \land q)$	$(p \lor q) \land \sim(p \land q)$	$\sim(p \lor q)$	$(p \lor q) \land \sim(p \land q) \to \sim(p \lor q)$
T	T	T	T	F	F	F	T
T	F	T	F	T	T	F	F
F	T	T	F	T	T	F	F
F	F	F	F	T	F	T	T

The second, simpler, statement is

$$(p \lor q) \to (p \land q)$$

Its truth table is:

p	q	$p \lor q$	$p \land q$	$(p \lor q) \to (p \land q)$
T	T	T	T	T
T	F	T	F	F
F	T	T	F	F
F	F	F	F	T

Since the last columns of the two truth tables are the same, the two statements are equivalent. We may therefore replace the more complex statement with the simpler one.

Supplementary Exercises

PROBLEM 6-27 Which of the following are valid propositions?

(a) George Washington was our second President.
(b) All cats are animals.
(c) Detroit is the capital of Michigan.
(d) Happy Birthday to you!
(e) San Diego is a beautiful city.
(f) The sun rises in the east.

PROBLEM 6-28 Negate each of the following statements:

(a) All lawyers are rich.
(b) Some movie stars are not famous.
(c) Today is not Monday.
(d) It isn't true that I didn't go fishing.

PROBLEM 6-29 Negate each of the following compound statements:

(a) I like chicken and you like fish.
(b) I will not go to work or you will dine alone.

PROBLEM 6-30 Using truth tables, prove the following properties.

(a) $p \vee q \leftrightarrow q \vee p$ Commutative property
(b) $p \vee p \leftrightarrow p$ and $p \wedge p \leftrightarrow p$ Idempotent properties
(c) $p \vee (p \wedge q) \leftrightarrow p$ Absorption law
(d) $p \wedge (q \wedge r) \leftrightarrow (p \wedge q) \wedge r$ Associative property
(e) $(p \vee q) \rightarrow r \leftrightarrow (p \rightarrow r) \wedge (q \rightarrow r)$

PROBLEM 6-31 Create the truth table for the following statements and identify them as a tautology, contradiction, or neither.

(a) $p \wedge q \rightarrow p \vee q$
(b) $\sim p \wedge q \leftrightarrow p \vee \sim q$
(c) $((p \vee q) \rightarrow r) \leftrightarrow r \wedge \sim q$

PROBLEM 6-32 Write each of the following arguments in symbols and test its validity:

(a) If I go to the prom, then if I rent a tux, it will be white. I will rent a tux. Therefore, if I go to the prom, the tux I rent will be white.
(b) Anyone who scored over 70 on all the quizzes, passes the course. Carlos passed the course. Therefore, Carlos scored over 70 on all the quizzes.

PROBLEM 6-33 Construct the truth tables for $p \wedge (\sim p \vee q)$ and $p \wedge q$. Are they equivalent?

PROBLEM 6-34 Write the converse and the contrapositive of the following statements:

(a) If I get paid Friday, I will go to the movies.
(b) If it is not raining, then I will go to the beach.

PROBLEM 6-35 Decide if the given pairs of statements are logically equivalent:

(a) $(\sim p \vee q) \wedge (\sim q \vee p)$ and $p \leftrightarrow q$
(b) $\sim(p \wedge q)$ and $(\sim p) \vee (\sim q)$
(c) $(p \wedge \sim q)$ and $\sim p \rightarrow q$

PROBLEM 6-36   Decide on the validity of the following arguments:

(a) If T. J. bought a boat, then either he sold his motorcycle or he borrowed from the credit union. T. J. did not borrow from the credit union. Therefore, if he has not sold his motorcycle, then he has not bought a boat.

(b) If it snows today, then the TV weatherperson is right or the newspaper is wrong. It did not snow today, or the TV person is wrong. Therefore the newspaper is right.

PROBLEM 6-37   Their is five errers in this sentense. See if you can find all of them.

Answers to Supplementary Exercises

6-27   (a), (b), (c), and (f) are propositions.

6-28   (a) Some lawyers are not rich.
    (b) All movie stars are famous.
    (c) Today is Monday.
    (d) It is true that I didn't go fishing.

6-29   (a) I do not like chicken or you do not like fish.
    (b) I will go to work and you will not dine alone.

6-31   (a) This is a tautology.
    (b) This is a contradiction.
    (c) This is neither.

6-32   (a)

$$p \to (q \to r)$$
$$q$$
$$\overline{\therefore p \to r}$$ It is valid.

    (b)
$$p \to q$$
$$q$$
$$\overline{\therefore p}$$ This is not valid.

6-33   This is a tautology.

6-34   (a) Converse: If I go to the movies, then I've been paid. Contrapositive: If I don't go to the movies, then I've not been paid.
    (b) Converse: If I go to the beach, then it is not raining. Contrapositive: If I don't go to the beach, then it is raining.

6-35   (a) equivalent   (b) equivalent
    (c) not equivalent.

6-36   (a) valid   (b) invalid.

6-37   Spelling errors: there, errors, sentence.
Grammatical error: "is" should be "are".
Logical error: there are four (not five) errors.

7 METHODS OF PROOF

THIS CHAPTER IS ABOUT

☑ **Problem-Solving Strategy**
☑ **Direct Proofs**
☑ **Proof of the Contrapositive**
☑ **Proof by Contradiction**
☑ **Proof by Counter-example**
☑ **Proof by Induction**

7-1. Problem-Solving Strategy

This chapter provides the details of some of the commonly used methods of solving problems and proving theorems. Since it is by no means an exhaustive study, we will concentrate on a few very basic techniques. To be a well-grounded student of the sciences you will need all of these techniques on many occasions, so you may find yourself referring back to this chapter frequently.

A **theorem**, or conjecture, is simply a collection of statements $H_1, H_2, \ldots, H_j$, collectively called **hypotheses**, that lead to—or imply—a conclusion C. The hypotheses are given information. Using logic symbolism (see Chapter 6), we can write a theorem as

$$(H_1 \wedge H_2 \wedge H_3 \wedge \cdots \wedge H_j) \to C \tag{7-1}$$

A theorem is **valid** if the collection of hypotheses does indeed imply the conclusion. A theorem is **invalid**, or a **fallacy**, if for some reason the hypotheses do not lead to the conclusion. To *prove a theorem* is to show that it is valid (or, as the case may be, that it is invalid).

Now, we're well aware that the word "prove" can sometimes send a student into paroxysms of panic. There is really very little to be afraid of. "Prove" only means to solve a problem, that is, to find a way out of a difficult situation by applying some previously tried steps and a bit of experimentation. Skill at doing proofs or solving problems comes from a thorough understanding of the basic principles of the discipline plus a lot of practice. It means you must study the problem to gain insight into what it's really asking for, and then expend effort to solve it, all the while focusing on the goal of the problem. In short, you need to decide where to start and what your goal is, and then fill in the details of the in-between. You will eventually find that there is a direct analogy between "doing proofs" and "writing programs," for they both address the fundamental task of problem solving.

There are three basic steps in the process of problem solving:

(1) *Discover*—read all the words of the problem, review technical terms, and identify the goal of the problem.
(2) *Experiment*—change the words of the problem into equations or formulas, replace expressions with equivalent items, consider alternative paths to the goal, and reduce the complexity of the problem possibly by subdividing the tasks.
(3) *Formulate*—put your ideas into a pattern, apply some logic, focus on the goal, and identify the best path to the goal.

In working any problem or doing any proof, it is important that you make each step follow logically from one or more of the preceding statements. If any of your hypotheses or intermediate statements are invalid, you have a fallacy, and the proof in question is also invalid. Sometimes it is very difficult to determine if and where a fallacy exists; if none is found, the proof is usually considered valid. To help avoid fallacies, you must consider all possible cases in a given problem and justify all conclusions you make. Avoiding ambiguity and unwarranted assumptions is crucial to effective problem solving.

Many different methods exist for doing proofs. In the following sections we will explore the most common ones:

(a) Direct proof
(b) Proof of the contrapositive
(c) Proof by contradiction
(d) Proof by counter-example
(e) Proof by induction

A Word of Encouragement

Working with proofs takes a lot of experimentation and often many false starts. In this chapter we will not explore how to prove specific theorems, but we *will* offer what processes are available to attack a proof, explain how they work, demonstrate that they do work, and let you absorb them into your scientific thinking process.

Learning proofs is somewhat like learning to drive. You can absorb all the general regulations, methods, situations, and solutions from a driver's manual, but the manual surely can't describe what to do at every instant you drive a vehicle. To be successful in driving, you must get out there and do it! At first the neophyte is petrified. Then a little experience is gained. Then more and varied situations are encountered, and more experience is gained, etc. Eventually an expert emerges. And so it is with learning proofs. Learn the basics first, then practice, practice, practice!

7-2. Direct Proofs

In a **direct proof** of $(H_1 \wedge H_2 \wedge H_3 \wedge \cdots \wedge H_j) \to C$, you must assume that the collection of H_i (where $i = 1, 2, 3, \ldots, j$) is true. Then, applying these and other facts available to you, you must build a chain of logical reasoning leading to the conclusion that C is true. Note that this type of proof does not show that C is true but shows that C *has to be true if all the H_i are true*.

note: If $(H_1 \wedge H_2 \wedge \cdots \wedge H_j) \to C$ is a tautology, then it is always true regardless of the truth value of any part (see chapter 6).

The facts available for a direct proof usually fall into two categories:

(a) purely logical facts
(b) facts that are related to the subject matter with which the problem deals

Another way to organize a direct proof is in the following fashion:

$$
\begin{array}{r}
p_1 \\
p_2 \\
p_3 \\
\vdots \\
\underline{p_n} \\
\therefore q
\end{array}
\qquad \text{(7-2)}
$$

where the p statements are the hypotheses and other intermediate steps and q is the conclusion. You can think of statement (7-2) as a kind of addition problem—you "add" the hypotheses together to get the conclusion, or sum effect, of the hypotheses. Statement (7-2) is the same as saying, "If p_1 and p_2 and p_3 and $\cdots p_n$, then q," which can be written symbolically as $p_1 \wedge p_2 \wedge p_3 \wedge \cdots \wedge p_n \to C$, the equivalent of statement (7-1). Let's do a few examples to practice the direct proof.

EXAMPLE 7-1 Prove the following theorem by direct proof:

If I work hard, I will be rich. If I am rich, then I will be happy. Thus if I work hard, I will be happy.

Solution First identify what the theorem is saying, then set it up in logic notation. The hypotheses are the implications

$$\{\text{I work hard}\} \rightarrow \{\text{I am rich}\} \quad \text{and} \quad \{\text{I am rich}\} \rightarrow \{\text{I am happy}\}$$
$$(p \rightarrow r) \qquad\qquad\qquad (r \rightarrow t)$$

The conclusion is the implication

$$\{\text{I work hard}\} \rightarrow \{\text{I am happy}\}$$
$$(p \rightarrow t)$$

The entire argument is now symbolically

$$(p \rightarrow r) \wedge (r \rightarrow t) \rightarrow (p \rightarrow t)$$

But this is the transitive tautology (see Chapter 6), so the theorem is obviously true.

EXAMPLE 7-2 Prove the following theorem by direct proof:

If a number is divisible by 8, it is divisible by 4.

Solution This is a direct $p \rightarrow q$ theorem. We must rely on mathematical facts to prove it, so we follow the three steps of problem solving: identify, experiment, and formulate.

IDENTIFY:		
(1) Let x be any number divisible by 8	Hypothesis.
(2) $x = k \cdot 8$ for some integer k	Definition of divisibility.
(3) $8 = 2 \cdot 4$	Known number fact.
(4) $x = k \cdot (2 \cdot 4)$	Replacement by equal.
(5) $x = (k \cdot 2)(4)$	Multiplication fact.
(6) $k \cdot 2$ is an integer m	Integer fact.
(7) $x = m \cdot 4$	Replacement by equal.
(8) x is divisible by 4	Conclusion. The theorem is true.

(Steps 2–7 bracketed as EXPERIMENT and FORMULATE)

EXAMPLE 7-3 Prove the following theorem by direct proof:

Running is healthy. If running is healthy then medical insurance premiums are lower. Therefore medical insurance premiums are lower.

Solution This theorem in symbolic form

$$\begin{array}{c} p \\ p \rightarrow q \\ \hline \therefore q \end{array}$$

is in the form of *modus ponens* (the method of asserting), and the argument is valid. The conclusion may or may not be true, but the argument to obtain the conclusion is valid, so the theorem itself is valid.

note: A theorem is a collection of statements that lead to a conclusion. Even though the conclusion is false, the hypotheses may validly lead to that conclusion. Thus the theorem itself may be valid.

EXAMPLE 7-4 Give a direct proof of the following mathematical theorem:

Let $m, x_1, x_2,$ and y be real numbers. If $m = \{\text{minimum of } x_1, x_2\}$ and $y < m$, then $y < x_1$ and $y < x_2$.

Solution This is a $p \rightarrow q$ theorem, and we must again apply general mathematical principles to prove it.

Identify hypothesis H_1:	**(1)** $m \le x_1$ and $m \le x_2$	Definition of minimum.
H_2:	**(2)** $y < m$	Hypothesis.
Experiment and Formulate:	**(3)** $y < m \le x_1$ and $y < m \le x_2$	Math fact—combine known statements.
	(4) $y < x_1$ and $y < x_2$	Property of inequality; conclusion. The theorem is valid.

7-3. Proof of the Contrapositive

If the conjecture you are trying to prove doesn't seem solvable by a direct proof but you feel that it is true, you may try some variants of that method. You saw in Chapter 6 that $p \rightarrow q$ and $\sim q \rightarrow \sim p$ are equivalent and that the form $\sim q \rightarrow \sim p$ is the *contrapositive* of $p \rightarrow q$. We may therefore establish the validity of $p \rightarrow q$ by establishing $\sim q \rightarrow \sim p$. To do so, we assume that the hypothesis $\sim q$ is true and proceed to show that p is false (i.e., that $\sim p$ is true).

EXAMPLE 7-5   Use the contrapositive to prove the following theorem:

Let x and y be integers. If $x + y \geq 57$, then $x \geq 29$ or $y \geq 29$.

Solution   Write the contrapositive statement:

If not $(x \geq 29$ or $y \geq 29)$, then not $(x + y \geq 57)$

To prove this, we can alter the left side by DeMorgan's Law (Chapter 6):

$\{$not $(x \geq 29$ or $y \geq 29)\}$ is the same as $\{($not $x \geq 29)$ and (not $y \geq 29)\}$

which is the same as $\{x < 29$ and $y < 29\}$, which is the same as $\{x \leq 28$ and $y \leq 28\}$.
   Similarly, the right side $\{$not $(x + y \geq 57)\}$ is the same as $\{x + y < 57\}$ or $\{x + y \leq 56\}$. Hence the contrapositive statement becomes

If $(x \leq 28$ and $y \leq 28)$, then $x + y \leq 56$.

Now applying one of the basic rules of inequalities:

$$a \leq b \text{ and } c \leq d \text{ implies } a + c \leq b + d$$

we have

$$x \leq 28 \text{ and } y \leq 28 \text{ implies } x + y \leq 28 + 28 = 56$$

which is true. The contrapositive is thus valid, so the original theorem is also valid for all integers x and y.

EXAMPLE 7-6   Use the contrapositive to prove:

If the product of two positive integers a and b is N, then $a^2 \leq N$ or $b^2 \leq N$.

Solution   Suppose the negation of the conclusion, i.e., that $\sim[(a^2 \leq N) \vee (b^2 \leq N)] = (a^2 > N) \wedge (b^2 > N)$. Then $a^2 > N$ and $b^2 > N$. Thus $a^2 b^2 > N^2$ or $a \cdot b > N$ (by math facts), which is a negation of the hypothesis that $a \cdot b = N$. Because we assumed $\sim q$ and have shown that $\sim p$ follows, we have proved by the contrapositive method that the theorem is valid.

EXAMPLE 7-7   Using the contrapositive, prove the theorem:

Let N be a positive integer. If N is a prime integer other than 2, then N is odd.

Solution   Identify what the problem is asking. We have statements

p: "N is a prime integer and $N \neq 2$."
q: "N is odd."

and the theorem is $p \rightarrow q$. We must take the contrapositive of this; assume $\sim q$ is true, then prove that $\sim p$ is true. The contrapositive $\sim q \rightarrow \sim p$ is

$\sim q$: "N is even."
$\sim p$: "N is not prime or $N = 2$." $\hspace{4em}$ **(7-3)**

Since we assume $\sim q$, that is, that N is even, then $N = 2 \cdot M$ for some positive integer M that is less than N. Now if $M = 1$, then $N = 2$, and $N = 2$ is a valid conclusion by statement (7-3). If $M > 1$, then N is not a prime number since N is the product of 2 and M by $\sim q$; that is, N is divisible by $M > 1$ and is therefore not prime. Thus we have a valid conclusion in statement (7-3), so the contrapositive is valid. Thus the original theorem is valid.

7-4. Proof by Contradiction

Another method of proving that an implication $p \to q$ is valid is the method of **contradiction**. In this technique we combine the assumption "q is false" together with the premise "p is true" in the form $p \wedge (\sim q)$ and, show that this leads to a contradiction. This method is often called the *indirect method* and is based on the tautology

$$[(p \to q) \wedge (\sim q)] \to (\sim p)$$

which states that if a statement p implies q and q is a false statement, then p must also be false.

We apply the proof by contradiction in the following manner: To prove $H_1 \wedge H_2 \wedge \cdots \wedge H_j \to C$ we negate q (the conclusion C) and add it to p (the list of hypotheses). We therefore get:

$$H_1 \wedge H_2 \wedge \cdots \wedge H_j \wedge (\sim C) \tag{7-4}$$

We already know that $H_1 \wedge H_2 \wedge \cdots \wedge H_j$ is a true statement because we assumed that all hypotheses $H_1, H_2, \ldots, H_j$ are true. Therefore, statement (7-4) will also be true only if $(\sim C)$ is true. It will be false only if $(\sim C)$ is false. Turning this around, we can find whether $(\sim C)$ is true or false by checking the validity of statement (7-4). If it is a contradiction (always false), then $(\sim C)$ is false, the original conclusion C is true (by the negation truth tables), and we have proved the theorem. If, however, statement (7-4) is not a contradiction, then $(\sim C)$ is true either all of the time (if statement 7-4 is a tautology) or some of the time (in certain cases of p and q). In these ways, we will have disproved the original theorem by finding that $(\sim C)$ is true (at least some of the time), and therefore that C is false (at least some of the time).

note: If you find C is false some of the time, you may have to rewrite the theorem by altering the hypotheses so they are more limiting, and thus make the theorem true all of the time.

EXAMPLE 7-8   Prove that if $xy \geq 4$, then $x \geq 2$ or $y \geq 2$.

Solution   Using the method of contradiction, we assume the negation of the conclusion and combine it with the hypothesis. We then have

$$\sim(x \geq 2 \text{ or } y \geq 2) \wedge (xy \geq 4)$$

or
$$(x < 2 \text{ and } y < 2) \wedge (xy \geq 4) \tag{7-5}$$

If our theorem is true, then statement (7-5) will be a contradiction. Calling on the laws of inequalities, we have

If $x < 2$ and $y < 2$, then $xy < 2 \cdot 2$ or $xy < 4$.

Substituting this into statement (7-5), we can simplify it to

$$(xy < 4) \wedge (xy > 4)$$

which is indeed a contradiction. Our negated conclusion must therefore be false, and the original conclusion must be true. Thus the theorem is valid.

EXAMPLE 7-9   Use proof by contradiction to show that if 2 divides N, then N is even.

Solution   Negate the conclusion to get $\sim(N$ is even$)$, which is equivalent to (N is odd). Combine this with the hypothesis to get

$$(2 \text{ divides } N) \wedge (N \text{ is odd})$$

Since N is odd, we may write it as $N = 2M + 1$. But a mathematics fact indicates that if 2 divides N, then $N = 2P$ where P is an integer. Thus

$$2M + 1 = 2P$$
$$2P - 2M = 1$$
$$2(P - M) = 1 \tag{7-6}$$

But since P and M are integers, so is $P - M$. Thus the left side of statement (7-6) is always even while the right side is never even (it is 1 and is always odd). This is a contradiction; thus the theorem is true.

Some problems will not require you to manipulate math symbols.

EXAMPLE 7-10   Use proof by contradiction to show that if an elementary school class has 25 students then there are at least 3 birthdays in some month.

Solution   The negation of the conclusion (that there are at least 3 birthdays in some month) is that each month has at most 2 birthdays among these students. Then in a year, if every month has the maximum of two birthdays, the class can contain a maximum of 24 persons. Combining this with the hypothesis that the class has 25 members provides a contradiction. The theorem is true.

Some theorems can become quite long and involved because they have a lot of interrelated statements. You can sometimes make a complicated theorem easier to prove by setting it up in the stacking format (see statement 7-2).

EXAMPLE 7-11   Prove the following conjecture by contradiction:

> If I made no coding errors, then my program will begin and will terminate. If I specified the initial conditions or my program terminates, then my teacher will be satisfied. I specified the initial conditions or I made no coding errors. Therefore my teacher will be satisfied.

Solution   First designate the simple statements as follows:

c: "I made no coding errors."
b: "My program will begin."
t: "My program will terminate."
i: "I specified the initial conditions."
s: "My teacher will be satisfied."

We can now write the conjecture symbolically in a stack:

$$c \rightarrow b \wedge t$$
$$i \vee t \rightarrow s$$
$$\underline{i \vee c}$$
$$\therefore s$$

Following the proof by contradiction procedure, we would first negate the conclusion and add it to the hypotheses to set it up in the form $H_1 \wedge H_2 \wedge H_3 \wedge (\sim C)$. We would then obtain:

$$(c \rightarrow b \wedge t) \wedge (i \vee t \rightarrow s) \wedge (i \vee c) \wedge (\sim s)$$

Our job is to show whether or not this rather long string of statements is a contradiction. The proof process says to modify and work with the statements to alter them into other equivalent forms. Then we continue to examine all the facts and see what we can finally conclude. (We number the steps for easy reference.)

(1) $c \rightarrow b \wedge t$	Hypothesis	
(2) $i \vee t \rightarrow s$	Hypothesis	
(3) $i \vee c$	Hypothesis	
(4) $\sim s$	Negation of conclusion—now a hypothesis	
(5) $\sim(i \vee t)$	Contrapositive of (2) using (4)—$(\sim i \vee t \rightarrow \sim s)$	
(6) $(\sim i) \wedge (\sim t)$	DeMorgan's Law on (5)	
(7) $\sim i$	Simplification rule on (6)—$(p \wedge q) \rightarrow p$	
(8) $(\sim t) \wedge (\sim i)$	Commutative law on (6)—see note below.	
(9) $\sim t$	Simplification rule on (8)	
(10) $i \vee c$	Statement (3)	
(11) c	Combine (7) and (10)	
(12) $b \wedge t$	From statement (1)	
(13) $t \wedge b$	Commutative law	

(14) t Simplification rule on (13)
(15) $(\sim t) \wedge t$ Combine (9) and (14)
(16) Contradiction   Statement (15)

In the above proof, statements (1)–(3) are the hypotheses; so is statement (4). Statements (5)–(14) are reworking the facts, and statement (15) is the conclusion.

note: Just because we reduced statement (6) to (7) doesn't mean we can't go back and use statement (6) again. Statement (6) always stems from the hypotheses (2) and (4), so when we simplify (6) to (7), we can "get another" (6) by calling up (2) and (4) again. That way, we can simplify it differently and still come up with other true statements (8) and (9). We just skipped repeating statements (2)–(5) to get another (6).

7-5. Proof by Counter-example

There are some cases when the truth of a particular statement might be suspect. If the statement takes the form "For every x, $P(x)$ is true," then one way to show this to be invalid is to find one case c for which $P(c)$ is false. The case c would be called a *counter-example* to the theorem, and would disprove the theorem.

EXAMPLE 7-12   Find a counter-example to the statement:

If a real number is not negative, then it must be positive.

Solution   Although the statement may seem to be correct, it is not. Consider the real number $x = 0$. It satisfies the hypothesis since it is not negative. But it is not positive either, so it denies the conclusion. Thus the statement is false.

EXAMPLE 7-13   Prove or disprove the statement:

Every positive integer may be expressed as the sum of the squares of three integers.

Solution   Not knowing whether the statement is valid or not, we begin by trying some cases.

$$1 = 1^2 + 0^2 + 0^2$$
$$2 = 1^2 + 1^2 + 0^2$$
$$3 = 1^2 + 1^2 + 1^2$$
$$4 = 2^2 + 0^2 + 0^2$$
$$5 = 2^2 + 1^2 + 0^2$$
$$6 = 2^2 + 1^2 + 1^2$$
$$7 = ? \qquad \text{It cannot be done!}$$

note: You should reserve this type of proof for those theorems you feel you can quickly find an exception to.

The values that may be used are 0, 1, 2, and 4, but no three value sums make 7. Thus, the value 7 is a counter-example to the theorem and the theorem is invalid.

EXAMPLE 7-14   Prove or disprove the theorem

If $x^2 = y^2$, then $x = y$.

Solution   This is a false statement because a counter-example exists when $x = 4$ and $y = -4$. The hypothesis is true since $4^2 = (-4)^2$ but the conclusion is false since $4 \neq -4$.

7-6. Proof by Induction

In science and mathematics we use two main paths of inquiry to discover new results—*deductive reasoning* and *inductive reasoning*. The preceding methods of proof have used the deductive approach: we accept specific statements as premises and axioms and use logical inference to develop these into new statements. In inductive reasoning, however, we search for facts by careful observation and experimentation and, on the basis of these facts, we arrive at a general statement that is valid for all cases. A centuries-old concept, the **proof by induction** technique—also known as **mathematical induction**—is a way to verify the truth of a general statement.

To develop the steps of an induction proof we will first examine a simple problem: Find the sum of the first n integers. The mathematical equivalent of this statement is: Find S for $S = 1 + 2 + 3 + 4 + \cdots + n$, where n is a positive integer.

There are two obvious facts we could infer from this problem:

(**a**) If $n = 1$, then sum $S = 1$.
(**b**) If we know the sum for a particular value of n, then we can easily find the sum for the next larger value of n by just adding one new integer to the previous sum; that is, for $n = 5$, if the sum is $S_5 = 1 + 2 + 3 + 4 + 5 = 15$, then $S_6 = S_5 + 6 = 15 + 6 = 21$ and, in general, $S_{k+1} = S_k + (k + 1)$.

Using some intuition and pattern recognition we can discover a third important fact: that S may be found for any case from the general result

(**c**) $S = 1 + 2 + 3 + \cdots + n = \dfrac{n(n + 1)}{2}$

$$\left(\text{For example, } S_6 = 1 + 2 + 3 + 4 + 5 + 6 = \frac{6(6 + 1)}{2} = \frac{42}{2} = 21, \text{ as above.} \right)$$

These three observations form the general outline of the proof by induction process. Stated simply, we can apply induction provided that

(**a**) We know the conjecture is true for the first value (usually $n = 1$);
(**b**) we are able to determine the answer at any one stage from the answer of the preceding stage; and
(**c**) we are able to display the general result.

If the general result is part of the conjecture, then mathematical induction provides a framework for confirming that the conjecture is true. Before stating the principle of mathematical induction in formal terms, let's consider an analogy. Suppose we set up a very long string of dominos by placing them close enough together to cause a chain-reaction fall when we push on the first one. We can be assured that all the dominos will fall over only if we have constructed the display to conform to the following two properties:

(**a**) We do push the first domino over and it strikes the second.
(**b**) If domino k of the string falls over, then domino $(k + 1)$ (the one next to it) will also fall over.

In this illustration, property (a) is equivalent to the validity of a mathematical statement for the case $n = 1$. Property (b) is equivalent to the mathematical property that the answer or validity proof at any stage be obtained from the answer or validity proof at the immediately preceding stage. Now for the official, formal definition:

THE PRINCIPLE OF MATHEMATICAL INDUCTION   Let n be an integer and let $P(n)$ be a statement that may be true or false for each n. To prove that $P(n)$ is true for all positive integers, we must show that

(**a**) $P(1)$ is true; and
(**b**) for all $k \geq 1$, the assumption that $P(k)$ is true implies that $P(k + 1)$ is true.

note: Step (a) corresponds to pushing over the first domino and step (b) corresponds to the falling of domino k that causes the fall of domino $(k + 1)$. The first value for k will be 1.

EXAMPLE 7-15   Use the principle of mathematical induction to prove that for each positive integer n, the sum of the first n integers is $\dfrac{n(n + 1)}{2}$.

Solution   The statement to be proved is

$$S(n): \quad 1 + 2 + 3 + \cdots + n = \frac{n(n + 1)}{2}$$

We must first establish the validity of $S(1)$ then, by assuming $S(k)$ is true, we must prove that $S(k + 1)$ follows from $S(k)$. First,

$$S(1): \quad 1 = \frac{1(1 + 1)}{2} = \frac{1 \cdot 2}{2} = 1$$

This is true, so the formula $S(n)$ holds for $n = 1$, and we can begin the rest of the proof. Next we assume the truth of $S(k)$, that is,

$$1 + 2 + 3 + \cdots + k = \frac{k(k + 1)}{2}$$

Once we show that the truth of $S(k + 1)$ follows from $S(k)$, we will have proved the theorem. The statement of $S(k + 1)$ is

$$1 + 2 + 3 + \cdots + k + (k + 1) = \frac{(k + 1)(k + 2)}{2} \tag{7-7}$$

By bracketing the terms on the left side, like so:

$$[1 + 2 + 3 + \cdots + k] + (k + 1) = \frac{(k + 1)(k + 2)}{2}$$

we can replace the portion in the brackets by its equal from the induction assumption $S(k)$:

$$\left[\frac{k(k + 1)}{2} \right] + (k + 1) = \frac{(k + 1)(k + 2)}{2} \tag{7-8}$$

Now all we have to do is prove eq. (7-8):

$$\frac{k^2 + k}{2} + \frac{2(k + 1)}{2} \overset{?}{=} \frac{(k + 1)(k + 2)}{2}$$

$$\frac{k^2 + k + 2k + 2}{2} \overset{?}{=}$$

$$\frac{k^2 + 3k + 2}{2} \overset{?}{=}$$

$$\frac{(k + 1)(k + 2)}{2} \overset{\checkmark}{=} \frac{(k + 1)(k + 2)}{2}$$

Therefore statement (7-7) is true, and the statement $S(n)$ is true for all n.

We can apply the formula we just proved by induction in Example 7-15. For example, suppose we are given the program segment

1. *ISUM* ← 0
2. FOR $I = 1$ TO N DO
   a. *ISUM* ← *ISUM* + *I*

In this segment, the variable *ISUM* calculates

$$1 + 2 + 3 + \cdots + n$$

Because of the above proof, we could replace this entire three-line segment by the one line

$$ISUM \leftarrow N * (N + 1)/2$$

EXAMPLE 7-16   Prove that the mathematical statement for the sum of the squares of the first n positive integers,

$$1^2 + 2^2 + 3^2 + 4^2 + \cdots + n^2 = \frac{n(n + 1)(2n + 1)}{6}$$

is a valid statement.

Solution Designate this proposition as

$$P(n): \quad 1^2 + 2^2 + 3^2 + \cdots + n^2 = \frac{n(n+1)(2n+1)}{6}$$

and validate the case for $n = 1$:

$$P(1): \quad 1^2 = \frac{1 \cdot (1+1)(2+1)}{6} = \frac{1 \cdot 2 \cdot 3}{6} = 1$$

This is true, so $P(n)$ is valid so far. Now assume the truth of statement $P(k)$

$$1^2 + 2^2 + 3^2 + \cdots + k^2 = \frac{k(k+1)(2k+1)}{6}$$

and we're ready to attack the statement $P(k+1)$ and show whether or not it is true:

$$P(k+1): \quad 1^2 + 2^2 + 3^2 + \cdots + k^2 + (k+1)^2 = \frac{(k+1)(k+2)(2(k+1)+1)}{6}$$

or

$$\underbrace{[1^2 + 2^2 + \cdots + k^2]}_{\text{Replace using } P(k)} + (k+1)^2 \stackrel{?}{=} \frac{(k+1)(k+2)(2k+3)}{6}$$

$$\frac{k(k+1)(2k+1)}{6} + (k+1)^2 \stackrel{?}{=} \frac{(k+1)(k+2)(2k+3)}{6}$$

$$\frac{k(k+1)(2k+1) + 6(k+1)^2}{6} \stackrel{?}{=} \frac{(k+1)(k+2)(2k+3)}{6}$$

$$\frac{(k+1)}{6}[2k^2 + k + 6(k+1)] \stackrel{?}{=} \frac{(k+1)(k+2)(2k+3)}{6}$$

$$\frac{(k+1)}{6}[2k^2 + 7k + 6] \stackrel{?}{=} \frac{(k+1)(k+2)(2k+3)}{6}$$

$$\frac{(k+1)(k+2)(2k+3)}{6} \stackrel{\checkmark}{=} \frac{(k+1)(k+2)(2k+3)}{6}$$

EXAMPLE 7-17 Prove or disprove the proposition

$P(n)$: For all positive integers n, $n^5 - n$ is an exact multiple of 10.

Solution Because it is set up as a "true for all values" statement, you might want to attempt to satisfy yourself whether this statement is true or false before you proceed to prove or disprove it. Try a few random values of n:

For $n = 5$, $5^5 - 5 = 3125 - 5 = 3120 = 312 \cdot 10$
For $n = 8$, $8^5 - 8 = 32768 - 8 = 32760 = 3276 \cdot 10$
For $n = 11$, $11^5 - 11 = 161051 - 11 = 161040 = 16104 \cdot 10$

This is *not* a proof, but it looks like the proposition might be true since each value is an exact multiple of 10. Thus, we move ahead with a mathematical induction proof to prove that the general statement is valid for every n:

$P(n)$: $n^5 - n$ is a multiple of 10

First, validate $P(1)$:

$P(1)$: $1^5 - 1 = 1 - 1 = 0$ is a multiple of 10. True.

Next assume $P(k)$:

$P(k)$: $k^5 - k$ is a multiple of 10.

Now we only need to show $P(k+1)$:

$P(k+1)$: $(k+1)^5 - (k+1)$ is a multiple of 10.

Expand and we get

$$k^5 + 5k^4 + 10k^3 + 10k^2 + 5k + 1 - (k+1) = k^5 - k + (5k^4 + 10k^3 + 10k^2 + 5k) + 1 - 1$$
$$= (k^5 - k) + 5k(k^3 + 2k^2 + 2k + 1)$$
$$= (k^5 - k) + 5k(k+1)^3$$
$$= \underbrace{(k^5 - k)}_{A} + \underbrace{5k(k+1)[(k+1)^2]}_{B}$$

The term labeled A is a multiple of 10 from the induction assumption $P(k)$. In the term labeled B, if k is even, then the $5k$ part has 10 as a factor; if k is odd, then the $5(k+1)$ part has 10 as a factor. Either way, this entire term B is also a multiple of 10. Therefore the whole expression is a multiple of 10 and the original proposition is true for all positive integers n.

Although mathematical induction seems to be only a tool for mathematicians, it serves a very useful purpose in computer science. For example, when a programmer writes a segment of program code to perform a particular task, it is often necessary to prove that the program will actually do what is intended. Thus we must prove the correctness of a program. Mathematical induction is a tool that helps us do this.

EXAMPLE 7-18   You are given the following program segment:

1. Input nonnegative integers x, y
2. $d \leftarrow x$
3. $t \leftarrow y$
4. WHILE $t > 0$ DO
   a. $d \leftarrow d - 1$
   b. $t \leftarrow t - 1$

Prove that this program segment will produce $d = x - y$ at completion.

Solution   To prove the claim, we could simply test it, but this procedure has its drawbacks; namely, How many tests do we do? and When are we done testing? A more sure way is to construct $x - d_n + t_n = y$, or $d_n - t_n = x - y$. This combination, called a **loop invariant**, uses the variables d_n and t_n, which change with n as the given program loop runs. If we can show that this combination is a valid expression for all values of n (that it is *invariant*), then when the loop is finished at the point when $t = 0$, the loop invariant statement is $x - d_f + 0 = y$, or $d_f - 0 = x - y$, or $d_f = x - y$, which is what we want the program to produce.

To prove that the loop invariant is valid, we use mathematical induction on the proposition

$$P(n): d_n - t_n = x - y$$

$P(1)$ should then give us

$$d_1 - t_1 = x - y$$

Checking the beginning of the program segment (steps 2 and 3) we find that we are given $d_1 = x$ and $t_1 = y$ so $d_1 - t_1 = x - y$. Thus our first requirement is indeed valid. Next we assume the truth of $P(k)$: $d_k - t_k = x - y$, the second requirement. Now, according to mathematical induction, $P(k+1)$ will equal $d_{(k+1)} - t_{(k+1)} = x - y$. But the loop in the program tells us that

$$d_{k+1} = d_k - 1 \quad \text{and} \quad t_{k+1} = t_k - 1$$

so
$$d_{k+1} - t_{k+1} = (d_k - 1) - (t_k - 1)$$
$$= d_k - t_k = x - y \qquad \text{by the induction assumption}$$

Thus $P(k+1)$ is true and $P(n)$ is true for all n. Since the program loop executes while t is positive, and since t decreases, the loop must eventually terminate. When it does, we will have $t_n = 0$ and $d_f = x - y$, which is what we set out to prove.

A Final Word

The nature of the way the hypotheses are stated may sometimes dictate the method of proof you choose to use. But whatever the theorem, in doing a proof you must simply *start* with some method. If it goes through, great! But if it doesn't (that is, if the steps constructed do not lead logically from the hypothesis to the conclusion), then try a different method. If that doesn't work, try yet another. You may often "feel your way through" a proof in this manner until you begin to suspect that the theorem is false. At that point, spend some time searching for some case in which the theorem is actually false.

But remember that *every* theorem is either true or false, so some method has to work to prove or disprove it. Which one is it? You may have to undertake a lot of hard, long, tedious trials, and you may never be able to do it. Mathematics is full of unproved theorems, some more than 300 years old. Geniuses have failed to find either a proof or a counter-example to them. But we keep trying. Someone else might have better luck.

SUMMARY

1. The main steps in problem solving are discovery, experimentation, and formulation.
2. In the direct method of proof, you assume the hypotheses are true, then combine them with other available facts in a chain of logical reasoning that leads to the conclusion. Symbolically, $p \rightarrow q$.
3. In the contrapositive method of proof, you negate the conclusion and combine it with available facts to lead to the negation of the hypothesis. Symbolically, $\sim q \rightarrow \sim p$.
4. In the proof by contradiction method, also called the indirect method, you assume your given hypotheses are true, then you negate the conclusion and combine it with the hypotheses to lead to a contradiction. Symbolically, $[(p \rightarrow q) \wedge (\sim q)] \rightarrow (\sim p)$.
5. In the method of counter-example, you find one specific case for the value(s) of the unknown(s) to show that the premise $P(x)$ is false.
6. You use the proof by induction technique to verify the truth of a general statement consisting of a function whose domain is the integers. You show that the conjecture of the general statement is true for an initial case (usually $n = 1$) and then verify the truth of the $(n + 1)^{\text{th}}$ case based on the assumed truth of the nth case.
7. You can apply the induction proof technique to prove the correctness of a computer program.

RAISE YOUR GRADES

Can you...?

☑ name the three main steps in problem solving
☑ give the details of each of these three steps
☑ define the process for a direct proof
☑ apply the direct proof process to a theorem
☑ define the process for a proof by the contrapositive method
☑ explain the process for a proof by contradiction
☑ apply the contradiction method to prove a theorem
☑ explain how a counter-example may be used in the proof of a theorem
☑ explain the process of proof by mathematical induction
☑ explain how induction is used to prove the correctness of programming segments and
☑ how the loop invariant is involved

SOLVED PROBLEMS

Direct Proofs

PROBLEM 7-1   Provide a proof of the statement: "If the tap is open, then the water is running. The water is not running. Thus the tap is not open."

Solution   Define the notation:

   p: "The tap is open."
   q: "The water is running."

We have the following theorem given:

$$p \rightarrow q$$
$$\frac{\sim q}{\therefore \sim p}$$

Its direct proof is:

(1) $p \rightarrow q$ Hypothesis
(2) $\sim q$ Hypothesis
(3) $\sim q \rightarrow \sim p$ Contrapositive of (1)
(4) $\sim p$ Combination of (2) and (3)

PROBLEM 7-2   Give a direct proof of the theorem, "Either wages increase or there is inflation. The cost of living increases or there is no inflation. The cost of living does not increase. Thus wages increase."

Solution   Remember the three-step problem-solving process: Discover (consider and use the hypotheses), experiment (change the statements to equivalent ones), and formulate (put the equivalent statements into a sequence of logical steps ending with the conclusion). First, then, let's put the theorem in symbolic notation:

   p: "Wages increase."
   q: "There is inflation."
   r: "Cost of living increases."

Thus we have

$$p \vee q$$
$$r \vee \sim q$$
$$\frac{\sim r}{\therefore p}$$

Experimenting and formulating, we find that the direct proof is:

(1) $r \vee \sim q$ Hypothesis
(2) $\sim r$ Hypothesis
(3) $\sim q$ Combination of (1) and (2) using *disjunctive simplification* (see the following note).
(4) $p \vee q$ Hypothesis
(5) p Combination of (3) and (4) using disjunctive simplification (see the following note).

note: We validate the disjunctive simplification as follows:

$$(u \vee v) \wedge (\sim v) \rightarrow u$$

u	v	$u \vee v$	$\sim v$	$(u \vee v) \wedge (\sim v)$	$(u \vee v) \wedge (\sim v) \rightarrow u$
T	T	T	F	F	T
T	F	T	T	T	T
F	T	T	F	F	T
F	F	F	T	F	T

PROBLEM 7-3   Give a direct proof that the product of odd integers is odd.

Solution   Let a and b be odd integers. Then $a = 2k + 1$ and $b = 2m + 1$, where k and m are integers, are valid representations of odd numbers. Then $a \cdot b = (2k + 1)(2m + 1) = 4km + 2k + 2m + 1$. But $4km$ is even, $2k$ is even, and $2m$ is even. Thus $a \cdot b$ is one greater than an even integer, so ab is odd.

PROBLEM 7-4   On most speedboats there is a device to measure engine hours of use. Say a boat rental agency rents a boat to a fisherman with the express instructions not to run the engine for more than 13 hours in a two-day period. He returns the boat two weeks later and the on-board device shows a total of 95 engine use hours. Prove that during that time, the fisherman used the boat for more than 13 hours in a two-day period.

Solution   If the boat had been used exactly the same number of hours each day in the two-week period, the use value for each day would be $D = \frac{95}{14} = 6\frac{11}{14}$ hours. Under these circumstances, two consecutive days would produce a use time T of

$$T = 2D = 2\left(\frac{95}{14}\right) = \frac{190}{14} = 13\frac{8}{14} \text{ hours}$$

which would be over the 13 hours allotted. Assume, however, that if in any one day, say, day A, the usage was *less* than $6\frac{11}{14}$ hours; then on some other day, say, day B, the use would have to be more than $6\frac{11}{14}$ hours. In this case, the total use on day B plus that on any consecutive day other than A would be more than T. Thus $T = 13\frac{8}{14}$ is the minimal two-day use value. It is more than 13, so the fisherman did indeed break the rules.

PROBLEM 7-5   There are 91 teenagers at a party. Prove that at least two of them were born in the same month of the same year.

Solution   The youngest person at the party could have age 13 years, 0 months, and the oldest could be 19 years, 11 months. This is a total of 84 different months in which these teens could have been born. Thus among 91 people at the party, at least two were born in the same month.

PROBLEM 7-6   Show that if a and x are real numbers and that if $a \cdot x = 0$ $(a \neq 0)$, then $x = 0$.

Solution   If $a \cdot x = 0$ and $a \neq 0$, then $1/a$ exists and

$$\frac{1}{a} \cdot (a \cdot x) = \frac{1}{a} \cdot 0$$

$$\left(\frac{1}{a} \cdot a\right) \cdot x = 0$$

$$1 \cdot x = 0$$

$$x = 0 \ \checkmark$$

Contrapositive Method

PROBLEM 7-7   Use the method of contrapositive to show that if $xy \geq 9$, then $x \geq 3$ or $y \geq 3$.

Solution   We have $p \to q$, with p: $xy \geq 9$ and q: $x \geq 3$ or $y \geq 3$. We need to show $\sim q \to \sim p$. First, we negate the conclusion:

$$\sim q: \ \sim(x \geq 3 \text{ or } y \geq 3) = (\sim x \geq 3) \text{ and } (\sim y \geq 3)$$
$$= (x < 3) \text{ and } (y < 3)$$

Now we have $\sim p$, which is $\sim(xy \geq 9) = xy < 9$. Taking our negated conclusion and applying the rules of inequality, we get

If $x < 3$ and $y < 3$, then $xy < 3 \cdot 3$, or $xy < 9$.

This is indeed the negated hypothesis $\sim p$, so we have shown that $\sim q \to \sim p$. This contrapositive is the same as the implication $p \to q$, so the theorem is true.

PROBLEM 7-8   Prove that if $m \cdot n \neq 0$, then $m \neq 0$ and $n \neq 0$.

Solution   Using the contrapositive method, we must show $\sim q \rightarrow \sim p$, that is, that

   If $\sim(m \neq 0$ and $n \neq 0)$ then $\sim(m \cdot n \neq 0)$.

This is equivalent to

$$(m = 0 \text{ or } n = 0) \rightarrow mn = 0 \qquad \text{(7-9)}$$

By the zero property of multiplication, $0 \cdot x = x \cdot 0 = 0$, we find statement (7-9) valid. Thus the original statement is also valid.

Contradiction Method

PROBLEM 7-9   Use contradiction to prove: "If every equation of the form $x + a = b$ has a solution, then it has at most one solution."

Solution   Identify the parts:

   p: "Every equation of the form $x + a = b$ has a solution."
   q: "It has at most one solution."

Identify what you need to do: In proof by contradiction you negate the conclusion, add it to the hypotheses, and show that this combination leads to a contradiction. The negated conclusion is $\sim q$: "It has more than one solution." We add this to the hypothesis, and our statement becomes:

   "Every equation of the form $x + a = b$ has a solution and it has more than one solution."

To decide whether or not this statement is a contradiction, we let x_1 and x_2 be two different solutions ($x_1 \neq x_2$). Then $x_1 + a = b$ is an identity and $x_2 + a = b$ is also an identity. But if $x_1 + a$ and $x_2 + a$ both equal b, then $x_1 + a = x_2 + a$. But then $x_1 = x_2$, which is a contradiction to our requirement that the equation has more than one solution. Thus the original theorem is true.

Counter-example Method

PROBLEM 7-10   Use a counter-example to prove the following statements are false:

(a) If $|x| < |y|$, then $x < y$.
(b) If $|x| + |y| = 1$, then $x + y = 1$.
(c) If $a \cdot b$ is divisible by 9, then a or b must be divisible by 9 (assume a and b are integers).

Solution   Find any exception to the given statement.

(a) Let $x = 4$, $y = -5$. Then $|x| = |4| = 4$ and $|y| = |-5| = 5$ so we do have $|x| < |y|$, but it is false that $x < y$.
(b) Let $x = -\frac{1}{4}$, $y = \frac{3}{4}$. Then $|x| + |y| = |-\frac{1}{4}| + |\frac{3}{4}| = \frac{1}{4} + \frac{3}{4} = 1$ but $x + y = -\frac{1}{4} + \frac{3}{4} = \frac{1}{2} \neq 1$.
(c) Let $a = 3$, $b = 3$ so that $ab = 9$. Then ab is divisible by 9 but neither a nor b is divisible by 9.

Induction Method

PROBLEM 7-11   Use induction to prove that $(ab)^n = a^n \cdot b^n$.

Solution   Designate this statement as

   $P(n)$: $(ab)^n = a^n \cdot b^n$

We must first show that the equation holds true for $n = 1$. Then, assuming $P(k)$ is true, we must show that the equation holds true for $P(k + 1)$. Plugging in $n = 1$, we get

   $P(1)$: $(ab)^1 = ab = a^1 \cdot b^1 = ab$ True

Then assume the proposition is true for the case $n = k$:

   $P(k)$: $(ab)^k = a^k \cdot b^k$

Then show that the case $n = k + 1$ follows from the induction assumption:

$$P(k + 1): (ab)^{k+1} \overset{?}{=} a^{k+1} \cdot b^{k+1}$$

$$\underbrace{(ab)^k} \cdot (ab) \overset{?}{=} a^{k+1} \cdot b^{k+1}$$

Substitute from
induction assumption

$$\downarrow$$

$$(a^k \cdot b^k)(ab) \overset{?}{=} a^{k+1} \cdot b^{k+1}$$

$$a^k \cdot a \cdot b^k \cdot b \overset{?}{=} a^{k+1} \cdot b^{k+1}$$

$$a^{k+1} \cdot b^{k+1} \overset{\angle}{=} a^{k+1} \cdot b^{k+1}$$

PROBLEM 7-12   Use induction to prove that $a^n - b^n$ is divisible by $a - b$.

Solution   Designate the general result as

   $P(n)$: $a^n - b^n$ is divisible by $a - b$

Then determine the truth of the general result for the case $n = 1$:

   $P(1)$: $a^1 - b^1$ is divisible by $a - b$

This is true since an expression is always divisible by itself. Assume the validity of the general proposition for the case $n = k$:

   $P(k)$: $a^k - b^k$ is divisible by $a - b$.

Show the validity of the general proposition for the case $n = k + 1$:

   $P(k + 1)$: $a^{k+1} - b^{k+1}$ is divisible by $a - b$.

Observe the following steps

$$a^{k+1} - b^{k+1} = a^{k+1} - a^k \cdot b + a^k \cdot b - b^{k+1}$$

$$= \underbrace{a^k(a - b)}_{\text{This is divisible by } a - b} + \underbrace{b(a^k - b^k)}_{\substack{\text{This is divisible by } a - b \text{ by the} \\ \text{induction hypothesis}}}$$

Therefore $a^{k+1} - b^{k+1}$ is divisible by $a - b$.

PROBLEM 7-13   Prove that the sum of the angles of a convex polygon with n sides is $(n - 2) \cdot 180°$.

Solution

note: Induction does not always have to begin with the case $n = 1$. If we begin at $n = b$, then we show the truth of $P(n)$ for all $n \geq b$.

   Begin at $n = 3$, (that is, with a triangle). If $n = 3$, the polygon is a triangle and $(n - 2) \cdot 180° = 180°$ which we know from plane geometry is the sum of the angles. Assume $P(k)$: The sum of the angles of a convex polygon with k sides is $(k - 2) \cdot 180°$. Show that $P(k + 1)$ is true: that the sum of the angles of a polygon with $(k + 1)$ sides is $(k - 1) \cdot 180°$.

   If the polygon has $(k + 1)$ sides, we can join two nonadjacent vertices to form a triangle A (see Figure 7-1). If we cut off triangle A, polygon B will now have k sides (since two sides have been replaced by one). By the induction hypothesis, the sum of the angles for polygon B is $(k - 2) \cdot 180°$. But triangle A has angle sum $180°$. Thus B with A re-attached has angle sum $(k - 2) \cdot 180° + 180° = (k - 1) \cdot 180°$. This is exactly proposition $P(k + 1)$, so we have proved the theorem.

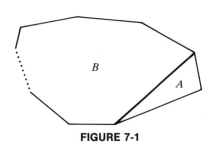

FIGURE 7-1

PROBLEM 7-14   Use mathematical induction to prove that $n! \geq 2^{n-1}$ for all positive integers n.

Solution   Designate the proposition to be proved as $P(n)$: $n! \geq 2^{n-1}$. Then

$$P(1):\ 1! \geq 2^{1-1}$$

$$1 \geq 2^0$$

$$1 \geq 1 \quad \text{True}$$

Assume the truth of $P(k)$: $k! \geq 2^{k-1}$. Show the truth of

$$P(k+1):\ (k+1)! \geq 2^{(k+1)-1}$$

$$\text{or}\quad (k+1)! \geq 2^k$$

Now, we know that

$$(k+1)! = (k+1)k!$$
$$\geq (k+1) \cdot 2^{k-1} \qquad \text{from the induction assumption.}$$
$$\geq 2 \cdot 2^{k-1} \qquad\ \text{Since } k+1 \geq 2.$$
$$\geq 2^k$$

$P(k+1)$ is established as true, thus $P(n)$ is true for all positive integers n.

PROBLEM 7-15   Prove the theorem: For all positive values of n, $n(n^2 + 5)$ is a multiple of 6.

Solution   Designate the conclusion as

$P(n)$: $n(n^2 + 5)$ is a multiple of 6.

Test it for the first value $n = 1$

$P(1)$: $1(1^2 + 5) = 6$ is a multiple of 6.   True

Next, assume

$P(k)$: $k(k^2 + 5)$ is a multiple of 6.

Then show

$P(k+1)$: $(k+1)((k+1)^2 + 5)$ is a multiple of 6.

To show this, expand the expression $P(k+1)$

$$(k+1)((k+1)^2 + 5) = (k+1)(k^2 + 2k + 1 + 5)$$
$$= (k+1)[(k^2 + 5) + (2k+1)]$$
$$= k(k^2 + 5) + k(2k+1) + (k^2 + 5) + (2k+1)$$
$$= k(k^2 + 5) + (2k^2 + k + k^2 + 5 + 2k + 1)$$
$$= k(k^2 + 5) + (3k^2 + 3k + 6)$$
$$= \underbrace{k(k^2 + 5)}_{\substack{\text{multiple of 6} \\ \text{by the induction} \\ \text{assumption}}} + \underbrace{3k(k+1)}_{\substack{\text{multiple of 6 since} \\ \text{either } k \text{ or } k+1 \\ \text{is even}}} + 6$$

This is a multiple of 6, so
$P(n)$ is true for all n.

PROBLEM 7-16   Verify the correctness of the program segment

1. Input nonnegative integers x, y
2. $S \leftarrow 3 * x$
3. $t \leftarrow y$
4. WHILE $t > 0$ DO
   a. $S \leftarrow S + 2 * y$
   b. $t \leftarrow t - 1$

to compute the value of the expression $S = 3x + 2y^2$.

Solution Construct the loop invariant by first examining the program segment. Since y (put into variable t) decreases in steps of 1 until it gets to zero, we keep track of the value S, designated S_n, as it passes through $S_n + 2 \cdot y \cdot t_n$. When t_n is zero, S should be $3x + 2y^2$. Our loop invariant is thus

$$S_n + 2 * y * t_n = 3x + 2y^2$$

Now prove that it is invariant for all n; then show that at termination its value is $3x + 2y^2$.

$$\text{Let } P(n): \quad S_n + 2 * y * t_n = 3x + 2y^2$$

$$\text{Then } P(1): \quad S_1 + 2 * y * t_1 = 3 * x + 2 * y * y$$
$$= 3x + 2y^2$$

To see if this is true, check the program. At the beginning of the program, $S_1 = 3x$ and $t_1 = y$ so $S_1 + 2yt_1 = 3x + (2y)(y) = 3x + 2y^2$ and the case of $P(1)$ holds true. Now assume

$$P(k): \quad S_k + 2 * y * t_k = 3x + 2y^2$$

We must show

$$P(k + 1): \quad S_{k+1} + 2 * y * t_{k+1} \stackrel{?}{=} 3x + 2y^2$$

But from the program segment we have $S_{k+1} = S_k + 2 * y$ and $t_{k+1} = t_k - 1$ (from the WHILE–DO loop) so making substitutions, we get

$$S_k + 2 * y + 2 * y * (t_k - 1) \stackrel{?}{=} 3x + 2y^2$$

or

$$S_k + \cancel{2 * y} + 2 * y * t_k - \cancel{2 * y} \stackrel{?}{=} 3x + 2y^2$$

But this is exactly the induction assumption, so it is true. Therefore the statement $P(n)$ is true for all n.

note: Since the loop variable t decreases and the loop terminates when $t = 0$, this segment will terminate at a value $t_f = 0$. At that point the loop invariant is

$$S_f + 2 * y * t_f = 3x + 2y^2$$

or

$$S_f + 0 = 3x + 2y^2$$

$$S_f = 3x + 2y^2$$

and the final value of the invariant is $3x + 2y^2$. Thus the segment does compute $3x + 2y^2$ for any nonnegative integers x and y.

Supplementary Exercises

PROBLEM 7-17   Prove that if integers a and b are divisible by 3, then $a + b$ is divisible by 3.

PROBLEM 7-18   Give a direct proof of the propostion: If y and z are even integers, then $y + z$ is an even integer.

PROBLEM 7-19   Use a direct proof to show that if n is odd, then n^2 is odd.

PROBLEM 7-20   Use a direct proof to show that if $x \leq 3$, then $2x + 1 \leq 7$.

PROBLEM 7-21   Using contradiction, prove that if m and n are odd integers then $m + n$ is an even integer.

PROBLEM 7-22   Using contradiction, prove that the product of two odd integers is odd.

PROBLEM 7-23   Using the method of the contrapositive, prove that

(a) If a number z is positive, then $z + 1$ is also positive.
(b) If y and z are even integers, then $y - z$ is an even integer.
(c) If $x \neq y$ and $z \neq 0$, then $xz \neq yz$.

PROBLEM 7-24   Find a counter-example to show that each of the following assertions is false:

(a) Every positive integer of the form $6k - 1$ is prime.
(b) If 2×2 matrices $\mathbf{A}$ and $\mathbf{B}$ satisfy $\mathbf{AB} = 0$, then $\mathbf{A} = 0$ or $\mathbf{B} = 0$.
(c) If a 2×2 matrix has all nonzero entries, then it has an inverse.
(d) $n^2 + n + 41$ is a prime number for each integer n.
(e) For rational numbers x, y, and z, $x + (y \cdot z) = (x + y) \cdot (x + z)$.

PROBLEM 7-25   Find a counter-example or prove the truth of each of the following:

(a) If $x < 1$, then $x^n < 1$ for all n.

(b) $\dfrac{2 \cdot 4 \cdot 6 \cdots (2n)}{1 \cdot 3 \cdot 5 \cdots (2n - 1)} \geq 2\sqrt{n}$ for all n.

(c) All prime numbers are odd.
(d) All people who study mathematics are tall or have blue eyes.
(e) If $n \geq 6$, then $n^2 \geq 36$.

PROBLEM 7-26   Use mathematical induction to prove that

(a) $1^3 + 2^3 + 3^3 + \cdots + n^3 = \dfrac{n^2(n + 1)^2}{4}$

(b) $1 + 5 + 9 + \cdots + (4n - 3) = n(2n - 1)$

(c) $\dfrac{1}{1 \cdot 2} + \dfrac{1}{2 \cdot 3} + \dfrac{1}{3 \cdot 4} + \cdots + \dfrac{1}{n(n + 1)} = \dfrac{n}{n + 1}$

PROBLEM 7-27   Use induction to prove that $n^3 - n$ is divisible by 3 for all $n \geq 0$.

PROBLEM 7-28   Use induction to prove that $7^n - 1$ is divisible by 6 for all positive integers n.

PROBLEM 7-29   Use mathematical induction to prove that

(a) $3^n \geq 1 + 2n$ for all n.
(b) $n^2 + n$ is an even integer for all n.

(c) $\sum_{i=1}^{n} 4i = 2n^2 + 2n$ for all positive n.

(d) The sum of the first n odd counting numbers is n^2.

(e) If $n \geq 2$, then $\dfrac{n^3 - n}{3}$ is a counting number.

Answers to Supplementary Exercises

7-24   (a) Use $k = 6$

(b) Use $\mathbf{A} = \begin{bmatrix} 1 & 0 \\ 0 & 0 \end{bmatrix}$, $\mathbf{B} = \begin{bmatrix} 0 & 0 \\ 1 & 0 \end{bmatrix}$

(c) Use $\mathbf{M} = \begin{bmatrix} 1 & 2 \\ 3 & 6 \end{bmatrix}$

(d) Use $n = 41$
(e) Use $x = 1$, $y = 2$, $z = 3$

7-25   (a) Use $x = -1$, $n = 2$
(b) Use $n = 2$
(c) $n = 2$ is prime
(d) Find a short, brown-eyed mathematician
(e) It is true. Prove it!

SETS

☑ **Sets, Membership, and Properties**
☑ **Operations on Sets**
☑ **The Algebra of Sets**
☑ **More on Operations with Sets**
☑ **Computer Representation of Sets**

8-1. Sets, Membership, and Properties

A. Definitions

A **set** is any well-defined collection of objects; the objects are called the *members* or *elements* of the set. For example, we could have the set of all Discrete Structures books, the set of all major league baseball players, or the set of positive integers greater than 100. A set may have any number of elements.

We indicate that a specific element is a member of a set with the notation $a \in S$, read as "*a* is an element of *S*." If an element *b* is not in the set *S* we write $b \notin S$, read "*b* is not an element of *S*."

Sets are fundamental to the study of mathematics and computer science, so you need to thoroughly understand their basic properties.

B. Notation

One way to denote a set is to list its members between braces; this method is called the *roster*, or *enumeration, notation*:

The set of Great Lakes: {Huron, Ontario, Michigan, Erie, Superior}
The set of positive integers less than six: $\{1, 2, 3, 4, 5\}$

note: This may also be written as $\{4, 2, 5, 1, 3\}$, for order inside the braces is not important.

A set of alphabet letters: $\{F, O, R, T, R, A, N\}$
A set of words: {BASIC, FORTRAN, COBOL, PASCAL}
A set that has no elements: {   }. This is denoted by $\varnothing$.

note: Do not confuse $\varnothing$, the **empty set**, with the set $\{0\}$, which contains the one element called zero.

Another way to denote a set is the *set-builder notation*, in which the conditions of membership are described in words and/or in mathematical notation. For example, the set $C = \{1, 2, 3, 4\}$ may be written as

$$C = \{x \mid x \text{ is a natural number less than 5}\}$$

which is read as "the set *C* of all elements *x* such that *x* is a natural number less than 5." (The vertical bar in the braces translates to "such that.")

EXAMPLE 8-1   Write the given set in the "other form," enumeration or set-builder:

(a) $D = \{1, 2, 3, \ldots, 9\}$
(b) $L = \{\text{red, amber, green}\}$
(c) $C = \{x \mid x \text{ is value of a common U.S. coin}\}$
(d) $E = \{e \mid e \text{ is a positive even integer less than 13}\}$

Solution

(a) $D = \{x \mid x \text{ is a positive single-digit integer}\}$
(b) $L = \{c \mid c \text{ is a color on a traffic light}\}$
(c) $C = \{1, 5, 10, 25, 50\}$
(d) $E = \{2, 4, 6, 8, 10, 12\}$

A few of the sets of numbers that are commonly referred to by mathematicians and that make up our system of numbers include

$\mathbb{N}$ = the natural numbers = $\{1, 2, 3, 4, \ldots\}$

$\mathbb{Z}$ = the integers = $\{\ldots, -3, -2, -1, 0, 1, 2, 3, \ldots\}$

$\mathbb{Q}$ = the rational numbers = $\left\{ \dfrac{p}{q} \,\middle|\, p, q \text{ are integers and } q \neq 0 \right\}$

$\mathbb{I}$ = the irrational numbers = $\left\{ x \,\middle|\, x \text{ cannot be expressed as } \dfrac{p}{q} \text{ where } p \text{ and } q \text{ are integers} \right\}$

$\mathbb{R}$ = the real numbers = $\{x \mid x \text{ is a member of } \mathbb{Q} \text{ or } \mathbb{I}\}$

C. Subsets

A set B is a **subset** of set A, written $B \subseteq A$, if every element of B is also an element of A. If B is not a subset of A, we write $B \nsubseteq A$.

EXAMPLE 8-2   Let $S = \{1, 2, 3\}$. Find all of the subsets of S.

Solution   Remember that a subset must be a set itself and that a set can have any number of elements. Therefore, we can make one-element and two-element combinations using the elements given in S. The one-element subsets are $\{1\} \subseteq S$, $\{2\} \subseteq S$, and $\{3\} \subseteq S$. The two-element subsets are $\{1, 2\} \subseteq S$, $\{1, 3\} \subseteq S$ and $\{2, 3\} \subseteq S$. In addition, there is the three-element subset $\{1, 2, 3\} \subseteq S$ (i.e., every set is a subset of itself), and the no-element set $\varnothing \subseteq S$ (i.e., the empty set is a subset of every set). We have listed eight subsets of S.

A set B is a *proper subset* of set A if B is a subset of A and A contains element(s) that are not in B. We write this $B \subset A$. For instance, in Example 8-2, $\{2, 3\}$ is a proper subset of S because S contains elements that are not in $\{2, 3\}$. The subset $\{3, 1, 2\}$, however, though a subset of S, is not a proper subset of S because S does not contain any elements that aren't also in the subset.

A *finite set* (a set with a finite number of elements) having n elements will have 2^n subsets. We will use this concept again in Section 8-3.

EXAMPLE 8-3   If $A = \{1, 2, 3\}$, $B = \{0, 1\}$, $C = \{1\}$, which of the following statements are true?

(a) $B \subseteq A$ (d) $1 \in A$ (g) $\varnothing \in B$
(b) $C \subseteq A$ (e) $\{1\} \subseteq A$ (h) $C \subseteq B$
(c) $\varnothing \subseteq C$ (f) $\{0\} \subseteq C$ (i) $C \subset B$

Solution

(a) This is F since the element 0 is in B but not in A.
(b) This is T since all the elements of C are also in A.

(c) This is T since the empty set is a subset of every set.
(d) This is T since the element 1 is a member of *A*.
(e) This is T since the members of the set {1} are also members of *A*.
(f) This is F since the member of {0} is not a member of *C*.
(g) This is F since $\varnothing$, the empty set, cannot be an element of some other set (it can only be a subset of some other set).
(h) This is T since all of the elements of *C* are elements of *B*.
(i) This is T since *C* is a subset of *B*, and *B* contains element(s) that are not in *C*.

D. Set equality

Two sets *A* and *B* are said to be **equal** if $A \subseteq B$ and $B \subseteq A$. That is, *A* and *B* are equal if each element of *A* is also an element of *B* and vice versa.

EXAMPLE 8-4   Decide if the following pairs of sets are equal:

(a) $S_1 = \{4, \frac{1}{3}, \frac{8}{5}, -5\}, S_2 = \{\frac{8}{5}, -5, 4, \frac{1}{3}\}$
(b) $T_1 = \{\pi, \frac{1}{2}, \frac{2}{9}\}, T_2 = \{\frac{1}{2}, \frac{2}{9}, \frac{22}{7}\}$
(c) $U_1 = \{x \mid x \text{ is a divisor of } 6\}, U_2 = \{1, 2, 3, 6\}$

Solution

(a) The sets S_1 and S_2 have exactly the same elements so $S_1 = S_2$.
(b) The element π in T_1 is different from the element $\frac{22}{7}$ in T_2. Thus $T_1 \neq T_2$.
(c) The divisors of 6 are exactly 1, 2, 3 and 6. Thus $U_1 = U_2$.

EXAMPLE 8-5   Is the set $A = \{x \mid 2x^2 - 5x - 12 = 0\}$ equal to the set $B = \{y \mid y^2 - \frac{5}{2}y - 6 = 0\}$?

Solution   Set *A* contains two values, the roots of the quadratic:

$$2x^2 - 5x - 12 = 0$$
$$(2x + 3)(x - 4) = 0$$
$$x = -\frac{3}{2}, 4$$

Set *B* contains two values, the roots of

$$y^2 - \frac{5}{2}y - 6 = 0$$
$$\left(y + \frac{3}{2}\right)(y - 4) = 0$$
$$y = -\frac{3}{2}, 4$$

Thus $A = B$.

E. Generating sequences

A **sequence** is an ordered listing of elements. Unlike those of a set, the elements in a sequence are in a specific order and are not set in braces. For example, the listing $1, \frac{3}{4}, \frac{1}{2}, \frac{1}{4}, 0$ is a finite sequence having five elements. The listing $1, \frac{1}{2}, \frac{1}{4}, \frac{1}{8}, \frac{1}{16}, \ldots$ is an infinite sequence.

Computer scientists often think of a sequence as a function subprogram which accepts positive integers as its parameters.

EXAMPLE 8-6   Form the sequence using the positive integers as the input values in the formula $S(n) = \frac{1}{n}$.

Solution   The sequence is $\dfrac{1}{1}, \dfrac{1}{2}, \dfrac{1}{3}, \dfrac{1}{4}, \dfrac{1}{5}, \cdots$

There is an important connection between sequences and sets. A sequence is a *generating sequence* for a set if a set is constructed using the exact elements of the sequence as its elements. (That is, to form a set from sequence, you take the elements of the sequence and stick them within braces as the elements of a set.) The sequence in Example 8-6 generates the set

$$\left\{ 1, \frac{1}{2}, \frac{1}{3}, \frac{1}{4}, \frac{1}{5}, \cdots \right\}$$

Because order does not matter in a set, this infinite set could also be written as

$$\left\{ \frac{1}{3}, \frac{1}{2}, 1, \frac{1}{8}, \frac{1}{7}, \frac{1}{6}, \cdots \right\}$$

The set-builder notation for this set is

$$\left\{ \frac{1}{n} \,\middle|\, n \text{ is a positive integer} \right\}$$

Thus we see that the sequence in Example 8-6 uniquely defines a set. But the converse is not true since the same set $\left\{ \dfrac{1}{n} \,\middle|\, n \text{ is a positive integer} \right\}$ could also generate the sequence $1, 1, \dfrac{1}{2}, \dfrac{1}{2}, \dfrac{1}{3}, \dfrac{1}{3}, \cdots$ (which is different from the earlier one).

8-2. Operations on Sets

A. Union and intersection

Just as we perform operations on numbers (add, subtract, multiply) to produce new numbers, we perform operations on sets to produce new sets. A simple way of combining two sets is to put all of their elements together to form a new set, called the **union** of the two sets. Let A and B be given sets. The union of A and B (written $A \cup B$) is

UNION OF *A* AND *B* $A \cup B = \{x \,|\, x \in A \text{ or } x \in B\}$

In this definition the "or" means x belongs to either set A or set B or possibly to both A and B. Note that the union operation on sets is very much like the "or" operation in logic (Section 6-1). A **Venn diagram** is a pictorial method of representing a set. It consists of a large rectangle, which indicates a universal set from which all elements come, and circles (or blobs), which represent the collection of elements in a given subset of the universal set. Shading emphasizes the set being defined. For example, Figure 8-1a illustrates a set A as a subset of the universal set, Figure 8-1b a set B, and Figure 8-1c the union of the subsets, $A \cup B$.

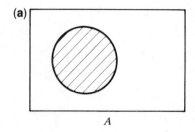

A

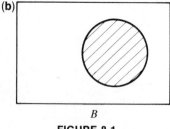

B

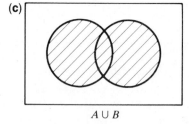

$A \cup B$

FIGURE 8-1

EXAMPLE 8-7   If $A = \{a, b, c, d, e, f\}$ and $B = \{b, a, d, f, o, r, m\}$, form the set $A \cup B$.

Solution   We need to find the set whose elements are letters that are in either the roster of set A or the roster of set B. Thus $A \cup B = \{a, b, c, d, e, f, o, r, m\}$.

note: Don't repeat a letter if it is found in both sets: just list all of the eligible candidates. The ordering of the elements is not important.

A second operation on sets acts like the "and" operation in logic (Section 6-1). If A and B are sets, then the **intersection** of A and B (written $A \cap B$) is

INTERSECTION OF *A* AND *B* $A \cap B = \{x \mid x \in A \text{ and } x \in B\}$

That is, to be a member of the intersection, an element must be in both set A and in set B simultaneously. Figure 8-2 illustrates the Venn diagrams for intersection.

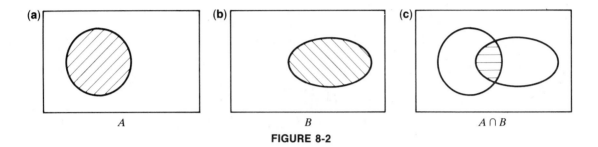

FIGURE 8-2

EXAMPLE 8-8   If $P = \{R, S, T, U, V, W\}$ and $Q = \{S, U, W, X\}$, then find **(a)** $P \cap Q$ and **(b)** $P \cup Q$.

Solution

(a) To form the set $P \cap Q$, find all of the elements that are in both P and Q. This is the set $\{S, U, W\}$.
(b) Recall that for the union of a set you simply list all of the elements that are in either P or Q. Thus $P \cup Q = \{R, S, T, U, V, W, X\}$.

If two sets A and B have no elements in common (that is, their intersection is the empty set), they are said to be **disjoint**:

DISJOINT SETS $A \cap B = \varnothing$

Figure 8-3 shows a Venn diagram of disjoint sets.

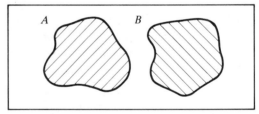

FIGURE 8-3   Disjoint sets: $A \cap B = \varnothing$.

B. Difference

The set **difference** is an operation that is similar to the arithmetic difference of numbers. If A and B are sets, then the difference of A and B (written $A - B$) is

DIFFERENCE OF SETS *A* AND *B* $A - B = \{x \mid x \in A \text{ and } x \notin B\}$ **(8-1)**

Alternately, the difference of B and A would be

DIFFERENCE OF SETS *B* AND *A* $B - A = \{y \mid y \in B \text{ and } y \notin A\}$ **(8-2)**

Essentially, you're subtracting from B all of the elements that intersect with A. The Venn diagrams of these two cases are shown in Figure 8-4.

note: Sometimes the set $A - B$ is called the **complement of set *B* relative to set *A***. It consists of everything that is not in set B but is in set A.

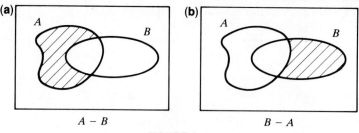

FIGURE 8-4

EXAMPLE 8-9   Let $S = \{1, 3, 5, 7\}$, $T = \{5, 6, 7, 8, 9\}$, and $W = \{8, 10, 12\}$. Find:

(a) $S \cup T$ **(c)** $S \cap W$ **(e)** $S - T$ **(g)** $W - T$
(b) $S \cap T$ **(d)** $S \cup W$ **(f)** $T - S$ **(h)** $T - W$

Solution

(a) The elements that are in S or in T are $\{1, 3, 5, 6, 7, 8, 9\}$.
(b) The elements that are in S and in T are $\{5, 7\}$.
(c) The elements that are in S and in W are $\{\ \} = \varnothing$. The sets S and W are disjoint.
(d) The elements that are in S or in W are $\{1, 3, 5, 7, 8, 10, 12\}$.
(e) The elements that are in S but are not in T are $\{1, 3\}$.
(f) The elements that are in T but are not in S are $\{6, 8, 9\}$.
(g) The elements that are in W but are not in T are $\{10, 12\}$.
(h) The elements that are in T but are not in W are $\{5, 6, 7, 9\}$.

Sometimes we want to refer to the complement of a set relative to the universal set. The universal set is the set from which all subsets we consider are taken (e.g., the rectangle in a Venn diagram). If A is a subset of the universal set U, then the **complement of A** (written $\bar{A}$) is

COMPLEMENT OF SET A
$$\bar{A} = U - A = \{x \mid x \in U \text{ and } x \notin A\}$$

or, more simply,
$$\bar{A} = \{x \mid x \notin A\}$$

For example, if $A = \{x \mid x \text{ is an integer and } x > 7\}$ then $\bar{A} = \{x \mid x \text{ is an integer and } x \leq 7\}$. Or if L is the set of alphabet letters in THE UNITED STATES OF AMERICA then $\bar{L}$ is the set $\{B, G, J, K, L, P, Q, V, W, X, Y, Z\}$. Figure 8-5 shows the Venn diagram of the complement.

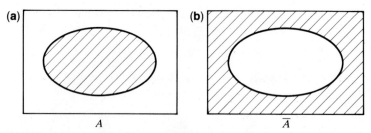

FIGURE 8-5   (a) Set A. (b) Complement of A relative to the universal set.

Note now that the set difference definitions of eqs. (8-1) and (8-2) could be written as

$$A - B = A \cap \bar{B}$$

and

$$B - A = B \cap \bar{A}$$

Recall that the union of two sets A and B is the set of elements in A or B or both. If we exclude the elements that are in both sets, we have the set A "exclusive or" B. This is called the **symmetric**

difference of *A* and *B*, and is defined by

$$A \oplus B = \{x \mid x \in A \text{ or } x \in B \text{ but not both}\}$$

It is also defined by

$$A \oplus B = \{x \mid (x \in A \text{ and } x \notin B) \text{ or } (x \in B \text{ and } x \notin A)\}$$

The Venn diagram of symmetric difference (Figure 8-6) clearly illustrates that, simply put,

$$A \oplus B = (A \cup B) - (A \cap B)$$

$$\uparrow \qquad\qquad \uparrow$$

elements in   remove the
A or B   elements in
or both   both A
and B

(a) 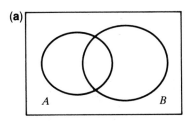   **(b)**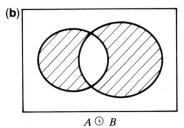

$$A \oplus B$$

FIGURE 8-6

This leads us to a final form of symmetric difference as:

SYMMETRIC DIFFERENCE OF SETS *A* AND *B* $A \oplus B = \{x \mid (x \in A \cup B) \text{ and } (x \notin A \cap B)\}$

EXAMPLE 8-10 Let the universal set be the single-digit positive integers. Let $A = \{1, 3, 5\}$, $B = \{3, 4, 5, 6, 7\}$, and $C = \{6, 7, 9\}$. Find **(a)** $\bar{A}$ **(b)** $\bar{B}$ **(c)** $\bar{C}$ **(d)** $A \oplus B$ **(e)** $B \oplus C$ **(f)** $\overline{A \cup B}$ **(g)** $\overline{B \cap C}$.

Solution

(a) $\bar{A}$ consists of the elements in the universal set that are not in A: $\bar{A} = \{2, 4, 6, 7, 8, 9\}$.
(b) $\bar{B}$ consists of the elements in the universal set that are not in B: $\bar{B} = \{1, 2, 8, 9\}$.
(c) $\bar{C}$ consists of the elements in the universal set that are not in C: $\bar{C} = \{1, 2, 3, 4, 5, 8\}$.
(d) $A \oplus B$ consists of the elements in A or B but not both: $A \oplus B = \{1, 4, 6, 7\}$.

    note: Check this by using

$$\begin{aligned} A \oplus B &= (A \cup B) - (A \cap B) \\ &= \{1, 3, 4, 5, 6, 7\} - \{3, 5\} \\ &= \{1, 4, 6, 7\} \end{aligned}$$

(e) $B \oplus C$ consists of the elements in B or C but not both: $B \oplus C = \{3, 4, 5, 9\} = (B \cup C) - (B \cap C)$.
(f) $\overline{A \cup B}$ consists of the elements in the universal set that are not in $A \cup B$: $\overline{A \cup B} = U - \{1, 3, 4, 5, 6, 7\} = \{2, 8, 9\}$.
(g) $\overline{B \cap C}$ consists of the elements in the universal set that are not in $B \cap C$: $\overline{B \cap C} = U - \{6, 7\} = \{1, 2, 3, 4, 5, 8, 9\}$.

    The operations on sets bear a striking similarity to the logical operations discussed in Chapter 6. We summarize them here using logic notation.

Union:	$(x \in A \cup B) \leftrightarrow (x \in A \lor x \in B)$
Intersection:	$(x \in A \cap B) \leftrightarrow (x \in A \land x \in B)$
Difference:	$(x \in A - B) \leftrightarrow (x \in A \land x \notin B)$
Complement (Universal set defined):	$(x \in \bar{A}) \leftrightarrow (x \notin A)$
Symmetric difference:	$(x \in A \oplus B) \leftrightarrow [(x \in A \cup B) \land (x \notin A \cap B)]$

EXAMPLE 8-11 Show that $\overline{A \cup B} \neq \overline{A} \cup \overline{B}$.

Solution This is done easily by using Venn diagrams and allowing sets A and B to intersect. Given sets A and B as in Figure 8-7a, we separately form Venn diagrams for $\overline{A} \cup \overline{B}$ and $\overline{A \cup B}$, as in Figures 8-7b and 8-7c, respectively. Examining the results of Figures 8-7b and 8-7c, it is clear that $\overline{A} \cup \overline{B} \neq \overline{A \cup B}$.

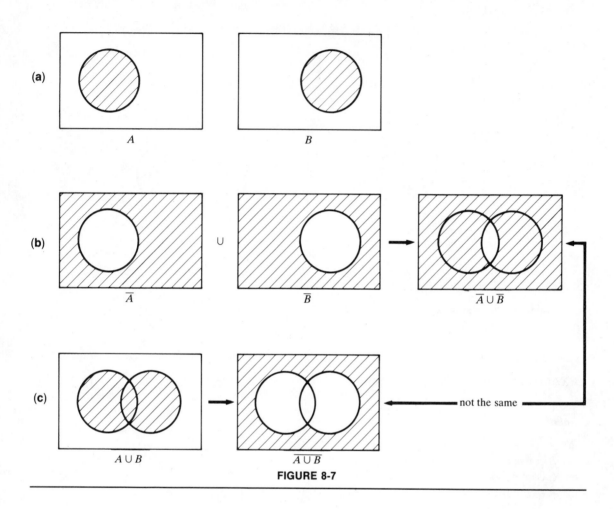

FIGURE 8-7

C. DeMorgan's laws

You may have noticed from Figure 8-7b of Example 8-11 that $\overline{A} \cup \overline{B}$ is the complement of the intersection of A and B. Thus we do have

$$\overline{A} \cup \overline{B} = \overline{A \cap B} \tag{8-3}$$

along with the companion identity

$$\overline{A} \cap \overline{B} = \overline{A \cup B} \tag{8-4}$$

Equations (8-3) and (8-4) are *DeMorgan's Laws* for sets.

EXAMPLE 8-12 Verify the first DeMorgan's Law (eq. 8-3) using the following sets:

$$U = \{x \mid x \text{ is a positive integer less than } 17\}$$

$$A = \{1, 2, 4, 8, 16\}$$

$$B = \{2, 4, 6, 8, 10, 12\}$$

Solution First calculate $\bar{A}$ and $\bar{B}$ and then $\bar{A} \cup \bar{B}$:

$$\bar{A} = \{3, 5, 6, 7, 9, 10, 11, 12, 13, 14, 15\}$$

$$\bar{B} = \{1, 3, 5, 7, 9, 11, 13, 14, 15, 16\}$$

$$\bar{A} \cup \bar{B} = \{1, 3, 5, 6, 7, 9, 10, 11, 12, 13, 14, 15, 16\}$$

Next find $A \cap B$ and then $\overline{A \cap B}$:

$$A \cap B = \{2, 4, 8\}$$

$$\overline{A \cap B} = \{1, 3, 5, 6, 7, 9, 10, 11, 12, 13, 14, 15, 16\}$$

Comparing the results shows that, indeed, $\bar{A} \cup \bar{B} = \overline{A \cap B}$.

8-3. The Algebra of Sets

A. Set properties

Let U be a universal set containing sets A, B, and C. The following is a list of some of the properties that exist for these sets. Their proofs, when needed, will be addressed in the problem section to follow.

(1) $\varnothing \subseteq A$ for all sets A.
(2) $A \subseteq A$ for all sets A.
(3) $\bar{\bar{A}} = A$ for all sets A.

> **note:** $\bar{A}$ means the complement of the complement. That is, given set A, $\bar{A}$ is the complement of A, and $\bar{\bar{A}} = (\overline{\bar{A}})$, the complement of the complement.

(4) $A \cup \bar{A} = U$ and $A \cap \bar{A} = \varnothing$ for all sets A.
(5) $A \cup U = U$ and $A \cap U = A$ for all sets A.

B. Set identities

Again let U be the universal set and A, B, and C be subsets in U. Then

(1) $\left.\begin{array}{l} A \cup A = A \\ A \cap A = A \end{array}\right\}$ **Idempotent Laws**

(2) $\left.\begin{array}{l} A \cup B = B \cup A \\ A \cap B = B \cap A \end{array}\right\}$ **Commutative Laws**

(3) $\left.\begin{array}{l} A \cup (B \cup C) = (A \cup B) \cup C \\ A \cap (B \cap C) = (A \cap B) \cap C \end{array}\right\}$ **Associative Laws**

(4) $\left.\begin{array}{l} A \cup (B \cap C) = (A \cup B) \cap (A \cup C) \\ A \cap (B \cup C) = (A \cap B) \cup (A \cap C) \end{array}\right\}$ **Distributive Laws**

Many of the above identities can be proven using Venn diagrams, others by a valid proof technique. These will be addressed in the problem section.

C. The power set

Earlier we discussed the set of all subsets of a given set A (Section 8-1C, Example 8-2). This set of subsets is called the **power set** of the original set A. We use the notation

$$\mathscr{P}(\{1, 2\}) = \{\varnothing, \{1, 2\}, \{1\}, \{2\}\}$$

$\uparrow\uparrow$
2 elements $2^2 = 4$ subsets in the power set

$$\mathscr{P}(\{A, B, C\}) = \{\varnothing, \{A, B, C\}, \{A, B\}, \{A, C\}, \{B, C\}, \{A\}, \{B\}, \{C\}\}$$

$\uparrow\uparrow$
3 elements $2^3 = 8$ subsets in the power set

> ***note:*** The power set contains only sets as its elements (that is, $1 \notin \mathscr{P}$ but $\{1\} \in \mathscr{P}$). A set with n elements has 2^n subsets in its power set.

EXAMPLE 8-13 Form the power set for the set $S = \{*, !, \circ, \Box\}$.

Solution Since S has 4 elements there are $2^4 = 16$ subsets in the power set. One subset has 4 elements, the set S itself; there are 4 three-element subsets

$$\{*, !, \circ\}, \{*, !, \Box\}, \{*, \circ, \Box\}, \{!, \circ, \Box\}$$

There are 6 two-element subsets:

$$\{*, !\}, \{*, \circ\}, \{*, \Box\}, \{!, \circ\}, \{!, \Box\}, \{\circ, \Box\}$$

There are 4 one-element subsets:

$$\{*\}, \{!\}, \{\circ\}, \{\Box\}$$

and 1 no-element subset, $\varnothing$. Thus the complete power set for $S = \{*, !, \circ, \Box\}$ is

$$\mathscr{P}(S) = \{\{*, !, \circ, \Box\}, \{*, !, \circ\}, \{*, !, \Box\}, \{*, \circ, \Box\}, \{!, \circ, \Box\}, \{*, !\}, \{*, \circ\}, \{*, \Box\},$$
$$\{!, \circ\}, \{!, \Box\}, \{\circ, \Box\}, \{*\}, \{!\}, \{\circ\}, \{\Box\}, \varnothing\}$$

D. Additional set properties

1. Cardinality

Let's denote the number of elements in a finite set A by $|A|$. For example, if $A = \{m, n, p, q\}$, then $|A| = 4$. (As a special case, $|\varnothing| = 0$.) This number of elements in a set is called the **cardinality** of A. If the cardinality of a set is an integer, we say that the set A is *finite*; otherwise we say A is *infinite*.

EXAMPLE 8-14 **(a)** Find a set that is finite and has cardinality six. **(b)** Find two infinite sets.

Solution

(a) The set $H = \{a, b, c, d, e, f\}$ is finite and has cardinality six.
(b) The set of all positive integers is infinite. The set of all rational numbers is infinite.

Using our previous results we can state the following facts about finite sets:

(1) If $A \subseteq B$, then $|A| \leq |B|$
(2) If $A \subset B$, then $|A| < |B|$
(3) If A and B are disjoint, then $|A \cup B| = |A| + |B|$
(4) In any case: $|A \cup B| = |A| + |B| - |A \cap B|$

> *note:* This is called the *principle of inclusion and exclusion*. It basically says: To count the elements in the union of two sets, count those in A, include the count of elements in B, and exclude the elements in the intersection, which have been counted twice.

EXAMPLE 8-15 In a box of books there are 21 paperbacks and 12 history books. Of those books, 7 are paperback history books. How many books are there?

Solution

$$\text{Total number books} = (\text{Number of paperbacks}) + (\text{Number of history books})$$
$$- (\text{Number of paperback history books})$$
$$= 21 + 12 - 7 = 26$$

2. Product set

You are probably familiar with an ordered pair from work in geometry and algebra. It is defined as a pair of objects (a, b) distinguishable by order. Two ordered pairs (m, n) and (p, q) are equal if and only if $m = p$ and $n = q$. This reliance upon the order of the elements distinguishes ordered pairs from sets of two elements. However, ordered pairs are useful when defining set multiplication.

　　If A and B are sets, then the **product** of A and B is a set of ordered pairs

PRODUCT OF SETS A AND B

$$A \times B = \{(x, y) \mid x \in A \text{ and } y \in B\}$$

If A or B is the empty set, then $A \times B$ is the empty set.

EXAMPLE 8-16　　Let $A = \{1, 2\}$ and $B = \{a, b, c\}$. Find $A \times B$.

Solution

$$A \times B = \{(1, a), (1, b), (1, c), (2, a), (2, b), (2, c)\}$$

We can visualize the set by plotting the ordered pairs in the Euclidian plane with the first member plotted horizontally and the second vertically (Figure 8-8). The cardinality of the product set is easy to find. If $|A| = m$ and $|B| = n$, then $|A \times B| = mn$. Here $m = 2$ and $n = 3$, so $mn = 6$.

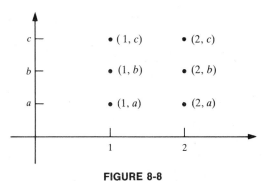

FIGURE 8-8

---

3. Partitions

The intuitive idea of a **partition** of a set is quite simple: split the set into nonoverlapping, nonempty subsets. For example, a restaurant partitions a pie into distinct pieces and a butcher partitions a side of beef into distinct nonempty portions. In the same way, we can partition a set into distinct subsets. Formally, we define a partition π of a set S as $\pi = \{P_1, P_2, \ldots, P_n\}$ where

(1) each P_i is a nonempty subset of S;
(2) $P_1 \cup P_2 \cup \cdots \cup P_n = S$;
(3) $P_i \cap P_j = \varnothing$ when $i \neq j$.

The sets P_i are called the *blocks of the partition*.

EXAMPLE 8-17　　Partition the set $S = \{0, 1, 2, 3, 4\}$ into **(a)** four blocks **(b)** two blocks.

Solution

(a) We need to create four nonempty subsets that exhaust all of S and that are nonintersecting. One possible solution is $\{\{0\}\ \{1, 2\}\ \{3\}\ \{4\}\}$. There are other partitions with four blocks.
(b) There are many partitions with two blocks, including $\{\{0, 1, 2\}\ \{3, 4\}\}$ and $\{\{0, 1, 3,\}\ \{2, 4\}\}$.

Computer languages often utilize the concept of partitions. In the Fortran computer language, for example, there are six basic data types: integer, logical, character, single precision real, double precision real, and complex. Every variable in a program must be of one and only one of these six types. Thus all variables in a program are partitioned into at most six blocks, although not all programs contain all six blocks. The concept of data type in Fortran is related to the set concept of partition. This is true in other computer languages as well.

8-4.　More on Operations with Sets

A.　Union and intersection of three sets

The associativity property for unions of sets (Section 8.3B) eliminates the need for parentheses in strings such as $A \cup B \cup C$, $A \cup B \cup C \cup D$, and so on. The formal definition of such unions follows naturally from the definition of union given earlier. For example, for the union of three sets we have

$$A \cup B \cup C = \{x \mid x \in A \text{ or } x \in B \text{ or } x \in C\} \tag{8-5}$$

The Venn diagram (Figure 8-9) is similarly straightforward.

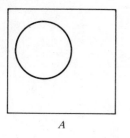 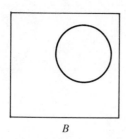 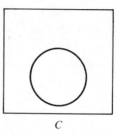 

$$\quad\quad A \quad\quad\quad\quad\quad\quad B \quad\quad\quad\quad\quad\quad C \quad\quad\quad\quad\quad\quad A \cup B \cup C$$

FIGURE 8-9

EXAMPLE 8-18 Let $A = \{p, q, r, s, t\}$, $B = \{q, s, v\}$, and $C = \{r, t, w, x\}$. Find $A \cup B \cup C$.

Solution According to eq. (8-5), you simply list all of the elements that belong to A, B, or C. Thus $A \cup B \cup C = \{p, q, r, s, t, v, w, x\}$.

Similarly, the intersection of more than two sets follows a natural extension of the earlier definition of intersection. For example, for three sets we have

$$A \cap B \cap C = \{x \mid x \in A \text{ and } x \in B \text{ and } x \in C\} \tag{8-6}$$

and the Venn diagram shown in Figure 8-10.

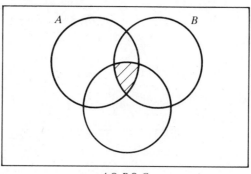

$$A \cap B \cap C$$

FIGURE 8-10

The cardinality of the set $A \cup B \cup C$ is obtained using the *Addition Principle*:

$$|A \cup B \cup C| = |A| + |B| + |C| - |A \cap B| - |B \cap C| - |A \cap C| + |A \cap B \cap C| \tag{8-7}$$

The proof of this cardinality formula follows from the formula for two sets (the principle of inclusion and exclusion):

$$|S \cup T| = |S| + |T| - |S \cap T|$$

Replace S by A and replace T by $(B \cup C)$ to get

$$\begin{aligned}
|A \cup (B \cup C)| &= |A| + |B \cup C| - |A \cap (B \cup C)| \\
&= |A| + |B| + |C| - |B \cap C| - |A \cap (B \cup C)| \\
&= |A| + |B| + |C| - |B \cap C| - |(A \cap B) \cup (A \cap C)| \quad \text{(Distributive Law)} \\
&= |A| + |B| + |C| - |B \cap C| \\
&\quad - \{|A \cap B| + |A \cap C| - |(A \cap B) \cap (A \cap C)|\} \quad \text{(principle of inclusion and exclusion)} \\
&= |A| + |B| + |C| - |B \cap C| - |A \cap B| - |A \cap C| + |A \cap B \cap C| \quad \text{(eq. 8-6)}
\end{aligned}$$

EXAMPLE 8-19   A survey was taken on spectator sports and the respondents were asked to check the sports they regularly watch, choosing from baseball, football, hockey. More than one check was allowed. The results were

37	Baseball
39	Football
21	Hockey
22	Baseball and Football
9	Baseball and Hockey
11	Football and Hockey
5	All three

How many people completed the survey?

Solution   Let set A be the baseball fans, set B be the football fans, and set C be the hockey fans. Then $|A| = 37$, $|B| = 39$, $|C| = 21$, $|A \cap B| = 22$, $|A \cap C| = 9$, $|B \cap C| = 11$, and $|A \cap B \cap C| = 5$ (see Figure 8-11). We must find the total number of respondents, or $|A \cup B \cup C|$. Using eq. (8-7), we have

$$\text{total respondents} = |A \cup B \cup C| = 37 + 39 + 21 - 22 - 11 - 9 + 5 = 60$$

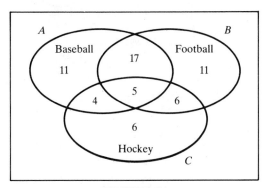

FIGURE 8-11

B. Union and intersection of several sets

In general, if $A_i, A_{i+1}, A_{i+2}, \ldots, A_j$ are sets, then the union of these sets is defined by

$$\bigcup_{k=i}^{j} A_k = \{x \mid x \in A_k \text{ for some } k \text{ in } i \le k \le j\} \tag{8-8}$$

and the intersection of these sets is defined by

$$\bigcap_{k=i}^{j} A_k = \{x \mid x \in A_k \text{ for all } k \text{ in } i \le k \le j\} \tag{8-9}$$

EXAMPLE 8-20   Suppose that for each integer k, $1 \le k \le 4$, the set A_k is defined by

$$A_k = \{x \mid k - 1 \le x \le 2k + 1 \text{ for integer } x\}$$

Find the union and intersection of the collection of sets A_k.

Solution   First form the sets:

$$A_1 = \{x \mid 0 \le x \le 3\} = \{0, 1, 2, 3\}$$
$$A_2 = \{x \mid 1 \le x \le 5\} = \{1, 2, 3, 4, 5\}$$
$$A_3 = \{x \mid 2 \le x \le 7\} = \{2, 3, 4, 5, 6, 7\}$$
$$A_4 = \{x \mid 3 \le x \le 9\} = \{3, 4, 5, 6, 7, 8, 9\}$$

Then using eq. (8-8) we have

$$\bigcup_{k=1}^{4} A_k = \{0,1,2,3,4,5,6,7,8,9\}$$

and from eq. (8-9) we find

$$\bigcap_{k=1}^{4} A_k = \{3\}$$

C. Venn diagrams for multiple sets

Using circles for sets, the Venn diagrams for many set combinations help to visualize the combination. We list a few of the possible sets in Figure 8-12.

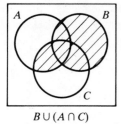 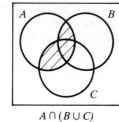 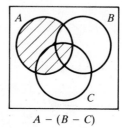 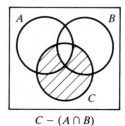

$B \cup (A \cap C)$ $\qquad$ $A \cap (B \cup C)$ $\qquad$ $A - (B - C)$ $\qquad$ $C - (A \cap B)$

FIGURE 8-12

D. The duality principle

If we have a valid equality involving sets, it can be changed to a *dual equality*, which is equally valid, by applying these rules:

(1) Replace each $\cup$ by $\cap$ and replace each $\cap$ by $\cup$.

(2) Replace $\varnothing$ with U and replace U with $\varnothing$.

DeMorgan's Laws are a particularly appropriate application. Using the first law

$$\overline{A \cup B} = \bar{A} \cap \bar{B}$$

the duality principle provides

$$\overline{A \cap B} = \bar{A} \cup \bar{B}$$

The set identities from Sections 8-3A and B give us, for example

$$A \cap \bar{A} = \varnothing \xrightarrow{\text{duality}} A \cup \bar{A} = U$$

$$A \cup (B \cap C) = (A \cup B) \cap (A \cup C) \xrightarrow{\text{duality}} A \cap (B \cup C) = (A \cap B) \cup (A \cap C)$$

8-5. Computer Representation of Sets

To represent a set in a computer we must first arrange the elements of the set into a sequence. Usually a set is listed without regard to the ordering of the elements; that is, the set $S = \{a, c, b, e, d\}$ is the same as the set $T = \{e, c, a, d, b\}$. The members of a sequence, however, depend on their given order. To gain an ordering in a set, then, we will identify the sequence $a, b, c, d, \ldots, n$ with the corresponding set $A = \{a, b, c, d, \ldots, n\}$; that is, think of the elements of the sequence a, b, c, etc. as the first, second, third, etc. elements of the set.

If we are working with a finite set of elements—if the universal set has n elements—we can form an n-element sequence whose members correspond to the elements of the set. Now, because computers only process data with 0's and 1's, we must set up our sequences using only 0's and 1's. So we let a 0 indicate that an element is excluded from a set and let a 1 indicate that an element is included in a set. Then any set can be represented as an n-element sequence of 0's and 1's. Put another way, a computer will use the function

$$f_S(a) = \begin{cases} 1 \text{ if } a \in S \\ 0 \text{ if } a \notin S \end{cases}$$

for each *a* in the universal set. Each subset would then be a sequence of *n* 0's and 1's, corresponding to exclusion or inclusion in the subset *S*.

EXAMPLE 8-21   Let the universe be the finite set

$$U = \{\text{New York, Detroit, Boston, Cleveland, Toronto, Milwaukee, Baltimore}\}$$

Represent **(a)** the subset *S* = {Detroit, Boston, Milwaukee}, **(b)** the subset *T* = {New York, Cleveland, Milwaukee}, and **(c)** the subset *V* = {Baltimore, Detroit, Toronto} as computer sequences.

Solution

(a) Relying on the order of the seven elements in the universal set, we establish a seven-element sequence of 0's and 1's to represent the subset *S* by inserting 0's for excluded elements and 1's for included elements:

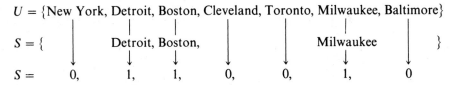

(b) Similarly, the set *T* will form 1, 0, 0, 1, 0, 1, 0.

(c) Here the elements of the set do not match the order of appearance in *U*, but since we only rely on the order of *U*'s elements, we simply match members and find that the sequence for this subset would be *V* = 0, 1, 0, 0, 1, 0, 1.

EXAMPLE 8-22   Let the universal set be the alphabet letters *A – N* (in alphabetic order). Represent the entire universe as a bit sequence and also represent the subset {*C, E, L, M*}.

Solution   The universal sequence would contain a 1 in every position:

A	B	C	D	E	F	G	H	I	J	K	L	M	N
1	1	1	1	1	1	1	1	1	1	1	1	1	1

The sequence for subset {*C, E, L, M*} would be

A	B	C	D	E	F	G	H	I	J	K	L	M	N
0	0	1	0	1	0	0	0	0	0	0	1	1	0

In the computer the bit sequence would be a one-dimensional array. It would be stored and accessed under the formats of the given language.

EXAMPLE 8-23   If the universal set is {1, 2, 3, 4, 5, 6}, set *A* is {1, 3, 4}, and set *B* is {2, 3, 4, 6}, form the bit arrays for **(a)** *A* and *B*, **(b)** *A* ∪ *B*, and **(c)** *A* ∩ *B*.

Solution

(a) The array for *A* = {1, 3, 4} is [1 | 0 | 1 | 1 | 0 | 0] and for *B* = {2, 3, 4, 6} is [0 | 1 | 1 | 1 | 0 | 1].

(b) For *A* ∪ *B* = {*x* | *x* ∈ *A* or *x* ∈ *B*} we compare the bits of *A* with the corresponding bits of *B* and use the rules

$$\begin{cases} 0 \text{ or } 0 \to 0 \\ 1 \text{ or } 0 \to 1 \\ 0 \text{ or } 1 \to 1 \\ 1 \text{ or } 1 \to 1 \end{cases} \quad \text{to get} \quad \boxed{1\ |\ 1\ |\ 1\ |\ 1\ |\ 0\ |\ 1}$$

which is the 6-bit array that represents *A* ∪ *B*.

(c) For $A \cap B = \{x \mid x \in A$ and $x \in B\}$ we compare the bits of A with the corresponding bits of B and use the rules

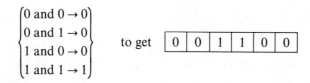

$$\left.\begin{array}{l} 0 \text{ and } 0 \to 0 \\ 0 \text{ and } 1 \to 0 \\ 1 \text{ and } 0 \to 0 \\ 1 \text{ and } 1 \to 1 \end{array}\right\} \quad \text{to get} \quad \boxed{0 \mid 0 \mid 1 \mid 1 \mid 0 \mid 0}$$

which is the 6-bit array that represents $A \cap B$.

Note that these "rules" correspond to the truth tables for "or" and "and" in Chapter 6. Other set operations may be similarly defined.

SUMMARY

1. A set is a collection of objects called elements.
2. The set with no elements is called the empty set.
3. A subset is a subcollection of the elements of a set.
4. A set may be described by enumeration, set-builder notation, or just specification.
5. Sets A and B are equal if they contain exactly the same elements.
6. The power set of a set is the collection of all the possible subsets of the set.
7. The product of two sets $A \times B$ is the collection of ordered pairs (x, y) with $x \in A$ and $y \in B$.
8. The cardinality of a set is the number of elements in the set.
9. The partition of a set is the "slicing" of the set into nonoverlapping parts.
10. The essential operations on sets are
    Union of two sets: $A \cup B =$ the set of all elements either in set A or in set B or both.
    Intersection of two sets: $A \cap B =$ the set of all elements in both A and B simultaneously.
    Complement of a set: $\bar{A} =$ the set of all elements in the universal set U which are not in set A.
11. The pictorial representation of the interaction between sets is called a Venn diagram.
12. Sets may be generated from sequences by forming the set elements from the sequence members.
13. The computer representation of sets is usually as a bit array. The format for specification changes with the computer language being used.

RAISE YOUR GRADES

Can you ...?

- ☑ define a finite set and an infinite set
- ☑ specify a set in three different formats
- ☑ define the five commonly used sets to describe our number system
- ☑ distinguish between a subset and a proper subset
- ☑ create the power set of a given set
- ☑ use a generating function to form a sequence, and then a set
- ☑ draw the Venn diagrams for the basic set operations
- ☑ describe the connection between set operations and logic operations
- ☑ recite the idempotent, commutative, associative, and distributive laws for set operations with union and intersection
- ☑ partition a given set
- ☑ write the formula for the Addition Principle
- ☑ use the Duality Principle to form new set equalities
- ☑ represent a set and its subsets in a bit array computer representation

SOLVED PROBLEMS

Sets, Membership, and Properties

PROBLEM 8-1   Let $S = \{3, 5, \{6, 7\}, 8\}$. Indicate true or false for each statement.

(a) $5 \in S$ (c) $\{6, 7\} \in S$ (e) $\varnothing \notin S$
(b) $3 \subseteq S$ (d) $\{5, 8\} \subseteq S$ (f) $\{3, 5\} \in S$

Solution

(a) True, 5 is an element in S.
(b) False; 3 is an element, not a subset.
(c) True; the set $\{6, 7\}$ is an element of S—it is in its listing.
(d) True; the set $\{5, 8\}$ is a subset of S.
(e) True; the empty set is not an element—it is a subset, however.
(f) False; the set $\{3, 5\}$ is not an element of S—it is a subset of S.

PROBLEM 8-2   Using the sets defined in Section 8-1B, indicate true or false for each statement.

(a) $5 \in \mathbb{Q}$ (c) $10 \in \mathbb{R}$ (e) $13.5 \in \mathbb{Q}$ (g) $\varnothing = \{0\}$
(b) $-3 \in \mathbb{N}$ (d) $\sqrt[3]{7} \in \mathbb{I}$ (f) $-\frac{1}{5} \in \mathbb{I}$

Solution

(a) True; 5 is the same as $\frac{5}{1}$—the form of a rational number.
(b) False; $\mathbb{N}$ contains only the positive integers.
(c) True; 10 is a real number.
(d) True; $\sqrt[3]{7}$ is irrational—its value is a nonending, nonrepeating decimal.
(e) True; 13.5 is the same as $\frac{27}{2}$—the form of a rational number.
(f) False; $-\frac{1}{5}$ is a rational number, not an irrational number.
(g) False; $\varnothing$ is the empty set, it has no elements. $\{0\}$ is a set with one element whose name is zero.

PROBLEM 8-3   Write the sets in roster notation from the given set-builder notation.

(a) $M = \{x \mid x$ is a month whose first letter is "J."$\}$ (c) $T = \{j \mid j \in \mathbb{Z}$ and $2 \leq j \leq 6\}$
(b) $V = \{v \mid v$ is a vowel.$\}$ (d) $A = \{y \mid y$ is an integer and $y^2 < 20\}$

Solution   List the members of the set.

(a) $M = \{$January, June, July$\}$ (c) $T = \{2, 3, 4, 5, 6\}$
(b) $V = \{a, e, i, o, u\}$ (d) $A = \{-4, -3, -2, -1, 0, 1, 2, 3, 4\}$

PROBLEM 8-4   Write the following sets in set builder notation.

(a) $A = \{3, 4, 5, 6, 7, 8\}$ (d) $D = \{-3, -2, -1, 0, 1, 2, 3\}$
(b) $B = \{2, 4, 8, 16, 32, 64, \ldots\}$ (e) The set of integers divisible by 5.
(c) $C = \{\$1, \$5, \$10, \$20, \$50\}$

Solution   Describe the conditions of membership.

(a) $A = \{x \mid x \in \mathbb{Z}$ and $3 \leq x \leq 8\}$
(b) $B = \{x \mid x = 2^n$ for $n \in \mathbb{N}\}$
(c) $C = \{x \mid x$ is a commonly used U.S. paper currency$\}$
(d) $D = \{t \mid t \in \mathbb{Z}$ and $t^2 \leq 9\}$
(e) $E = \{i \mid i = 5j$ and $j \in \mathbb{Z}\}$

PROBLEM 8-5   (a) Create the sequence using the set **N** in the formula $S(n) = n^2$. (b) Form a set from this sequence.

Solution

(a) $S(1) = 1$, $S(2) = 4$, $S(3) = 9$, $S(4) = 16$, etc., so the sequence is $1, 4, 9, 16, 25, 36, \ldots$
(b) The corresponding set is $\{1, 4, 9, 16, 25, 36, \ldots\}$

PROBLEM 8-6   Prove that the empty set $\varnothing$ is a subset of every set B.

Solution   To show that a given set A is a subset of B we must show that every element that is in A is also in B. Here we must show that every element in $\varnothing$ is also in B, that is, if $x \in \varnothing$, then $x \in B$. But in this implication, the hypothesis is always false (there are *no* elements in $\varnothing$). Hence the implication is true.

Operations on Sets

PROBLEM 8-7   Use Venn diagrams to find an example of sets A, B, C to show that

(a) $A \cup B = A \cup C$ but $B \neq C$ (b) $A \cap B = A \cap C$ but $B \neq C$

Solution

(a) Construct A to contain both B and C; as in Figure 8-13.

   note: $A \cup B = A$ and $A \cup C = A$, but $B \neq C$.

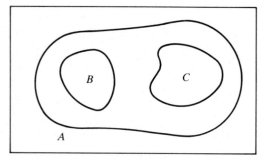

FIGURE 8-13

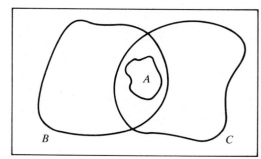

FIGURE 8-14

(b) Construct B and C to be different but have each one contain A, as in Figure 8-14.

   note: $A \cap B = A$ and $A \cap C = A$, but $B \neq C$.

PROBLEM 8-8   Let $A = \{2, 4, 6, 8, 10\}$, $B = \{1, 3, 8, 15\}$, and $C = \{3, 6, 9, 12, 15\}$. Find the following sets.

(a) $A \cup (B \cap C)$ (b) $A \cap (B \cup C)$ (c) $B - C$ (d) $A \oplus B$.

Solution

(a) $(B \cap C) = \{3, 15\}$ so $A \cup (B \cap C) = \{2, 3, 4, 6, 8, 10, 15\}$
(b) $(B \cup C) = \{1, 3, 6, 8, 9, 12, 15\}$ so $A \cap (B \cup C) = \{6, 8\}$
(c) $B - C = B \cap \bar{C} = \{x \mid x \in B \text{ and } x \notin C\} = \{1, 8\}$
(d) $A \oplus B = (A - B) \cup (B - A) = \{x \mid (x \in A \text{ and } x \notin B) \text{ or } (x \in B \text{ and } x \notin A)\} = \{2, 4, 6, 10\} \cup \{1, 3, 15\} = \{1, 2, 3, 4, 6, 10, 15\}$

PROBLEM 8-9   Let $P = \{n \mid n \text{ is odd and } 1 \leq n \leq 13\}$ and $Q = \{n \mid n \text{ is a prime and } 1 \leq n \leq 13\}$. Compute (a) $P \cap Q$, (b) $P \cup Q$, (c) $P - Q$, (d) $(Q - P) \times (P - Q)$.

Solution   First write P and Q in roster form: $P = \{1, 3, 5, 7, 9, 11, 13\}$, $Q = \{1, 2, 3, 5, 7, 11, 13\}$. Then

(a) $P \cap Q = \{1, 3, 5, 7, 11, 13\}$ (c) $P - Q = \{9\}$ [and $Q - P = \{2\}$]
(b) $P \cup Q = \{1, 2, 3, 5, 7, 9, 11, 13\}$ (d) $(Q - P) \times (P - Q) = \{(2, 9)\}$.

PROBLEM 8-10   Let $U = \{a, b, c, d, e, f, g\}$, $A = \{a, c, e\}$, $B = \{c, d, e, f\}$, $C = \{a, d, f, g\}$, and $D = \{b, e, g\}$. Compute (a) $A \cap C$ (b) $(B \cup D) \cup A$ (c) $(C - A) - D$ (d) $B \oplus C$ (e) $\bar{A} \cap D$ (f) $\overline{A - B}$.

Solution

(a) $A \cap C = \{a\}$
(b) $(B \cup D) \cup A = \{b, c, d, e, f, g\} \cup \{a, c, e\} = \{a, b, c, d, e, f, g\} = U$
(c) $(C - A) - D = \{d, f, g\} - \{b, e, g\} = \{d, f\}$
(d) $B \oplus C = \{a, c, e, g\}$
(e) $\bar{A} \cap D = \{b, d, f, g\} \cap \{b, e, g\} = \{b, g\}$
(f) $\overline{A - B} = \overline{\{a, c, e\} - \{c, d, e, f\}} = \overline{\{a\}} = \{b, c, d, e, f, g\}$

The Algebra of Sets

PROBLEM 8-11   Use DeMorgan's Laws and other set identities to prove that

$$(A - B) \cup (B - A) \cup (A \cap B) = A \cup B$$

Solution   Work on the left side: Since $A - B = A \cap \bar{B}$ by definition of the difference of sets, we have

$$(A \cap \bar{B}) \cup (B \cap \bar{A}) \cup (A \cap B)$$

Use the distributive laws twice:

$$[(A \cap \bar{B}) \cup B] \cap [(A \cap \bar{B}) \cup \bar{A}] \cup (A \cap B)$$

$$[(A \cup B) \cap (\bar{B} \cup B)] \cap [(A \cup \bar{A}) \cap (\bar{B} \cup \bar{A})] \cup (A \cap B)$$

But $\bar{B} \cup B = U$ and $A \cup \bar{A} = U$, so

$$[(A \cup B) \cap U] \cap [U \cap (\bar{B} \cup \bar{A})] \cup (A \cap B)$$

By the identity $S \cap U = S$ we now have

$$[A \cup B] \cap [\bar{B} \cup \bar{A}] \cup (A \cap B)$$

By the second DeMorgan's Law (eq. 8-4), the middle term is changed:

$$(A \cup B) \cap (\overline{A \cap B}) \cup (A \cap B)$$

Using $S \cup \bar{S} = U$ on the last pair we have

$$(A \cup B) \cap U$$

or $A \cup B$, which is the right-hand side of the identity we set out to prove.

PROBLEM 8-12   Construct the Venn diagram for the identity of Problem 8-11.

Solution   Break each part of the left-hand side into its own Venn diagram, then solve for their union, as in Figure 8-15.

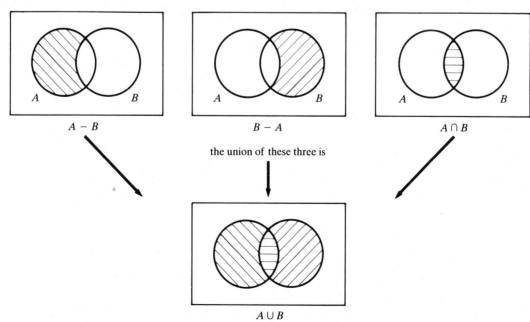

the union of these three is

$$A \cup B$$

FIGURE 8-15

PROBLEM 8-13 Form the power set for $\{\alpha, \beta, \gamma, \delta\}$.

Solution The power set has 2^n elements when the original set has n elements. Thus we have $2^4 = 16$ sets in the power set:

$$\mathscr{P}(\{\alpha, \beta, \gamma, \delta\}) = \{\varnothing, \{\alpha\}, \{\beta\}, \{\gamma\}, \{\delta\}, \{\alpha, \beta\}, \{\alpha, \gamma\}, \{\alpha, \delta\}, \{\beta, \gamma\}, \{\beta, \delta\}, \{\gamma, \delta\},$$
$$\{\alpha, \beta, \gamma\}, \{\alpha, \beta, \delta\}, \{\alpha, \gamma, \delta\}, \{\beta, \gamma, \delta\}, \{\alpha, \beta, \gamma, \delta\}\}$$

PROBLEM 8-14 If $A = \{1, 2, 3\}$ and $B = \{$apples, bananas, carrots$\}$, find **(a)** the cardinality of A **(b)** the cardinality of B **(c)** the cardinality of $A \times B$ **(d)** the cardinality of the power set of B.

Solution Recall that cardinality is the number of elements in the set.

(a) $|A| = 3$
(b) $|B| = 3$
(c) $|A \times B| = 3 \cdot 3 = 9$
(d) $|\mathscr{P}(B)| = 2^3 = 8$

PROBLEM 8-15 Using the sets A and B of Problem 8-14, list the members of $A \times B$.

Solution The members are element pairs whose first element comes from A and whose second element comes from B: $\{1, $apples$\}, \{1, $bananas$\}, \{1, $carrots$\}, \{2, $apples$\}, \{2, $bananas$\}, \{2, $carrots$\}, \{3, $apples$\}, \{3, $bananas$\}, \{3, $carrots$\}$.

PROBLEM 8-16 Form a partition of the positive integers into five mutually disjoint sets.

Solution This can be done in many ways, but one easy partition is:

$$P_1 = \{1, 6, 11, 16, \ldots\} = \{x \,|\, x = 5n - 4 \text{ for } n \in \mathbf{N}\}$$
$$P_2 = \{2, 7, 12, 17, \ldots\} = \{x \,|\, x = 5n - 3 \text{ for } n \in \mathbf{N}\}$$
$$P_3 = \{3, 8, 13, 18, \ldots\} = \{x \,|\, x = 5n - 2 \text{ for } n \in \mathbf{N}\}$$
$$P_4 = \{4, 9, 14, 19, \ldots\} = \{x \,|\, x = 5n - 1 \text{ for } n \in \mathbf{N}\}$$
$$P_5 = \{5, 10, 15, 20, \ldots\} = \{x \,|\, x = 5n \text{ for } n \in \mathbf{N}\}\}$$

note: $P_i \cap P_j = \varnothing$ and $P_1 \cup P_2 \cup P_3 \cup P_4 \cup P_5 = N$, so the qualifications for partitions are met.

More on Operations with Sets

PROBLEM 8-17 Let U be the set of uppercase alphabet letters. Let $\Delta = \{A, B, C, D, E\}$, $\square = \{C, E, G, I, K, M, O\}$, $* = \{A, D, G, J, M, P\}$, $? = \{E, J, O, T, X\}$. For each of the following list its elements and state its cardinality.

(a) $(\Delta \cap *) \cup ?$ **(b)** $(\square \cap *) \cap \Delta$ **(c)** $(\Delta \cup \square) \cup (* \cup ?)$

Solution

(a) $\Delta \cap * = \{A, D\}$; $(\Delta \cap *) \cup ? = \{A, D\} \cup \{E, J, O, T, X\} = \{A, D, E, J, O, T, X\}$; cardinality of 7.
(b) $(\square \cap *) = \{G, M\}$; $(\square \cap *) \cap \Delta = \{G, M\} \cap \{A, B, C, D, E\} = \varnothing$; cardinality of zero.
(c) The union of all four subsets is $\{A, B, C, D, E, G, I, J, K, M, O, P, T, X\}$; cardinality of 14.

PROBLEM 8-18 Draw the Venn diagrams for **(a)** $\bar{A} \cap (B \cup C)$ **(b)** $(A - B) \cap (A \cup B)$ **(c)** $A \cup (\overline{A \cup B})$.

Solution

(a) Put three disks labeled A, B, and C into the universe. We want points that are in either B or C (in $B \cup C$) and that are simultaneously outside of A (in $\bar{A}$). See Figure 8-16.

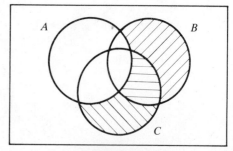

FIGURE 8-16

(b) We want points that are simultaneously in $A \cap \bar{B}$ and in $A \cup B$. It turns out that these are the points in $A \cap \bar{B} = A - B$. See Figure 8-17.

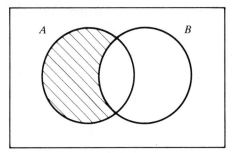

FIGURE 8-17

(c) The points are to be either in A or outside of $A \cup B$. This turns out to be equal to $\bar{B} - A$. See Figure 8-18.

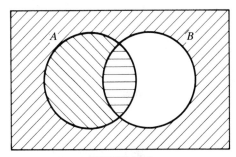

FIGURE 8-18

PROBLEM 8-19 Prove the conclusion of part (c) of Problem 8-18: $A \cup (\overline{A \cup B}) = \bar{B} - A$.

Solution Using the DeMorgan's Law, $\overline{A \cup B} = \bar{A} \cap \bar{B}$, we have

$$A \cup (\bar{A} \cap \bar{B}) = (A \cup \bar{A}) \cap (A \cup \bar{B})$$
$$= U \cap (A \cup \bar{B})$$
$$= A \cup \bar{B} = \overline{\bar{A} \cap B} = \overline{B - A}$$

PROBLEM 8-20 Given the following data counts

32	students are taking Calculus
37	students are taking Spanish
41	students are taking Biology
16	students are taking both Calculus and Spanish
5	students are taking both Calculus and Biology
8	students are taking both Spanish and Biology
	No students are in all three courses

(a) How many students are in at least one of these courses?
(b) How many students are taking only Biology?
(c) How many students are taking only Calculus?

Solution We need to draw the Venn diagram with no triple overlap (since there are no students taking all three courses), as in Figure 8-19. Use the data to fill in the numbers in the parts where two courses overlap and then deduct the value of the overlapping parts from the given totals for the courses to get the remaining parts:

students taking only Calculus $\rightarrow 32 - 16 - 5 = 11$

students taking only Spanish $\rightarrow 37 - 16 - 8 = 13$

students taking only Biology $\rightarrow 41 - 8 - 5 = 28$

Now we can answer the questions.

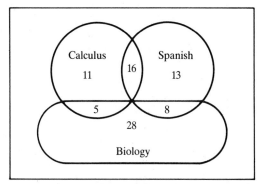

FIGURE 8-19

(a) To find how many students are in at least one of these courses, add the numbers in all the subsets to get the total count: $11 + 16 + 13 + 5 + 8 + 28 = 81$.

(b) Those taking only Biology = 28.

(c) Those taking only Calculus = 11.

Computer Representation of Sets

PROBLEM 8-21   Let the universal set be $U = \{a, b, c, d, e, f, g, h, i, j\}$. Let $A = \{b, e, g, j\}$ and $B = \{a, b, c, h, i, j\}$. **(a)** Represent the sets A and B as sequences of 0's and 1's. **(b)** Form the bit arrays for A, B, $A \cup B$, $A \cap B$, $A - B$, $A \oplus B$.

Solution

(a) The sequence for A is

$$0, \quad 1, \quad 0, \quad 0, \quad 1, \quad 0, \quad 1, \quad 0, \quad 0, \quad 1$$
$$\quad\quad b \quad\quad\quad\quad\quad e \quad\quad\quad g \quad\quad\quad\quad\quad j$$

and for B is

$$1, \quad 1, \quad 1, \quad 0, \quad 0, \quad 0, \quad 0, \quad 1, \quad 1, \quad 1$$
$$a \quad\, b \quad\, c \quad\quad\quad\quad\quad\quad\quad h \quad\, i \quad\, j$$

(b) The bit arrays for A and B will of course be:

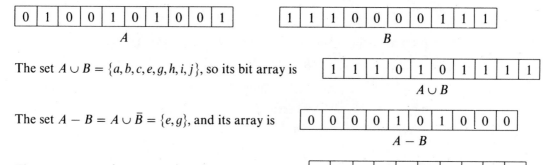

The set $A \cup B = \{a, b, c, e, g, h, i, j\}$, so its bit array is

The set $A - B = A \cup \bar{B} = \{e, g\}$, and its array is

The set $A \oplus B = \{a, c, e, g, h, i\}$, whose bit array is

Supplementary Exercises

PROBLEM 8-22   Describe the sets **(a)** $\mathbb{Z} - \mathbb{N}$, **(b)** $\mathbb{R} - \mathbb{Q}$.

PROBLEM 8-23   Use set notation to describe the following sets:

(a) The set of odd positive integers having a single digit.

(b) The set of squares of all nonnegative integers.

(c) The set of states whose first letter is Z.

(d) The set of rationals that are reciprocals of positive integers.

PROBLEM 8-24   Indicate if the given statement is true or false.

(a) $\{1, 4\} \subseteq \{1, 2, 4, 6, 8\}$

(b) $\{3, 8, 2\} = \{2, 3, 8\}$

(c) $\{0\} \subseteq \varnothing$

(d) $\{u\} \in \{\{t\}, \{u\}, \{v\}\}$

(e) $-3.6284 \in \mathbb{Q}$

(f) $\{a\} \in \{a, b, c\}$

(g) $A \cup B \subseteq A \cap B$

PROBLEM 8-25   Define $S_i = \{i - 1, i, i + 1\}$. Compute (a) $S_1 \cup S_2$ and (b) $S_1 \cap S_2$.

PROBLEM 8-26   Form the power set for the set $\{X, Y\}$

PROBLEM 8-27   Form the Venn diagrams for (a) $\bar{A} \cup B$, (b) $\overline{\overline{A \cup B}}$, (c) $\bar{A} - B$, (d) $A \cup (B - C)$.

PROBLEM 8-28   Let A and B be subsets of a universal set U. Prove that if $A \subseteq B$, then $\bar{B} \subseteq \bar{A}$.

PROBLEM 8-29   Use the laws of set operations to prove that $\overline{\bar{A} \cup B} = A - B$

PROBLEM 8-30   Use the laws of set operations to simplify (a) $(A \cap B) \cup (A - B)$ and (b) $(A - B) \cap B$.

PROBLEM 8-31   Prove that $A - B$ and $B - A$ are disjoint sets.

PROBLEM 8-32   A new way to define a set is to define it inductively. This is done with three axioms as follows:

(1) The numbers 5 and 7 are in S.
(2) If the number x and y are in S, then the numbers $x - y$ and $x + y$ are in S.
(3) Only numbers formed in this way are in S.

Prove that this set S is the set of integers.

PROBLEM 8-33   In Figure 8-20 region 1 is expressed by $A - B - C$ and region 5 is $A \cap B \cap C$. Express all of the other regions in terms of sets A, B, and C.

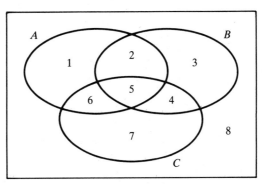

FIGURE 8-20

PROBLEM 8-34   Let $S = \{+, -\}$ and $B = \{00, 01, 10, 11\}$. Describe $S \times B$.

PROBLEM 8-35   Prove that if A and B are disjoint, then $A \oplus B = A \cup B$.

PROBLEM 8-36   In a computer science class of 26 students, 16 knew Fortran, 9 knew PASCAL, and 3 knew neither of these languages. How many students knew both languages?

PROBLEM 8-37   A biologist studying marsh birds reported the following counts:

43	were mallards
36	were female birds
27	were migratory
12	were female mallards
9	were female migratory
8	were migratory mallards
4	were female migratory mallards

(a) How many birds were there altogether in the count?
(b) How many male nonmigratory mallards were there?
(c) How many female nonmallards were there?
(d) How many birds were males?

Answers to Supplementary Exercises

8-22   **(a)** The negative integers and zero.   **(b)** The irrational numbers.

8-23   **(a)** $\{1, 3, 5, 7, 9\}$ **(b)** $\{y \mid y = x^2 - 1 \text{ for } x \in \mathbb{N}\}$ **(c)** $\varnothing$ **(d)** $\left\{ t \mid t = \dfrac{1}{n} \text{ for } n \in \mathbb{N} \right\}$

8-24   **(a)** T   **(b)** T   **(c)** F   **(d)** T   **(e)** T   **(f)** F   **(g)** F

8-25   **(a)** $(0, 1, 2, 3\}\}$ **(b)** $\{1, 2\}$

8-26 $\mathscr{P}(X, Y\}) = \{\varnothing, \{X\}, \{Y\}, \{X, Y\}\}$

8-27   **(a)**
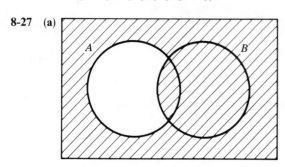
(b)
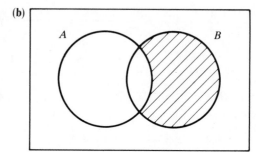

(c)
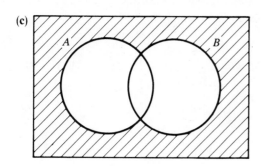
(d)
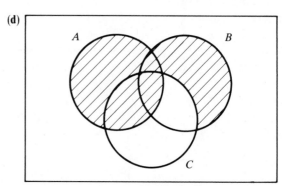

8-30   **(a)** A **(b)** $\varnothing$

8-33   Region 2 is $(A \cap B) - C$, region 3 is $B - A - C$, region 4 is $(B \cap C) - A$, region 6 is $(A \cap C) - B$, region 7 is $C - A - B$, and region 8 is $\overline{A \cup B \cup C}$.

8-34 $\{(+, 00), (+, 01), (+, 10), (+, 11), (-, 00), (-, 01), (-, 10), (-, 11)\}$

8-36   2

8-37   **(a)** 81   **(b)** 27   **(c)** 24   **(d)** 45

9 *BOOLEAN ALGEBRA*

THIS CHAPTER IS ABOUT

☑ **Switching Systems**
☑ **Boolean Algebra**
☑ **Boolean Functions and Boolean Expressions**
☑ **Logic Gates**
☑ **Applications—Binary Addition**
☑ **Minimizing Expressions—Karnaugh Maps**

Boolean algebra is a very important tool in computer science and is quite different from the ordinary algebra you have previously studied. It all started in the nineteenth century when British mathematician George Boole wanted to develop tools for logical calculus, in which structures and rules would apply to logic symbols in a way similar to how ordinary algebra rules apply to numbers. There are many Boolean algebras. Though the one we study in this chapter is a simple one, it is one that underlies the design of switching systems used in telephone networks, logic circuits, computers, and so on.

9-1. Switching Systems

A. Definitions

A **switch** is a device with only two states: "open" or "closed." When a switch is closed, electricity is conducted; when it's open, no electricity is conducted. A **switching system** consists of an **energy source** (such as a battery or a wall plug) and an **energy user** (such as a motor or light bulb) connected by a number of **switches**, designated by x, y, z, etc. There are two ways to connect the switches in a system: in a series (Figure 9-1a) or in parallel (Figure 9-1b).

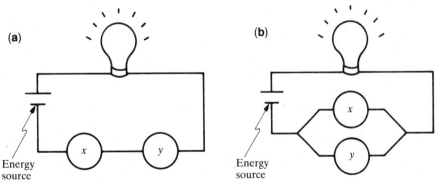

FIGURE 9-1 (a) A series connection. (b) A parallel connection.

In a series connection (Figure 9-1a), the power must go through both x and y to get to the bulb. Thus, if both switches x and y are closed, then the light will go on—and we say the system is closed. But if only one of the switches, or neither of them, is closed, then the light will not go on—and we say the system is open. The four possible states of a series connection are shown in Table 9-1.

In a parallel connection (Figure 9-1b), the power can branch off to either x or y to get to the bulb. That is, if either switch x or switch y is closed, then the light will go on and the system is closed. But if both x and y are open, then the light will not go on, and the system is open. Table 9-2 summarizes the states of a parallel connection.

One additional system configuration is important. Each switch has a corresponding **complementary switch**, $\bar{x}$, whose state is always opposite to the state of switch x. Table 9-3 summarizes the states of a complementary switch.

TABLE 9-1   Series Connection

Switch x	Switch y	State of system
open	open	open
open	closed	open
closed	open	open
closed	closed	closed

TABLE 9-2   Parallel Connection

Switch x	Switch y	State of system
open	open	open
open	closed	closed
closed	open	closed
closed	closed	closed

TABLE 9-3 Complementary Switches

Switch x	Switch $\bar{x}$
open	closed
closed	open

You may already be anticipating what's coming!

B. Notation

Given the various states of a series connection, a parallel connection, and a complementary switch, we adopt the following rules to simplify the notation:

(1) We substitute 0 for "open."
(2) We substitute 1 for "closed."
(3) We let $x \cdot y$ mean that switches x and y are connected in series.
(4) We let $x + y$ mean that switches x and y are connected in parallel.
(5) We let $\bar{x}$ represent the complement of switch x.

We now have three operations called multiplication ($\cdot$), addition ($+$), and complementation ($^-$) which operate on the two-element set $S = \{0, 1\}$. With this new notation, Tables 9-1, 9-2, and 9-3 can be rewritten in an easier-to-read format as Tables 9-4, 9-5, and 9-6, respectively.

TABLE 9-4   Series Connection

x	y	$x \cdot y$
0	0	0
0	1	0
1	0	0
1	1	1

TABLE 9-5 Parallel Connection

x	y	$x + y$
0	0	0
0	1	1
1	0	1
1	1	1

TABLE 9-6 Complementary Switches

x	$\bar{x}$
0	1
1	0

note: If you haven't already noticed, you should see now that this open/closed system of switches corresponds nicely to the binary, on/off system of computers. Notice, too, the similarity of Tables 9-4, 9-5, and 9-6 to the truth tables you studied in Chapter 6. We'll explore their relationship in the next section.

C. Laws

Now that we have a set of elements and some operations that we can perform on them, we need some rules to express the various ways we can manipulate these elements and the operations. Given the two-element set $S = \{0, 1\}$, let a, b, and c be elements of S; that is, the values of a, b, c are either 0 or 1. The variables a, b, and c are called **Boolean variables**. Let the operations of addition ($+$), multiplication ($\cdot$), and complementation ($^-$) be given as in Tables 9-4, 9-5, and 9-6. Then the set S

satisfies the following laws:

(1) $\left.\begin{array}{l} a + b = b + a \\ a \cdot b = b \cdot a \end{array}\right\}$ **Commutative Laws**

(2) $\left.\begin{array}{l} (a + b) + c = a + (b + c) \\ (a \cdot b) \cdot c = a \cdot (b \cdot c) \end{array}\right\}$ **Associative Laws**

(3) $\left.\begin{array}{l} a + (b \cdot c) = (a + b) \cdot (a + c) \\ a \cdot (b + c) = (a \cdot b) + (a \cdot c) \end{array}\right\}$ **Distributive Laws**

(4) $\left.\begin{array}{l} a + 0 = a \\ a \cdot 1 = a \end{array}\right\}$ **Identity Laws**

(5) $\left.\begin{array}{l} a + \bar{a} = 1 \\ a \cdot \bar{a} = 0 \end{array}\right\}$ **Complementation Laws**

EXAMPLE 9-1   Verify the above laws using Tables 9-4, 9-5, and 9-6:

(a) $a \cdot b = b \cdot a$ **(d)** $a + 0 = a$

(b) $(a + b) + c = a + (b + c)$ **(e)** $a \cdot \bar{a} = 0$

(c) $a \cdot (b + c) = (a \cdot b) + (a \cdot c)$

Solution   Let a, b, and c take on all their possible combinations of values. Use the tables to evaluate the expressions. (This is very similar to what you did with logic tables in Chapter 6.)

note: In our operations on expressions containing Boolean variables we will assume that multiplication takes precedence over addition and that complementation takes precedence over both addition and multiplication in the order of operations.

(a)

a	b	$a \cdot b$	$b \cdot a$
0	0	0	0
0	1	0	0
1	0	0	0
1	1	1	1

same

(b)

a	b	c	$a + b$	$(a + b) + c$	$b + c$	$a + (b + c)$
0	0	0	0	0	0	0
0	1	0	1	1	1	1
1	0	0	1	1	0	1
1	1	0	1	1	1	1
0	0	1	0	1	1	1
0	1	1	1	1	1	1
1	0	1	1	1	1	1
1	1	1	1	1	1	1

same

(c)

a	b	c	$b + c$	$a \cdot (b + c)$	$a \cdot b$	$a \cdot c$	$(a \cdot b) + (a \cdot c)$
0	0	0	0	0	0	0	0
0	1	0	1	0	0	0	0
1	0	0	0	0	0	0	0
1	1	0	1	1	1	0	1
0	0	1	1	0	0	0	0
0	1	1	1	0	0	0	0
1	0	1	1	1	0	1	1
1	1	1	1	1	1	1	1

same

(d)

a	0	$a + 0$
0	0	0
1	0	1

same

(e)

a	$\bar{a}$	$a \cdot \bar{a}$	0
0	1	0	0
1	0	0	0

same

The above combination of set S, elements 0 and 1, and the operations $+$, $\cdot$, and $^-$ has provided us with one particular example of a Boolean algebra, which we explore in the next section.

9-2. Boolean Algebra

A   Definitions

Before we formally define a Boolean algebra, we need two more definitions. A **binary operation** on a set A is an operation that uses a pair of elements from the set A and assigns to them an element from set A. In set notation, we use a member of $A \times A$ and assign a member of A to correspond with it. The operations of $+$ and $\cdot$ as we explored them in Section 9-1 are both binary operations. (For example, $0 + 1 \leftrightarrow 1$ and $1 \cdot 0 \leftrightarrow 0$ each show two elements in equivalency with a single element that is also a member of their own set.)

A **unary operation** on a set B is an operation that assigns an element of B to another (or the same) element of B. Thus members (or elements) of B are paired (associated) with members of B. The operation of $^-$ as we explored it in Section 9-1 is a unary operation. (For example, $\bar{1} \leftrightarrow 0$ and $\bar{0} \leftrightarrow 1$ each show one element in equivalency with a single element that is also a member of the same set.)

Now we're ready for the definition. A **Boolean algebra** (such as $[S, +, *, ^-, 0, 1]$) consists of a set (S), two binary operations ($+$ and $*$), a unary operation ($^-$), and two special distinct elements of S (called the **identities** of the Boolean algebra) (0 and 1) and satisfies the following axioms for every choice of x, y, and z in S:

$x + y = y + x$	$x * y = y * x$	Commutative Axioms
$x + (y + z) = (x + y) + z$	$x * (y * z) = (x * y) * z$	Associative Axioms
$x + (y * z) = (x + y) * (x + z)$	$x * (y + z) = (x * y) + (x * z)$	Distributive Axioms
$x + 0 = x$	$x * 1 = x$	Identity Axioms
$x + \bar{x} = 1$	$x * \bar{x} = 0$	Complementation Axioms

note: The Boolean algebra we defined in Section 9-1 ($B = [S, +, \cdot, ^-, 0, 1]$) happened to have a 2-element set $S = \{0, 1\}$ whose two elements were also its identities (those two distinct elements that satisfy the axioms of complementation and identity as listed above). There can, however, be other elements in the set S besides the identities, as Example 9-4 will show. But to be a Boolean algebra, *all* the elements of S must satisfy the ten axioms and there must be two distinct elements in set S—the identities—that cover the complementation and identity parts of the ten axioms.

B. Implementation

Boolean algebras come in many forms, depending on their use in programming, logic, set notation, and so on. The following examples explore some types of Boolean algebras.

EXAMPLE 9-2   In computer programming, a programmer determines if a set of conditions (a programming statement) is true (T) or false (F) using the operations OR, AND, NOT. Define binary and unary operations to form a Boolean algebra for computer programming.

Solution   We need a sextuple whose general form looks like $[S, +, *, ^-, 0, 1]$. Let $S = \{T, F\}$. Let the binary operations be defined by the logic tables for addition and multiplication:

x	y	x OR y		x	y	x AND y
F	F	F		F	F	F
F	T	T		F	T	F
T	F	T		T	F	F
T	T	T		T	T	T

and a unary operation defined by the logic table for complementation:

x	NOT x
F $\cdot$	T
T	F

Let 0 be represented by F and 1 be represented by T. The Boolean algebra now is $[S, \text{OR}, \text{AND}, \text{NOT}, \text{F}, \text{T}]$.

note: There is a direct connection between this Boolean algebra and the earlier one described in Section 9-2A: T $\leftrightarrow$ 1, F $\leftrightarrow$ 0, OR $\leftrightarrow$ +, AND $\leftrightarrow$ *, NOT $\leftrightarrow$ $^-$. Since we have this direct correlation, we know that this second Boolean algebra meets all of the criteria the first one does.

note: By applying logic notation to the same principles of truth tables, a parallel Boolean algebra would be $[S, \vee, \wedge, \sim, 0, 1]$, with $S = \{0, 1\}$.

EXAMPLE 9-3 Using the Boolean algebra of Example 9-2, determine the value (T or F) of the compound statement in

IF ($(A > 4$ OR $B = 2$) AND (NOT $C < 7$)) THEN...

if the statement $A > 4$ is true, $B = 2$ is false, and $C < 7$ is true.

Solution We need to replace the statements with their respective truth values and determine the entire statement's value from the tables we created in Example 9-2 for this Boolean algebra. Plugging in the given truth values, the programming statement reads:

IF ((T OR F)   AND   (NOT T)) THEN...

From the OR table   From the NOT table
this is T.   this is F.

We now have

IF (T AND F) THEN...

From the AND table
this is F.

We now have

IF (F) THEN...

Under the given conditions, the compound statement is false.

EXAMPLE 9-4 Set notation can also be used as a basis for a Boolean algebra. Define operations so that the power set $\mathscr{P}(X)$ of a nonempty set X becomes the basis for a Boolean algebra, and give the Boolean algebra.

recall: The power set of X contains all the subsets of X. Thus if $X = \{1, 2, 3\}$, $\mathscr{P}(X) = \{\varnothing, \{1\}, \{2\}, \{3\}, \{1, 2\}, \{2, 3\}, \{1, 3\}, \{X\}\}$.

Solution We want a sextuple in the form $[S, +, *, ^-, 0, 1]$. Given the nonempty set X, let $Y = \mathscr{P}(X) =$ the set of all subsets of X, and let Y represent set S. For the binary operations we will define the $+$ operation to be union (that is, for subsets A and B in Y, let $A + B = A \cup B$). Define the $*$ operation to be intersection (that is, for subsets A and B in Y, let $A * B = A \cap B$). Define the unary operation to be set complementation (that is, for C in Y, let $\bar{C} = X - C$). (Put another way, these operations mean that for any A, B, C in Y, the operations of $A \cup B$, $A \cap B$, and $X - C$ will produce an element of set Y). For the identities, let the 0 set be $\varnothing$, and let the 1 set be all of X. Thus the Boolean algebra for the power set of a nonempty set X is $[Y, A \cup B, A \cap B, X - C, \varnothing, X]$.

To verify that this is a Boolean algebra, we would have to verify the ten axioms in the definition to prove that the binary and unary operations hold for all values of Y. This is done readily using the basic properties of union, intersection, and complementation (Section 8-3). The $\varnothing$ and X will satisfy the identity and complementation axioms for this Boolean algebra.

9-3. Boolean Functions and Boolean Expressions

A. Definitions

Suppose the student senate at school S has an electronic voting device. The 13 senators have aligned themselves in voting blocs: 6 in bloc A, 4 in bloc B, and 3 in bloc C. They always vote by blocs and each bloc uses a switch to vote: closing it means a "yes" vote, opening it means a "no" vote. (Thus if only bloc A votes yes, there is a total of 6 yes votes and 7 no votes; if both blocs A and C vote yes, there is a total of 9 yes votes and 4 no votes, and so on.) If the total of yes votes is a majority vote, a bell rings to indicate that the motion passed.

Let the circuit for this electronic device be designed so that a 1 indicates a final yes vote (the bell rings) and a 0 indicates a final no vote (the bell doesn't ring). Table 9-7 shows the possible voting situations and outcomes. Using Boolean algebra, we will eventually design a switching system to build the circuit in its simplest form (see Example 9-26). But first we need a few new ideas.

TABLE 9-7   State of Electronic Voting Device

A	B	C	Yes votes	No votes	State of bell
0	0	0	0	13	0
0	0	1	3	10	0
0	1	0	4	9	0
0	1	1	7	6	1
1	0	0	6	7	0
1	0	1	9	4	1
1	1	0	10	3	1
1	1	1	13	0	1

A **Boolean function** on a Boolean algebra B is a function from $B \times B \times B \times \cdots \times B$ to B. That is, some n-tuple of elements is matched with a single element. For example, the first, second, third, and sixth columns in Table 9-7 form a Boolean function—call it f—because the values of the first three columns, read across, produce a corresponding value in the sixth column. [Thus $f(0,0,0) = 0$ and $f(0,1,1) = 1$ and $f(1,1,0) = 1$, etc.] Put another way, a triple of input values produces a single output value; a triple vote tally produces a single voting outcome which does or does not ring the bell.

A **Boolean expression** is a combination of Boolean variables and Boolean operators. If the variables were x and y and the operators $\wedge$ and $\vee$, then some Boolean expressions would be $x \wedge y$, $x \vee (x \wedge y)$, and $\bar{x} \vee y$.

More precisely, we define a Boolean expression recursively: If a Boolean algebra B has variables $x_1, x_2, \ldots, x_n$, then a **Boolean expression on B** is a combination of the variables that meet two criteria: **(a)** Each variable is a Boolean expression. **(b)** For Boolean expressions f and g, the combinations $f + g$, $f * g$, and $\bar{f}$ are also Boolean expressions.

EXAMPLE 9-5   Using the notation above, list some Boolean expressions. Let the variables be x, y, and z.

Solution   Following the rules of the definition we would have this partial list:

x	y	z	$\bar{x}$	$\bar{y}$	$\bar{z}$
$x + y$	$y + z$	$x + z$	$\overline{(x+y)}$	$\overline{(y+z)}$	$\overline{(x+z)}$
$x * y$	$y * z$	$x * z$	$\overline{(x*y)}$	$\overline{(y*z)}$	$\overline{(x*z)}$
$(x+y)*\bar{z}$	$y*z+x$	$x*(z+\bar{x})$	$x*\bar{y}*\bar{z}$	$\bar{x}+y*\bar{z}*z$	$(\bar{y}*z)*x+\bar{z}$

Each Boolean expression is related to a Boolean function. Thus the expression $\bar{x} + y * z$ is associated with the function $f : B \times B \times B \to B$ defined by $f(x, y, z) = \bar{x} + y * z$. Conversely, each Boolean function may be associated with a Boolean expression. In fact, many expressions can be associated with a single function because many expressions may look different but actually be equivalent.

Two Boolean expressions f and g are **equivalent** when they produce the same values on a truth table; then we write $f = g$. When two expressions are equivalent, they represent the same Boolean function. The "simplest" expression is usually the one we are interested in obtaining.

EXAMPLE 9-6   Show the equivalence of the Boolean expressions $(x * y) + (\bar{x} * y) + (x * \bar{y})$ and $x + y$.

Solution   Use Tables 9-4, 9-5, and 9-6 to build a truth table to compare the values of the two expressions.

x	y	$x+y$	$x*y$	$\bar{x}$	$\bar{x}*y$	$\bar{y}$	$x*\bar{y}$	$(x*y)+(\bar{x}*y)$	$(x*y)+(\bar{x}*y)+(x*\bar{y})$
0	0	0	0	1	0	1	0	0	0
0	1	1	0	1	1	0	0	1	1
1	0	1	0	0	0	1	1	0	1
1	1	1	1	0	0	0	0	1	1

same

The two expressions are equivalent, so they represent the same function. Notice, however, how much simpler the expression $(x + y)$ is than the other.

EXAMPLE 9-7   Show the equivalence of the Boolean expressions $\overline{(x + \bar{y})} * \bar{z}$ and $\bar{x} * y * \bar{z}$.

Solution   Create the case table and compare the final values.

x	y	z	$\bar{y}$	$x + \bar{y}$	$\overline{x + \bar{y}}$	$\bar{z}$	$\overline{(x + \bar{y})} * \bar{z}$	$\bar{x}$	$\bar{x} * y$	$\bar{x} * y * \bar{z}$
0	0	0	1	1	0	1	0	1	0	0
0	1	0	0	0	1	1	1	1	1	1
1	0	0	1	1	0	1	0	0	0	0
1	1	0	0	1	0	1	0	0	0	0
0	0	1	1	1	0	0	0	1	0	0
0	1	1	0	0	1	0	0	1	1	0
1	0	1	1	1	0	0	0	0	0	0
1	1	1	0	1	0	0	0	0	0	0

$\underset{\text{same}}{\underline{\quad\quad\quad\quad\quad\quad\quad\quad\quad\quad}}$

The two expressions are equivalent, so they represent the same function. In this case, however, the expression $\bar{x} * y * \bar{z}$ is a simpler version of the function because it is more straightforward.

A Boolean expression is called a **tautology** if its value is 1 for all inputs; it is called a **contradiction** if its value is 0 for all inputs. In terms of the associated Boolean function, the rule is a tautology if $f \equiv 1$ and a contradiction if $f \equiv 0$.

EXAMPLE 9-8   (a) Show that the function $f(a, b) = \bar{a} * (a + b) * (a + \bar{b})$ is a contradiction. (b) Show that the function $g(a, b) = a + b + (\bar{a} * \bar{b})$ is a tautology.

Solution   Form the complete case table for both functions.

a	b	$\bar{a}$	$\bar{b}$	$\bar{a} * \bar{b}$	$a + b$	$a + \bar{b}$	$\bar{a} * (a + b)$	$\bar{a} * (a + b) * (a + \bar{b})$	$a + b + (\bar{a} * \bar{b})$
0	0	1	1	1	0	1	0	0	1
0	1	1	0	0	1	0	1	0	1
1	0	0	1	0	1	1	0	0	1
1	1	0	0	0	1	1	0	0	1

(**a**) a contradiction   (**b**) a tautology

In order to understand the concept of the simplest expression of a function, consider the following: Let E be a Boolean expression. A variable is called a **literal** of E if the variable or its complement appears in E. Now, let f be a Boolean function and let E be a Boolean expression that represents f. We say that E is a **minimal representation** (the simplest expression) of f if

(**1**) E is the sum of product terms.
(**2**) The number of product terms that occur in E is less than or equal to the number of product terms in any other expression representing the function f.
(**3**) When the number of product terms in E is equal to the number of product terms in some other representation, then the number of literals in E is less than or equal to the number of literals in the other representation.

Thus in the process of obtaining a minimal representation we must first minimize the number of product terms and then minimize the number of literals that are used. To do this, however, we first need a few laws of Boolean expressions.

B. Laws of Boolean expressions

There are many useful identities that may be used when simplifying Boolean expressions. Let x, y, and z be Boolean variables. Then

(**1**) $\left. \begin{array}{l} x \wedge y = y \wedge z \\ x \vee y = y \vee x \end{array} \right\}$ Commutative Laws

(2) $(x \wedge y) \wedge z = x \wedge (y \wedge z)$
   $(x \vee y) \vee z = x \vee (y \vee z)$ } Associative Laws

(3) $x \wedge (y \vee z) = (x \wedge y) \vee (x \wedge z)$
   $x \vee (y \wedge z) = (x \vee y) \wedge (x \vee z)$ } Distributive Laws

(4) $x \wedge x = x$
   $x \vee x = x$ } Idempotent Laws

(5) $x \wedge (x \vee y) = x$
   $x \vee (x \wedge y) = x$ } Absorption Laws

note: These laws are defined using the symbols $\wedge$ and $\vee$ from logic notation, but we could just have easily defined them with the * or · and + notation. Because they represent the same binary operations, these notations are completely interchangeable with Boolean algebras and functions.

To prove these laws are valid, for all Boolean expressions, we have only the axioms of the definition of a Boolean algebra to work with. Thus the proofs are somewhat tricky because of the lack of ammunition.

EXAMPLE 9-9   Prove the idempotent laws are valid in a Boolean algebra. (Use $B = [S, +, *, ^-, 0, 1]$)

Solution

$x \vee x = (x \vee x) * 1$ Identity Axiom

   $= (x \vee x) * (x \vee \bar{x})$ Complement Axiom

   $= x \vee (x \wedge \bar{x})$ Distributive Axiom

   $= x \vee 0$ Complement Axiom

   $= x$ Identity Axiom

$x \wedge x = (x \wedge x) + 0$ Identity Axiom

   $= (x \wedge x) + (x \wedge \bar{x})$ Complement Axiom

   $= x \wedge (x \vee \bar{x})$ Distributive Axiom

   $= x \wedge 1$ Complement Axiom

   $= x$ Identity Axiom

---

note: We could have saved some work in Example 9-9 if we had used the **dual principle**: If we make the swaps $\vee \leftrightarrow \wedge$ and $0 \leftrightarrow 1$ in a given statement, we arrive at a new and equally valid statement called the statement's **dual equality**.

Two other important laws are consequences of the idempotent laws and use the principle of dual equality:

$$\overline{x \vee y} = \bar{x} \wedge \bar{y}$$
$$\overline{x \wedge y} = \bar{x} \vee \bar{y}$$ } DeMorgan's Laws

---

EXAMPLE 9-10   Verify the DeMorgan's Laws for Boolean variables.

Solution   In order to show $\overline{x \vee y} = \bar{x} \wedge \bar{y}$, it suffices to show that $\bar{x} \wedge \bar{y}$ behaves like the complement of $x \vee y$ (i.e., that $(x \vee y) \wedge (\bar{x} \wedge \bar{y}) = 0$ and $(x \vee y) \vee (\bar{x} \wedge \bar{y}) = 1$). For the first case we have

$(x \vee y) \wedge (\bar{x} \wedge \bar{y}) = x \wedge (\bar{x} \wedge \bar{y}) \vee y \wedge (\bar{x} \wedge \bar{y})$ Distributive Law

   $= (x \wedge \bar{x}) \wedge \bar{y} \vee (y \wedge \bar{y}) \wedge \bar{x}$ Commutative and Associative Laws

   $= 0 \wedge \bar{y} \vee 0 \wedge \bar{x}$ Complement Axioms

   $= 0 \vee 0$ Complement Laws

   $= 0$ Identity Axiom

For the second case we have

$$(x \vee y) \vee (\bar{x} \wedge \bar{y}) = ((x \vee y) \vee \bar{x}) \wedge ((x \vee y) \vee \bar{y}) \qquad \text{Distributive Law}$$
$$= (y \vee (x \vee \bar{x})) \wedge (x \vee (y \vee \bar{y})) \qquad \text{Commutative and Associative Laws}$$
$$= (y \vee 1) \wedge (x \vee 1) \qquad \text{Complement Axiom}$$
$$= 1 \wedge 1 \qquad \text{Complement Laws}$$
$$= 1 \qquad \text{Identity Axiom}$$

Two other simple laws will be needed later. These are the **Identity Laws**. For any Boolean variable z,

$$\left. \begin{array}{l} z \vee 1 = 1 \\ z \wedge 0 = 0 \end{array} \right\} \quad \text{Identity Laws}$$

EXAMPLE 9-11 Prove the Identity Law $z \vee 1 = 1$ for any Boolean variable z.

Solution

$$z \vee 1 = z \vee (z \vee \bar{z}) \qquad \text{Complement Axiom}$$
$$= (z \vee z) \vee \bar{z} \qquad \text{Associative Axiom}$$
$$= z \vee \bar{z} \qquad \text{Idempotent Law}$$
$$= 1 \qquad \text{Complement Axiom}$$

EXAMPLE 9-12 Prove the Identity Law $z \wedge 0 = 0$ for any Boolean variable z.

Solution Use the dual principle. Since we know $z \vee 1 = 1$, the dual is obtained by replacing $\vee \leftrightarrow \wedge$ and $0 \leftrightarrow 1$. We get $z \wedge 0 = 0$.

Armed with all of the preceding laws, we are now ready to simplify Boolean expressions to their minimal representations.

EXAMPLE 9-13 Simplify $x \vee (\bar{x} \wedge y)$ to become $x \vee y$ (i.e., show that they are equivalent).

Solution

$$x \vee (\bar{x} \wedge y) = (x \vee \bar{x}) \wedge (x \vee y) \qquad \text{Distributive Law}$$
$$= 1 \wedge (x \vee y) \qquad \text{Complement Axiom}$$
$$= x \vee y \qquad \text{Identity Axiom}$$

EXAMPLE 9-14 Use the laws to show that the Boolean expression $x \vee (\bar{x} \wedge y) \vee (\bar{x} \wedge \bar{y})$ is equivalent to 1. (This makes it a tautology.)

Solution

$$x \vee (\bar{x} \wedge y) \vee (\bar{x} \wedge \bar{y}) = (x \vee \bar{x}) \wedge (x \vee y) \vee (\bar{x} \wedge \bar{y}) \qquad \text{Distributive Law}$$
$$= 1 \wedge (x \vee y) \vee (\bar{x} \wedge \bar{y}) \qquad \text{Complement Axiom}$$
$$= (x \vee y) \vee (\bar{x} \wedge \bar{y}) \qquad \text{Identity Axiom}$$
$$= ((x \vee y) \vee \bar{x}) \wedge (x \vee y \vee \bar{y}) \qquad \text{Distributive Law}$$
$$= ((x \vee \bar{x}) \vee y) \wedge (x \vee (y \vee \bar{y})) \qquad \text{Associative Law}$$
$$= (1 \vee y) \wedge (x \vee 1) \qquad \text{Complement Axiom}$$
$$= 1 \vee (y \wedge x) \qquad \text{Distributive Axiom}$$
$$= 1 \qquad \text{Identity Law}$$

9-4. Logic Gates

Logic gates are the building blocks of the electronic circuits in computers, calculators, telephone systems, etc. A gate changes one or more incoming signals into a new, single signal. There are several kinds of gates that circuit designers utilize in various combinations. The three basic gates are the OR gate, the AND gate, and the inverter (or complementation) gate. They correspond directly to the three basic operations of a Boolean algebra: $+$, $\cdot$, and $^-$. For example, if binary inputs x and y are given, then the OR gate yields an output of $x + y$ and the AND gate yields an output of $x \cdot y$. If a unary input x is given, the inverter yields the complement value $\bar{x}$. Thus logic gates are the physical realization of the defined operations of a Boolean algebra. The standard symbols, with their input–output (truth) tables, are shown in Figure 9-2.

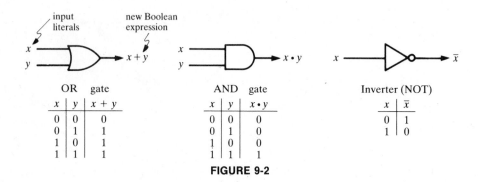

OR	gate	
x	y	$x + y$
0	0	0
0	1	1
1	0	1
1	1	1

AND	gate	
x	y	$x \cdot y$
0	0	0
0	1	0
1	0	0
1	1	1

Inverter (NOT)	
x	$\bar{x}$
0	1
1	0

FIGURE 9-2

note: The terms truth table, input–output table, case table, and logic table all refer to the same table of data; we will use these terms interchangeably when dealing with Boolean algebras and their applications.

agreement: For the rest of the chapter we will adopt the following simplifying notation: When the binary operations $+$ and $*$, $\vee$ and $\wedge$, or $+$ and $\cdot$ appear in an expression, we will use $+$ for addition and eliminate the multiplication symbols by using simple juxtaposition. Thus the expressions

$$x \vee (x \wedge \bar{y} \wedge z) \vee y, \quad x + x * \bar{y} * z + y, \quad \text{and} \quad x + x \cdot \bar{y} \cdot z + y$$

will all simply be written as

$$x + x\bar{y}z + y$$

When enough logic gates are correctly combined to produce a desired output for a given input arrangement, we obtain a **logic circuit**. And when this circuit is etched into a silicon chip, we obtain an integrated circuit chip, those (usually very complicated) chips that run our wrist watches, microwave ovens, VCRs, computers, etc.

As we pass literals through the various gates of a logic circuit, the expression that comes out each gate becomes more and more complex because the gate it just passed through has affected its signal. Put another way, when we pass an expression through a gate it becomes a new Boolean expression. For example, in Figure 9-3 the three literals pass through a series of five gates to eventually produce the expression $\overline{z}(x + y) + x$.

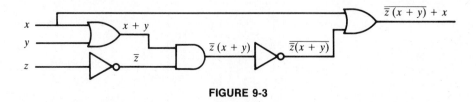

FIGURE 9-3

EXAMPLE 9-15 Write the Boolean expression represented by each of the circuits in Figure 9-4.

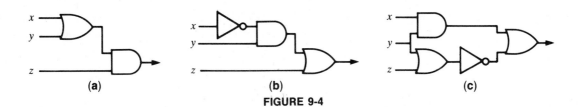

FIGURE 9-4

Solution Work from the left (input) side and pass the literals through the various gates as the diagrams show. Keep track of how the generated expressions change as they go through successive gates, then record the final output.

(a) We first have the literals x and y going through an OR gate; this produces $x + y$. Then this output meets up with the literal z at an AND gate; this gives us the final output $(x + y)z$.

(b) First x goes through an inverter to become $\bar{x}$, which goes in with y through an AND gate to produce $\bar{x}y$. This output is now input with z in an OR gate to give $(\bar{x}y) + z$ as the final output.

(c) Here the y signal is split, so we first get xy on top through the AND gate and $(y + z)$ on the bottom through the OR gate. The lower signal is inverted to $\overline{(y + z)}$ and finally combined with the upper signal xy to give us $xy + \overline{(y + z)}$.

EXAMPLE 9-16 Find the expression for the circuit in Figure 9-5 and form its logic table.

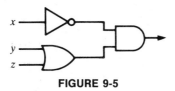

FIGURE 9-5

Solution At the top left the x going through the inverter gate produces $\bar{x}$, and at the lower left the y and z through the OR gate produces $y + z$. These become the inputs for the final AND gate, so the result is $\bar{x}(y + z)$. The logic table is:

x	y	z	$\bar{x}$	$y + z$	$\bar{x}(y + z)$
0	0	0	1	0	0
0	1	0	1	1	1
1	0	0	0	0	0
1	1	0	0	1	0
0	0	1	1	1	1
0	1	1	1	1	1
1	0	1	0	1	0
1	1	1	0	1	0

Examine the circuit, the Boolean expression, and the logic table we produced in Example 9-16. They are all related in a very important way. Since the expression $\bar{x}(y + z)$ describes the output of the given circuit, and since the logic table describes the output of the expression $\bar{x}(y + z)$, we can say that the logic table describes the outputs of the given circuit for all possible inputs x, y, z. We read the table this way: If the inputs x, y, z produce a value of 1 in the final column of the table, the circuit in Figure 9-5 is closed and therefore conducts electricity given those inputs. If the inputs x, y, z produce a final value of 0, the entire circuit is open and therefore doesn't conduct electricity given those inputs.

Let's say a light bulb were attached to the end of the circuit in Figure 9-5. A look at the table in Example 9-16 reveals that the bulb would only go on (have a value of 1) when the values of x, y, z were either (0, 1, 0), (0, 0, 1), or (0, 1, 1), respectively. Recall from our work with functions that a triple across a logic table can be written as a Boolean function (Section 9-3). Thus our table in Example 9-16 produces the function $f(x, y, z) = \bar{x}(y + z)$, or $f(0, 0, 0) = 0$, $f(0, 1, 0) = 1$, and so on.

Now, if we express a variable with a notation 1 and its complement with a notation 0, then a triple of numbers from the logic table may be expressed as a product of the variables represented in it. Thus the triple $(0, 1, 0)$ becomes $\bar{x}y\bar{z}$, $(0, 0, 1)$ becomes $\bar{x}\bar{y}z$, and $(0, 1, 1)$ becomes $\bar{x}yz$. Using this notation, we could then say the bulb will go on when we have $\bar{x}y\bar{z}$ OR $\bar{x}\bar{y}z$ OR $\bar{x}yz$. That is, we can say that the circuit is closed $[f(x, y, z) = 1]$ when:

$$f(x, y, z) = \bar{x}y\bar{z} + \bar{x}\bar{y}z + \bar{x}yz \tag{9-1}$$

Here's the important result: Eq. (9-1) is another, equivalent, expression that describes the circuit in Figure 9-5. Only, instead of getting it by examining the circuit itself, we got it by examining the circuit's logic table created in Example 9-16. (You will prove in the supplementary exercises that $\bar{x}(y + z)$ is equivalent to eq. 9-1).

These relationships between the expression, the physical logic circuit, the logic table, and the Boolean function are crucial to the building of efficient circuits. We will return to them in Section 9-6.

EXAMPLE 9-17   Create another circuit equivalent to the circuit of Example 9-16. Sketch it and obtain the truth table.

Solution   To create a circuit equivalent to that of Example 9-16, we must first find an expression that is equivalent to the expression of the original circuit. Using the distributive law we modify the original function $f(x, y, z) = \bar{x}(y + z)$ to obtain $g(x, y, z) = \bar{x}y + \bar{x}z$. Then we draw the circuit for g by studying the expression and using the appropriate gates to represent the additions and multiplications. The circuit is shown in Figure 9-6.

This circuit is considered to be more complicated than that of Example 9-16 because it uses 4 gates instead of 3. Thus the original circuit created by $f(x, y, z) = \bar{x}(y + z)$ is more efficient. A logic table of the new expression g, however, reveals that this new circuit does indeed produce exactly the same outputs for the same inputs as the circuit in Example 9-16:

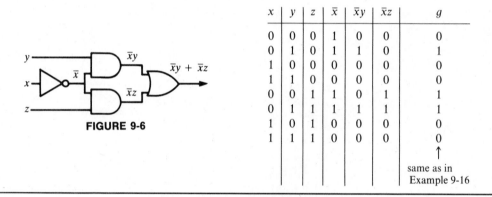

x	y	z	$\bar{x}$	$\bar{x}y$	$\bar{x}z$	g
0	0	0	1	0	0	0
0	1	0	1	1	0	1
1	0	0	0	0	0	0
1	1	0	0	0	0	0
0	0	1	1	0	1	1
0	1	1	1	1	1	1
1	0	1	0	0	0	0
1	1	1	0	0	0	0

$\uparrow$
same as in
Example 9-16

FIGURE 9-6

In addition to the three basic gates defined earlier, circuit designers use three other gates, shown in Figure 9-7:

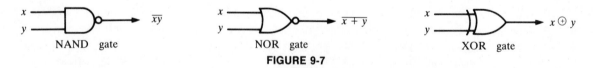

NAND gate   NOR gate   XOR gate

FIGURE 9-7

- The NAND gate (from "not–and"), defined by $z = \overline{xy}$.
- The NOR gate (from "not–or"), defined by $z = \overline{x + y}$.
- The XOR gate (from "exclusive or"), defined by $z = x \oplus y = x\bar{y} + \bar{x}y$.

The XOR gate is not used as extensively as the others since it is easy to produce the same result of an XOR gate by combining the AND, OR, and NOT gates, as in the circuit in Figure 9-8. This circuit may be simplified from five gates to four gates by applying laws to produce the equivalency expression

$$x \oplus y = x\bar{y} + \bar{x}y = (x + y)(\overline{xy}) \tag{9-2}$$

The circuit of this expression is shown in Figure 9-9.

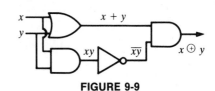

FIGURE 9-8   FIGURE 9-9

EXAMPLE 9-18 Verify the equivalency expression eq. (9-2).

Solution Use the previously given laws:

$$
\begin{aligned}
(x + y)(\overline{xy}) &= x(\overline{xy}) + y(\overline{xy}) && \text{Distributive Law} \\
&= x(\overline{x} + \overline{y}) + y(\overline{x} + \overline{y}) && \text{DeMorgan's Law} \\
&= x\overline{x} + x\overline{y} + y\overline{x} + y\overline{y} && \text{Distributive Law} \\
&= 0 + x\overline{y} + y\overline{x} + 0 && \text{Complementation Law} \\
&= x\overline{y} + y\overline{x}
\end{aligned}
$$

Table 9-8 summarizes the Boolean function values associated with the six basic gates used in circuit construction.

TABLE 9-8 Summary of Logic Values of Logic Gates

x	y	NOT $\overline{x}$	OR $x + y$	NOR $\overline{x + y}$	AND xy	NAND $\overline{xy}$	XOR $x \oplus y$
0	0	1	0	1	0	1	0
0	1	1	1	0	0	1	1
1	0	0	1	0	0	1	1
1	1	0	1	0	1	0	0

The two gates AND and OR may also have three or more input values. In this case, their truth tables are relatively easy to construct, and the circuit sketches have the multiple "wires" coming in from the left and going out as a single "wire" to the right (see Figure 9-10).

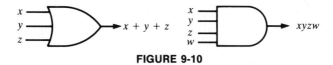

FIGURE 9-10

A set of gates $\{g_1, g_2, \ldots, g_k\}$ is said to be **functionally complete** if it is possible to construct a logic circuit that computes any given Boolean function f using only the gates $g_1, g_2, \ldots, g_k$. We present without proof an important result:

THEOREM: The set of gates {AND, OR, NOT} is functionally complete.

In other words, the circuit of any Boolean function f can be drawn using only AND, OR, and NOT gates. As illustrations of this theorem, look back at all the circuits studied thus far. They use only the gates AND, OR, and NOT. Some additional results, however, are very useful.

THEOREM: The set of gates {AND, NOT} and the set of gates {OR, NOT} are each functionally complete.

Proof: For the first case, using only AND and NOT we need to produce the same value that an OR gate does. But

$$
\begin{aligned}
x + y &= \overline{\overline{x}} + \overline{\overline{y}} && \text{(by Involution law)} \\
&= \overline{\overline{x}\,\overline{y}} && \text{(by DeMorgan's law)}
\end{aligned}
$$

Thus the OR gate can be replaced by one AND and three NOT gates, as in Figure 9-11.

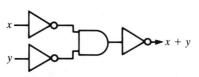

FIGURE 9-11 AND and NOT gates can create the same effect as an OR gate.

For the second case, we want to replace AND by a combination of OR and NOT. But

$$xy = \overline{\overline{x}\,\overline{y}} \qquad \text{(by Involution law)}$$
$$= \overline{\overline{x} + \overline{y}} \qquad \text{(by DeMorgan's law)}$$

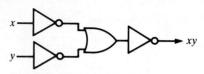

FIGURE 9-12 OR and NOR gates can create the same effect as an AND gate.

So we can indeed use OR and NOR gates to produce the same result as an AND gate, as Figure 9-12 illustrates.

Though none of the three basic gates is functionally complete by itself, the NAND gate does have this remarkable property.

EXAMPLE 9-19 Show that the set of gates {NAND} is functionally complete.

Solution We will have proved this if we show that each of the gates OR and NOT can be created from a combination of NAND gates. (Since we've already proved, above, that the set {OR, NOT} is functionally complete, if we can create OR and NOT gates with only NAND gates, the NAND gate will also be functionally complete.) Since by definition

$$x \text{ NAND } y = \overline{xy}$$

we can write

$$x \text{ NAND } x = \overline{xx} = \overline{x}$$

and

$$(x \text{ NAND } x) \text{ NAND } (y \text{ NAND } y) = (\overline{x}) \text{ NAND } (\overline{y}) = \overline{\overline{x}\,\overline{y}} = x + y$$

These show that $\overline{x}$ (the NOT gate) and $x + y$ (the OR gate) may be written as NAND gate combinations. Thus {NAND} is functionally complete. The circuits for these are shown in Figures 9-13a and 9-13b, respectively.

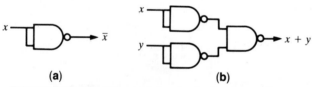

(a)   **(b)**

FIGURE 9-13 NAND gates can create the same effect as (a) a NOT gate and (b) an OR gate.

note: Often circuit designers, for the sake of construction simplicity, will alter the final design to use only the NAND gate. This, of course, is a consequence of the above properties.

9-5. Applications—Binary Addition

A. The half adder

In a computer, the circuit to perform binary arithmetic is a most important function. Consider how addition of two binary numbers goes. When a computer adds the numbers 13 and 6 in binary arithmetic, for example, we get

$$\begin{array}{r} \text{carries} \to 1\,1 \\ 13_{10} = 1101 \\ 6_{10} = \underline{\quad 110} \\ 10011 \to \text{result is } 19_{10} \end{array}$$

Thus the only rules we need for binary addition are

$$0 + 0 = 0 \qquad 0 + 1 = 1 + 0 = 1 \qquad 1 + 1 = 10$$

Notice that these cover all the possibilities of adding two single-bit binary numbers together. (The addition of $1 + 1$ plus a carry 1, which equals 11, is performed as two separate additions: $1 + 1 = 10$, $10 + 1 = 11$.) But also notice that we have one instance in which the answer takes up two bit places instead of one ($1 + 1 = 10$). To allow a computer to deal with the possibility of a two-bit answer, all the answers to binary additions are recorded as a pair of bits c and s, in which s is the **sum number** (the right-hand bit) and c is the **carry number** (the left-hand bit). Thus all the possible values of c and s would be as in Table 9-9. A circuit that adds two single-bit binary numbers is called a **half adder** (so called only because two half adders make a full adder, discussed in the next section). The output values of a half adder are exactly those outputs described in Table 9-9.

TABLE 9-9

x	y	Carry c	Sum s
0	0	0	0
0	1	0	1
1	0	0	1
1	1	1	0

Note that the carry column of Table 9-9 is identical to the AND output, while the sum column is identical to the XOR output. Thus we can say that the value for c is xy and the value for s is $x\bar{y} + y\bar{x}$.

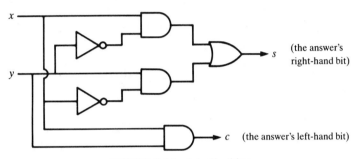

FIGURE 9-14   A half adder.

The circuit design that will add x to y and give us the correct two-bit answer is shown in Figure 9-14. This is a half adder and uses six gates. (Notice that this circuit produces *two* outputs: the bit that is put down in the sum and the bit that is carried to the next column.) The circuit for the half adder is produced in its simplest form using the equivalency of the XOR gate to produce the sum output:

$$s = x \oplus y = x\bar{y} + y\bar{x} = (x + y)(\overline{xy})$$

and the expression $c = xy$ to produce the carry output. This allows us to use only four gates to obtain the efficient half adder shown in Figure 9-15.

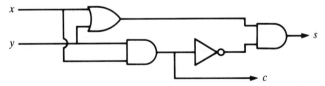

FIGURE 9-15   A simplified half adder.

B. The full adder

The addition of two single-bit binary numbers is done with the half adder. But suppose—as is usually the case—we want to add numbers that are two bits or more longer? In this case, addition in any of the other columns requires *three* input values: the bits x and y being added and a potential carry bit, call it c_{i-1}, from the column to the right of x and y. The **full adder**, which adds three single-bit binary numbers, is constructed from two half adders as in Figure 9-16.

Observe that we send the carry bit c_{i-1} and bit x through the first half adder, obtaining a sum bit s' and a new carry bit c'. The sum bit s' meets up with the value of y in a second half adder, which, in turn, produces a final sum bit s and a carry bit c''. The carry bits c' and c'' (one of which will always be

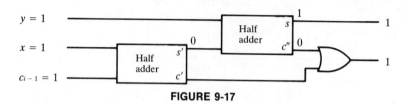

FIGURE 9-16   A full adder.

a zero) are then added together to get a final value of the carry bit c_i, the carry bit of the entire addition. Figure 9-17 shows the full adder at work adding $1 + 1$ plus a carry bit of 1.

FIGURE 9-17

Now, Figure 9-16 obviously isn't a complete circuit design, since the half adders aren't gates but are themselves a series of gates. To design the full adder circuit efficiently, we must have a table of all possible outputs. From the binary addition rules, we find that the addition of any three single-bit numbers produces the final outputs c and s as described in Table 9-10. (The variable of the carry bit c_{i-1} has been replaced by the general variable z.)

Table 9-10 clearly shows that the output for s is 1 when only one of x, y, z is 1 or when all of x, y, z are 1. This is the same as the exclusive–or form, $x \oplus y \oplus z$ (see Problem 9-36). Hence, we may use two XOR circuits to obtain the desired s outputs. To obtain the desired c outputs, we first see that Table 9-10 shows $c = 1$ when we have

TABLE 9-10

x	y	z	c	s
0	0	0	0	0
0	1	0	0	1
1	0	0	0	1
1	1	0	1	0
0	0	1	0	1
0	1	1	1	0
1	0	1	1	0
1	1	1	1	1

$$x y \bar{z} + \bar{x} y z + x \bar{y} z + x y z$$

This may be simplified in the following way:

$$
\begin{aligned}
x y \bar{z} + \bar{x} y z + x \bar{y} z + x y z &= x y (\bar{z} + z) + \bar{x} y z + x \bar{y} z && \text{Distributive law on 1st and 4th terms.} \\
&= x y + \bar{x} y z + x \bar{y} z + x y z && \text{Add in a repetition of an original term.} \\
&= x y + \bar{x} y z + x z (\bar{y} + y) && \text{Distributive law on last two terms.} \\
&= x y + x z + \bar{x} y z + x y z && \text{Add in an original term.} \\
&= x y + x z + y z (\bar{x} + x) && \text{Distributive law on last two terms.} \\
&= x y + x z + y z \\
&= x y + z (x + y) && \text{Distributive law again.}
\end{aligned}
$$

The full adder circuit is now easy to construct using ten gates (Figure 9-18).

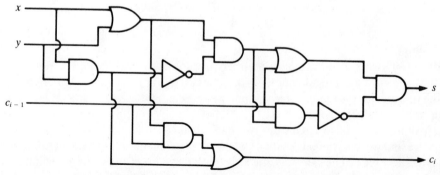

FIGURE 9-18   The circuit of a full adder.

C. Digital display

Many commonplace display devices use a seven-segment display to produce the ten decimal digits from 0 through 9:

Segment identifier

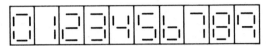

Formation of digits using the seven-segment display

If we have binary coded decimal input, then all the digits may be formed from 4-bit input signals; call the bits of these 4-bit signals W, X, Y, Z, respectively. Then the output and input patterns would be as shown in Table 9-11.

TABLE 9-11

Digit to be displayed	Input values				Output needed						
	W	X	Y	Z	a	b	c	d	e	f	g
0	0	0	0	0	1	1	1	1	1	1	0
1	0	0	0	1	0	1	1	0	0	0	0
2	0	0	1	0	1	1	0	1	1	0	1
3	0	0	1	1	1	1	1	1	0	0	1
4	0	1	0	0	0	1	1	0	0	1	1
5	0	1	0	1	1	0	1	1	0	1	1
6	0	1	1	0	0	0	1	1	1	1	1
7	0	1	1	1	1	1	1	0	0	0	0
8	1	0	0	0	1	1	1	1	1	1	1
9	1	0	0	1	1	1	1	0	0	1	1

The design of the circuit to perform the desired operations would follow a block-design in which four inputs create eight outputs:

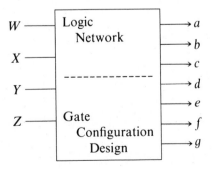

Even in everyday devices, the logic circuits of this chapter are common usage. Look carefully at a digital wristwatch. The numbers are probably formed exactly as above. This logic network is built into the computer chip that runs the watch.

9-6. Minimizing Expressions—Karnaugh Maps

In the last section we used our knowledge of Boolean algebra to design logic circuits. Since there are many ways to design a circuit to do a particular job, we are vitally interested in a design that accomplishes the goal with a minimal effort (in terms of cost or materials or space). If a particular circuit is described by a Boolean expression, we want to find another, simpler, expression that is equivalent to the original one. In turn, this simpler expression will obviously produce a simpler circuit. In this section we will investigate a special tool for minimizing Boolean expressions.

A. The disjunctive normal form

If a Boolean expression is generated by variables $x_1, x_2, \ldots, x_n$, then a **minterm** is a product of n distinct variables. There are 2^n possible minterms generated by n variables. For variables x_1 and x_2 the four possible minterms are

$$x_1 x_2 \qquad \bar{x}_1 x_2 \qquad x_1 \bar{x}_2 \qquad \bar{x}_1 \bar{x}_2$$

For variables x, y, and z the eight possible minterms are

$$xyz \qquad \bar{x}yz \qquad x\bar{y}z \qquad xy\bar{z} \qquad \bar{x}\bar{y}z \qquad \bar{x}y\bar{z} \qquad x\bar{y}\bar{z} \qquad \bar{x}\bar{y}\bar{z}$$

note: To be a minterm of a Boolean expression of n variables, a term *must* contain all n variables. Thus if we have $f(x, y, z)$, the terms xy and $y\bar{z}$ aren't minterms, but xyz and $\bar{x}y\bar{z}$ are.

A Boolean expression is said to be in **disjunctive normal form** if it is written as the sum of minterms. Every Boolean expression can be written in this form, abbreviated **d.n.f.**

EXAMPLE 9-20 Change the expression $f(x_1, x_2) = (x_1 + \bar{x}_2)x_2$ into disjunctive normal form.

Solution Simply expand it using the proper rules of Boolean algebra to produce

$$f(x_1, x_2) = (x_1 + \bar{x}_2)x_2 = x_1 x_2 + x_2 \bar{x}_2 = x_1 x_2 + 0 = x_1 x_2$$

Thus the expression $(x_1 + \bar{x}_2)x_2$ is equivalent to the d.n.f. $x_1 x_2$.

EXAMPLE 9-21 Change the expression $g(x, y, z) = (x + y)z$ to disjunctive normal form.

Solution There are two methods. The first deals directly with the expression and, using the rules, modifies it to the desired form:

$$g(x, y, z) = xz + yz$$

$$= x \cdot 1 \cdot z + 1 \cdot y \cdot z$$

Though this is a simpler form, it is not d.n.f. because it is not written as the sum of minterms.

$$= x \cdot (y + \bar{y}) \cdot z + (x + \bar{x}) \cdot y \cdot z$$

The repeated term is cast away, since $x \vee x = x$.

$$= xyz + x\bar{y}z + xyz + \bar{x}yz$$

$$= xyz + x\bar{y}z + \bar{x}yz$$

The second method involves creating the logic table for g and using it to identify the minterms that produce the desired output. The d.n.f. is then written directly from this table:

$$g(x, y, z) = xz + yz$$

x	y	z	xz	yz	$xz + yz$
0	0	0	0	0	0
0	1	0	0	0	0
1	0	0	0	0	0
1	1	0	0	0	0
0	0	1	0	0	0
0	1	1	0	1	1
1	0	1	1	0	1
1	1	1	1	1	1

At each point where the function value is zero, we have zero contribution. Where the function value is 1 we have a three-variable term that contributes to the desired output (1). Thus this minterm is a term of the d.n.f. Writing the minterms—and thus the d.n.f.—from the table is a simple task: If a 1 occurs in the x, y, or z column, the minterm has the variable x, y, or z in it; if a 0 occurs in the x, y, or z column, the minterm has the variable $\bar{x}$, $\bar{y}$, or $\bar{z}$ in it. From this table we see there are only three minterms that produce a value of 1, so the d.n.f. is:

$$g(x, y, z) = \bar{x}yz + x\bar{y}z + xyz$$

Recall that this is exactly what we did in Section 9-4, where we wrote the expression of a circuit from the logic table of Example 9-16. We were actually producing the d.n.f. of that circuit!

note: All equivalent expressions of the same Boolean function lead to the same d.n.f., since they all produce the same logic table.

EXAMPLE 9-22 From the logic table given, find the d.n.f.

x	y	$f(x, y)$
0	0	1
0	1	1
1	0	0
1	1	1

Solution There are three minterms that produce the desired output 1. We get

$$f(x, y) = 1 \cdot \bar{x}\bar{y} + 1 \cdot \bar{x}y + 0 \cdot x\bar{y} + 1 \cdot xy$$
$$= \bar{x}\bar{y} + \bar{x}y + xy$$

Now we finally have all the tools we need to attack the problem of getting the simplest Boolean expression so that circuit design is optimal.

B. Karnaugh maps

We know that every Boolean expression corresponds to a unique logic table and that this generates a unique function, the disjunctive normal form, that is a sum of products (minterms). We are now ready to find the simplest expression equivalent to the d.n.f. generated by the truth table. To do so, we use the **Karnaugh map**, a rectangular array of boxes in which the tops and sides are labeled with Boolean variables and their complements. The map represents the d.n.f. of a Boolean expression by identifying the minterms of the d.n.f. An entry of 1 in a particular box indicates the presence of a corresponding minterm in the expression. For example, the following Karnaugh maps (read sort of like a multiplication table) "map" the expressions $f(x, y) = xy + xy$ and $g(x, y) = xy$, respectively:

We will explore how to map expressions that have two variables and expressions that have three variables.

1. Map in two variables

An expression in two variables requires a collection of four boxes labeled in the following appropriate way:

The boxes are now associated with the four possible minterms for the two variables: $\bar{x}\bar{y}$, $\bar{x}y$, $x\bar{y}$, and xy. We will say that two boxes are **adjacent** if they share a common side; two adjacent boxes correspond to a single variable (that of the row or column they share). Thus if a 1 appears in the upper left and lower left boxes we have a correspondence with the variable $\bar{y}$. Therefore, if a d.n.f. has the two minterms $\bar{x}\bar{y}$ and $x\bar{y}$, its Karnaugh map shows two adjacent boxes $\bar{x}\bar{y}$ and $x\bar{y}$, and we can *replace* the two minterms in the d.n.f. by the value of $\bar{y}$. (This works out with the laws: $\bar{x}\bar{y} + x\bar{y} = \bar{y}(x + \bar{x}) = \bar{y} \cdot 1 = \bar{y}$.) Similarly, a 1 in the lower left and lower right boxes indicates a correspondence with the variable x, so the sum $x\bar{y} + xy$ can be replaced by the simpler value x.

Now consider the Boolean expression (the disjunctive normal form) that was generated from the logic table in Example 9-22. We have $f(x, y) = \bar{x}\bar{y} + \bar{x}y + xy$. The Karnaugh map for this expression has a 1 in the boxes corresponding to these minterms. We therefore get

	$\bar{y}$	y
$\bar{x}$	1	1
x		1

We want to use this map to reduce the function to a simplified form, so now we check for adjacent pairs. Circle adjacent pairs to cover all the 1's you can, but don't use more circles than necessary.

	$\bar{y}$	y
$\bar{x}$	1	1
x		1

(Note that overlapping is allowed.) Write down the new simplified expression for $f(x, y)$ as the sum of the new single-variable expressions corresponding to the circles drawn. From the top circle we get $\bar{x}$; from the right circle we get y. The simplified expression is therefore

$$f(x, y) = \bar{x} + y$$

We prove this in the next example.

EXAMPLE 9-23   As a check of the above process, form the logic table for the newly derived expression and compare it to the logic table from Example 9-22. Then compare the logic circuits for the original expression and the simplified expression.

Solution   The function in question is $f(x, y) = \bar{x} + y$. It produces the same logic table as the original function in Example 9-22, so the functions are indeed equivalent:

x	y	$\bar{x}$	$\bar{x} + y$
0	0	1	1
0	1	1	1
1	0	0	0
1	1	0	1

└── same as in Example 9-22

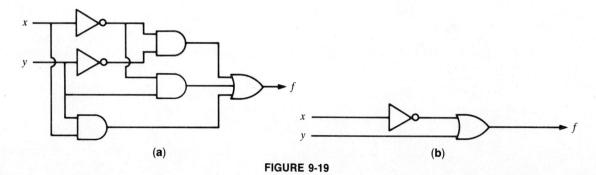

(a)   **(b)**

FIGURE 9-19

Figures 9-19a and 9-19b show the circuits for $f(x, y) = \bar{x}\bar{y} + \bar{x}y + xy$ (requiring six gates) and for $f(x, y) = \bar{x} + y$ (requiring only two gates), respectively.

---

note: Sometimes the d.n.f. cannot be simplified, as in $f(x, y) = x\bar{y} + \bar{x}y$. The Karnaugh map for this function shows no adjacent pairs, so we can't reduce it further.

2. Map in three variables

Since there are eight possible minterms in an expression in three variables we have a more complicated situation. The map will contain eight boxes, and we will label them as follows:

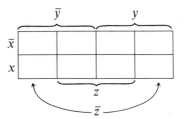

If we fill in all the box labels, our general Karnaugh map describes all eight minterms like this:

$\bar{x}\bar{y}\bar{z}$	$\bar{x}\bar{y}z$	$\bar{x}yz$	$\bar{x}y\bar{z}$
$x\bar{y}\bar{z}$	$x\bar{y}z$	xyz	$xy\bar{z}$

For this type of map we define two boxes to be adjacent if the minterms they correspond to differ in only a single variable (that is, if the minterms share two out of three variables). This definition of adjacency includes boxes that "wrap around" the sides of the map (think of the map as being wrapped around your favorite soft-drink can. The ends would meet and be adjacent.):

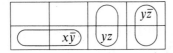

Then an adjacent pair of boxes may be described with an expression that has one fewer variable than either of the original minterms. Thus all four of the following expressions result from adjacency:

$$x\bar{y}\bar{z} + x\bar{y}z = x\bar{y}(\bar{z} + z) = x\bar{y} \cdot 1 = x\bar{y}$$

$$\bar{x}yz + xyz = yz(\bar{x} + x) = yz \cdot 1 = yz$$

$$\bar{x}y\bar{z} + \bar{x}\bar{y}\bar{z} = \bar{x}\bar{z}(y + \bar{y}) = \bar{x}\bar{z} \cdot 1 = \bar{x}\bar{z}$$

$$xy\bar{z} + x\bar{y}\bar{z} = x\bar{z}(y + \bar{y}) = x\bar{z} \cdot 1 = x\bar{z}$$

There are also groups of four boxes that can be represented by an expression in only one variable:

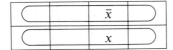

The process for simplifying a Boolean expression containing minterms in three variables follows a process similar to that used with two variables. For each minterm in the expression, enter a 1 in the corresponding box. Then enclose in ovals those adjacent cells having 1's in them, in either quadruples or pairs. Try to join larger groups of cells first; then join the smaller groups, being sure that all possible 1's are covered. Overlapping of ovals is allowed, but try to minimize the number of ovals you use.

EXAMPLE 9-24 Given the Boolean expression

$$f(x, y, z) = xyz + \bar{x}\bar{y}\bar{z} + \bar{x}yz + x\bar{y}z + \bar{x}\bar{y}z$$

draw the Karnaugh map and find the simplifed expression.

Solution Since the expression is already in d.n.f., we only need to enter the 1's in the boxes corresponding to the minterms:

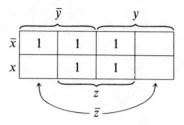

Then we look for adjacent quadruples and pairs. We have one large oval and one small oval that overlaps it:

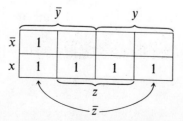

The large one corresponds to z (all of the minterms in that quadruple share a z), and the small one corresponds to $\bar{x}\bar{y}$ (those two minterms share $\bar{x}\bar{y}$). Thus the simplified expression is

$$f(x, y, z) = z + \bar{x}\bar{y}$$

EXAMPLE 9-25 Suppose a circuit is to be designed and the engineer finds the Boolean expression for it to be

$$f(x, y, z) = (x(y + z)) + \bar{y}\bar{z}$$

Form the Karnaugh map and obtain a simplified expression.

Solution Since the function is not in disjunctive normal form, we need to get the logic table first and then form the d.n.f. The logic table is

x	y	z	$\bar{y}$	$\bar{z}$	$\bar{y}\bar{z}$	$y + z$	$x(y + z)$	f
0	0	0	1	1	1	0	0	1
0	1	0	0	1	0	1	0	0
1	0	0	1	1	1	0	0	1
1	1	0	0	1	0	1	1	1
0	0	1	1	0	0	1	0	0
0	1	1	0	0	0	1	0	0
1	0	1	1	0	0	1	1	1
1	1	1	0	0	0	1	1	1

Which produces five minterms. Form the minterms according to the entry of 1 or 0 in the x, y, and z columns. The d.n.f. of the given function is therefore

$$f(x, y, z) = \bar{x}\bar{y}\bar{z} + x\bar{y}\bar{z} + xy\bar{z} + x\bar{y}z + xyz$$

The expression's Karnaugh map is

from which we form two ovals:

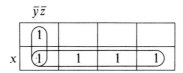

The large one corresponds to x; the small one corresponds to $\bar{y}\bar{z}$. Thus the simplified form is

$$f(x, y, z) = x + \bar{y}\bar{z}$$

For the engineer this will result in a great saving of circuitry.

Now we're ready to tie all of this together. Remember the student senate voting machine of Section 9-3A? We're going to make the circuit for that machine. We repeat the values of Table 9-7 here for reference:

A	B	C	Yes votes	No votes	State of bell
0	0	0	0	13	0
0	0	1	3	10	0
0	1	0	4	9	0
0	1	1	7	6	1
1	0	0	6	7	0
1	0	1	9	4	1
1	1	0	10	3	1
1	1	1	13	0	1

EXAMPLE 9-26 Form the simplified function and draw the circuit of the student senate voting machine from Section 9-3A.

Solution There are four minterms in the d.n.f., corresponding to the four 1's in the "State of bell" column. Thus we get

$$f(A, B, C) = \bar{A}BC + A\bar{B}C + AB\bar{C} + ABC$$

The Karnaugh map produces three adjacent pairs:

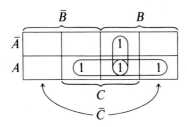

so the d.n.f. reduces to the simplified form

$$f(A, B, C) = AC + AB + BC$$

The circuit is shown in Figure 9-20.

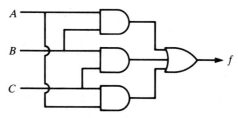

FIGURE 9-20 The circuit for the student senate voting machine.

SUMMARY

1. A switching system is an adaptation of a Boolean algebra.
2. A Boolean algebra is a sextuple that consists of a set S, two binary operations, a unary operation, and two distinct elements of S (identities) and that satisfies ten axioms.

3. A Boolean expression is a combination of Boolean variables and operators.
4. In a Boolean function an *n*-tuple of elements is matched with a single element.
5. Boolean expressions are equivalent if they have the same truth table.
6. In simplifying Boolean expressions the following laws are important tools: Commutative, Associative, Distributive, Idempotent, Absorption, DeMorgan's, and Identity.
7. Logic gates are the physical realization of the Boolean operations. They combine to form circuit networks.
8. The disjunctive normal form allows a Boolean expression to be written as minterms. It may be formed from the logic table.
9. Karnaugh maps are used to write the disjunctive normal form into a simplified Boolean expression.

RAISE YOUR GRADES

Can you ...?

☑ design a circuit to perform a particular operation
☑ define a Boolean algebra
☑ construct a Boolean expression and function
☑ recognize a Boolean tautology and contradiction
☑ define the Boolean laws
☑ construct the logic tables for the six logic gates
☑ define a functionally complete set of gates
☑ create a disjunctive normal form from a logic table
☑ form a Karnaugh map in two variables and three variables
☑ enter a d.n.f. into a Karnaugh map and output its simplified form

SOLVED PROBLEMS

Boolean Algebra

PROBLEM 9-1 Use a truth table to show the equivalence of the Boolean expressions $\overline{\overline{x} + y}$ and $x * \overline{y}$.

Solution

x	y	$\overline{x}$	$\overline{x} + y$	$\overline{\overline{x} + y}$	$\overline{y}$	$x * \overline{y}$
0	0	1	1	0	1	0
0	1	1	1	0	0	0
1	0	0	0	1	1	1
1	1	0	1	0	0	0

             └── same ──┘

PROBLEM 9-2 Define a set S to contain all divisors of 30. (Thus $S = \{1, 2, 3, 5, 6, 10, 15, 30\}$.) Define the binary operations as follows: for x and y in S, let

$$x + y = \text{lcm}(x, y) \quad \text{(the least common multiple of } x \text{ and } y)$$
$$x * y = \text{gcd}(x, y) \quad \text{(the greatest common divisor of } x \text{ and } y)$$

Define the unary operation

$$\overline{x} = 30/x$$

For example

$$3 + 10 = \text{lcm}(3, 10) = 30 \qquad 3 * 10 = \text{gcd}(3, 10) = 1 \qquad \bar{3} = 10$$
$$3 + 15 = \text{lcm}(3, 15) = 15 \qquad 3 * 15 = \text{gcd}(3, 15) = 3 \qquad \bar{5} = 6$$
$$6 + 15 = \text{lcm}(6, 15) = 30 \qquad 6 * 15 = \text{gcd}(6, 15) = 3 \qquad \overline{15} = 2$$

This is the beginning of a valid Boolean algebra. Form the truth (logic) tables for the binary operations of this new Boolean algebra we are creating, and then find the identities. (You are proving that the defined operations lcm and gcm are valid binary operations for the new Boolean algebra.)

Solution Put the tables into cross-reference form.

lcm (+)	1	2	3	5	6	10	15	30
1	1	2	3	5	6	10	15	30
2	2	2	6	10	6	10	30	30
3	3	6	3	15	6	30	15	30
5	5	10	15	5	30	10	15	30
6	6	6	6	30	6	30	30	30
10	10	10	30	10	30	10	30	30
15	15	30	15	15	30	30	15	30
30	30	30	30	30	30	30	30	30
	↑							

gcd (*)	1	2	3	5	6	10	15	30
1	1	1	1	1	1	1	1	1
2	1	2	1	1	2	2	1	2
3	1	1	3	1	3	1	3	3
5	1	1	1	5	1	5	5	5
6	1	2	3	1	6	2	3	6
10	1	2	1	5	2	10	5	10
15	1	1	3	5	3	5	15	15
30	1	2	3	5	6	10	15	30
								↑

x	1	2	3	5	6	10	15	30
$\bar{x}$	30	15	10	6	5	3	2	1

To find the identities, examine the columns of the tables for the lcm and gcm operations. The identity for the lcm operation would be lcm$(i, y) = y$, for any y. In our new Boolean algebra, the identity for lcm is $i = 1$, because—according to the lcm table—the lcm of 1 and any value in S produces the original value (see arrow at column). The identity for the gcd operation would be, for any y, gcd$(i, y) = y$. In our new Boolean algebra the identity for gcd is $i = 30$ because—according to the gcd table—the gcd of 30 and any value in S produces the original value (see arrow in column). Thus the complete Boolean algebra is $[S, \text{lcm}, \text{gcd}, \bar{\ }, 1, 30]$.

PROBLEM 9-3 For the Boolean algebra described in Problem 9-2, verify the axioms defined in the following expressions:

(a) $3 + (6 * 10) = (3 + 6) * (3 + 10)$ Distributive
(b) $5 * (10 * 15) = (5 * 10) * 15$ Associative
(c) $30 * (2 + 6) = (30 * 2) + (30 * 6)$ Distributive

Solution Use the tables from Problem 9-2 and the order of operations:

(a) $3 + (6 * 10) = 3 + 2 = 6; (3 + 6) * (3 + 10) = 6 * 30 = 6$
(b) $5 * (10 * 15) = 5 * 5 = 5; (5 * 10) * 15 = 5 * 15 = 5$
(c) $30 * (2 + 6) = 30 * 6 = 6; (30 * 2) + (30 * 6) = 2 + 6 = 6$

PROBLEM 9-4 Let $D_{18} =$ the set of divisors of $18 = \{1, 2, 3, 6, 9, 18\}$. Define the binary operations to be lcm and gcd. Define the unary operation to be $\bar{x} = 18/x$. Show that $B = [D_{18}, \text{lcm}, \text{gcd}, \bar{\ }, 1, 18]$ is not a Boolean algebra.

Solution First let's create the complements table (using $\bar{x} = 18/x$) and the binary operations tables to identify all the values our supposed algebra generates:

lcm (+)	1	2	3	6	9	18
1	1	2	3	6	9	18
2	2	2	6	6	18	18
3	3	6	3	6	9	18
6	6	6	6	6	18	18
9	9	18	9	18	9	18
18	18	18	18	18	18	18

gcd (*)	1	2	3	6	9	18
1	1	1	1	1	1	1
2	1	2	1	2	1	2
3	1	1	3	3	3	3
6	1	2	3	6	3	6
9	1	1	3	3	9	9
18	1	2	3	6	9	18

x	1	2	3	6	9	18
$\bar{x}$	18	9	6	3	2	1

Now, recall that to be a Boolean algebra, B must satisfy ten axioms. Let's start with the Complements axioms; here $x + \bar{x} = 18$ and $x * \bar{x} = 1$ must hold for all values of x in the set D_{18}. But by inspecting the tables we see that the complementary values 3 and 6 do not produce the required results:

$$\begin{Bmatrix} 3 + 6 \neq 18 \\ 6 + 3 \neq 18 \end{Bmatrix} \quad \text{and} \quad \begin{Bmatrix} 3 * 6 \neq 1 \\ 6 * 3 \neq 1 \end{Bmatrix}$$

Thus this is not a Boolean algebra.

PROBLEM 9-5 Define the set $P = \{00, 01, 10, 11\}$. Define a binary operation $+$ by adding positional digits (that is, $01 + 11 = 10$ since $0 + 1 = 1$ and $1 + 1 = 1$, using the tables of the series and parallel switch connections) and define the binary operation $*$ by multiplying positional digits (that is, $10 * 01 = 00$ since $1 * 0 = 0$ and $0 * 1 = 0$, again with the switch connection tables). The unary operation $\bar{\ }$ is by digit complementation (that is, $\overline{11} = 00$ and $\overline{01} = 10$).

(a) Construct the tables for $+$, $*$, and $\bar{\ }$.
(b) Find the identities and verify the complement properties for this Boolean algebra.
(c) Evaluate $11 + (01 * 11) * (10 + 11)$

Solution

(a) Construct row and column reference tables for the binary and unary operations:

$+$	00	01	10	11
00	00	01	10	11
01	01	01	11	11
10	10	11	10	11
11	11	11	11	11

$*$	00	01	10	11
00	00	00	00	00
01	00	01	00	01
10	00	00	10	10
11	00	01	10	11

	$(\bar{\ })$
00	11
01	10
10	01
11	00

(b) From the above tables it is easy to see that the identity for $+$ is 00 since for any element x, we have

$$x + 00 = 00 + x = x$$

The identity for $*$ is 11 since for any element x we have

$$x * 11 = 11 * x = x \quad \text{and} \quad x * \bar{x} = 00$$

The complement properties $x + \bar{x} = 11$ and $x * \bar{x} = 00$ are verified by

x	$\bar{x}$	$x + \bar{x}$	$x * \bar{x}$
00	11	11	00
01	10	11	00
10	01	11	00
11	00	11	00

(c) $11 + (01 * 11) * (10 + 11) = 11 + (01) * (11)$ Inside parentheses first.

$\qquad\qquad\qquad\qquad\qquad = 11 + 01$ Multiplication takes precedence over addition.

$\qquad\qquad\qquad\qquad\qquad = 11$

Boolean Functions and Boolean Expressions

PROBLEM 9-6 Use the laws and axioms of Boolean expressions to show that the expression

$$x \lor (\bar{y} \land z) \lor \bar{z} \land (x \lor (\bar{y} \land z))$$

is equivalent to

$$x \vee (\bar{y} \wedge z)$$

Solution

$$
\begin{aligned}
x \vee (\bar{y} \wedge z) \vee \bar{z} \wedge (x \vee (\bar{y} \wedge z)) &= 1 \wedge (x \vee (\bar{y} \wedge z)) \vee \bar{z} \wedge (x \vee (\bar{y} \wedge z)) && \text{Identity Axiom} \\
&= (1 \vee \bar{z}) \wedge (x \vee (\bar{y} \wedge z)) && \text{Distributive Law} \\
&= 1 \wedge (x \vee (\bar{y} \wedge z)) && \text{Identity Law} \\
&= x \vee (\bar{y} \wedge z) && \text{Identity Law}
\end{aligned}
$$

PROBLEM 9-7 Show that $(x \wedge y) \vee (\bar{x} \wedge y) = y$ and write its dual equality.

Solution Use the laws of Boolean algebra.

$$
\begin{aligned}
(x \wedge y) \vee (\bar{x} \wedge y) &= (y \wedge x) \vee (y \wedge \bar{x}) && \text{Commutative Law} \\
&= (y \vee (x \wedge \bar{x}) && \text{Distributive Law} \\
&= y \vee 0 && \text{Complementation Law} \\
&= y && \text{Identity Law}
\end{aligned}
$$

To find the dual equality, you swap $\wedge \leftrightarrow \vee$ for every operation in the statement. The dual equality is therefore

$$(x \vee y) \wedge (\bar{x} \vee y) = y$$

PROBLEM 9-8 Verify the Absorption Laws: Let a and b be elements of a Boolean algebra B. Then

$$a + (a * b) = a \quad \text{and} \quad a * (a + b) = a.$$

Solution As usual, this requires manipulating the statements with the laws.

$$
\begin{aligned}
a + (a * b) &= a * 1 + (a * b) && \text{Identity Axiom } (a = a * 1) \\
&= a * (1 + b) && \text{Distributive Axiom} \\
&= a * 1 && \text{Identity Axiom } (1 + b = 1 \text{—see Example 9-11}) \\
&= a && \text{Identity Axiom}
\end{aligned}
$$

The expression $a * (a + b) = a$ is true because it is the dual equality of the expression we just proved.

PROBLEM 9-9 Verify the Involution Law: For every element a in a Boolean algebra B, $\bar{\bar{a}} = a$.

Solution Since $\bar{\bar{a}} = \overline{(\bar{a})}$ we are looking for a complement of $\bar{a}$. But we know from complementation rules that $\bar{a} + a = 1$ and $a * \bar{a} = 0$. So a is one complement of $\bar{a}$. Since in a Boolean algebra the complement of an element is a unique element, then a is the only complement of $\bar{a}$. Thus $\bar{\bar{a}} = a$.

PROBLEM 9-10 Use the laws of Boolean algebra to establish the validity of

(a) $(x + \bar{y})y = xy$ **(b)** $\overline{xy} + z = \bar{x} + (\bar{y} + z)$

Solution

(a)
$$
\begin{aligned}
(x + \bar{y})y &= xy + \bar{y}y && \text{Distributive Law} \\
&= xy + 0 && \text{Complementation Law} \\
&= xy && \text{Identity Law}
\end{aligned}
$$

(b)
$$
\begin{aligned}
\overline{(xy)} + z &= (\bar{x} + \bar{y}) + z && \text{DeMorgan's Law} \\
&= \bar{x} + (\bar{y} + z) && \text{Associative Law}
\end{aligned}
$$

PROBLEM 9-11 Use a logic table to establish the validity of $\bar{\bar{x}} = x$.

Solution

x	$\bar{x}$	$\bar{\bar{x}}$
0	1	0
1	0	1

$\llcorner$ same $\lrcorner$

PROBLEM 9-12   For the given table, find the Boolean expression of the circuit the table describes.

(a) x	y	f
0	0	1
0	1	1
1	0	1
1	1	0

(b) x	y	z	g
0	0	0	1
0	1	0	0
1	0	0	0
1	1	0	1
0	0	1	0
0	1	1	1
1	0	1	1
1	1	1	0

Solution   Remember to use x, y, or z for every value of 1 and $\bar{x}$, $\bar{y}$, or $\bar{z}$ for every value of 0. Identify the pairs or triples that close the circuit (that create a final value of 1).

(a) We have three pairs that produce a value of 1 for the circuit. For the $x = 0$, $y = 0$ case, we have $\bar{x}\bar{y}$; for the $x = 0$, $y = 1$ case, we have $\bar{x}y$; and for the $x = 1$, $y = 0$ case, we have $x\bar{y}$. Thus

$$f(x, y) = \bar{x}\bar{y} + \bar{x}y + x\bar{y}$$

(b) We have four triples that produce the desired value of 1. For the $x = 0$, $y = 0$, $z = 0$ case, we have $\bar{x}\bar{y}\bar{z}$; for the $x = 1$, $y = 1$, $z = 0$ case, we have $xy\bar{z}$; for the $x = 0$, $y = 1$, $z = 1$ case we have $\bar{x}yz$; and for the $x = 1$, $y = 0$, $z = 1$ case we have $x\bar{y}z$. Thus

$$g(x, y, z) = \bar{x}\bar{y}\bar{z} + xy\bar{z} + \bar{x}yz + x\bar{y}z$$

Logic Gates

PROBLEM 9-13   Write the Boolean expression associated with the circuits given in Figure 9-21a and 9-21b.

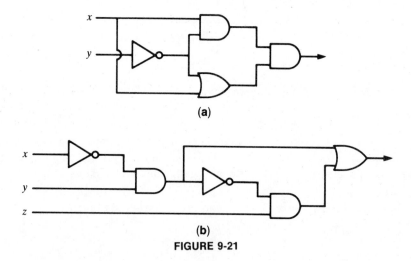

(a)

(b)

FIGURE 9-21

Solution   Trace the variables x and y through the gates, keeping track of the expressions created as they pass through them.

(a) In Figure 9-21a, the y variable goes through a complement gate to become $\bar{y}$, then splits off to join up with x in an AND gate at the top and an OR gate at the bottom. The outputs from these two gates, $(x\bar{y})$ and $(x + \bar{y})$, respectively, then pass through another AND gate to become the final output of $(x\bar{y})(x + \bar{y})$.

(b) In Figure 9-21b, x becomes $\bar{x}$ through a complement gate, and then produces $\bar{x}y$ through an AND gate. This output splits, going straight ahead to become $(\overline{\bar{x}y})$, then combines with z in an AND gate, producing $(\overline{\bar{x}y})z$. This output goes up to join the second output of $\bar{x}y$ in an OR gate. The final output is

$$(\overline{\bar{x}y})z + \bar{x}y$$

PROBLEM 9-14   Find the Boolean expression that corresponds to the gate implementation in Figure 9-22.

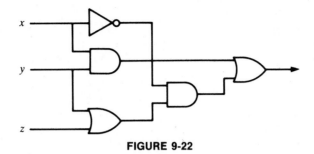

FIGURE 9-22

Solution   Break down the outputs on the first set of three gates at the left: the top gate produces $\bar{x}$, the middle gate produces xy, and the bottom gate produces $y + z$. These outputs become the inputs in stage two, in which $\bar{x}$ and $y + z$ pass through an AND gate to give $\bar{x}(y + z)$. This output then combines at the final stage with xy to produce

$$(xy) + \bar{x}(y + z)$$

PROBLEM 9-15   Build a gate circuit to represent the Boolean expression $x\bar{y} + y$.

Solution   You will need to send y through a complement gate to get $\bar{y}$, then combine that with x in an AND gate to obtain $x\bar{y}$. Finally, pass this through an OR gate with y to obtain the desired result of $x\bar{y} + y$. Figure 9-23 shows the circuit.

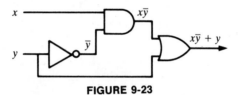

FIGURE 9-23

PROBLEM 9-16   Build a gate circuit to represent the Boolean expression $x\bar{y}z + \bar{x}y$.

Solution   This expression is more complicated than that of Problem 9-15 so its circuit will involve more gates. It's easiest to form the complements and the product terms before putting the product terms through an OR gate to add them together (this corresponds to the order of operations requirements). Thus the final circuit will be as shown in Figure 9-24.

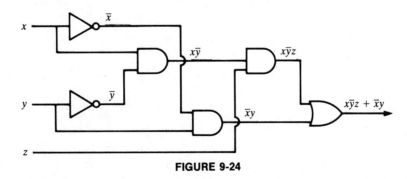

FIGURE 9-24

PROBLEM 9-17   Draw the gate circuit for the function that has the following logic table:

x	y	Output
0	0	1
0	1	0
1	0	1
1	1	0

Solution   We have a closed circuit when x and y are open or when x is closed and y is open (that is, when the output is 1). Thus the Boolean expression of the circuit is

$$f(x, y) = \bar{x}\bar{y} + x\bar{y}$$

and the circuit itself requires five gates (Figure 9-25).

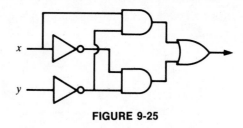

FIGURE 9-25

Minimizing Expressions—Karnaugh Maps

PROBLEM 9-18   Form the logic table for the output expression $f(x, y, z)$ of Problem 9-13(b). Then write the disjunctive normal form of the function.

Solution   The final function of Problem 9-13(b) was $f(x, y, z) = (\overline{\bar{x}y})z + \bar{x}y$. Recall that the disjunctive normal form (d.n.f.) is written as the sum of the minterms and that we get the minterms from the logic table of a function. The logic table of our function looks like this;

x	y	z	$\bar{x}$	$\bar{x}y$	$\overline{\bar{x}y}$	$\overline{\bar{x}y}z$	$f(x, y, z)$		Minterms
0	0	0	1	0	1	0	0		
0	1	0	1	1	0	0	1	$\longrightarrow$	$\bar{x}y\bar{z}$
1	0	0	0	0	1	0	0		
1	1	0	0	0	1	0	0		
0	0	1	1	0	1	1	1	$\longrightarrow$	$\bar{x}\bar{y}z$
0	1	1	1	1	0	0	1	$\longrightarrow$	$\bar{x}yz$
1	0	1	0	0	1	1	1	$\longrightarrow$	$x\bar{y}z$
1	1	1	0	0	1	1	1	$\longrightarrow$	xyz

Because we are only concerned with the combinations of x, y, z that produce a 1 (that is, that create a closed system), we pull out those minterms and add them together. Thus the d.n.f. is

$$f(x, y, z) = \bar{x}y\bar{z} + \bar{x}\bar{y}z + \bar{x}yz + x\bar{y}z + xyz$$

PROBLEM 9-19   Use a Karnaugh map to simplify the d.n.f. you found in Problem 9-18. Draw the new circuit.

Solution   The function in question is $f(x, y, z) = \bar{x}y\bar{z} + \bar{x}\bar{y}z + \bar{x}yz + x\bar{y}z + xyz$. Since this is a three-variable expression, we need a map with eight boxes. Place a 1 in the appropriate box for each of the minterms in the d.n.f., then look for adjacent quadruples and pairs. The Karnaugh map produces one quadruple and one pair:

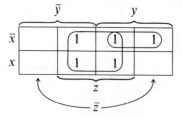

This allows us to reduce the d.n.f. from six terms to two terms. (The terms $\bar{x}\bar{y}z$, $\bar{x}yz$, $x\bar{y}z$, and xyz simplify to z, and the terms $\bar{x}y\bar{z}$ and $\bar{x}yz$ simplify to $\bar{x}y$.) Thus the function becomes

$$f(x, y, z) = z + \bar{x}y$$

Figure 9-26 shows this simple circuit; notice how much more efficient it is than the circuit in Problem 9-13(b) (Figure 9-21b), which represents an equivalent expression.

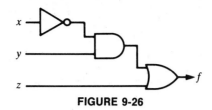

FIGURE 9-26

PROBLEM 9-20   The secret formula for making Smozo-Cola is kept in a safe whose lock is opened electronically using three switches. Each of the top three big-wigs of the company has a unique key for a switch. If at least two of the three keys are turned, the safe will open. Design the system for the lock and derive the Boolean function for the system.

Solution   Let a turned key represent a closed connection; that is, represent it by 1. An open connection (no key) will be a 0. If x, y, and z denote the switches (keys), then the schematic is as shown in Figure 9-27.

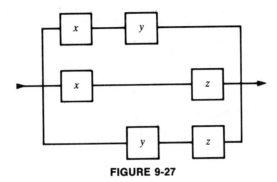

FIGURE 9-27

The circuit is closed (the lock will open) if the function has the property that $f(1, 1, 0) = 1$, $f(1, 0, 1) = 1$, $f(0, 1, 1) = 1$, and $f(1, 1, 1) = 1$. The complete representation of the function then is

$$f(x, y, z) = xy\bar{z} + x\bar{y}z + \bar{x}yz + xyz$$

By inspecting the design in Figure 9-27, however, we can accomplish this task with the even simpler function

$$g(x, y, z) = \quad xy \quad + \quad xz \quad + \quad yz$$

closed or closed or closed
x and y x and z y and z

PROBLEM 9-21   It is common in residential wiring to have two single-throw switches combine to control a single light. For example, switches are sometimes located at the ends of a hall or at the top and bottom of stairs to turn on a single hall or stairwell light. Construct the wiring diagram and form the truth table.

Solution   The switches A and B may be either "up" or "down". Say that if both are up or both are down we have a connection; if one is up and one is down we don't have a connection. The truth table is as follows, and the diagram is shown in Figure 9-28.

A	B	light
up	down	off
up	up	on
down	down	on
down	up	off

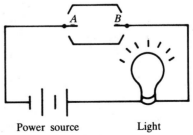

Power source   Light

FIGURE 9-28

PROBLEM 9-22   Use the Karnaugh map to simplify the expressions

(a) $f(x, y) = x\bar{y} + \bar{x}y + xy$ **(b)** $g(x, y, z) = x\bar{y}\bar{z} + xyz + \bar{x}\bar{y}\bar{z} + \bar{x}\bar{y}z$

Solution

(a) This expression has two variables, so we need a 4-box Karnaugh map. Enter the 1's to correspond to the minterms, circle any adjacent pairs, and reduce the expression using the correspondences of the adjacent pairs.

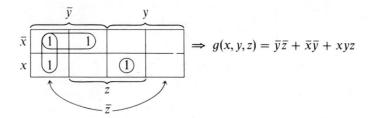

$$\Rightarrow f(x, y) = x + y$$

(b) This expression has three variables, so it requires an 8-box Karnaugh map. Look for possible quadruples first, then for pairs. Simplify as necessary according to the adjacency correspondences.

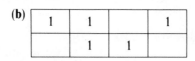

$$\Rightarrow g(x, y, z) = \bar{y}\bar{z} + \bar{x}\bar{y} + xyz$$

PROBLEM 9-23   Write the simplified Boolean expression from the given Karnaugh map.

(a)

1	1	1	1
1	1		1

(b)

1	1		1
	1	1	

Solution   Draw the ovals around adjacent quadruples and pairs and write the corresponding expressions.

(a)

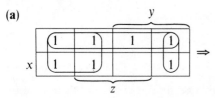

$$f(x, y, z) = \bar{y} + \bar{x} + y\bar{z}$$

(b)

$$g(x, y, z) = \bar{x}\bar{z} + \bar{y}z + xz$$

PROBLEM 9-24   Without using a logic table, change the Boolean expression $B(x, y, z) = y(x + z) + \bar{x}z$ to the disjunctive normal form.

Solution   Use the laws.

$$B(x, y, z) = y(x + z) + \bar{x}z$$
$$= xy + yz + \bar{x}z \qquad\qquad \text{Distributive Law}$$
$$= xy \cdot 1 + yz \cdot 1 + \bar{x}z \cdot 1 \qquad \text{Identity Axiom}$$
$$= xy(z + \bar{z}) + yz(x + \bar{x}) + \bar{x}z(y + \bar{y}) \qquad \text{Complement Axiom}$$
$$= xyz + xy\bar{z} + xyz + \bar{x}yz + \bar{x}yz + \bar{x}\bar{y}z$$
$$= xyz + xy\bar{z} + \bar{x}yz + \bar{x}\bar{y}z$$

PROBLEM 9-25   The circuit in Figure 9-29, using nine input gates, produces an output in the form of a disjunctive normal form. Find the form; reduce it to simpler form and redesign the circuit.

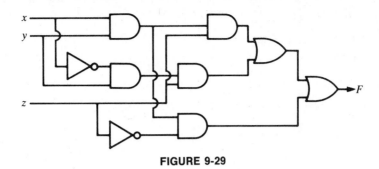

FIGURE 9-29

Solution Beginning across the top row, at the upper left we get xy; then z is put in through another AND gate to get xyz. In the second row at the left, $\bar{x}y$ is formed and then joins up with z at the center to form $\bar{x}yz$. Meanwhile, at the upper right we have an OR gate that produces $xyz + \bar{x}yz$. Then at the bottom row of activity we combine $\bar{z}$ with xy to get $xy\bar{z}$, which meets $xyz + \bar{x}yz$ at another OR gate to get the final result

$$F = xyz + \bar{x}yz + xy\bar{z}$$

We'll use algebra and the Boolean laws to obtain the reduction:

$$
\begin{aligned}
F = xyz + \bar{x}yz + xy\bar{z} &= (xyz + \bar{x}yz) + (xy\bar{z} + xyz) \\
&= (x + \bar{x})yz + xy(\bar{z} + z) \\
&= 1 \cdot yz + xy \cdot 1 = yz + xy \\
&= y(z + x)
\end{aligned}
$$

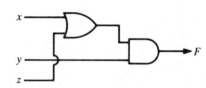

The simplified circuit is shown in Figure 9-30.

FIGURE 9-30

PROBLEM 9-26   Use algebraic techniques to simplify the four-variable Boolean expression

$$G = w\bar{x}y\bar{z} + \bar{w}\bar{x}yz + w\bar{x}yz + \bar{w}\bar{x}y\bar{z}$$

Solution   We must look for terms in which three entries are the same and the fourth entries are complements; this is true of the first and last terms and is true of the inner two terms. Joining these two pairs together, we get

$$
\begin{array}{ll}
G = (w\bar{x}y\bar{z} + \bar{w}\bar{x}y\bar{z}) + (\bar{w}\bar{x}yz + w\bar{x}yz) & \text{Reorder terms and group.} \\
 = (w + \bar{w})(\bar{x}y\bar{z}) + (\bar{w} + w)(\bar{x}yz) & \text{Distributive Law} \\
 = 1 \cdot \bar{x}y\bar{z} + 1 \cdot \bar{x}yz & \text{Complement Axiom} \\
 = \bar{x}y\bar{z} + \bar{x}yz & \text{Identity Axiom} \\
 = \bar{x}y(\bar{z} + z) & \text{Distributive Law} \\
 = \bar{x}y \cdot 1 & \\
 = \bar{x}y & \text{Complement Axiom}
\end{array}
$$

Supplementary Exercises

PROBLEM 9-27   Following the format of Problem 9-2 construct the Boolean algebra $\{D_{110}, \text{lcm}, \text{gcd}, ^-, 1, 110\}$ where $\bar{x} = 110/x$ and D_{110} are all the divisors of 110; that is, $D_{110} = \{1, 2, 5, 10, 11, 22, 55, 110\}$. Construct the tables for the binary and unary operations and prove the identities.

PROBLEM 9-28   Find the dual equalities for **(a)** $a + \bar{a} * b = a + b$ and **(b)** $(a * 1) * (0 + \bar{a}) = 0$.

PROBLEM 9-29 Verify the equality of the expressions.

(a) $(x \wedge y) \vee (\bar{x} \wedge z) \vee (y \wedge z) = (x \wedge y) \vee (\bar{x} \wedge z)$
(b) $(x \vee y) \wedge (x \vee \bar{y}) = x$
(c) $(y \wedge (x \vee \bar{z})) \vee x = x \vee y \wedge z$

PROBLEM 9-30 Using DeMorgan's Laws, verify that $(\overline{\bar{x} \wedge \bar{y} \vee z}) = x \wedge y \wedge \bar{z}$.

PROBLEM 9-31 Verify that $(u \wedge v) \vee (\bar{u} \wedge v) \vee (u \wedge \bar{v}) = u \vee v$.

PROBLEM 9-32 Draw the circuits for Problem 9-29(c).

PROBLEM 9-33 Let $A = x * \bar{y} + x * y * \bar{z} + \bar{x} * y * \bar{z}$. Consider (**1**) A, (**2**) $A + x$, (**3**) $A + \bar{z}$, (**4**) $A + x * \bar{z}$. Which two of these are equivalent?

PROBLEM 9-34 Which of the following expressions are tautologies and which are contradictions?

(a) $\bar{x} \vee (x \vee y)$ (c) $(x \vee y) \wedge (\bar{x} \wedge \bar{y})$
(b) $\overline{x \wedge y} \vee x$ (d) $(\overline{x \wedge \bar{x} \vee y}) \vee (x \wedge y)$

PROBLEM 9-35 Verify the equivalency of the two expressions in Example 9-16 (That is, show $f(x, y, z) = \bar{x}(y + z)$ is equivalent to $g(x, y, z) = \bar{x}y\bar{z} + \bar{x}\bar{y}z + \bar{x}yz$.

PROBLEM 9-36 Verify that the output of $x \oplus y \oplus z$ is the sum digit of the full adder.

PROBLEM 9-37 Show that the circuits in Figure 9-31 are equivalent by constructing logic tables.

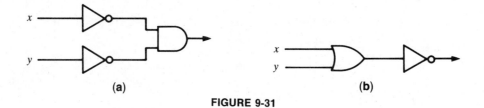

(a)   (b)

FIGURE 9-31

PROBLEM 9-38 Use the given table to find the disjunctive normal form for $f(x, y, z)$.

x	y	z	f
0	0	0	1
0	1	0	0
1	0	0	0
1	1	0	1
0	0	1	1
0	1	1	0
1	0	1	0
1	1	1	1

PROBLEM 9-39 Use algebra techniques to get the d.n.f. for $f(x, y, z) = (x\bar{z} + yz)(\bar{z} + x\bar{y})$.

PROBLEM 9-40 Reconstruct the half adder circuit using only NOR gates.

PROBLEM 9-41 Form a logic table and a simplified Boolean expression for the function

$$f(x, y, z) = \begin{cases} 1 & \text{if } y = z \\ 0 & \text{otherwise} \end{cases}$$

PROBLEM 9-42   Form a logic table and a simplified Boolean expression for the function

$$f(x, y, z) = \begin{cases} 1 & \text{if } x = 0, y = z = 1 \\ 1 & \text{if } x = y = z = 1 \\ 0 & \text{otherwise} \end{cases}$$

Answers to Supplementary Exercises

9-28 **(a)** $a * (\bar{a} + b) = a * b$
(b) $(a + 0) + (1 * \bar{a}) = 1$

9-38 $f(x, y, z) = \bar{x}\bar{y}\bar{z} + xy\bar{z} + \bar{x}\bar{y}z + xyz$

9-32

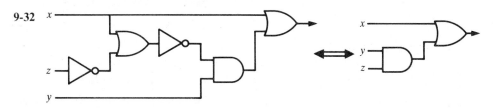

9-33 **(1)** and **(4)** are equivalent

9-39 $f(x, y, z) = xy\bar{z}$

9-34 **(a)**, **(b)**, and **(d)** are tautologies; **(c)** is a contradiction.

9-41

x	y	z	f
0	0	0	1
0	1	0	0
1	0	0	1
1	1	0	0
0	0	1	0
0	1	1	1
1	0	1	0
1	1	1	1

$= \bar{y}\bar{z} + yz$

9-35 Reconstruct g to be $g(x, y, z)$
$= \bar{x}y\bar{z} + \bar{x}yz + \bar{x}\bar{y}z + \bar{x}yz$
$= \bar{x}y(\bar{z} + z) + \bar{x}z(\bar{y} + y)$
$= \bar{x}y + \bar{x}z$
$= \bar{x}(y + z)$

9-36

x	y	z	$x \oplus y$	$(x \oplus y) \oplus z$
0	0	0	0	0
0	1	0	1	1
1	0	0	1	1
1	1	0	0	0
0	0	1	0	1
0	1	1	1	0
1	0	1	1	0
1	1	1	0	1

9-42

x	y	z	f
0	0	0	0
0	1	0	0
1	0	0	0
1	1	0	0
0	0	1	0
0	1	1	1
1	0	1	0
1	1	1	1

$= yz$

9-37 $\bar{x} * \bar{y}$ and $\overline{x + y}$ are same by DeMorgan's Laws.

10 COMBINATORICS— THE ART OF COUNTING

THIS CHAPTER IS ABOUT

- ☑ **A Counting Principle**
- ☑ **Permutations and Combinations**
- ☑ **The Binomial Theorem**

10-1. A Counting Principle

Counting is not always as easy as 1, 2, 3,.... When we're counting results (things, people, whatever) in a multistage process, we need some special techniques, from which we can define a counting principle. In this chapter, we'll investigate some of those techniques.

A. Tree diagrams

Suppose you are required to take a quiz that consists of 3 true/false questions. You haven't studied the material, so you must guess on every question. In how many different ways could you form an answer set? That is, how would you count the number of possible answers?

One way to find out is to construct a **tree diagram**, a graphical depiction of the possible solution sets in a multistage counting process. Figure 10-1 shows its basic construction. From "begin" to "end" there are 8 paths, which represent the 8 possible solution sets. Follow the paths to find the solution sets. That is: The first question presents you with 2 choices, true (T) or false (F). For each of these responses, you have 2 additional choices T or F for the second question, which gives you a total of 4 choices on the first two questions: TT, TF, FT, or FF. For each of these 4 choices you have 2 additional choices on the third question, for a total of $2 \times 2 \times 2 = 8$ choices: TTT, TTF, TFT, TFF, FTT, FTF, FFT, FFF.

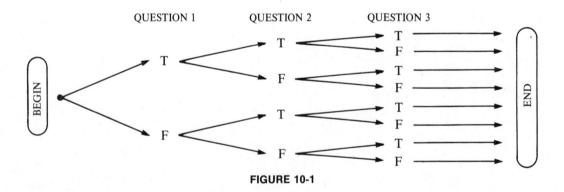

FIGURE 10-1

EXAMPLE 10-1 There are 3 roads connecting Waterville to Clay City and 2 roads connecting Clay City to Mud Flats. A truck driver starting in Waterville must make a delivery in Clay City and end up in Mud Flats. How many different routes can she take?

Solution Construct a tree diagram, as in Figure 10-2. Label the 3 routes from Waterville to Clay City a, b, c, and the 2 routes from Clay City to Mud Flats p, q. Then follow the paths of the routes, which are easily listed by road pairs: ap, aq, bp, bq, cp, cq, for a total of $3 \times 2 = 6$ ways.

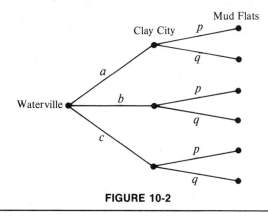

FIGURE 10-2

B. The multiplication rule

The tree diagram is just a picture of a basic technique that may be applied to situations in which counting is required.

EXAMPLE 10-2 A hiking club with 8 members needs to choose a chairperson, secretary, and treasurer for the group. In how many different ways can this be done?

Solution The tree diagram would be overly cumbersome here, but we can still use its construction technique to solve the problem. At step 1, the chairperson can be any of the 8 persons. So this step has 8 choices. Once that office has been filled, any one of the remaining 7 persons could be chosen as secretary. For each of the original 8 choices there are now 7 additional possibilities for a total of $8 \times 7 = 56$ ways of electing the first two officers. Now, for each of those 56 choices any one of the remaining 6 people could be chosen for the treasurer's post. That gives $8 \times 7 \times 6 = 336$ ways of choosing the 3 officers from among the 8 club members.

If the items we are counting are selected in stages, and if we know how many choices we have at each stage, we multiply the choice numbers of the various stages to get the total number of ways of doing the counting. The formal statement of this technique is called the **Multiplication Rule**:

MULTIPLICATION RULE
Suppose N objects are to be selected in a sequence of events. If the first event can occur in k_1 ways, the second in k_2 ways, the third in k_3 ways, and so forth, then the number of ways all N of the events occur is

$$k_1 \times k_2 \times k_3 \times \ldots \times k_N \text{ ways}$$

The multiplication rule is, in fact, a **counting principle**.

EXAMPLE 10-3 Many years ago the automobile license plates in Michigan consisted of 2 letters followed by 4 digits (e.g., MN 5216). How many different letter/digit groupings could be produced in this pattern?

Solution Consider the 6 entries to be selected in 6 stages. The first entry, a letter of the alphabet, may be chosen in 26 ways. The second entry may also be chosen in 26 ways (assuming there is no restriction on the letters' being the same). The next 4 stages involve digits. A digit may be chosen from 0, 1, 2, 3, 4, 5, 6, 7, 8, 9, so there are 10 choices for each of the 4-digit entries. Thus a 6-entry plate may be constructed in

$$26 \times 26 \times 10 \times 10 \times 10 \times 10 = 6\,760\,000 \text{ ways}$$

EXAMPLE 10-4 Not long ago the number of cars in Michigan was approaching 6.5 million and rising. The state officials needed to change the format for the plates to avoid repetition of letter/number combinations. Now the license plates have 3 letters and 3 numbers. How many different plates are now possible?

Solution For each of the 3 letters there are 26 possible choices, and for each of the 3 digits there are 10 possible choices. The total number of ways in which plates can be constructed now is

$$26 \times 26 \times 26 \times 10 \times 10 \times 10 = 17\,576\,000 \text{ ways}$$

EXAMPLE 10-5 At a choose-it-yourself sandwich bar you have the following choices:

Type of bread:	Whole wheat, rye, sourdough, white
Type of spread:	Butter, mayonnaise, mustard
Type of meat:	Ham, beef, turkey, salami
Type of cheese:	American, Swiss
Extras:	Lettuce, tomato, sliced pickles, chopped olives

If you are limited to 1 item from each category, how many visits could you make and have a different sandwich each time?

Solution There are 5 main choices, so this is a 5-stage process, and the number of choices at each stage must be counted. The total number of different sandwiches is

$$4 \times 3 \times 4 \times 2 \times 4 = 384$$

10-2. Permutations and Combinations

A. Simple permutations

We use a special case of the multiplication rule when we want to count the number of possible ordered arrangements of a given collection of items.

EXAMPLE 10-6 Suppose we have 4 balls—red, blue, green, and yellow—which we want to arrange in a row. How many different arrangements are possible?

Solution We may pick any of the 4 colored balls for the first position. Then, there are 3 choices left for the second position, 2 choices for the third, and 1 choice for the fourth. Thus the total number of arrangements is $4 \times 3 \times 2 \times 1 = 24$ arrangements (see Figure 10-3).

- An ordered arrangement of a set of objects is called a **permutation** of those objects.

EXAMPLE 10-7 In how many different ways could 6 people line up to board a bus?

Solution We can think of this as if we were filling 6 waiting spots with 6 different people. We can choose any of the 6 people to put in spot 1, which leaves 5 choices for spot 2, 4 for spot 3, etc. Thus the total number of ordered arrangements, or permutations, is

$$6 \times 5 \times 4 \times 3 \times 2 \times 1 = 6! = 720 \text{ ways}$$

These two examples illustrate a general law:

- The number of permutations of n objects taken n at a time is $n!$, denoted

$$_nP_n = n!$$ (10-1)

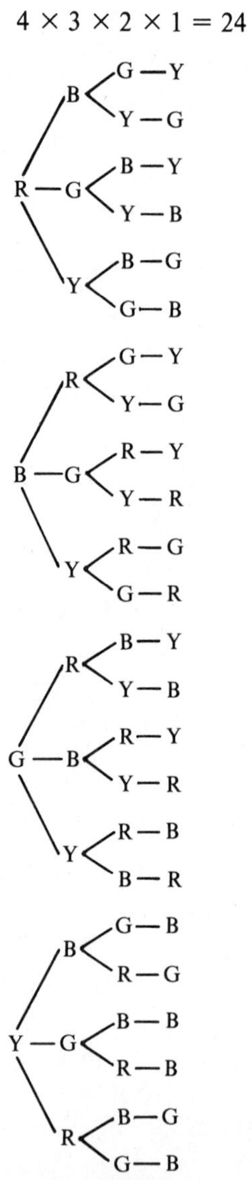

$4 \times 3 \times 2 \times 1 = 24$

FIGURE 10-3

note: The phrase "*n* objects taken *n* at a time" means that we use ALL the objects, *n* of *n*.

But, if we want to choose only *r* of the *n* original distinguishable objects and put these *r* objects into an arrangement, then we are finding the permutation of *n* objects taken *r* at a time, denoted by $_nP_r$.

EXAMPLE 10-8   If we are given the letters in the word DEPOT, how many 3-letter code groupings (not necessarily words) could we form?

Solution   We are to fill three slots, _ _ _ , with the letters from DEPOT. Any one letter could go first. Thus there are 5 choices for the first slot. Then, of the remaining letters, we have 4 choices for the second slot and 3 choices for the third slot. Hence there are

$$5 \times 4 \times 3 = 60 \text{ total choices}$$

That is, we have the permutation of 5 things taken 3 at a time.

• The general formula for computing the number of permutations of *n* objects taken *r* at a time is

$$_nP_r = n(n-1)(n-2)\cdots(n-r+2)(n-r+1) \tag{10-2a}$$

EXAMPLE 10-9   If we use the digits 1, 2, 3, 4, 5, 6, 7 in forming various 4-digit numbers, how many numbers can be formed if no repetitions are allowed?

Solution   We have 7 objects that we are taking 4 at a time and we are concerned about their arrangement; i.e., this is the permutation of $n = 7$ objects taken $r = 4$ at a time. We can do this logically: We know that we have 7 choices for the first digit, 6 choices for the second digit, 5 for the third digit, and 4 for the final digit. Thus there are $7 \times 6 \times 5 \times 4 = 840$ different 4-digit numbers. Or, we can use formula (10-2a) for $_7P_4$, which is the product of the digits from $n = 7$ going DOWN to $(n - r + 1) = (7 - 4 + 1) = 4$. Hence

$$_7P_4 = 7(7-1)(7-2)(7-3) = 7 \times 6 \times 5 \times 4 = 840$$

EXAMPLE 10-10   Suppose we have the same conditions as in Example 10-9 except that repetition of the digits is allowed. Now how many numbers can be formed?

Solution   We go back to the counting procedure for multistage processes and use the multiplication rule. The choices are 7-fold for each of the four slots, so we would have

$$7 \times 7 \times 7 \times 7 = 2401 \text{ numbers}$$

The formula for the number of permutations of *n* objects taken *r* at a time may be algebraically rewritten as follows:

$$_nP_r = n(n-1)(n-2)\cdots(n-r+2)(n-r+1)$$

$$= n(n-1)(n-2)\cdots(n-r+2)(n-r+1)\left(\frac{(n-r)(n-r-1)\cdots 3\cdot 2\cdot 1}{(n-r)(n-r-1)\cdots 3\cdot 2\cdot 1}\right)$$

or

NUMBER OF PERMUTATIONS (r of n)

$$_nP_r = \frac{n(n-1)(n-2)\cdots 3\cdot 2\cdot 1}{(n-r)(n-r-1)\cdots 3\cdot 2\cdot 1} = \frac{n!}{(n-r)!} \tag{10-2b}$$

EXAMPLE 10-11   Evaluate (a) $_4P_2$, (b) $_6P_3$, and (c) $_5P_5$.

Solution   We need only put the numbers into the formula $_nP_r = \dfrac{n!}{(n-r)!}$:

(a) $_4P_2 = \dfrac{4!}{(4-2)!} = \dfrac{4!}{2!} = \dfrac{4\cdot 3\cdot 2\cdot 1}{2\cdot 1} = 4\cdot 3 = 12$

(b) $_6P_3 = \dfrac{6!}{(6-3)!} = \dfrac{6!}{3!} = \dfrac{6 \cdot 5 \cdot 4 \cdot 3 \cdot 2 \cdot 1}{3 \cdot 2 \cdot 1} = 6 \cdot 5 \cdot 4 = 120$

(c) $_5P_5 = \dfrac{5!}{(5-5)!} = \dfrac{5!}{0!} = \dfrac{5!}{1} = 5! = 5 \cdot 4 \cdot 3 \cdot 2 \cdot 1 = 120$

  note: Remember that 0! is defined as 1.

EXAMPLE 10-12   A softball team is made up of 9 players. Find the number of different batting orders the manager can make if

(a) The first 4 batters are fixed in the line-up.
(b) Only the pitcher, who bats last, is fixed in the line-up.

Solution

(a) Four slots are already filled. The manager must permute the last 5 players into the remaining 5 slots. This is

$$_5P_5 = \dfrac{5!}{(5-5)!} = \dfrac{5!}{0!} = 5! = 120$$

(b) Only 1 slot is filled, the other 8 players get permuted into the 8 remaining slots. This is

$$_8P_8 = 8! = 40\,320$$

EXAMPLE 10-13   If 10 basketball teams are in a tournament, in how many ways could the potential first, second, and third place teams be listed?

Solution   This is an ordered arrangement of 10 items taken 3 at a time. This is

$$_{10}P_3 = \dfrac{10!}{(10-3)!} = \dfrac{10!}{7!} = \dfrac{10 \cdot 9 \cdot 8 \cdot 7 \cdot 6 \cdot 5 \cdot 4 \cdot 3 \cdot 2 \cdot 1}{7 \cdot 6 \cdot 5 \cdot 4 \cdot 3 \cdot 2 \cdot 1} = 10 \cdot 9 \cdot 8 = 720$$

EXAMPLE 10-14   Find the number of ways the first 4 cards can be dealt from a deck of 52 cards.

Solution   We have 52 items and we are choosing an ordered set of 4. We have

$$_{52}P_4 = \dfrac{52!}{(52-4)!} = \dfrac{52!}{48!} = 52 \cdot 51 \cdot 50 \cdot 49 = 6\,497\,400$$

B. Permutations with complications

So far we've assumed that the objects we were arranging were distinguishable one from the other. But an arrangement can contain some objects that are exactly alike. If we have some objects that are not distinguishable, we have fewer arrangements.

EXAMPLE 10-15   Dennis' boat carries 3 identical blue flags as well as 1 red flag and 1 green flag. If he displays all 5 flags on a vertical pole, how many distinguishable arrangements are possible?

Solution   If the flags were all different colors, he could make $_5P_5 = 5! = 120$ different arrangements. But the 3 blue flags can themselves be put into $3! = 3 \times 2 \times 1 = 6$ different arrangements without anyone's being able to tell them apart. For example, BRBBG could be displayed in 6 different ways by switching the blue ones around ($B_1RB_2B_3G$, $B_2RB_1B_3G$, etc.). Hence we need to divide out these repetitions. We divide by 3! to get

$$\dfrac{5!}{3!} = \dfrac{5 \cdot 4 \cdot 3 \cdot 2 \cdot 1}{3 \cdot 2 \cdot 1} = 5 \cdot 4 = 20 \text{ different arrangements}$$

In general, we have the following rule:

- If in a collection of N objects there are k_1 of one kind, k_2 of another kind, k_3 of a third kind, etc., then the number of distinguishable permutations of the N objects is

$$_N P_{k_i} = \frac{N!}{(k_1)!(k_2)!\cdots(k_i)!} \tag{10-3}$$

EXAMPLE 10-16 We have 3 black disks, 2 white disks, and 1 orange disk. We want to form a line of 6 disks. How many distinguishable permutations could be formed?

Solution We have $N = 6, k_1 = 3, k_2 = 2, k_3 = 1$. Thus we have

$$\frac{N!}{k_1!k_2!k_3!} = \frac{6!}{3!2!1!} = \frac{6\cdot5\cdot4\cdot3\cdot2\cdot1}{3\cdot2\cdot1\cdot2\cdot1\cdot1} = 60 \text{ distinguishable lines}$$

EXAMPLE 10-17 How many distinguishable 9-letter groupings can be formed from the letters of TENNESSEE?

Solution We have 9 letters, but there are 2 N's, 2S's, and 4 E's besides the lonesome T. Thus we have

$$\frac{9!}{2!2!4!1!} = \frac{9\cdot8\cdot7\cdot6\cdot5\cdot4\cdot3\cdot2\cdot1}{2\cdot1\cdot2\cdot1\cdot4\cdot3\cdot2\cdot1\cdot1} = 9\cdot4\cdot7\cdot3\cdot5 = 3780 \text{ groupings}$$

C. Combinations

When a baseball manager makes out his batting line-up, he assigns specific positions in the line-up to specific players. This is an ordered arrangement, or permutation. But sometimes the order of the persons (or objects) is not important. For the basketball coach to pick 5 starters from among the 10 players does not involve ordering. To pick a set of 3 volunteers from a group of 9 people does not involve ordering them. If we need to select 4 books from an assortment of 15 books for our Book-of-the-Month premium, we are not concerned about the order of the books in the selection. These "unordered" selections are called *combinations*.

- When we select subsets of r objects from a set of n objects, without regard to order, we are forming **combinations of** n objects taken r at a time, denoted $_n C_r$.

Thus the basketball coach needs combinations of 10 things taken 5 at a time, $_{10}C_5$. The volunteers chosen from a group of people make up combinations of 9 things taken 3 at a time, $_9C_3$. The books chosen from an assortment are combinations of 15 things taken 4 at a time, $_{15}C_4$.

The number of permutations of n things taken r at a time, $_n P_r$, takes into account the order in which the r things are selected. But we are not concerned with order in combinations, so we divide out the ordered sets to find the number of unordered sets. Since each combination corresponds to $r!$ permutations of the r objects such that $_n P_r = (r!)(_n C_r)$, we get

NUMBER OF COMBINATIONS (r of n)
$$_n C_r = \frac{_n P_r}{r!} = \frac{\frac{n!}{(n-r)!}}{r!} = \frac{n!}{r!(n-r)!} \tag{10-4}$$

EXAMPLE 10-18 Calculate the combinations in **(a)** $_{10}C_5$, **(b)** $_9C_3$, and **(c)** $_{15}C_4$.

Solution

(a) $_{10}C_5 = \dfrac{10!}{5!(10-5)!} = \dfrac{10!}{5!5!} = \dfrac{10\cdot9\cdot8\cdot7\cdot6}{5\cdot4\cdot3\cdot2\cdot1} = 2\cdot3\cdot2\cdot7\cdot3 = 252$

(b) $_9C_3 = \dfrac{9!}{3!(9-3)!} = \dfrac{9!}{3!6!} = \dfrac{9\cdot8\cdot7}{3\cdot2\cdot1} = 3\cdot4\cdot7 = 84$

(c) $_{15}C_4 = \dfrac{15!}{4!11!} = \dfrac{15 \cdot 14 \cdot 13 \cdot 12}{4 \cdot 3 \cdot 2 \cdot 1} = 15 \cdot 7 \cdot 13 = 1365$

EXAMPLE 10-19   A child has a penny (P), a nickel (N), a dime (D), a quarter (Q), and a half-dollar (H). How many different 2-coin payouts can be made? List the payouts.

Solution   The order of the two coins is not important. So we need to find the combinations of 5 things taken 2 at a time:

$$_{5}C_2 = \frac{5!}{2!3!} = \frac{5 \cdot 4 \cdot 3 \cdot 2 \cdot 1}{2 \cdot 1 \cdot 3 \cdot 2 \cdot 1} = \frac{5 \cdot 4}{2 \cdot 1} = 10 \text{ two-coin payouts}$$

PN PD PQ PH ND NQ NH DQ DH QH

EXAMPLE 10-20   Suppose we construct a 12-sided polygon by picking 12 points on a circle and connecting them consecutively. Then, defining a diagonal as a line segment joining two nonconsecutive vertices, we draw all its diagonals. How many diagonals could we draw?

Solution   The total number of lines that could be put in as diagonals for n points is $_{n}C_2$. But if we include all the possibilities, we are including lines between adjacent points, which are not diagonals but are edges. So we deduct the number of edges, which is $n = 12$. Thus for n points we have $_{n}C_2 - n$. For a 12-sided polygon we have

$$_{12}C_2 - 12 = \frac{12!}{2!10!} - 12 = \frac{12 \cdot 11}{2} - 12 = 66 - 12 = 54 \text{ diagonals}$$

In Section 10-3 we'll learn how to find the terms of the expansion of the binomial expression $(x + y)^n$ for positive integer n. The coefficients in this expansion are called **binomial coefficients,** and they turn out to be $_{n}C_k$ for various values of k. In that context they are written using the notation

$$_{n}C_k = \binom{n}{k} = \frac{n!}{k!(n-k)!}$$

EXAMPLE 10-21   Find the value of **(a)** $\binom{6}{2}$ and **(b)** $\binom{9}{6}$.

Solution

(a) $\binom{6}{2} = {_{6}C_2} = \dfrac{6!}{2!4!} = \dfrac{6 \cdot 5}{2 \cdot 1} = 15$ **(b)** $\binom{9}{6} = {_{9}C_6} = \dfrac{9!}{6!3!} = \dfrac{9 \cdot 8 \cdot 7}{3 \cdot 2 \cdot 1} = 3 \cdot 4 \cdot 7 = 84$

EXAMPLE 10-22   Dear Aunt Emily offers her cat Brunhilde a choice of any 4 of 6 different flavors of Kitty Treats—tuna, liver, beef, egg, milk, and cheese. In how many ways can the choices be made?

Solution   In any problem involving choosing we always need to decide whether the order of the choices is important or immaterial. If order is important, we have a permutation problem; if the order of the choices is immaterial, we have a combination problem. Here, Brunhilde is not concerned over the order in which the treats are chosen—she just picks 'em and starts crunchin'. So we have to determine the number of combinations of 6 things taken 4 at a time:

$$_{6}C_4 = \frac{6!}{4!2!} = \frac{6 \cdot 5}{2 \cdot 1} = 15$$

note: From Examples 10-21 and 10-22 you may have observed that

$$_{6}C_2 = \frac{6!}{2!4!} = 15 = \frac{6!}{4!2!} = {_{6}C_4}$$

Thus

$$\binom{6}{2} = \binom{6}{4}$$

In general $_nC_r = {_n}C_{n-r}$. An algebraic proof is:

$$_nC_{n-r} = \frac{n!}{(n-(n-r))!(n-r)!} = \frac{n!}{r!(n-r)!} = {_n}C_r$$

A combinatorial proof is: Suppose S is a set with cardinality n and r is a positive integer less than or equal to n. If P is a subset of S with cardinality r, then $S-P$ is a unique subset of cardinality $n-r$. Similarly, if Q is a subset of S at cardinality $n-r$, then there is a unique subset $S-Q$ of cardinality $n-(n-r)=r$. We then have an exact matching (a one-to-one correspondence) between subsets of S of cardinality r with subsets of S of cardinality $n-r$. Since there are $_nC_r$ subsets of cardinality r and $_nC_{n-r}$ subsets of cardinality $n-r$ and they are matched one-to-one, we conclude that $_nC_r = {_n}C_{n-r}$.

EXAMPLE 10-23 A student has 4 English books, 3 math books, and 5 psychology books. In how many ways can these books be arranged on a shelf if the books in the same subject must be kept together?

Solution Here, order is very important, since an interchange of just two of the books on the shelf results in a different arrangement. Thus we have permutation problems to solve. First look at the subject groups. The 4 English books can be permuted in $_4P_4 = 4! = 24$ ways. For each of these ways the math books are permuted in $_3P_3 = 3! = 6$ ways. And for each of these 24×6 ways for English and math we have $_5P_5 = 5! = 120$ ways for the psychology books. So far we have

$$_4P_4 \times {_3}P_3 \times {_5}P_5 = 24 \times 6 \times 120 = 17\,280 \text{ ways}$$

But the books, although kept in groups by subject matter, may have permutations within groups, e.g.,

$$\boxed{\text{Eng}}\ \boxed{\text{Math}}\ \boxed{\text{Psych}} \quad \text{or} \quad \boxed{\text{Psych}}\ \boxed{\text{Eng}}\ \boxed{\text{Math}} \quad \text{or}\cdots$$

There are $_3P_3 = 3! = 6$ ways of doing this. Thus our final answer is

$$6(24 \times 6 \times 120) = 6 \times 17\,280 = 103\,680 \text{ arrangements}$$

EXAMPLE 10-24 The PTA executive board, consisting of 9 women and 7 men, must have a subcommittee consisting of exactly 3 women and 3 men. In how many ways can this subcommittee be selected?

Solution Since the order in which the people are selected is not at all important, we have a combination problem. We have two nonoverlapping sets (men and women), so we work on each segment separately. We must choose 3 women from the available 9 women. This is done in $_9C_3$ ways. For each of these ways we can now choose the 3 men from the available 7 in $_7C_3$ ways. The total number of subcommittees then is

$$_9C_3 \cdot {_7}C_3 = \frac{9!}{3!6!} \cdot \frac{7!}{3!4!} = \frac{9 \cdot 8 \cdot 7}{3 \cdot 2 \cdot 1} \cdot \frac{7 \cdot 6 \cdot 5}{3 \cdot 2 \cdot 1} = (3 \cdot 4 \cdot 7)(7 \cdot 5) = 2940 \text{ ways}$$

10-3. The Binomial Theorem

A. Finding and using the binomial theorem

The binomial theorem is a formula to find the expansion of the product $(x + y)^n$ for positive integer n. In order to obtain the formula, we'll first look at a few simple cases and examine the patterns that emerge:

$$(x + y)^1 = x + y$$
$$(x + y)^2 = x^2 + 2xy + y^2$$
$$(x + y)^3 = x^3 + 3x^2y + 3xy^2 + y^3$$
$$(x + y)^4 = x^4 + 4x^3y + 6x^2y^2 + 4xy^3 + y^4$$
$$(x + y)^5 = x^5 + 5x^4y + 10x^3y^2 + 10x^2y^3 + 5xy^4 + y^5$$

From these cases we make some observations:

(1) The number of terms in the expansion of $(x + y)^n$ is $n + 1$.
(2) The first term is always x^n; the last term is always y^n.
(3) The coefficient of the second term is always n.
(4) In every term, the sum of the exponents on x and y is n.
(5) The exponents on x begin at n on the left and decrease in steps of 1 to an exponent of 0 at the right end. The exponents on y begin at 0 on the left end and increase in steps of 1 to an exponent of n on the right end.
(6) The sequence of coefficients in any expansion read from left to right is the same as the sequence of coefficients read from right to left.
(7) The coefficient of the first term is $1 = {}_nC_0 = \binom{n}{0}$.

The coefficient of the second term is $n = {}_nC_1 = \binom{n}{1}$.

The coefficient of the third term is $\dfrac{n(n-1)}{2} = {}_nC_2 = \binom{n}{2}$.

The coefficient of the fourth term is $\dfrac{n(n-1)(n-2)}{3 \cdot 2 \cdot 1} = {}_nC_3 = \binom{n}{3}$.

$$\vdots$$

The coefficient of the last term is $1 = {}_nC_n = \binom{n}{n}$.

And we know the notation for binomial coefficients.

$$_nC_k = \binom{n}{k} = \frac{n!}{k!(n-k)!}$$

so that, for example, we write

$$_nC_2 = \binom{n}{2} = \frac{n!}{2!(n-2)!} \quad \text{and} \quad {}_nC_3 = \binom{n}{3} = \frac{n!}{3!(n-3)!}$$

These observations allow us to form a conclusion, which is called the **Binomial Theorem**:

$$(x+y)^n = \binom{n}{0}x^n + \binom{n}{1}x^{n-1}y + \binom{n}{2}x^{n-2}y^2 + \cdots + \binom{n}{n-1}xy^{n-1} + \binom{n}{n}y^n \quad \textbf{(10-5)}$$

or

BINOMIAL THEOREM

$$(x+y)^n = \sum_{k=0}^n \binom{n}{k}x^{n-k}y^k$$

EXAMPLE 10-25   Find the binomial expansion of $(x + y)^6$.

Solution   We substitute into formula (10-5) using $n = 6$:

$$(x+y)^6 = \binom{6}{0}x^6 + \binom{6}{1}x^5y + \binom{6}{2}x^4y^2 + \binom{6}{3}x^3y^3 + \binom{6}{4}x^2y^4 + \binom{6}{5}xy^5 + \binom{6}{6}y^6$$

We have computed some of these coefficients before; e.g., $\binom{6}{2} = \binom{6}{4} = 15$. The others are

$$\binom{6}{0} = \binom{6}{6} = \frac{6!}{6!0!} = \frac{6!}{6!1} = 1$$

$$\binom{6}{1} = \binom{6}{5} = \frac{6!}{5!1!} = 6$$

$$\binom{6}{3} = \frac{6!}{3!3!} = \frac{6 \cdot 5 \cdot 4}{3 \cdot 2 \cdot 1} = 5 \cdot 4 = 20$$

Then

$$(x + y)^6 = x^6 + 6x^5y + 15x^4y^2 + 20x^3y^3 + 15x^2y^4 + 6xy^5 + y^6$$

note: All seven of the observations made above hold for this example.

Suppose we want to pick out a specific term of the expansion, say the *i*th term, where $i = 1, 2, 3, \dots$. Observe that the exponent on y is 1 less than the term number and the lower number in the coefficient notation is the same as the exponent on y. The exponent sum is still n and the upper number in the coefficient notation is n:

$$(x + y)^n = \overbrace{\binom{n}{0}x^ny^0}^{\substack{\text{1st} \\ \text{term}}} + \overbrace{\binom{n}{1}x^{n-1}y^1}^{\substack{\text{2nd} \\ \text{term}}} + \overbrace{\binom{n}{2}x^{n-2}y^2}^{\substack{\text{3rd} \\ \text{term}}} + \overbrace{\binom{n}{3}x^{n-3}y^3}^{\substack{\text{4th} \\ \text{term}}} + \overbrace{\binom{n}{4}x^{n-4}y^4}^{\substack{\text{5th} \\ \text{term}}} + \cdots \quad \textbf{(10-6)}$$

Hence the *i*th term is

ith TERM OF A BINOMIAL EXPANSION $\qquad \binom{n}{i-1}x^{n-(i-1)}y^{i-1}$

EXAMPLE 10-26   What is the 5th term in the expansion of $(x + y)^{12}$?

Solution   Let $i = 5$. We have $n = 12$. Then the 5th term is

$$\binom{n}{i-1}x^{n-(i-1)}y^{i-1} = \binom{12}{5-1}x^{12-5+1}y^{5-1}$$

$$= \binom{12}{4}x^8y^4 = \frac{12!}{4!8!}x^8y^4 = \frac{12 \cdot 11 \cdot 10 \cdot 9}{4 \cdot 3 \cdot 2 \cdot 1}x^8y^4$$

$$= 495x^8y^4$$

EXAMPLE 10-27   Find the expansion of $(a^2 + 3b^2)^4$.

Solution   Here we need to substitute a^2 for x and $(3b^2)$ for y everywhere in the expansion of $(x + y)^4$, so we expand $(x + y)^4$ first:

$$(x + y)^4 = \binom{4}{0}x^4 + \binom{4}{1}x^3y + \binom{4}{2}x^2y^2 + \binom{4}{3}xy^3 + \binom{4}{4}y^4$$

$$= x^4 + 4x^3y + 6x^2y^2 + 4xy^3 + y^4$$

Then we do the substitution:

$$(a^2 + 3b^2)^4 = (a^2)^4 + 4(a^2)^3(3b^2) + 6(a^2)^2(3b^2)^2 + 4(a^2)(3b^2)^3 + (3b^2)^4$$

$$= a^8 + 4a^6(3b^2) + 6a^4(9b^4) + 4a^2(27b^6) + 81b^8$$

$$= a^8 + 12a^6b^2 + 54a^4b^4 + 108a^2b^6 + 81b^8$$

EXAMPLE 10-28   Find the expansion of $(a - 2b)^5$.

Solution   We need to substitute a for x and $-2b$ for y in the expansion of $(x + y)^5$ after expanding.

If $\qquad (x + y)^5 = x^5 + 5x^4y + 10x^3y^2 + 10x^2y^3 + 5xy^4 + y^5$

then

$$(a - 2b)^5 = a^5 + 5(a)^4(-2b) + 10(a)^3(-2b)^2 + 10(a)^2(-2b)^3$$

$$+ 5(a)(-2b)^4 + (-2b)^5$$

$$= a^5 + 5a^4(-2b) + 10a^3(4b^2) + 10a^2(-8b^3) + 5a(16b^4) + (-32b^5)$$

$$= a^5 - 10a^4b + 40a^3b^2 - 80a^2b^3 + 80ab^4 - 32b^5$$

EXAMPLE 10-29 Find the expansion of $(1 - t^2)^6$.

Solution We need to use the expansion of $(x + y)^6$ and substitute $x = 1$ and $y = -t^2$.

If

$$(x + y)^6 = x^6 + 6x^5y + 15x^4y^2 + 20x^3y^3 + 15x^2y^4 + 6xy^5 + y^6$$

then

$$(1 - t^2)^6 = (1)^6 + 6(1)^5(-t^2) + 15(1)^4(-t^2)^2 + 20(1)^3(-t^2)^3$$
$$+ 15(1)^2(-t^2)^4 + 6(1)(-t^2)^5 + (-t^2)^6$$
$$= 1 - 6t^2 + 15t^4 - 20t^6 + 15t^8 - 6t^{10} + t^{12}$$

B. Pascal's triangle

The binomial coefficients $\binom{n}{k}$ have a significant property in their position in a special number display. If we write the coefficients $\binom{n}{k}$ for $k = 0, 1, 2, 3, \ldots$, on the nth line, we get **Pascal's Triangle**:

Line 0						1						$(x + y)^0$
Line 1					1		1					$(x + y)^1$
Line 2				1		2		1				$(x + y)^2$
Line 3			1		3		3		1			$(x + y)^3$
Line 4		1		4		6		4		1		$(x + y)^4$
Line 5	1		5		10		10		5		1	$(x + y)^5$
•		1	6	15		20		15	6	1		•
•	1	7	21	35		35		21	7	1		•
•		•	•	•	•	•	•	•	•	•		•

In Pascal's Triangle, the numbers in line n are the coefficients in the expansion $(x + y)^n$. And this display *shows* some properties of these coefficients:

- Any number (*except* 1) in line n can be obtained by adding together the two numbers closest to it in the line $n - 1$ above it.
- The sum of the numbers in any row is 2^n.
- The triangle is symmetric with respect to a vertical line down the middle.

All of these properties are provable—in fact, we'll do one. This property is sometimes called "**rabbit ears**": Draw the "rabbit ears" from a number, say the first 10 in line 5. The sum of the two numbers at the ends of the "ears," $4 + 6$, is the chosen number, 10.

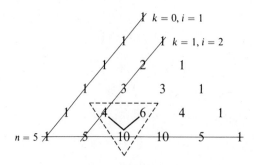

Proof: Pick any number in line n and let it be the ith one from the left end. Then its coefficient designation is $\binom{n}{k} = \binom{n}{i - 1}$. Thus the first 10 above is in line 5, position $i = 3$, so its designation is

$\binom{5}{3-1} = \binom{5}{2}$. Then the first 4 above is in line $n - 1 = 4$ and position $i = 2$, so its designation is $\binom{4}{1}$; and 6 is in line $n - 1 = 4$ and position $i = 3$, so its designation is $\binom{4}{2}$. Now using these coefficient designations, we can set up our rabbit ears property as an equation:

$$\binom{n}{k} = \binom{n-1}{k-1} + \binom{n-1}{k}$$

We express each of these in factorials:

$$\binom{n}{k} = \frac{n!}{(n-k)!k!}$$

$$\binom{n-1}{k-1} = \frac{(n-1)!}{(n-k)!(k-1)!}$$

$$\binom{n-1}{k} = \frac{(n-1)!}{(n-1-k)!k!}$$

Then

$$\frac{n!}{(n-k)!k!} = \frac{(n-1)!}{(n-k)!(k-1)!} + \frac{(n-1)!}{(n-1-k)!k!}$$

$$= \frac{(n-1)!}{(n-k)!(k-1)!}\left(\frac{k}{k}\right) + \frac{(n-1)!}{(n-1-k)!k!}\left(\frac{n-k}{n-k}\right)$$

$$= \frac{(n-1)!k}{(n-k)!k!} + \frac{(n-1)!(n-k)}{(n-k)!k!}$$

$$= \frac{(n-1)!}{k!}\left[\frac{k}{(n-k)!} + \frac{n-k}{(n-k)!}\right]$$

$$= \frac{(n-1)!n}{(n-k)!k!} \qquad \text{[Remember that } n! = n(n-1)!\text{]}$$

$$= \frac{n!}{(n-k)!k!} \qquad \text{QED}$$

SUMMARY

1. The counting principle allows multistage processes to be evaluated for a number of possible occurrences:

   - If stage 1 of an N-stage process may be done in k_1 ways and stage 2 following it may be done in k_2 ways, etc., then the combination of stages may be done in $k_1 \times k_2 \times \cdots \times k_N$ ways.
   - A tree diagram is a graphical way of depicting a multistage process.

2. If a set of objects n is put into an ordered arrangement, the ordering is called a permutation P. The number of permutations of n objects taken n at a time is

   $$_nP_n = n!$$

   The number of permutations of n objects taken r at a time is

   $$_nP_r = \frac{n!}{(n-r)!}$$

The number of distinguishable permutations of N objects taken k_i at a time, where k_1 is one kind, k_2 is another kind, etc., is

$$_N P_{k_i} = \frac{N!}{(k_1)!(k_2)! \cdots (k_i)!}$$

3. If an arrangement of a set of n objects is made without regard to ordering it is called a combination. The number of combinations of n objects taken r at a time is

$$_n C_r = \frac{n!}{r!(n-r)!}$$

4. The number of permutations of n objects taken r at a time is more than the number of combinations of n objects taken r at a time: The connection is $_n P_r = (r!)(_n C_r)$.

5. The coefficients in the expansion of the power of a binomial are found by evaluating combination values:

If $\binom{n}{k} = {_n C_k} = \dfrac{n!}{k!(n-k)!}$, then

$$(x+y)^n = \binom{n}{0}x^n + \binom{n}{1}x^{n-1}y + \binom{n}{2}x^{n-2}y^2 + \cdots + \binom{n}{n-1}xy^{n-1} + \binom{n}{n}y^n$$

These same coefficients may be found by examining the correct row of Pascal's triangle.

RAISE YOUR GRADES

Can you...?

☑ form the tree diagram for a counting problem
☑ use the multiplication rule for counting objects
☑ decide, in a given problem, whether the permutation technique or the combination technique should be used
☑ find the number of permutations of n objects taken r at a time
☑ find the number of distinguishable permutations of N objects if some of the objects are identical
☑ find the number of combinations of n objects taken r at a time
☑ relate the number of permutations to the number of combinations of a collection of objects
☑ use the Binomial Theorem to find the binomial expansion $(x+y)^n$
☑ explain the correlation between combinations and binomial coefficients
☑ construct Pascal's triangle and explain how the rows are formed and their significance

SOLVED PROBLEMS

A Counting Principle

PROBLEM 10-1 You are preparing to compete in the league championship track meet and you are allowed to pick exactly 1 running event and 1 field event. You have 3 specialities in running—the 100-meter, 200-meter, and 400-meter races—and 2 specialties in the field events—the long jump (LJ) and the triple jump (TJ). How many different event selections do you have?

Solution Form a tree diagram, as in Figure 10-4, and count the possible selections. There are six possible selections.

Or, use the multiplication rule: If you break the problem down into choices at the various stages, you get

Total choices = 3 × 2 = 6

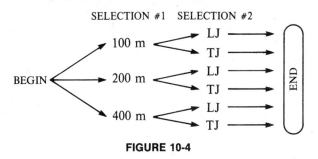

FIGURE 10-4

PROBLEM 10-2 In how many ways can you form a 3-digit number using only the digits 4, 5, 6, 7, 8 with no repetitions?

Solution You have 5 choices for the first digit. Once this digit has been chosen, you have 4 choices for the second digit; then 3 choices for the third digit. So you can make 5 × 4 × 3 = 60 number choices.

PROBLEM 10-3 In how many different ways could you arrange the letters in the word WORD without repetitions?

Solution You have to fill 4 slots, each with a different letter. You can pick any letter to start, so you have 4 choices for the first slot. Then there are 3 choices for the second slot, 2 choices for the third slot, and 1 choice (the only letter left) for the final slot. So you have

$$4 \times 3 \times 2 \times 1 = 24 \text{ choices}$$

PROBLEM 10-4 You toss one coin and roll one 6-sided die. How many different outcomes are possible? List them.

Solution There are 2 outcomes, H and T, for the coin toss and 6 outcomes, 1 through 6, for the die roll, so there are 2 × 6 = 12 outcomes. They are

$$\{H1, H2, H3, H4, H5, H6, T1, T2, T3, T4, T5, T6\}$$

PROBLEM 10-5 If an ID number has 1 letter followed by 4 digits, how many ID numbers could be created?

Solution There are 26 choices for the letter and 10 choices for each digit, so the total is

$$26 \times 10 \times 10 \times 10 \times 10 = 260\,000$$

PROBLEM 10-6 If 7 horses run a race, how many possibilities are there for the first 3 finishes, win, place, and show?

Solution Any one of the 7 horses could win. Any one of the remaining 6 horses could place second, and any one of the remaining 5 could "show" the crowd that she is third. Thus there are 7 × 6 × 5 = 210 finish orders.

PROBLEM 10-7 Your casual wardrobe consists of 4 different pairs of jeans, 5 different sweatshirts, and 3 different pairs of shoes. How many outfits are possible?

Solution 4 × 5 × 3 = 60 choices.

PROBLEM 10-8 We want to form an identifier consisting of three alphabet letters. How many of these contain either an X or a Y?

Solution We get the answer by counting the total number of identifiers that could be made using any letters: 26 · 26 · 26 = 17 576 and subtract all those that *do not* contain X and Y: 24 · 24 · 24 = 13 824, for a total of 17 576 − 13 824 = 3752 that *do* contain an X or Y.

PROBLEM 10-9 Gordon has 7 ties, 8 shirts, and 5 pairs of trousers. Assuming everything coordinates, how many different outfits can be formed?

Solution   Use the multiplication rule: For each of the 7 ties there are 8 shirts for a total of $7 \cdot 8 = 56$ different tie-shirt combos. For each of these there are 5 trouser choices for a total of $7 \cdot 8 \cdot 5 = 280$ choices for an outfit.

Permutations and Combinations

PROBLEM 10-10   Evaluate (a) $_7P_2$, (b) $_8P_4$, and (c) $_6P_1$.

Solution   You know that $_nP_r = \dfrac{n!}{(n-r)!}$, so

(a) $_7P_2 = \dfrac{7!}{(7-2)!} = \dfrac{7!}{5!} = \dfrac{7 \cdot 6 \cdot 5 \cdot 4 \cdot 3 \cdot 2 \cdot 1}{5 \cdot 4 \cdot 3 \cdot 2 \cdot 1} = 7 \cdot 6 = 42$

(b) $_8P_4 = \dfrac{8!}{(8-4)!} = \dfrac{8!}{4!} = 8 \cdot 7 \cdot 6 \cdot 5 = 1680$

(c) $_6P_1 = \dfrac{6!}{(6-1)!} = \dfrac{6!}{5!} = 6$

EXAMPLE 10-11   A bookstore owner has just bought 8 different bestsellers at a hefty discount, and he wants to display 5 of them on his bargain shelf. How many different displays could he make?

Solution   Because changing the order of the books is the only way to make a different display, this is a permutation problem. This is the permutation of 8 objects taken 5 at a time, $_8P_5$. Thus

$$_8P_5 = \frac{8!}{(8-5)!} = \frac{8!}{3!} = 8 \cdot 7 \cdot 6 \cdot 5 \cdot 4 = 6720 \text{ ways}$$

PROBLEM 10-12   You have 6 different coins (a penny, a nickel, a dime, a quarter, a half-dollar, and an odd shilling). (a) If you arrange some subset of 4 of them in a row, how many different arrangements are possible? (b) How many different arrangements are possible if you use all six of the coins?

Solution

(a) You're concerned with order, so you want to permute 6 objects taken 4 at a time:

$$_6P_4 = \frac{6!}{(6-4)!} = \frac{6!}{2} = 6 \cdot 5 \cdot 4 \cdot 3 = 360 \text{ arrangements}$$

(b) $_6P_6 = 6! = 720$ arrangements

PROBLEM 10-13   How many 3-letter sets of initials (repetitions not allowed) are possible?

Solution   You are permuting 3 objects from 26 objects:

$$_{26}P_3 = \frac{26!}{(26-3)!} = \frac{26!}{23!} = 26 \cdot 25 \cdot 24 = 15\,600 \text{ sets}$$

PROBLEM 10-14   How many ways are there to order the 5 letters P, Q, R, S, T?

Solution $_5P_5 = 5! = 120$ ways

PROBLEM 10-15   How many different letter arrangements can you make with the letters in MISSISSIPPI?

Solution   If the letters were all different, you would have $N! = 11!$ permutations. But the letter I occurs 4 times, the letter S occurs 4 times, and P occurs twice, so $k_1 = 4$, $k_2 = 4$, and $k_3 = 2$. Thus there are

$$\frac{N!}{k_1! k_2! k_3} = \frac{11!}{4!4!2!} = \frac{11 \cdot 10 \cdot 9 \cdot 8 \cdot 7 \cdot 6 \cdot 5 \cdot 4 \cdot 3 \cdot 2 \cdot 1}{4 \cdot 3 \cdot 2 \cdot 1 \cdot 4 \cdot 3 \cdot 2 \cdot 1 \cdot 2 \cdot 1}$$

$$= 11 \cdot 10 \cdot 9 \cdot 7 \cdot 5$$

$$= 34\,650 \text{ distinguishable permutations}$$

PROBLEM 10-16 How many distinguishable permutations can you make with the letters in COLLEGE?

Solution There are 7 objects, but there are two L's and two E's. Thus there are

$$\frac{7!}{2!2!} = \frac{7 \cdot 6 \cdot 5 \cdot 4 \cdot 3 \cdot 2 \cdot 1}{2 \cdot 1 \cdot 2 \cdot 1} = 7 \cdot 6 \cdot 5 \cdot 3 \cdot 2 = 1260 \text{ distinguishable permutations}$$

PROBLEM 10-17 When you're dealing a poker hand, you don't care about the order in which the cards are dealt. How many different hands of 5 cards can you deal from a standard 52-card deck?

Solution Since order is not important, this is a combination problem. Since there are 52 objects taken 5 at a time, there are

$$_nC_r = \frac{n!}{r!(n-r)!}$$

$$_{52}C_5 = \frac{52!}{5!(52-5)!} = \frac{52!}{5!47!} = \frac{52 \cdot 51 \cdot 50 \cdot 49 \cdot 48}{5 \cdot 4 \cdot 3 \cdot 2 \cdot 1} = 2\,598\,960 \text{ different hands}$$

PROBLEM 10-18 How many different subcommittees of 3 persons can be chosen from a committee of 10 persons?

Solution Since the subcommittees can be made up of any 3 persons without regard to order, the number of different subcommittees is

$$_{10}C_3 = \frac{10!}{3!7!} = \frac{10 \cdot 9 \cdot 8}{3 \cdot 2 \cdot 1} = 10 \cdot 3 \cdot 4 = 120$$

PROBLEM 10-19 You have 8 kinds of flowers and you need to make up bouquets consisting of 3 different kinds of flowers. How many different kinds of bouquets can you make?

Solution
$$_8C_3 = \frac{8!}{3!5!} = \frac{8 \cdot 7 \cdot 6}{3 \cdot 2 \cdot 1} = 56 \text{ bouquets}$$

PROBLEM 10-20 In Audrey's town all of the streets are laid out in a grid pattern. She lives at 3rd Street and B Avenue and her school is at 6th Street and F Avenue as shown on the following map:

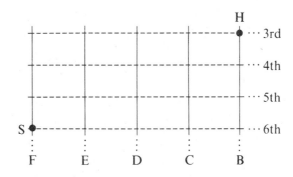

On her trip to school she natually walks only west and south in some combination. How many different paths are available from home H to school S?

Solution Audrey must go south three blocks and west four blocks, so the path is a collection of seven letters, three of which are S and four of which are W, such as WSWWSSW. There are seven slots to fill but when four slots are filled with W, the other three must be filled with S. Thus the total number of paths is

$$_7C_4 = \frac{7!}{4!3!} = \frac{7 \cdot 6 \cdot 5}{3 \cdot 2 \cdot 1} = 35$$

PROBLEM 10-21 If Audrey must also walk home and again uses the "conservation of energy" by going only east and north, how many paths are there to walk from school S to home H. In a 180-day school year could she use a different "round trip path" each day?

Solution There are also exactly 35 paths from school to home (the reverse of each of the home to school paths). The round trip now can be constructed from any one of 35 going-to-school paths followed by any one of 35 returning-to-home paths for a total number of $35 \cdot 35 = 1225$ different round-trip paths. This could take almost 7 years to cover all of these possible paths.

PROBLEM 10-22 How many strings of length 8 can be formed from the letters $\{b, e, n, r, s, t\}$ in which the letter e must appear exactly twice?

Solution We have eight slots to fill but two of them must have e in them. The number of positions for them is ${}_8C_2 = \frac{8!}{6!2!} = \frac{8 \cdot 7}{2 \cdot 1} = 28$. The remaining six positions are filled with any of the other five letters. There are $5 \cdot 5 \cdot 5 \cdot 5 \cdot 5 \cdot 5 \cdot = 5^6$ ways of doing this. The total number of strings then is $28 \cdot 5^6 = 437\,500$.

PROBLEM 10-23 In a state lottery called LOTTO Pamela, the player, pays \$1 to pick six numbers from the list 1–40. If all six computer-chosen numbers are matched identically by her picks, the jackpot is hers! What are her chances of winning?

Solution Since she picks any six of the forty without regard for order, there are ${}_{40}C_6$ different number combinations. Computing:

$$
\begin{aligned}
{}_{40}C_6 &= \frac{40!}{6!34!} \\
&= \frac{40 \cdot 39 \cdot 38 \cdot 37 \cdot 36 \cdot 35}{6 \cdot 5 \cdot 4 \cdot 3 \cdot 2 \cdot 1} \\
&= 2 \cdot 39 \cdot 38 \cdot 37 \cdot 35 \\
&= 3\,838\,380
\end{aligned}
$$

Her chances are 1 in ${}_{40}C_6$, or about 1 in 4 million.

PROBLEM 10-24 In SUPER LOTTO Pamela picks six numbers from 1–44. What are her chances of winning?

Solution Now we need to find

$$
\begin{aligned}
{}_{44}C_6 &= \frac{44!}{6!38!} \\
&= \frac{44 \cdot 43 \cdot 42 \cdot 41 \cdot 40 \cdot 39}{6 \cdot 5 \cdot 4 \cdot 3 \cdot 2 \cdot 1} \\
&= 44 \cdot 43 \cdot 7 \cdot 41 \cdot 13 \\
&= 7\,059\,052
\end{aligned}
$$

Her chances are now 1 in about 7 million.

PROBLEM 10-25 How many different basketball teams of 5 girls can a coach form from a squad of 15 girls?

Solution Since the coach is not choosing positions, she doesn't distinguish among the team members; i.e., order is not important. Thus there are

$$
{}_{15}C_5 = \frac{15!}{5!10!} = \frac{15 \cdot 14 \cdot 13 \cdot 12 \cdot 11}{5 \cdot 4 \cdot 3 \cdot 2 \cdot 1} = 3003 \text{ teams}
$$

PROBLEM 10-26 How many ways are there of getting exactly 3 heads in the toss of 5 coins?

Solution There are

$$
{}_5C_3 = \frac{5!}{3!2!} = \frac{5 \cdot 4}{2 \cdot 1} = 10 \text{ ways of getting 3 heads}
$$

(out of a total of $2^5 = 32$ ways the coins could fall).

PROBLEM 10-27   In how many ways can a committee of 3 Democrats and 4 Republicans be chosen from a group of 5 Democrats and 8 Republicans?

Solution   The 3 Democrats are chosen from the available 5 in $_5C_3$ ways. The 4 Republicans are chosen from the available 8 in $_8C_4$ ways. Thus the total number of ways is

$$_5C_3 \cdot {_8C_4} = \frac{5!}{3!2!} \cdot \frac{8!}{4!4!} = \frac{5 \cdot 4}{2 \cdot 1} \cdot \frac{8 \cdot 7 \cdot 6 \cdot 5}{4 \cdot 3 \cdot 2 \cdot 1} = 10 \cdot 70 = 700 \text{ ways}$$

PROBLEM 10-28   In a league of 10 softball teams, how many games will be played in a season if each team plays each other team once?

Solution   There are 10 teams and each game involves 2 teams, so

$$_{10}C_2 = \frac{10}{2!8!} = \frac{10 \cdot 9}{2 \cdot 1} = 45 \text{ games}$$

PROBLEM 10-29   Pepe's Pizza Palace offers 3 choices of salad, 12 kinds of pizza, and 4 different drinks. How many different selections can you make for a salad, a pizza, and a drink?

Solution   This is a 3-stage counting problem. You have 3 salad choices. Then, for each salad choice, you have 12 choices for pizza. Then for each of these two-item selections, you have 4 choices for the drink. The total number of selections is $3 \times 12 \times 4 = 144$.

PROBLEM 10-30   Izzy's Ice Cream Shoppe carries 31 flavors. How many different double-dip ice cream cones could you make, assuming that the dips are of different flavors and you don't care which flavor is on top?

Solution   There are two ways to approach this problem. By the counting principle, you have 31 choices for the first flavor and 30 choices for the second flavor, which gives you a total of $31 \times 30 = 930$ different cones. But if you use this total, you are double-counting the cones with the flavors flipped (e.g., chocolate atop strawberry is the same as strawberry atop chocolate for all practical purposes). So you must divide by 2 to get $930/2 = 465$ cones.

Approaching the question as a combination problem, you are choosing 2 flavors from among the 31 available, so there are

$$_{31}C_2 = \frac{31!}{2!29!} = \frac{31 \cdot 30}{2 \cdot 1} = 31 \cdot 15 = 465 \text{ cones}$$

The Binomial Theorem

PROBLEM 10-31   Working with the notation for binomial coefficients, evaluate

(a) $\binom{5}{2}$ **(b)** $\binom{8}{6}$ **(c)** $\binom{12}{11}$ **(d)** $\binom{4}{4}$

Solution   By definition $\binom{n}{k} = {_nC_k} = \frac{n!}{k!(n-k)!}$, so

(a) $\binom{5}{2} = \frac{5!}{2!3!} = \frac{5 \cdot 4}{2 \cdot 1} = 10$ **(c)** $\binom{12}{11} = \frac{12!}{11!1!} = 12$

(b) $\binom{8}{6} = \frac{8!}{6!2!} = \frac{8 \cdot 7}{2 \cdot 1} = 28$ **(d)** $\binom{4}{4} = \frac{4!}{4!0!} = 1$

PROBLEM 10-32   Create the expansion of $(x + y)^7$.

Solution   Use the expanded formula (10-5) from the Binomial Theorem, with $n = 7$:

$$(x+y)^n = \binom{n}{0}x^n + \binom{n}{1}x^{n-1}y + \binom{n}{2}x^{n-2}y^2 + \cdots + \binom{n}{n-1}xy^{n-1} + \binom{n}{n}y^n$$

$$(x+y)^7 = \binom{7}{0}x^7 + \binom{7}{1}x^6y + \binom{7}{2}x^5y^2 + \binom{7}{3}x^4y^3 + \binom{7}{4}x^3y^4 + \binom{7}{5}x^2y^5 + \binom{7}{6}xy^6 + \binom{7}{7}y^7$$

Then find the coefficients for each term in the expansion by using the formula

$$\binom{n}{k} = \frac{n!}{k!(n-k)!}$$

The coefficients of the first and seventh terms are particularly easy to find:

$$\binom{7}{0} = \binom{7}{7} = \frac{7!}{7!} = 1 \qquad \begin{array}{l}\text{[The coefficient of the first term in the expansion of } (x+y)^n \text{ is always 1.}\\ \text{By the symmetry property the coefficient of the last term will also be 1.]}\end{array}$$

And the coefficients of the second and sixth terms are just as easy to find:

$$\binom{7}{1} = \binom{7}{6} = \frac{7!}{6!} = 7 \qquad \begin{array}{l}\text{[The coefficient of the 2nd term is always } n; \text{ and since the left-to-right}\\ \text{sequence is the same as the right-to-left, the coefficient of the 6th}\\ \text{term in a 7-term expansion also equals } n.]\end{array}$$

Now you only have to calculate the coefficients of the third and fourth terms, since by the symmetry property the coefficient of the fifth term will be the same as that of the third:

$$\binom{7}{2} = \frac{7!}{2!(7-2)!} = \frac{7!}{2!5!} = 7 \cdot 3 = 21 = \binom{7}{5}$$

$$\binom{7}{3} = \frac{7!}{3!(7-3)!} = \frac{7!}{3!4!} = 7 \cdot 5 = 35$$

Thus $\qquad (x+y)^7 = x^7 + 7x^6y + 21x^5y^2 + 35x^4y^3 + 35x^3y^4 + 21x^2y^5 + 7xy^6 + y^7$

PROBLEM 10-33 What is the ninth term of $(2a - b)^{14}$?

Solution The ith term of $(x+y)^n$ is

$$\binom{n}{i-1}x^{n-(i-1)}y^{i-1}$$

So if $i = 9$, $n = 14$, $x = 2a$, $y = -b$, the solution is

$$\binom{14}{8}(2a)^6(-b)^8 = \frac{14!}{8!6!}(2)^6a^6b^8 = 7 \cdot 3 \cdot 11 \cdot 3(64)a^6b^8 = 192\,192a^6b^8$$

PROBLEM 10-34 Find the expansion of $(a^3 - 2b)^5$.

Solution Begin with the expansion of $(x+y)^5$, then substitute a^3 for x and $(-2b)$ for y: Since

$$(x+y)^5 = \binom{5}{0}x^5 + \binom{5}{1}x^4y + \binom{5}{2}x^3y^2 + \binom{5}{3}x^2y^3 + \binom{5}{4}xy^4 + \binom{5}{5}y^5$$

$$= x^5 + 5x^4y + 10x^3y^2 + 10x^2y^3 + 5xy^4 + y^5$$

Then

$$(a^3 - 2b)^5 = (a^3)^5 + 5(a^3)^4(-2b) + 10(a^3)^3(-2b)^2 + 10(a^3)^2(-2b)^3 + 5(a^3)(-2b)^4 + (-2b)^5$$

$$= a^{15} - 10a^{12}b + 40a^9b^2 - 80a^6b^3 + 80a^3b^4 - 32b^5$$

PROBLEM 10-35 Use the binomial expansion to find the value $(1.02)^{15}$, accurate to three decimal places.

Solution First, note that $1.02 = 1 + .02$. Then, since the first four terms of $(x+y)^{15}$ are

$$(x+y)^{15} = x^{15} + 15x^{14}y + 105x^{13}y^2 + 455x^{12}y^3 + \cdots$$

You have

$$(1 + .02)^{15} \simeq 1^{15} + 15(1)^{14}(.02) + 105(1)^{13}(.02)^2 + 455(1)^{12}(.02)^3 + \cdots$$

$$= 1 + 15(.02) + 105(.0004) + 455(.000\,008) = 1 + .30 + .042 + .003\,64 + \cdots$$

$$= 1.345\,64\ldots \cong 1.346$$

note: Using a calculator to get the accurate result, you find $(1.02)^{15} = 1.345\,83 \cong 1.346$. Not bad!

PROBLEM 10-36 Find the expansion of $(1 - 2i)^4$.

Solution Using $(x + y)^4$ with $x = 1$ and $y = -2i$, you have

$$(x + y)^4 = x^4 + 4x^3y + \left(\frac{4!}{2!2!}\right)x^2y^2 + 4xy^3 + y^4$$

$$= x^4 + 4x^3y + 6x^2y^2 + 4xy^3 + y^4$$

Then

$$(1 - 2i)^4 = 1^4 + 4(1)^3(-2i) + 6(1)^2(-2i)^2 + 4(1)(-2i)^3 + (-2i)^4$$

$$= 1 - 8i + 24i^2 - 32i^3 + 16i^4$$

$$= 1 - 8i - 24 + 32i + 16 \qquad (\textbf{\textit{recall}}: i^2 = -1)$$

$$= -7 + 24i$$

PROBLEM 10-37 Find the expansion of $(2x^{1/2} - y^{1/2})^6$.

Solution First write out $(a + b)^6$, then replace a by $(2x^{1/2})$ and b by $(-y^{1/2})$:

$$(a + b)^6 = a^6 + 6a^5b + \left(\frac{6!}{2!4!}\right)a^4b^2 + \left(\frac{6!}{3!3!}\right)a^3b^3 + \left(\frac{6!}{4!2!}\right)a^2b^4 + 6ab^5 + b^6$$

$$= a^6 + 6a^5b + 15a^4b^2 + 20a^3b^3 + 15a^2b^4 + 6ab^5 + b^6$$

$$(2x^{1/2} - y^{1/2})^6 = (2x^{1/2})^6 + 6(2x^{1/2})^5(-y^{1/2}) + 15(2x^{1/2})^4(-y^{1/2})^2 + 20(2x^{1/2})^3(-y^{1/2})^3$$

$$+ 15(2x^{1/2})^2(-y^{1/2})^4 + 6(2x^{1/2})(-y^{1/2})^5 + (-y^{1/2})^6$$

$$= 64x^3 - 192x^{5/2}y^{1/2} + 240x^2y - 160x^{3/2}y^{3/2} + 60xy^2 - 12x^{1/2}y^{5/2} + y^3$$

PROBLEM 10-38 Prove that $_nC_0 + {_nC_1} + {_nC_2} + \cdots + {_nC_n} = 2^n$.

Solution Write the general formula for $(x + y)^n$:

$$(x + y)^n = {_nC_0}x^n + {_nC_1}x^{n-1}y + {_nC_2}x^{n-2}y^2 + \cdots + {_nC_n}y^n$$

Now set $x = 1$, $y = 1$ to get

$$(1 + 1)^n = {_nC_0}(1)^n + {_nC_1}(1)^{n-1}(1) + {_nC_2}(1)^{n-2}(1)^2 + \cdots + {_nC_n}(1)^n$$

or

$$2^n = {_nC_0} + {_nC_1} + {_nC_2} + \cdots + {_nC_n}$$

Try it on sample values:

$$n = 2: \quad {_2C_0} + {_2C_1} + {_2C_2} = 1 + 2 + 1 = 4 = 2^2$$

$$n = 3: \quad {_3C_0} + {_3C_1} + {_3C_2} + {_3C_3} = 1 + 3 + 3 + 1 = 8 = 2^3$$

$$n = 4: \quad {_4C_0} + {_4C_1} + {_4C_2} + {_4C_3} + {_4C_4} = 1 + 4 + 6 + 4 + 1 = 16 = 2^4$$

Supplementary Exercises

PROBLEM 10-39 How many seven-digit telephone numbers can be formed if neither 0 nor 1 can be used as the first or second digit?

PROBLEM 10-40 Given the digits 1 through 9, you wish to form four-digit numbers whose digits are all different. How many are there?

PROBLEM 10-41 Of the above four-digit numbers, how many are less than 5000?

PROBLEM 10-42 Of the above four-digit numbers, how many are odd?

PROBLEM 10-43   If six people enter a bus and there are eight vacant seats, in how many ways can they be seated?

PROBLEM 10-44   Find n if $_nP_5 = 6 \cdot {_nP_4}$.

PROBLEM 10-45   The Greek alphabet has 24 letters. How many three-letter fraternity names could be formed if (**a**) repetitions are not permitted, (**b**) repetitions are permitted?

PROBLEM 10-46 $_nC_{(n-r)} = {_nC_r}$: True or False?

PROBLEM 10-47   From an ordinary deck of playing cards, how many ways are there of drawing three black cards?

PROBLEM 10-48   How many distinct numbers greater than 6000 without repeated digits can be formed with 1, 2, 3, 8, 9?

PROBLEM 10-49   In how many ways can you divide 9 books between two of your friends if one is to receive 5 books and the other 4 books?

PROBLEM 10-50   A post office keeps supplies of 8 different airmail stamps. In how many ways can a person buy 3 different stamps?

PROBLEM 10-51   How many seven-letter code words may be formed from the letters POPOVER?

PROBLEM 10-52   If 4 red, 3 black, 6 green, and 3 yellow disks are to be arranged in a row, what is the number of possible arrangements of the disks by color?

PROBLEM 10-53   In how many ways can a panel of 8 judges reach a majority decision?

PROBLEM 10-54   If twelve motorcyclists enter a race, in how many ways could the first three place winners be listed?

PROBLEM 10-55   A class of 22 students wants to elect a president, secretary, and treasurer. In how many ways can this be done?

PROBLEM 10-56   If a family has 6 children, in how many different orders could 4 boys and 2 girls have been born?

PROBLEM 10-57   If you invest $\$A$ at $p\%$ compounded annually, then the amount after n years is $A(1 + \frac{p}{100})^n$. Suppose you invest $\$1000$ at 10%. How much will your investment be worth after 5 years?

PROBLEM 10-58   Expand (**a**) $(3x + y)^4$, (**b**) $(x^2 + 2y^2)^3$, (**c**) $(5w - 4v)^4$, (**d**) $(a^3 - b^2)^5$.

PROBLEM 10-59   Find the first four terms of $(2a^2 + 3b^2)^9$.

PROBLEM 10-60   Find the fifth term of the expansion of $(5u^2 - 7v)^8$.

PROBLEM 10-61   Prove that if everyone in Smalltown, USA—population 25 000—has 3 initials, two people must have the same initials.

PROBLEM 10-62   Find n if $_nP_6 = 720$.

PROBLEM 10-63   In how many ways can 5 people line up to buy bus tickets?

PROBLEM 10-64   Find the number of ways three men and three women can use one restroom one at a time.

PROBLEM 10-65   Expand $(2 + 3i)^4$.

PROBLEM 10-66   If an automobile manufacturer makes four body models in six colors with three different engines, how many cars does a dealer need to stock to have one of each kind on hand?

PROBLEM 10-67   A programmer is instructed to write a program to sort a list of 12 items into some order. How many different orders can be produced?

PROBLEM 10-68   If a computer operator receives programs from seven students and must run these in some pre-specified order, how many different orders could be made for running the programs?

PROBLEM 10-69   Eight people have an investment club. The positions of president and treasurer are rotated each year. In how many years will the office holder slates be used up?

Answers to Supplementary Exercises

10-39 64×10^5

10-40   3024

10-41   1344

10-42   1680

10-43   20 160

10-44   10

10-45   (a) 12 144   (b) 13 824

10-46   True

10-47   15 600

10-48   5760

10-49   126

10-50   56

10-51   1260

10-52   33 633 600

10-53   93

10-54   1320

10-55   9240

10-56 $_6C_2 = 15$

10-57   \$1610.51

10-58   (a) $81x^4 + 108x^3y + 54x^2y^2 + 12xy^3 + y^4$
   (b) $x^6 + 6x^4y^2 + 12x^2y^4 + 8y^6$
   (c) $625w^4 - 2000w^3v + 2400w^2v^2 - 1280wv^3 + 256v^4$
   (d) $a^{15} - 5a^{12}b^2 + 10a^9b^4 - 10a^6b^6 + 5a^3b^8 - b^{10}$

10-59 $512a^{18} + 6912a^{16}b^2 + 41\,472a^{14}b^4 + 145\,152a^{12}b^6 + \cdots$

10-60 $105\,043\,750u^8v^4$

10-61   17 576 different initial combinations

10-62 $n = 6$

10-63   120

10-64   20

10-65 $-119 - 120i$

10-66   72

10-67 $12! = 479\,001\,600$

10-68 $7! = 5040$

10-69   28

EXAM 2 (Chapters 6–10)

1. Establish the validity of the statement $p \vee q \leftrightarrow \sim(\sim p \wedge (\sim q))$.

2. Determine if the following statement is a tautology, a contradiction, or neither:

$$[\sim q \wedge (p \rightarrow q)] \rightarrow \sim p$$

3. Write the following argument in symbolic notation and test its validity:

> "I will go skiing if and only if I have some money. If I have some money, then I will rent a video movie. I will not go skiing or I will buy a pizza. Therefore, I will rent a movie or I will buy a pizza."

4. Use the principle of mathematical induction to prove that the proposition

$$P(n): 3 + 7 + 11 + 15 + \cdots + (4n - 1) = n(2n + 1)$$

is valid for all positive integers n.

5. Let $P = \{2, 4, 6, 8\}$, $Q = \{1, 2, 5, 6, 7\}$, $R = \{1, 3, 5, 8, 9\}$, and $S = \{1, 6, 7, 9, 10\}$. Find

(a) $(P \cap Q) \cap S$ (b) $(P \cup R) \cap Q$ (c) $(P - Q)$ (d) $R \oplus S$

6. Prove that if A and B are disjoint sets, then $A \cup B = A \oplus B$.

7. Let the universal set be $U = \{$single-digit positive integers$\}$. Let $A = \{x \in U \mid x$ is a multiple of $2\}$, $B = \{x \in U \mid x^2 - 10 > 0\}$. Find

(a) $A \cup B$ (b) $A \cap B$ (c) $\overline{A}$ (d) $\overline{B}$

8. Show the equivalence of the Boolean expressions

$$x * z + x * y * \overline{z} = x * (y + z)$$

9. Define $f(x, y, z)$, given that the following table defines a Boolean expression:

x	y	z	$f(x, y. z)$
0	0	0	0
0	1	0	1
1	0	0	0
1	1	0	1
0	0	1	1
0	1	1	1
1	0	1	0
1	1	1	0

10. Use a Karnaugh map to find a simplified expression for the result of problem 9. Draw the logic circuit.

11. (a) Linn has 6 blouses, 8 skirts, and 5 sweaters. Assuming all coordinate, how many different outfits can she create?

(b) Benny, Joshua, and five other children line up to board a bus. In how many ways can they form a single line?

12. Find the first four terms of the expression of $(a^2 + 2b^2)^7$.

Solutions to Exam 2

1. Form the truth table of the complete statement:

p	q	$p \vee q$	$\sim p$	$\sim q$	$\sim p \wedge (\sim q)$	$\sim(\sim p \wedge (\sim q))$
T	T	T	F	F	F	T
T	F	T	F	T	F	T
F	T	T	T	F	F	T
F	F	F	T	T	T	F

2. Form the truth table:

p	q	$\sim q$	$p \rightarrow q$	$\sim q \wedge (p \rightarrow q)$	$\sim p$	$(\sim q \wedge (p \rightarrow q)) \rightarrow \sim p$
T	T	F	T	F	F	T
T	F	T	F	F	F	T
F	T	F	T	F	T	T
F	F	T	T	T	T	T

It is a tautology.

3. Assign a symbol to each statement of the argument:

p: "I will go skiing"
q: "I have some money"
r: "I will rent a video movie"
s: "I will buy a pizza"

You can then represent the argument by the statement

$$[(p \leftrightarrow q) \wedge (q \rightarrow r) \wedge (\sim p \vee s)] \rightarrow (r \vee s)$$

The complete truth table will have $2^4 = 16$ rows. You should create it and verify that when $p, q, r,$ and s are all false, we have a false implication. Therefore the argument is not valid.

4. First, show $P(1)$ is true: $3 = 1(2 + 1) = 3$
Next, assume $P(k)$ is true: It is true that $3 + 7 + 11 + 15 + \cdots + (4k - 1) = k(2k + 1)$
Finally, show $P(k + 1)$ is true:

$$[3 + 7 + 11 + 15 + \cdots + (4k - 1)] + (4k + 3) = (k + 1)(2k + 3)$$

$$k(2k + 1) + (4k + 3) = (k + 1)(2k + 3)$$

$$2k^2 + k + 4k + 3 = (k + 1)(2k + 3)$$

$$(k + 1)(2k + 3) = (k + 1)(2k + 3)$$

5. (a) We need elements in all of $P, Q,$ and S simultaneously: $(P \cap Q) \cap S = \{6\}$
(b) $P \cup R = \{1, 2, 3, 4, 5, 6, 8, 9\} \rightarrow (P \cup R) \cap Q = \{1, 2, 5, 6\}$
(c) We need elements in P and not in Q: $P - Q = \{4, 8\}$
(d) We need elements in R or in S but not in both: $R \oplus S = \{3, 5, 6, 7, 8, 10\}$

6. First, show $A \cup B \subseteq A \oplus B$. If $x \in A \cup B$, then $x \in A$ or $x \in B$ or both. But x cannot be in both since A and B are disjoint. Thus $x \in A \oplus B$.
Then show $A \oplus B \subseteq A \cup B$. If $x \in A \oplus B$, then $x \in A$ or $x \in B$, but not in both. But then $x \in A \cup B$. Thus x is an element of both $A \cup B$ and $A \oplus B$ and $A \cup B = A \oplus B$.

7. We have sets $A = \{2, 4, 6, 8\}$ and $B = \{4, 5, 6, 7, 8, 9\}$.

(a) $A \cup B = \{2, 4, 5, 6, 7, 8, 9\}$ **(c)** $\bar{A} = \{1, 3, 5, 7, 9\}$
(b) $A \cap B = \{4, 6, 8\}$ **(d)** $\bar{B} = \{1, 2, 3\}$

8. Set up a truth table:

x	y	z	$x*z$	$x*y$	$\bar{z}$	$x*y*\bar{z}$	$y+z$	$(x*z)+(x*y*\bar{z})$	$x*(y+z)$
0	0	0	0	0	1	0	0	0	0
0	1	0	0	0	1	0	1	0	0
1	0	0	0	0	1	0	0	0	0
1	1	0	0	1	1	1	1	1	1
0	0	1	0	0	0	0	1	0	0
0	1	1	0	0	0	0	1	0	0
1	0	1	1	0	0	0	1	1	1
1	1	1	1	1	0	0	1	1	1

↑ —— same —— ↑

9. The function will be the sum of the minterms that produce a value of 1 in the $f(x,y,z)$ column. Thus $f(x,y,z) = \bar{x}y\bar{z} + xy\bar{z} + \bar{x}\bar{y}z + \bar{x}yz$.

10. Enter 1's in the boxes corresponding to the minterms:

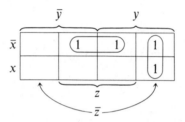

Thus the simplified version of the function is $f(x,y,z) = \bar{x}z + y\bar{z}$. Its logic circuit is shown in Figure E2-1.

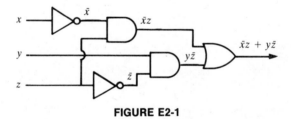

FIGURE E2-1

11. **(a)** Linn is limited to one item from each category; she therefore has $6 \cdot 8 \cdot 5 = 240$ different outfits.
(b) This is a permutation of seven items. There are $7! = 5040$ ways the children can form a single line.

12. We need to substitute a^2 for x and $2b^2$ for y in the binomial expansion of $(x+y)^7$:

$$(x+y)^7 = \binom{7}{0}x^7 + \binom{7}{1}x^6y + \binom{7}{2}x^5y^2 + \binom{7}{3}x^4y^3 + \cdots$$

$$= x^7 + 7x^6y + 21x^5y^2 + 35x^4y^3 + \cdots$$

$$= (a^2)^7 + 7(a^2)^6(2b^2) + 21(a^2)^5(2b^2)^2 + 35(a^2)^4(2b^2)^3 + \cdots$$

$$= a^{14} + 14a^{12}b^2 + 84a^{10}b^4 + 280a^8b^6 + \cdots$$

11 PROBABILITY

THIS CHAPTER IS ABOUT

☑ **Elementary Properties and Laws**
☑ **Probabilities for Compound Events**
☑ **Probabilities for Dependent Events**
☑ **Binomial Probabilities**

11-1. Elementary Properties and Laws

A. Definition of probability

In the classical definition of probability, we define the **probability of an event** as the value obtained when the number of outcomes favorable to the event is divided by the total number of possible outcomes.

EXAMPLE 11-1 Suppose a bag contains 2 red, 2 white, 1 blue, and 1 yellow marble. We blindly draw one out. What is the probability that it is red? blue? red or white?

Solution The total possible number of outcomes is 6 since there are six marbles in the bag. The number of favorable outcomes for the event of drawing a red one is 2 since there are two red marbles in there. Thus $P(\text{red}) = \frac{2}{6}$. Similarly, $P(\text{blue}) = \frac{1}{6}$ because there is only one blue marble, or a one-in-six chance of drawing it out. For $P(\text{red or white})$ we need to determine the number of favorable outcomes. The total number of red or white marbles is 4. Thus $P(\text{red or white}) = \frac{4}{6}$.

In many situations we can use this classical law. For example, the probability of drawing on ace from a card deck is $\frac{4}{52}$. The probability of rolling a 1 on a single die is $\frac{1}{6}$. But the definition has its limitations:

(1) It allows us to deal only with a finite number of outcomes.
(2) It depends on the events' being "equally likely."

To understand the concept of equally likely events, consider the outcomes for rolling two dice. We could list the outcomes as the set $\{2, 3, 4, 5, 6, 7, 8, 9, 10, 11, 12\}$ since these numbers represent the possible spot totals on the two dice. This set of 11 numbers, however, is not an equally likely set of outcomes since it is clearly more likely that the result 7 will appear than the result 12 will appear. Using the classical law to find $P(7) = \frac{1}{11}$ and $P(12) = \frac{1}{11}$ is therefore clearly wrong. To apply the classical law correctly, we must reconstruct the set of possible outcomes as 36 equally likely events consisting of number pairs representing the spots shown on the dice. Then any such *pair* (not their spot total) is equally likely to occur. Since many experiments can have more than one set of outcomes, it is helpful to use a set of outcomes in which the events are equally likely.

EXAMPLE 11-2 If a box contains 70 red poker chips, 20 white chips, and 10 blue chips, find $P(\text{red})$, $P(\text{white})$, and $P(\text{blue})$ if one chip is drawn. Are they equally likely?

Solution   Since we have 100 chips in the box, it is easy to see that $P(\text{red}) = \frac{70}{100}$, $P(\text{white}) = \frac{20}{100}$, and $P(\text{blue}) = \frac{10}{100}$, and that the chances of drawing a particular color are not equally likely.

note: If the set of outcomes is simply listed as the possible outcomes (red, white, blue) then the probabilities of $P(\text{red}) = \frac{1}{3}$, $P(\text{white}) = \frac{1}{3}$, and $P(\text{blue}) = \frac{1}{3}$ would be wrong because the events red, white, blue are not equally likely.

The set of outcomes of equally likely events is:

$$\underbrace{\{\text{red, red},\ldots,\text{red}}_{\text{70 of them}},\ \underbrace{\text{white, white},\ldots,\text{white}}_{\text{20 of them}},\ \underbrace{\text{blue, blue},\ldots,\text{blue}\}}_{\text{10 of them}}$$

100 equally likely events
(thus we use 100 as the denominator according to the classical definition)

That is, you're equally likely to draw any chip, but you're not equally likely to draw a red, white, or blue one.

B.   Sample spaces

In our study of probability we define anything that has outcomes (consequences) to be called an **experiment**. An experiment for which we describe all possible outcomes is called a **random experiment**, and the set of all possible outcomes is called the **sample space**. We usually denote the sample space by **S** and include in braces { } the collection of its outcomes.

EXAMPLE 11-3   You flip a fair coin once. What is the sample space?

Solution   There can be only two possible outcomes: heads or tails. Thus **S** = {heads, tails}.

note: In probability, "fair" means that a coin, die, or other object has not been "loaded," or tampered with, so it will produce a random, unbiased result every time it is thrown. In this text we will assume all objects are fair, whether stated or not.

EXAMPLE 11-4   A fair six-sided die is rolled and the number on top noted. What is the sample space?

Solution   **S** = $\{1, 2, 3, 4, 5, 6\}$ because there are only six possible outcomes.

EXAMPLE 11-5   Two coins, a penny and a nickel, are tossed and the results noted as an ordered pair (penny, nickel). Find the sample space.

Solution   There are four possible outcomes: **S** = {HH, HT, TH, TT}.

EXAMPLE 11-6   Two dice are rolled and the sum of their spots is observed. What is the sample space?

Solution   There are eleven possible sums as outcomes: **S** = $\{2, 3, 4, 5, 6, 7, 8, 9, 10, 11, 12\}$. This is not an equally likely sample space.

EXAMPLE 11-7   An experiment consists of drawing a marble from a bucket that contains yellow, green, and blue marbles. What is the sample space? Is this an equally likely sample space?

Solution   The sample space is **S** = {yellow, green, blue}, but since we don't know how many of each color are in the bucket, we cannot be sure if it is an equally likely sample space.

EXAMPLE 11-8   Suppose two marbles are drawn one at a time from the bucket of Example 11-7. List the sample space.

Solution   The set of all possible two-marble combinations that could be drawn is **S** = {YY, YG, YB, GY, GG, GB, BY, BG, BB}. (Note that even though you can end up with a green and yellow marble in your hand, the order in which they might be drawn makes the GY and YG draws totally different.)

C. Events

In a given random experiment, an **event** is a subset A of the sample space **S**. This subset may contain no outcomes, exactly one outcome, or more than one outcome. Each outcome is called a **simple event**, and more than one outcome is a **compound event**. Thus if subset A contains only one element, A is a simple event. If subset A has more than one element, it is called a compound event.

EXAMPLE 11-9   Suppose we roll a four-sided die twice and record the results as number pairs. There will be 16 ordered pairs in the sample space: (i, j), $1 \leq i \leq 4, 1 \leq j \leq 4$. Describe the event for which (**a**) the sum 2 is obtained (**b**) the sum 6 is obtained (**c**) the sum 9 is obtained.

Solution

(**a**) In only one outcome in the sample space will the sum 2 occur; in pair $(1, 1)$. Thus $A = \{(1, 1)\}$ —a simple event.

(**b**) The sum 6 is obtained from the subset $A = \{(2, 4), (3, 3), (4, 2)\}$ —a compound event.

(**c**) The sum 9 is impossible. Thus $A = \{\quad\} = \varnothing$

EXAMPLE 11-10   The Daily Four Game Lottery allows the player to pick four digits from 0 to 9. A match of all four numbers in any order produces a win. The sample space has $10 \cdot 10 \cdot 10 \cdot 10 = 10\,000$ possible equally likely combinations. If a player chooses the digits 1, 1, 2, and 2, what is the subset corresponding to the event?

Solution   This is the compound event $A = \{(1, 1, 2, 2), (1, 2, 1, 2), (1, 2, 2, 1), (2, 1, 1, 2), (2, 1, 2, 1), (2, 2, 1, 1)\}$.

D. Probability for an event

The likelihood that an outcome of event A will occur is called the **probability** of A, denoted $P(A)$. It is always a number in the unit interval, $0 \leq P(A) \leq 1$. If the event is certain, then $P(A) = 1$; if the event is impossible, then $P(A) = 0$.

EXAMPLE 11-11   Suppose a sack contains ten pennies and ten dimes. Find the probability that on a single draw (**a**) you will pick a quarter (**b**) you will pick a penny or a dime.

Solution

(**a**) There are no quarters in there, so it is impossible to draw one; $P(\text{quarter}) = 0$.

(**b**) The sack has only pennies and dimes, so to pick one or the other is a surety; $P(\text{penny or dime}) = 1$.

1. Probability for equally likely events

In some random experiments, each of the possible outcomes is equally likely to occur. In rolling a single fair die, each face is an equally likely occurrence. In flipping a single fair coin, each face is equally likely to come up. The **probability for equally likely events** is

PROBABILITY FOR EQUALLY LIKELY EVENTS
$$P(A) = \frac{m}{n}$$

where m is the number of outcomes in A; n is the number of outcomes in the sample space.

EXAMPLE 11-12   Find the following probabilities:

(**a**) getting "heads" when tossing a coin

(**b**) getting "both heads" when tossing a dime and a quarter

(**c**) drawing a king from a deck of cards

(**d**) drawing a spade from a deck of cards

(**e**) drawing the ace of diamonds from a deck of cards

Solution

(a) The sample space has 2 outcomes, H or T. Only 1 (H) is favorable. Thus $P(\text{head}) = \frac{1}{2}$.

(b) The sample space now has 4 outcomes: $S = \{HH, HT, TH, TT\}$ and only 1 is favorable. Thus

$$P(HH) = \frac{\text{number of outcomes favorable}}{\text{number of outcomes in sample space}} = \frac{1}{4}.$$

(c) The sample space has 52 outcomes and 4 of them are "kings." Thus $P(\text{king}) = \frac{4}{52} = \frac{1}{13}$.

(d) From the 52 possible outcomes, 13 of them are now in A, so $P(\text{spade}) = \frac{13}{52} = \frac{1}{4}$.

(e) There is only one event in A, that is, get the ace of diamonds, so $P(\text{ace of diamonds}) = \frac{1}{52}$.

EXAMPLE 11-13   In Example 11-9 we rolled a four-sided die twice. The sample space had 16 outcomes. What is the probability that the sum is (a) two (b) six (c) nine?

Solution

(a) Since we could get a sum of two in only one way, $A = \{(1, 1)\}$, we have $P(\text{two}) = \frac{1}{16}$.

(b) There were three ways to get a sum of six, and the event is compound; $|A|$ (the cardinality of set A) is 3, so $P(\text{six}) = \frac{3}{16}$.

(c) Since $A = \varnothing$, $P(\text{nine}) = 0$ (Surely impossible!)

EXAMPLE 11-14   In the Daily Four Game Lottery of Example 11-10 what is the probability of winning if I use only the numbers 1, 2, 3, and 4?

Solution   We know $|S| = 10\,000$. We need $|A|$. But to fill the four slots for an event in A, we have four choices for each slot. Thus the number of possible combinations in the event (the number of different ways my four digits can be ordered) is $4 \cdot 4 \cdot 4 \cdot 4 = 256$. Thus $P(\text{winning}) = \dfrac{256}{10000} = 0.0256 = 0.256\%$

(Notice that when I choose 4 different digits my event has 256 possible chances of winning. In Example 11-10, however, the player's choice of 4 digits included two duplications. The event for this choice of numbers had only 6 chances of winning!)

---

2. **Probability for events not necessarily equally likely—relative frequency**

   If we repeat an experiment N times and let $|A|$ be the number of times an outcome belongs to subset A, then $\dfrac{|A|}{N}$ denotes the **relative frequency** of the occurrence of A. For large N, this value approaches a fixed number p in $[0, 1]$ and stabilizes there. The value p is the probability $P(A)$ for event A.

EXAMPLE 11-15   You are given a box of 100 thumbtacks. You want to find the probability that when a tack is tossed onto a table it will land "point up" 🔼 compared to the alternative of "point down" 🔽. When you shake the box and dump it out the 100 tacks fall onto the table and assume their position— point up or point down. How would you find the required probability?

Solution   Count the number of tacks that have fallen point up; call it n_1. Find $\dfrac{n_1}{100}$. This is our first estimate of p. Now repeat this procedure *many* times, obtaining successively $\dfrac{n_1 + n_2}{200}, \dfrac{n_1 + n_2 + n_3}{300}, \ldots$ This sequence will converge to the final value for p–the probability that a given tack will fall point up.

> *note:* This process is sometimes called the **Law of Large Numbers**:
> If the number of times that an experiment is repeated is increased, the ratio of the number of successful occurrences to the number of trials will approach the theoretical probability for the outcome of a single trial of that event.

EXAMPLE 11-16   You wish to test the "honesty" of a coin (that is, is it constructed so that each side has an equal chance of landing face up?). Describe the process.

Solution Toss the coin *many* times (thousands upon thousands). Record the number of times the coin comes up heads. The number of heads divided by the number of tosses gives the relative frequency of the occurrence of heads. This should approach $\frac{1}{2}$ if the coin is fair.

11-2. Probabilities for Compound Events

A. Complements and mutually exclusive events

The **complement of an event** A, denoted $\bar{A}$, is the set of all sample points in the sample space that do not belong to event A.

PROBABILITY OF COMPLEMENT OF EVENT A	$P(\bar{A}) = 1 - P(A)$

EXAMPLE 11-17 If the probability of drawing an ace from an ordinary deck is $\frac{4}{52}$, what is the probability of drawing a non-ace?

Solution A non-ace is the complement of an ace since {Ace} $\cup$ {Non-ace} = whole deck = universal set. Thus

$$P(\text{non-ace}) = 1 - P(\text{ace}) = 1 - \frac{4}{52} = \frac{48}{52}$$

EXAMPLE 11-18 If three coins are tossed simultaneously, what is the probability of at least one head appearing?

Solution Let A denote the event of no head appearing. Then $\bar{A}$ is the set of all the rest of the possible occurrences, that is, the set of those occurrences in which at least one head appears. The sample space is $S = \{HHH, HHT, HTH, HTT, THT, TTH, TTT\}$. Now $P(A) = \frac{1}{8}$ since no heads appear only in outcome, TTT. Thus $P(\bar{A}) = 1 - \frac{1}{8} = \frac{7}{8}$.

If two events are defined in such a way that the occurrence of one precludes the occurrence of the other, the events are said to be **mutually exclusive**. Thus, if one occurs, the other cannot, and vice versa. One way to describe the relationship is to indicate $(A \cap B) = \varnothing$. In terms of probability, this would be written

PROBABILITY OF MUTUALLY EXCLUSIVE EVENTS	$P(A \text{ and } B) = 0$

That is, if A and B are mutually exclusive, the probability that they will both occur is zero.

EXAMPLE 11-19 Describe several mutually exclusive events.

Solution

(1) The occurrence of heads and the occurrence of tails on one flip of a coin.
(2) The occurrence of all of the dot totals 1, 2, 3, 4, 5, and 6 on a single roll of a die.
(3) The occurrence of missing the basketball free-throw and the occurrence of making it.

EXAMPLE 11-20 Determine if the following events are mutually exclusive:

(a) Drawing an ace; drawing a heart.
(b) A randomly selected student is male; a randomly selected student is over 21.
(c) Two dice rolled show a sum that is even; the dice show a sum of 11.

Solution

(a) These are not mutually exclusive since one card in the desk is both an ace and a heart at the same time (the Ace of Hearts).

(b) These are not mutually exclusive since the student could be male and over 21 at the same time.
(c) These are mutually exclusive since the even sum and the sum of 11 cannot happen at the same time.

B. Special rule of addition

Recall that when events A and B are mutually exclusive, then the probability that they both occur simultaneously is zero (i.e., it is impossible). Thus $P(A \cap B) = 0$; the Venn diagram for these sets A and B would show the two sets to be disjoint. It follows, then, that under these circumstances the probability that event A *or* event B (but not both) would happen should equal the probability that A would happen plus the probability that B would happen. This is indeed true and is known as the **special addition rule** of probabilities. In short,

SPECIAL ADDITION RULE OF MUTUALLY EXCLUSIVE EVENTS $P(A \text{ or } B) = P(A) + P(B)$

EXAMPLE 11-21   Use the special addition rule to find the probability that on a single roll of a die, the outcome is 5 or 6.

Solution   Since the outcomes 5 and 6 are mutually exclusive (you can't have both on one roll), we need

$$P(5 \text{ or } 6) = P(5) + P(6)$$

$$= \frac{1}{6} + \frac{1}{6} = \frac{2}{6} = \frac{1}{3}$$

EXAMPLE 11-22   Refer again to Example 11-18 where three coins were tossed simultaneously. Use an extended special addition rule to verify the probability of at least one head appearing.

Solution   To obtain the probability of at least one head appearing we would need to obtain the probability of exactly one head, then exactly two heads, then exactly three heads, appearing. Seeing that these events are mutually exclusive, we could use a special addition rule. Now $P(1 \text{ head}) = \frac{3}{8}$ since the successes HTT, THT, and TTH are from the sample space of 8 possible occurrences. Also $P(2 \text{ heads}) = \frac{3}{8}$ using the favorable occurrences HHT, HTH, and THH. Finally $P(3 \text{ heads}) = \frac{1}{8}$ using the outcome HHH. Thus the rule extends to

$$P(1 \text{ head or 2 heads or 3 heads}) = P(1 \text{ head}) + P(2 \text{ heads}) + P(3 \text{ heads}) = \frac{3}{8} + \frac{3}{8} + \frac{1}{8} = \frac{7}{8}$$

which agrees with the answer of Example 11.18.

We can extend this addition rule to any number of mutually exclusive events. If events $A, B, C, D, .., K$ are all mutually exclusive, then

EXTENDED ADDITION RULE OF MUTUALLY EXCLUSIVE EVENTS $P(A \text{ or } B \text{ or } C \text{ or } D \text{ or} \cdots \text{or } K) = P(A) + P(B) + P(C) + \cdots + P(K)$

EXAMPLE 11-23   Suppose $P(A) = 0.2$ and $P(B) = 0.5$ and A and B are mutually exclusive. Find (a) $P(\bar{A})$ (b) $P(\bar{B})$ (c) $P(A \text{ or } B)$ (d) $P(A \text{ and } B)$.

Solution

(a) $P(\bar{A}) = 1 - P(A) = 1 - 0.2 = 0.8$
(b) $P(\bar{B}) = 1 - P(B) = 1 - 0.5 = 0.5$
(c) $P(A \text{ or } B) = P(A) + P(B) = 0.2 + 0.5 = 0.7$
(d) $P(A \text{ and } B) = 0$ since A and B are mutually exclusive.

C. Independence of events and the multiplication rule

Consider the situation for the compound event in which both A and B occur. If we toss a dime and a quarter, what is the probability that we obtain heads on the dime and heads on the quarter? This would correspond to the outcome {HH} as a subset of the sample space {HH, HT, TH, TT}. Thus the probability is $\frac{1}{4}$. By the classical definition of probability, we have one favorable result among four possible outcomes. But if we separately consider both $P(A)$ and $P(B)$—that is, P(heads on dime) and P(heads on quarter)—independently in their respective flip, each has value $\frac{1}{2}$. (Put another way, there's a 50% chance of the dime's landing heads up and a 50% chance of the quarter's landing heads up. Each probability, taken independently, has a value of $\frac{1}{2}$. By multiplying these independent values, we again get the result $\frac{1}{4}$. Thus the case P(heads on dime and heads on quarter) = P(heads on dime) · P(heads on quarter), or

$$P(A \text{ and } B) = P(A) \cdot P(B) \tag{11-1}$$

The simple multiplication rule of eg. (11-1) works only under special circumstances, called **independence of events**. Two events A and B are independent events if the occurrence of one of them does not affect the probability given to the occurrence of the other event. (For example, the occurrence of heads on the dime has no influence on the occurrence of heads on the quarter, so the two events are considered independent.)

EXAMPLE 11-24 Suppose you roll two six-sided dice, one red and one green, and record the results as number pairs: (red number, green number). What is the sample space?

Solution Each die has six possible outcomes, and the number that comes up on one die does not affect the number that comes up on the other die. Therefore the event A (the occurrence for the red die) and event B (the occurrence for the green die) are independent. There will be 36 points in the sample space:

(1, 1)	(1, 2)	(1, 3)	(1, 4)	(1, 5)	(1, 6)
(2, 1)	(2, 2)	(2, 3)	(2, 4)	(2, 5)	(2, 6)
(3, 1)	(3, 2)	(3, 3)	(3, 4)	(3, 5)	(3, 6)
(4, 1)	(4, 2)	(4, 3)	(4, 4)	(4, 5)	(4, 6)
(5, 1)	(5, 2)	(5, 3)	(5, 4)	(5, 5)	(5, 6)
(6, 1)	(6, 2)	(6, 3)	(6, 4)	(6, 5)	(6, 6)

EXAMPLE 11-25 Using the data from Example 11-24, suppose event A is that the red die shows 6 and event B is that the green die shows 6. If we roll the dice together what is the probability that two 6's occur?

Solution By classical definition only one outcome is favorable (6, 6) out of 36 possible outcomes, so P(two 6's) = $\frac{1}{36}$. But on the red die, $P(6) = \frac{1}{6}$ and on the green die, $P(6) = \frac{1}{6}$, and their outcomes are independent of each other. Therefore the multiplication rule holds and

$$P(6 \text{ and } 6) = P(6) \cdot P(6) = \frac{1}{6} \cdot \frac{1}{6} = \frac{1}{36}$$

EXAMPLE 11-26 Using the 6-by-6 grid of sample space outcomes from Example 11-24, find the following: (a) $P(7)$, (b) $P(9)$, (c) P(sum less than 4), (d) P(doubles).

Solution

(a) $P(7) = \dfrac{\text{number of times the sum of 7 occurs}}{36} = \dfrac{6}{36}$ since the sum 7 comes from (1, 6), (2, 5), (3, 4), (4, 3), (5, 2), and (6, 1).

(b) $P(9) = \dfrac{\text{number of times the sum of 9 occurs}}{36} = \dfrac{4}{36}$ since the sum 9 comes from (6, 3), (5, 4), (4, 5), and (3, 6).

(c) P(sum less than 4) = P(3 or 2) = $P(3) + P(2) = \dfrac{2}{36} + \dfrac{1}{36} = \dfrac{3}{36} = \dfrac{1}{12}$

(d) $P(\text{doubles}) = \dfrac{\text{number of times doubles occur}}{36} = \dfrac{6}{36} = \dfrac{1}{6}$ since doubles comes from (1, 1), (2, 2), (3, 3), (4, 4) 5, 5), and (6, 6).

But multiplication doesn't always work, so be careful.

EXAMPLE 11-27 Find the probability that on a roll of two dice we have a sum of 5 and "doubles."

Solution Since "doubles" means that the two dice show the same number and these sums are always even, the occurrence of 5 and "doubles" simultaneously is impossible. Hence $P(5 \text{ and doubles}) = 0$. But if we blindly went to the multiplication rule and calculated $P(5 \text{ and doubles}) = P(5) \cdot P(\text{doubles}) = \frac{4}{36} \cdot \frac{6}{36} = \frac{1}{9} \cdot \frac{1}{6} = \frac{1}{54}$, we get an incorrect result.

EXAMPLE 11-28 Show that simple multiplication does not work for $P(8 \text{ and doubles})$.

Solution The answer by looking into the sample space shows (4, 4) to be the only one of the 36 points applicable. Thus $P(8 \text{ and doubles}) = \frac{1}{36}$. But the multiplication rule would have $P(8 \text{ and doubles}) = P(8) \cdot P(\text{doubles}) = \frac{5}{36} \cdot \frac{6}{36} = \frac{5}{216}$, which is not correct. The events "8" and "doubles" are not independent.

We can extend the multiplication rule to any number of independent events. If $A, B, C, \ldots, L$ are independent events, then

EXTENDED MULTIPLICATION RULE $\quad P(A \text{ and } B \text{ and } C \text{ and } \cdots \text{ and } L) = P(A) \cdot P(B) \cdot P(C) \cdot \cdots \cdot P(L)$

EXAMPLE 11-29 Using the red and green dice of Example 11-24, we define the following events:

$$A = \{\text{sum is 3}\} = \{(1, 2), (2, 1)\}$$

$$B = \{\text{sum is more than 9}\} = \{(6, 4), (5, 5), (4, 6), (6, 5), (5, 6), (6, 6)\}$$

$$C = \{\text{sum is 10}\} = \{(6, 4), (5, 5), (4, 6)\}$$

$$D = \{\text{set of doubles}\} = \{(1, 1), (2, 2), (3, 3), (4, 4), (5, 5), (6, 6)\}$$

Define the sets **(a)** $A \cap B$, **(b)** $A \cup C$, **(c)** $B \cap C$, **(d)** $A \cap D$, **(e)** $B \cap D$.

Solution A simple way to tackle this problem is to use the sample space and circle the occurrences of events $A, B, C,$ and D:

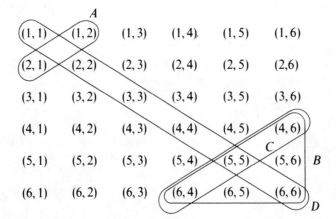

The results are now easy to see by inspecting the diagram.

(a) $A \cap B = \varnothing$
(b) $A \cup C = \{(2,1),(1,2),(6,4),(5,5),(4,6)\}$
(c) $B \cap C = C$
(d) $A \cap D = \varnothing$
(e) $B \cap D = \{(5,5),(6,6)\}$

EXAMPLE 11-30   Use the 6-by-6 grid of results for rolling two dice to find **(a)** $P(7 \text{ or } 8)$, **(b)** $P(2 \text{ and } 12)$, **(c)** $P(2 \text{ or } 12)$.

Solution

(a) Count all the occurrences of the sum 7 or 8. There are six and five occurrences, respectively. Thus

$$P(7 \text{ or } 8) = \frac{6+5}{36} = \frac{11}{36}$$

(b) $P(2 \text{ and } 12) = 0$ since those are nonoverlapping.
(c) $P(2 \text{ or } 12) = P(2) + P(12) = \frac{1}{36} + \frac{1}{36} = \frac{1}{18}$

11-3. Probabilities for Dependent Events

Lack of independence is called **dependence** and is illustrated by the following situation. Consider again the experiment of rolling two dice and recording two events: rolling "8" or rolling "doubles." We know from the grid sample space of Example 11-24 that $P(8) = \frac{5}{36}$ and $P(\text{doubles}) = \frac{6}{36}$. In the roll, does the occurrence of 8 affect the probability of getting a double?

If we know 8 has occurred, then it came from the subset $\{(2,6),(3,5),(4,4),(5,3),(6,2)\}$ and a double does occur in here. Thus we observe that the chances of getting a double knowing that an 8 has been rolled is easy. It is $\frac{1}{5}$—the pair $(4,4)$ from the above five. This is written $P(\text{double} \mid 8) = \frac{1}{5}$, and is read "The probability of getting a double given than an 8 has occurred is one fifth." Earlier we saw that $P(\text{double}) = \frac{1}{6}$ so the occurrence of 8 did effect the probability of the double. Hence these two occurrences, that of "8" and that of "double," are dependent events.

This situation is called **conditional probability** and we denote it by $P(B \mid A)$, read "the probability that event B will occur, given that event A has already occurred." It is defined by

CONDITIONAL PROBABILITY $$P(B \mid A) = \frac{P(A \text{ and } B)}{P(A)}$$ **(11-2)**

EXAMPLE 11-31   Use the formula for conditional probability to check out the situation that the occurrence of "8" affects the probability of the occurrence of "doubles."

Solution $P(8 \text{ and double})$ is $\frac{1}{36}$ since only one possible outcome can be both 8 and double—(the $(4,4)$ outcome). Also $P(8) = \frac{5}{36}$ as noted earlier. Thus

$$P(\text{double} \mid 8) = \frac{\frac{1}{36}}{\frac{5}{36}} = \frac{1}{5}$$

EXAMPLE 11-32   A box contains five yellow and four green ping pong balls. What is the probability that when two are selected, both will be green if **(a)** we replace (put back) the first one after its selection **(b)** we don't replace the first one after its selection.

Solution

(a) If we replace the ball, then the two parts of the event are independent and we don't need to use conditional probability; so

$$P(\text{green and green}) = P(\text{green}) \cdot P(\text{green}) = \frac{4}{9} \cdot \frac{4}{9} = \frac{16}{81} = .197$$

(b) If we don't replace the ball, the result of the second draw is dependent upon the result of the first draw, so we must use the conditional probability formula, eq. (11-2). By cross-multiplying, we change its form to be

$$P(A) \cdot P(B \mid A) = P(A \text{ and } B)$$

We want A to be green, B to be green. Thus

$$P(\text{green and green}) = P(\text{green}) \cdot P(\text{green} \mid \text{green}) = \frac{4}{3} \cdot \frac{3}{8} = \frac{1}{6} = .166$$

since on the first draw there are four eligible green among nine balls and then on the second draw (there was no replacement) there are three eligible green among the remaining eight balls. Notice that in this case we have a greater chance of drawing two green balls if the first ball drawn is replaced in the box.

note: Sampling without replacement is equivalent to drawing the whole set at once (think of it as a very quick operation—first one, then immediately another, then another, etc.).

EXAMPLE 11-33   A bucket contains 20 pennies, 10 nickles, and 5 dimes. You grab three coins. **(a)** What is the probability that all are nickels? **(b)** that they are all different?

Solution

(a) The probability for a triple event of choosing three nickels is

$$P(\text{nickel, nickel, nickel}) = P(\text{nickel}) \cdot P(\text{nickel} \mid \text{nickel}) \cdot P(\text{nickel} \mid 2 \text{ nickels})$$

That is, the probability of choosing three nickels in a row is equivalent to the probability of choosing one nickel times the probability of choosing a second nickel given that you have chosen the first nickel, times the probability of choosing a third nickel given that you have chosen two nickels. Plugging in the numbers, we get

$$P(n, n, n) = P(n) \cdot P(n \mid n) \cdot P(n \mid 2n)$$

$$= \frac{10}{35} \cdot \frac{9}{34} \cdot \frac{8}{33} = \frac{720}{39270}$$

$$= .018$$

(b) The probability for this triple event of choosing one of each coin is

$$P(\text{penny, nickel, dime}) = P(\text{penny}) \cdot P(\text{nickel} \mid \text{penny}) \cdot P(\text{dime} \mid \text{nickel and penny})$$

$$= \frac{20}{35} \cdot \frac{10}{34} \cdot \frac{5}{33} = \frac{1000}{39270}$$

$$= .025$$

EXAMPLE 11-34   Draw three cards without replacement from a standard deck. Let AKQ denote the event of drawing an Ace, a King, and a Queen, in that order. Find $P(\text{AKQ})$.

Solution   There are four of each card value in a standard deck, and your chances of drawing a different card increase as more cards are pulled from the deck:

$$P(\text{AKQ}) = \quad P(\text{A}) \quad \cdot \quad P(\text{K} \mid \text{A}) \quad \cdot \quad P(\text{Q} \mid \text{AK})$$

$$= \quad \frac{4}{52} \quad \cdot \quad \frac{4}{51} \quad \cdot \quad \frac{4}{50}$$

$$\uparrow \qquad\qquad \uparrow \qquad\qquad \uparrow$$

| 4 aces among the 52 cards | 4 kings among the remaining 51 cards | 4 queens among the remaining 50 cards |

Recall that events A and B are *independent* if the probability for event B is not affected by whether or not event A has occurred. Under this circumstance

$$P(B \mid A) = P(B) \quad \text{and} \quad P(A \mid B) = P(A)$$

The conditional probability formula then becomes

$$P(A) \cdot P(B \mid A) = P(A \text{ and } B)$$

or

$$P(A) \cdot P(B) = P(A \text{ and } B) \qquad \textbf{(11-3)}$$

Eq. (11-3) corresponds with our earlier formula of the multiplication rule for independent events (eq. 11-1).

We saw earlier that if events A and B are *mutually exclusive* (if one of them happens, the other cannot), then the special addition rule gives us

$$P(A \text{ or } B) = P(A) + P(B)$$

(Figure 11-1). If the events are not mutually exclusive, however, we cannot find the probability of one or the other by simply adding their respective probabilities because there might be overlapping members of each event; by adding $P(A)$ and $P(B)$ we actually cover the overlapped region twice (Figure 11-2). Thus we need to deduct the repetition of the $A \cap B$ region from the total probability. This gives us the general formula for the probability of the occurrence of two intersecting, not mutually exclusive, events:

$$P(A \text{ or } B) = P(A) + P(B) - P(A \text{ and } B) \qquad \textbf{(11-4)}$$

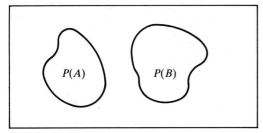

A and B are mutually exclusive

FIGURE 11-1

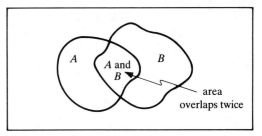

FIGURE 11-2

EXAMPLE 11-35   To get into Prestige University a student must pass at least one of two entrance exams. If the probability of passing Test I is $\frac{3}{5}$ and the probability of passing Test II is $\frac{1}{2}$, while the probability of passing both tests is $\frac{1}{5}$, find the probability that a student will satisfy the requirement.

Solution   Use eq. (11-4):

$$P(\text{I or II}) = P(\text{I}) + P(\text{II}) - P(\text{I and II})$$

$$= \frac{3}{5} + \frac{1}{2} - \frac{1}{5} = \frac{6}{10} + \frac{5}{10} - \frac{2}{10}$$

$$= \frac{9}{10}$$

EXAMPLE 11-36   A box contains 30 small motors of which 6 are known to be defective. You select 3 motors from the box. What is the probability that none of those you select is defective?

Solution   For the first motor to be nondefective, you must select one of the 24 good ones among the 30, so $P(\text{good on first draw}) = \frac{24}{30}$. If this is done, then there are 23 good ones among the 29 left, so $P(\text{good on second draw} \mid \text{good on first}) = \frac{23}{29}$. If these are both done then $P(\text{good on third} \mid \text{good on first and second}) = \frac{22}{28}$. Thus

$$P(3 \text{ good}) = \tfrac{24}{30} \cdot \tfrac{23}{29} \cdot \tfrac{22}{28} = .4985$$

11-4. Binomial Probabilities

A. Bernoulli experiment

If an experiment can result in one of only two possible outcomes, it is called a **Bernoulli experiment**. We refer to the two outcomes as "success" and "failure" and denote them by S and F, respectively. The notation $P(S) = p$ denotes the probability of success, and $P(F) = q$ denotes the probability of failure. Since one of the two outcomes must occur, $p + q = 1$.

EXAMPLE 11-37 Interpret the probability of success and failure of the following Bernoulli experiments: **(a)** the flip of an honest coin **(b)** the test of a light bulb **(c)** the shooting of a basketball free-throw.

Solution

(a) Since flipping a coin can result in exactly two outcomes, heads or tails, we call one of them (say, heads) a success and the other a failure. We can write

$$P(\text{success}) = P(\text{heads}) = p = 0.5$$
$$P(\text{failure}) = P(\text{tails}) = q = 0.5$$

(b) If we have randomly selected a light bulb and we flip the switch, it either lights or it doesn't. If it works we will call it a success; else a failure. The values of $p = P(\text{success})$ and $q = P(\text{failure})$ are usually not known, but a "track record" of that manufacturer's quality could be researched and reported as a probability that a light bulb you buy from that company will work.

(c) Any given free-throw shot results in success or failure. A player would have a season record or career record of the number of successes in n total tries. This ratio would serve as the value of $p = P(\text{success})$. Since the experiment has only two possible outcomes,

$$q = P(\text{failure}) = 1 - p$$

B. Binomial probabilities

A **binomial experiment** is a Bernoulli experiment that is repeated n times and that meets three criteria:

(1) Each trial is independent of the others (no one outcome affects the others);
(2) The probability for success p (and thus the probability of failure q) does not change from trial to trial;
(3) The number of successes X in the n trials is an integer in $1 \leq X \leq n$.

EXAMPLE 11-38 Suppose you are required to take a four-question multiple choice quiz, with each question having three choices. You haven't studied, you're tired, and you want to go to the beach. So you just circle any ol' letter—an A here, a C there, maybe a B on this one, whatever—and leave. Now you're on your way to the beach and some questions pop into your mind:

(a) what is the probability that I got all four wrong?
(b) what is the probability that I got all four correct?
(c) how many did I probably get correct?

Investigate these questions.

Solution We want to find out how the probability is distributed amongst the various possible outcomes. To do this we will make some calculations and thereby construct the *binomial probability distribution*. In this problem we proceed as follows: If x is the number of questions you got correct, then x may have values 0, 1, 2, 3, or 4. Since only one of the three choices for the questions is correct, you have $\frac{1}{3}$ chance of guessing it correctly. Also, you have $\frac{2}{3}$ chance that you guessed wrong. Since the choices were made independently, you use the multiplication rule (eq. 11-1) to find the probabilities:

(a) $P(0 \text{ correct}) = P(4 \text{ wrong})$

$$= P(\text{wrong and wrong and wrong and wrong})$$

$$= P(\text{wrong}) \cdot P(\text{wrong}) \cdot P(\text{wrong}) \cdot P(\text{wrong})$$

$$= \frac{2}{3} \cdot \frac{2}{3} \cdot \frac{2}{3} \cdot \frac{2}{3} = \frac{16}{81} = .198$$

So you have just under 20% chance that the quiz was a complete flop!

(b) Now calculate your chances of getting them all right:

$$P(4 \text{ correct}) = P(\text{correct and correct and correct and correct})$$

$$= P(\text{correct}) \cdot P(\text{correct}) \cdot P(\text{correct}) \cdot P(\text{correct})$$

$$= \frac{1}{3} \cdot \frac{1}{3} \cdot \frac{1}{3} \cdot \frac{1}{3} = \frac{1}{81} = .012$$

Oh no! you have only a 1% chance that they are all correct!

(c) Getting exactly one correct answer could occur in four ways—(first correct; others wrong) or (second correct; others wrong) or, etc. Thus

$$P(1 \text{ correct}) = 4[P(\text{correct and wrong and wrong and wrong})]$$

$$= 4[P(c) \cdot P(w) \cdot P(w) \cdot P(w)]$$

$$= 4\left[\frac{1}{3} \cdot \frac{2}{3} \cdot \frac{2}{3} \cdot \frac{2}{3}\right] = \frac{32}{81} = .395$$

Getting two correct could occur in $_4C_2 = 6$ ways. Thus

$$P(2 \text{ correct}) = 6[P(c \text{ and } c \text{ and } w \text{ and } w)]$$

$$= 6[P(c) \cdot P(c) \cdot P(w) \cdot P(w)]$$

$$= 6\left[\frac{1}{3} \cdot \frac{1}{3} \cdot \frac{2}{3} \cdot \frac{2}{3}\right]$$

$$= \frac{24}{81} = .296$$

In turn, there would be four ways of getting three correct and one wrong, so

$$P(3 \text{ correct}) = {}_4C_3[P(c \text{ and } c \text{ and } c \text{ and } w)]$$

$$= 4[P(c) \cdot P(c) \cdot P(c) \cdot P(w)]$$

$$= 4\left[\frac{1}{3} \cdot \frac{1}{3} \cdot \frac{1}{3} \cdot \frac{2}{3}\right]$$

$$= \frac{8}{81} = .098$$

note: Recall that $_nC_x$ is a shorthand way to refer to the various x-element combinations that n elements can create and that the value of $_nC_x$ is an integer such that

$$_nC_x = \frac{n!}{x!(n-x)!}$$

(see Section 10-2C). In this example, $_nC_x$ is an integer where n is the number of trials, x is the number of successes, and C is the number of ways we can get x successes out of n trials. Thus there are

$$_4C_2 = \frac{4!}{2!(4-2)!} = \frac{4!}{2!2!} = \frac{4 \cdot 3}{2!} = 6$$

ways to get 2 questions correct on a 4-question test.

Now let's get an overall view of the probabilities of your various possible outcomes. We summarize these probabilities in a table called a binomial probability distribution:

x(number correct)	$P(x)$
0	.198
1	.395
2	.296
3	.098
4	.012
TOTAL	1.000

Using the table you can now answer lots of specific questions. For example, what is the probability that you got 2 or more answers correct? Add the probabilities for $x = 2, 3$, and 4 to get $P(x \geq 2) = .406$. What is the probability that you got fewer than 3 correct? We get

$$P(x \leq 2) = P(x = 0) + P(x = 1) + P(x = 2)$$
$$= .198 + .395 + .296 = .889$$

That is, you have an 89% chance that you got only 0, 1, or 2 correct. Oh well, the beach is more fun!

We can define a **binomial probability formula** that will calculate the probability that the variable X of a binomial experiment has a specific value x:

BINOMIAL PROBABILITY FORMULA

$$P(x) = {}_nC_x(p)^x(1 - p)^{n-x} \tag{11-5}$$

or,

$$P(x) = \binom{n}{x}(p)^x(q)^{n-x} \tag{11-6}$$

note: The notation $\binom{n}{x}$ is the same as ${}_nC_x$ (see Section 10-2C).

EXAMPLE 11-39 Use the binomial probability formula (eq. 11-5) to check the results of Example 11-38.

Solution In that experiment $n = 4$ (4 questions), $p = \frac{1}{3}$ (chance of success if you guess), $q = 1 - p = \frac{2}{3}$ (chance of failure if you guess). So $P(x)$, your chances of getting x questions correct, would be:

$$P(0) = {}_4C_0\left(\frac{1}{3}\right)^0\left(\frac{2}{3}\right)^{4-0} = 1 \cdot 1\left(\frac{2}{3}\right)^4 = \frac{16}{81}$$

$$P(1) = {}_4C_1\left(\frac{1}{3}\right)^1\left(\frac{2}{3}\right)^{4-1} = 4\left(\frac{1}{3}\right)\left(\frac{2}{3}\right)^3 = \frac{32}{81}$$

$$P(2) = {}_4C_2\left(\frac{1}{3}\right)^2\left(\frac{2}{3}\right)^{4-2} = 6\left(\frac{1}{3}\right)^2\left(\frac{2}{3}\right)^2 = \frac{24}{81}$$

$$P(3) = {}_4C_3\left(\frac{1}{3}\right)^3\left(\frac{2}{3}\right)^{4-3} = 4\left(\frac{1}{3}\right)^3\left(\frac{2}{3}\right)^1 = \frac{8}{81}$$

$$P(4) = {}_4C_4\left(\frac{1}{3}\right)^4\left(\frac{2}{3}\right)^{4-4} = 1\left(\frac{1}{3}\right)^4\left(\frac{2}{3}\right)^0 = \frac{1}{81}$$

$$\frac{81}{81} = 1$$

EXAMPLE 11-40 Suppose you take a 10-question true–false test and guess at every one. Calculate the probability of getting 8 or more correct.

Solution In the binomial probability formula (eq. 11-5) we use $n = 10$ (ten trials of the guessing event), $p = \frac{1}{2}$ (an even chance of being correct on a guess), and $q = 1 - p = \frac{1}{2}$, and calculate for $x = 8, 9$, and 10:

$$P(8) = {}_{10}C_8 \left(\frac{1}{2}\right)^8 \left(\frac{1}{2}\right)^{10-8} \qquad P(9) = {}_{10}C_9 \left(\frac{1}{2}\right)^9 \left(\frac{1}{2}\right)^{10-9} \qquad P(10) = {}_{10}C_{10} \left(\frac{1}{2}\right)^{10} \left(\frac{1}{2}\right)^{10-10}$$

$$= \frac{10!}{8!2!} \cdot \left(\frac{1}{2}\right)^8 \left(\frac{1}{2}\right)^2 \qquad\qquad = \frac{10!}{9!1!} \cdot \left(\frac{1}{2}\right)^9 \left(\frac{1}{2}\right)^1 \qquad\qquad = \frac{10!}{10!0!} \cdot \left(\frac{1}{2}\right)^{10} \left(\frac{1}{2}\right)^0$$

$$= \frac{10 \cdot 9}{2 \cdot 1} \left(\frac{1}{2}\right)^{10} = \frac{45}{1024} \qquad = 10 \left(\frac{1}{2}\right)^{10} = \frac{10}{1024} \qquad = 1 \cdot \left(\frac{1}{2}\right)^{10} = \frac{1}{1024}$$

$$P(8 \text{ or more correct}) = P(8 \text{ or } 9 \text{ or } 10) = P(8) + P(9) + P(10)$$

$$= \frac{45}{1024} + \frac{10}{1024} + \frac{1}{1024} = \frac{56}{1024} = .0546$$

EXAMPLE 11-41 You toss 5 honest coins onto a table. What is the probability that you get 3 or more heads?

Solution This is a binomial experiment since each coin (independently) has only two possible outcomes. Thus $n = 5$, $p = \frac{1}{2}$, $q = \frac{1}{2}$, and we compute for $x = 3, 4$, and 5.

$$P(3 \text{ heads}) = {}_5C_3 \left(\frac{1}{2}\right)^3 \left(\frac{1}{2}\right)^{5-3} = \frac{5!}{3!2!} \left(\frac{1}{2}\right)^3 \left(\frac{1}{2}\right)^2 = 10 \cdot \left(\frac{1}{2}\right)^5 = \frac{10}{32}$$

$$P(4 \text{ heads}) = {}_5C_4 \left(\frac{1}{2}\right)^4 \left(\frac{1}{2}\right)^{5-4} = \frac{5!}{4!1!} \left(\frac{1}{2}\right)^4 \left(\frac{1}{2}\right)^1 = 5 \left(\frac{1}{2}\right)^5 = \frac{5}{32}$$

$$P(5 \text{ heads}) = {}_5C_5 \left(\frac{1}{2}\right)^5 \left(\frac{1}{2}\right)^{5-5} = \frac{5!}{5!} \left(\frac{1}{2}\right)^5 \left(\frac{1}{2}\right)^0 = 1 \left(\frac{1}{2}\right)^5 = \frac{1}{32}$$

$$P(3 \text{ or more heads}) = P(3) + P(4) + P(5) = \frac{10 + 5 + 1}{32} = \frac{16}{32} = \frac{1}{2}$$

EXAMPLE 11-42 In Example 11-41 what is the probability of at most 2 heads?

Solution This is the complement of the question in the above example. At most 2 heads includes 0 heads, 1 head, or 2 heads. Thus $P(\text{at most } 2) = 1 - P(3 \text{ or more}) = 1 - \frac{1}{2} = \frac{1}{2}$.

Most elementary statistics books or math handbooks contain prepared tables about computing binomial probabilities. Table 11-1 is a portion of such a table for the $n = 8$ case (8 trials) and a selection of different p values; it gives the probability for x successes in 8 trials if the probability of success in a single trial is p.

TABLE 11-1

n	x	$p = .10$	$p = .30$	$p = .50$	$p = .80$
8	0	.430	.058	.004	0
	1	.383	.198	.031	0
	2	.149	.296	.109	.001
	3	.033	.254	.219	.009
	4	.005	.136	.273	.046
	5	0	.047	.219	.147
	6	0	.010	.109	.294
	7	0	.001	.031	.336
	8	0	0	.004	.168

EXAMPLE 11-43 Use Table 11-1 to find the probability of getting exactly 6 correct on an 8-item true–false test written in Sanskrit.

Solution Chances are, you can't read the test, so you guess with probability $p = \frac{1}{2}$ on each item. Use $n = 8$ and $p = \frac{1}{2}$ and find the value $p(x = 6)$. This value from Table 11-1 is .109.

EXAMPLE 11-44 If 30% of the student body smokes, what is the probability that when 8 students are randomly selected, (a) exactly four are smokers (b) at the most 4 are smokers?

Solution

(a) Use $n = 8$ and $p = .30$ and calculate (look in Table 11-1) for $x = 4$. We find $P(x = 4) = .136$.
(b) Use $n = 8$ and $p = .30$ and obtain values for $x = 0, 1, 2, 3$, and 4, the occurrences that at most 4 students are smokers.

$$P(\text{at most 4 smokers}) = P(0) + P(1) + P(2) + P(3) + P(4)$$
$$= .058 + .198 + .296 + .254 + .136$$
$$= .942$$

EXAMPLE 11-45 When symbols are transmitted in binary form over a certain fiber-optic cable, the probability of error on any one of them is $p = .001$. If 30 symbols are transmitted, what is the probability of no error?

Solution This is binomial, with $n = 30$, $p = .001$, $q = .999$, and $x = 0$. Thus we can use eq. (11-5):

$$P(\text{zero errors}) = {}_{30}C_0(.001)^0(.999)^{30-0}$$
$$= \frac{30!}{0!30!}(.001)^0(.999)^{30}$$
$$= 1 \cdot 1 \cdot (.999)^{30} = .9704$$

C. Repeated trials—expected values

Suppose you are playing a die rolling game with the following rules: If an even number shows on the die, you get the number shown in dollars. If an odd number shows, you pay the number shown in dollars. We want to determine the amount you expect to win or lose when playing this game. According to the following chart,

Die face	Expected payoff	Probability that this face shows
1	−1	1/6
2	+2	1/6
3	−3	1/6
4	+4	1/6
5	−5	1/6
6	+6	1/6

on the average you would expect to finish with $(-1)(\frac{1}{6}) + 2(\frac{1}{6}) - 3(\frac{1}{6}) + 4(\frac{1}{6}) - 5(\frac{1}{6}) + 6(\frac{1}{6})$ dollars for each game played. This amount is

$$-\frac{1}{6} + \frac{2}{6} - \frac{3}{6} + \frac{4}{6} - \frac{5}{6} + \frac{6}{6} = \frac{1}{2} \text{ dollars}$$

Thus playing the game 50 times should net you $50 \cdot (\frac{1}{2}) = \$25.00$. That is, \$25.00 is your *expected value* after 50 games.

Put in general terms, if X is a random variable with values $X_1, X_2, \ldots, X_n$, and if the probability of X_i occurring is denoted by $P(X_i)$, then the **expected value of X** is defined by

$$E(X) = x_1 P(x_1) + x_2 P(x_2) + \cdots + x_n P(x_n)$$

EXPECTED VALUE OF X

$$= \sum_{i=1}^{n} x_i P(x_i)$$

(11-7)

EXAMPLE 11-46 A sack contains 2 quarters, 4 dimes, and 2 nickels. A trained monkey draws two coins which have equal likelihood (i.e., each one has the same chance of being selected). How much money do you expect to receive?

Solution We need to find the probability for all the different values in the sample space of values $S = \{(N, N), (N, D), (N, Q), (D, D), (D, Q), (Q, Q)\}$. Define x to be the monetary value, so x can assume the number $\{(10, 15, 30, 20, 35, 50\}$. We make the following chart:

Value x	$P(x)$	
.10	$\frac{2}{8} \cdot \frac{1}{7} = \frac{1}{28}$	Probability of N on first draw followed by N on second draw.
.15	$2 \cdot \frac{2}{8} \cdot \frac{4}{7} = \frac{8}{28}$	Two ways of getting 15 cents (ND or DN). Probability of N followed by the probability of D.
.20	$\frac{4}{8} \cdot \frac{3}{7} = \frac{6}{28}$	Probability of D on the first draw followed by D on the second.
.30	$2 \cdot \frac{2}{8} \cdot \frac{2}{7} = \frac{4}{28}$	Two ways of getting 30 cents (NQ or QN). Probability of N followed by the probability of Q.
.35	$2 \cdot \frac{2}{8} \cdot \frac{4}{7} = \frac{8}{28}$	Two ways of getting 35 cents (DQ or QD). Probability of Q followed by the probability of D.
.50	$\frac{2}{8} \cdot \frac{1}{7} = \frac{1}{28}$	Probability of Q on first draw followed by Q on second draw.

Now, to find the expected value plug the numbers into eq. (11-7):

$$E(X) = \sum x \cdot P(x)$$

$$= .10\left(\frac{1}{28}\right) + .15\left(\frac{8}{28}\right) + .20\left(\frac{6}{28}\right) + .30\left(\frac{4}{28}\right) + .35\left(\frac{8}{28}\right) + .50\left(\frac{1}{28}\right)$$

$$= \frac{.10 + 1.20 + 1.20 + 1.20 + 2.80 + .50}{28} = \frac{7.00}{28} = .25$$

SUMMARY

1. The probability for an event is defined to be the ratio of the number of outcomes favorable to the total number of possible outcomes.
2. The set of all possible outcomes for an experiment is called its sample space.
3. The values assigned to probabilities fall into the interval $0 \leq P(A) \leq 1$, with the endpoint zero corresponding to impossibility and 1 corresponding to certainty.
4. The Law of Large Numbers is used to find probabilities for events when exact values are not available experimentally.
5. If events A and B are mutually exclusive, then the probability that they will both occur is zero.
6. If events A and B are mutually exclusive, then $P(A \text{ or } B) = P(A) + P(B)$.
7. If events A and B are independent, then $P(A \text{ and } B) = P(A) \cdot P(B)$.
8. The conditional probability of an event B given another event A, denoted $P(B|A)$ is defined by

$$P(B|A) = \frac{P(A \text{ and } B)}{P(A)}$$

9. If an event is a binomial event (i.e., with only two possible outcomes), then the probability of x successes in n trials of the event for which the probability of success in a single trial is p is given by

$$P(x) = {}_nC_x(p)^x(1 - p)^{n - x}$$

RAISE YOUR GRADES

Can you...?

☑ create a sample space for an experiment
☑ form the probability definition for an event
☑ explain simple and compound events
☑ explain the Law of Large Numbers
☑ explain the relationship between an event and its complement
☑ describe mutually exclusive events
☑ describe independent events and dependent events
☑ recite the addition and multiplication rules of probability
☑ explain conditional probability
☑ describe a Bernoulli experiment
☑ define a binomial experiment and use the binomial probability formula
☑ calculate the expected value for an experiment

SOLVED PROBLEMS

Elementary Properties and Laws

PROBLEM 11-1   Suppose an urn contains 15 blue marbles, 12 yellow marbles, and 8 brown marbles. If we pick one marble, what is the probability that it is (**a**) blue (**b**) yellow (**c**) brown (**d**) blue or yellow (**e**) yellow and brown (**f**) blue or yellow or brown?

Solution   Divide the number of favorable outcomes by the total number of possible outcomes.

(**a**) There are 15 favorable outcomes (blue marbles) among the sample space (total) of 35 marbles. Thus
$$P(\text{blue}) = \frac{15}{35}.$$

(**b**) Similarly, $P(\text{yellow}) = \dfrac{\text{number of yellows}}{\text{number of marbles}} = \dfrac{12}{35}$

(**c**) $P(\text{brown}) = \dfrac{\text{number of browns}}{\text{number of marbles}} = \dfrac{8}{35}$

(**d**) With this option, our chances of success are improved, for we are seeking either a blue marble (15 chances) or a yellow marble (12 chances). Thus we have a total of 27 ways to be successful in this draw. That is,

$$P(\text{blue or yellow}) = \frac{\text{number of blues or yellows}}{\text{number in sample space}} = \frac{27}{35}$$

(**e**) $P(\text{yellow and brown}) = 0$; there is no one marble that is both yellow *and* brown.
(**f**) $P(\text{blue or yellow or brown}) = 1$ since all possible outcomes are favorable.

PROBLEM 11-2   Suppose we had two four-sided dice, with 1, 2, 3, 4 spots, respectively, on the faces. What is the sample space (the set of all possible outcomes) for these two dice?

Solution We record the possible results as number pairs imagining, say, that one die is black and one is white. Record the pairs as (black number, white number). The sample space is:

$$(1,1) \quad (1,2) \quad (1,3) \quad (1,4)$$
$$(2,1) \quad (2,2) \quad (2,3) \quad (2,4)$$
$$(3,1) \quad (3,2) \quad (3,3) \quad (3,4)$$
$$(4,1) \quad (4,2) \quad (4,3) \quad (4,4)$$

PROBLEM 11-3 Using the sample space and the two dice described in Problem 11-2, what is the probability of tossing (**a**) a total of 6 (**b**) a total less than 4?

Solution

(**a**) There are three ways to get a total of 6 among the 16 sample points: (4, 2), (3 3), (2, 4). So $P(6) = \frac{3}{16}$.
(**b**) To have a total less than 4 we have to have a total of 3 or 2. There are 3 outcomes favorable to this: (2, 1), (1, 2), and (1, 1). Thus $P(\text{less than 4}) = \frac{3}{16}$.

PROBLEM 11-4 A bag contains ping pong balls numbered 1 thru 7. You reach in and draw one ball. What is the probability its number is odd?

Solution Since there are four odd numbers among the seven, $P(\text{odd}) = \frac{4}{7}$.

Probabilities for Compound and Dependent Events

PROBLEM 11-5 If the bag from Problem 11-4 is sampled twice with replacement, what is the probability that (**a**) both are even (**b**) both are odd (**c**) one is even and one is odd?

Solution Recall that replacing a drawn ball makes the samplings independent. Therefore you can use the multiplication rule (eq. 11.1).

(**a**) $P(\text{even and even})$ and $P(\text{even} \cdot P(\text{even}) = \frac{3}{7} \cdot \frac{3}{7} = \frac{9}{49}$
(**b**) $P(\text{odd and odd}) = P(\text{odd}) \cdot P(\text{odd}) = \frac{4}{7} \cdot \frac{4}{7} = \frac{16}{49}$
(**c**) This case is the complement of the combination of the other two cases; that is

$$P(\text{even and odd}) = 1 - P(\text{both even}) - P(\text{both odd}) = 1 - \frac{9}{49} - \frac{16}{49} = \frac{24}{49}$$

As a check, we work from the multiplication formula, considering the independent draws

$$P(\text{even and odd}) = P(\text{even}) \cdot P(\text{odd}) = \frac{3}{7} \cdot \frac{4}{7} = \frac{12}{49}$$

But there are two ways of pulling out an even and an odd, (draw even then odd or odd then even) so we need to multiply by 2 to get $P(\text{even and odd}) = \frac{24}{49}$, which is what we got by the complement method.

PROBLEM 11-6 City H is proposing a city income tax on all persons employed in the city. A poll of 250 randomly selected residents of the city and its surrounding townships produced the following results:

City residents voted 80 in favor and 120 opposed.
Outside-of-city residents voted 10 in favor and 40 opposed.

If one of these persons were to be randomly selected for further interview, what is the probability that she would (**a**) favor the tax, (**b**) oppose the tax if she were a city resident, (**c**) oppose the tax if she were an outside-of-city resident?

Solution   Form a small summary table:

	City (C)	Non-city (N)	Total
Favor (F)	80	10	90
Oppose (O)	120	40	160
Total	200	50	250

(a) Use the total sample space of 250 polled people:

$$P(F) = \frac{\text{number who favor}}{\text{number in sample}} = \frac{90}{250} = .36$$

(b) Here the sample space is reduced to the 200 city residents, and we have a conditional probability. That is, for this question, the number of people opposing the tax law is dependent upon their also living in the city. Thus we use eq. (11-2):

$$P(O\,|\,C) = \frac{\text{number in opposition and in city}}{\text{number in city}} = \frac{120}{200} = .60$$

(c) The sample space is now reduced to the non-city residents, so we still use eq. (11-2):

$$P(O\,|\,N) = \frac{\text{number opposed and not in the city}}{\text{number in N}} = \frac{40}{50} = .80$$

PROBLEM 11-7   Twenty-five balloons are attached to a dart board. Five ballons contain slips of paper for prizes and twenty contain blank slips. You throw darts until you pop two balloons. What is the probability that you receive **(a)** zero prizes **(b)** exactly one prize **(c)** two prizes?

Solution

(a) On the successive balloon pops you need a "no prize" followed by a "no prize." This is conditional (the probability of the second outcome depends on the result of the first outcome) so we rearrange eq. (11-2) and get for $P(\text{zero prizes})$:

$P(\text{no prize and no prize}) = P(\text{no prize}) \cdot P(\text{no price}\,|\,\text{no prize})$

$$= \frac{\text{number of blank slips}}{\text{number of balloons}} \cdot \frac{\text{number of blank slips left after the first pop}}{\text{number of balloons left after the first pop}}$$

$$= \frac{20}{25} \cdot \frac{19}{24} = \frac{19}{30}$$

(b) There are two ways to get one prize:

$P(\text{one prize}) = P[(\text{no prize on first and prize on second}) \text{ or } (\text{prize on first and no prize on second})]$

These two conditional events are mutually exclusive, so we can use both the multiplication and addition rules:

$P(\text{one prize}) = P(\text{no prize}) \cdot P(\text{prize}\,|\,\text{no prize}) + P(\text{prize}) \cdot P(\text{no prize}\,|\,\text{prize})$

$$= \frac{20}{25} \cdot \frac{5}{24} + \frac{5}{25} \cdot \frac{20}{24} = \frac{10}{30}$$

recall: $P(\text{prize}\,|\,\text{no prize})$ means the probability of popping a prize balloon if you've already popped a no-prize balloon.

(c) $P(\text{two prizes}) = P(\text{prize on first and prize on second})$

$$P(\text{prize}) \cdot P(\text{prize}\,|\,\text{prize})$$

$$= \frac{5}{25} \cdot \frac{4}{24} = \frac{1}{30}$$

note: Within two balloon pops you *must* obtain one of these outcomes; this is verified by adding the probabilities: $\frac{19}{30} + \frac{10}{30} + \frac{1}{30} = \frac{30}{30} = 1$.

PROBLEM 11-8   Two cards are drawn from an ordinary deck, with the first card replaced before the second is drawn. What is the probability that (**a**) both cards are hearts (**b**) we will have either two hearts or two clubs (**c**) the cards will be of different suits?

Solution

(**a**) Since replacement is made, the two draws are independent. Thus

$$P(\text{heart and heart}) = \frac{13}{52} \cdot \frac{13}{52} = \frac{1}{16}$$

(**b**) $P(2 \text{ clubs}) = P(2 \text{ hearts})$, so $P(2 \text{ clubs or 2 hearts}) = \frac{1}{16} + \frac{1}{16} = \frac{1}{8}$.

(**c**) $P(\text{different suits}) = P(\text{pick any suit followed by a pick of a different suit})$
$$= P(\text{pick any}) \cdot P(\text{different one})$$
$$= 1 \cdot \frac{39}{52} = \frac{3}{4} \qquad (\textit{note}: 3 \text{ other suits} \cdot 13 \text{ cards per suit} = 39)$$

PROBLEM 11-9   Assume there are equal numbers of male and female students at your school and that the probability is $\frac{1}{5}$ that a male student is a computer science major and is $\frac{1}{20}$ that a female student is a computer science major. What is the probability that (**a**) a randomly selected student will be a male computer science major (**b**) a randomly selected student is a computer science major (**c**) a computer science major, randomly picked, is male?

Solution

(**a**) We need $P(\text{male and C.S.}) = P(\text{male}) \cdot P(\text{C.S.} \mid \text{male}) = \frac{1}{2} \cdot \frac{1}{5} = \frac{1}{10}$

(**b**) We need

$$P[(\text{male and C.S.}) \text{ or } (\text{female and C.S.})] = P(\text{male and C.S.}) + P(\text{female and C.S.})$$
$$= \frac{1}{10} + \frac{1}{2} \cdot \frac{1}{20} = \frac{1}{10} + \frac{1}{40} = \frac{1}{8}$$

(**c**) Use parts (a) and (b) and eq. (11-2) to find

$$P(\text{male} \mid \text{C.S.}) = \frac{P(\text{male and C.S.})}{P(\text{C.S.})} = \frac{\frac{1}{10}}{\frac{1}{8}} = \frac{4}{5}$$

PROBLEM 11-10   A dish contains 5 butterscotch balls and 3 peppermint swirls. What is the probability when 3 are selected at random that they are all butterscotch balls if (**a**) we select with replacement (**b**) we select without replacement?

Solution

(**a**) With replacement the draws are independent, so

$$P(\text{B and B and B}) = P(\text{B}) \cdot P(\text{B}) \cdot P(\text{B}) = \frac{5}{8} \cdot \frac{5}{8} \cdot \frac{5}{8} = \frac{125}{512} = .244$$

(**b**) Without replacement we have conditional probability, so

$$P(\text{B and B and B}) = P(\text{B}) \cdot P(\text{B} \mid \text{B}) \cdot P(\text{B} \mid \text{B and B})$$
$$= \frac{5}{8} \cdot \frac{4}{7} \cdot \frac{3}{6} = \frac{60}{336} = \frac{5}{28} = .179$$

PROBLEM 11-11   On a slot machine there are three reels which each contain the numbers 0, 1, 2, 3 and cherries and a plum. You insert a coin and pull the lever and the reels spin independently, coming to rest in one of the six positions mentioned. Find the probabilities of (**a**) all three cherries (**b**) one plum and two numbers (**c**) three different digits (**d**) no cherries.

Solution

(a) The probability of cherries on any wheel is $\frac{1}{6}$ so

$$P(\text{cherries and cherries and cherries}) = P(\text{cherries}) \cdot P(\text{cherries}) \cdot P(\text{cherries})$$

$$= \frac{1}{6} \cdot \frac{1}{6} \cdot \frac{1}{6} = \frac{1}{216}$$

(b) $P(\text{plum}) = \frac{1}{6}$ and $P(\text{number}) = \frac{4}{6}$, so with three arrangements of plum and two numbers (PNN or PNP or NNP) we have

$$P(\text{plum and two numbers}) = 3 \cdot \frac{1}{6} \cdot \frac{4}{6} \cdot \frac{4}{6} = \frac{2}{9}$$

(c) To get three different digits we have $P(\text{first digit}) = \frac{4}{6}$; $P(\text{digit} \,|\, \text{digit}) = \frac{3}{5}$; $P(\text{digit} \,|\, 2 \text{ digits}) = \frac{2}{4}$. So

$$P(3 \text{ different digits}) = \frac{4}{6} \cdot \frac{3}{5} \cdot \frac{2}{4} = \frac{1}{5}$$

(d) $P(\text{no cherries}) = \dfrac{5}{6} \cdot \dfrac{5}{6} \cdot \dfrac{5}{6} = \dfrac{125}{216}$

PROBLEM 11-12   When rolling two dice, the sum of the spots on top has outcomes $2, 3, 4, \ldots, 12$. What is the probability of each of these outcomes?

Solution   Since the roll of each dice is independent, we have the 36-point sample space of Example 11-29. Counting the number of times a particular sum occurs gives us

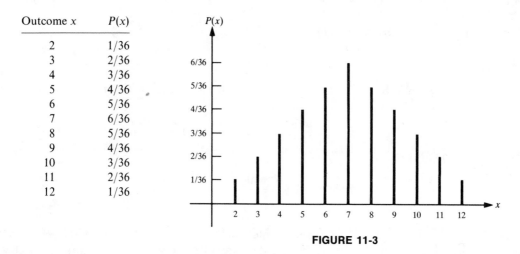

Outcome x	$P(x)$
2	1/36
3	2/36
4	3/36
5	4/36
6	5/36
7	6/36
8	5/36
9	4/36
10	3/36
11	2/36
12	1/36

FIGURE 11-3

This is called a *probability distribution* and can be drawn as in Figure 11-3 to display some of its interesting characteristics. Note that when rolling two dice the most probable outcome is 7 and the least probable is 2 or 12.

Binomial Probabilities

PROBLEM 11-13   When you go pheasant hunting the chance that you will down a particular bird on one shot has been calculated on long-term records to be .75. On today's hunt you shot at 6 birds. What is the probability you got 5 of them?

Solution   This is a binomial experiment (hit 'em or miss 'em) with $n = 6$, $p = .75$, and $x = 5$. So the binomial probability formula (eq. 11-5) produces

$$P(5 \text{ birds down}) = {}_6C_5(.75)^5(.25)^{6-5} = \frac{6!}{5!1!} \cdot \left(\frac{3}{4}\right)^5 \left(\frac{1}{4}\right)^1$$

$$= 6 \cdot \frac{3^5}{4^6} = .356$$

PROBLEM 11-14   In archery you get a bull's-eye 80% of the time. In shooting 8 arrows, what are the chances of your getting 7 or 8 bull's-eyes?

Solution   This is a binomial experiment with $n = 8$, $p = .80$, $q = .20$, and $x = 7$ or 8. We could compute using the binomial probability formula and a calculator:

$$P(7) = {}_8C_7(.80)^7(.20)^1 = .33554432$$

$$P(8) = {}_8C_8(.80)^8(.20)^0 = .16777216$$

or use Table 11-1 on page 289 to get

$$P(7) = .336 \quad \text{and} \quad P(8) = .168$$

Hence your chances of getting 7 or 8 bull's-eyes is the sum

$$P(7 \text{ or } 8) = .336 + .168 = .504$$

PROBLEM 11-15   Using the binomial probability formula, find the probability of $0, 1, 2$, and 3 heads when tossing three coins.

Solution   Each coin toss is independent of the others so we can treat this as three trials of a Bernoulli experiment with the two possible outcomes, heads or tails. We have $n = 3$, $p = \frac{1}{2}$, and $q = \frac{1}{2}$ and will compute for $x = 0, 1, 2, 3$.

$$P(0) = {}_3C_0\left(\frac{1}{2}\right)^0\left(\frac{1}{2}\right)^3 = 1 \cdot \left(\frac{1}{2}\right)^3 = \frac{1}{8}$$

$$P(1) = {}_3C_1\left(\frac{1}{2}\right)^1\left(\frac{1}{2}\right)^2 = 3 \cdot \left(\frac{1}{2}\right)^3 = \frac{3}{8}$$

$$P(2) = {}_3C_2\left(\frac{1}{2}\right)^2\left(\frac{1}{2}\right)^1 = 3 \cdot \left(\frac{1}{2}\right)^3 = \frac{3}{8}$$

$$P(3) = {}_3C_3\left(\frac{1}{2}\right)^3\left(\frac{1}{2}\right)^0 = 1 \cdot \left(\frac{1}{2}\right)^3 = \frac{1}{8}$$

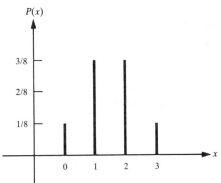

FIGURE 11-4

The graphic representation of the results is shown in Figure 11-4.

PROBLEM 11-16   You are rolling one die and the roll is considered a success if you get one spot, call it "an ace." Failure, then, is obtaining a 2, 3, 4, 5 or 6, called "non-ace." In four rolls, what are the probabilities of zero through four aces?

Solution   The rolls are independent so we have a binomial probability experiment with $n = 4$, $p = \frac{1}{6}$ (probability of getting an ace on a single roll), and $q = 1 - p = \frac{5}{6}$ (probability of a non-ace on a single roll), and we compute for $x = 0, 1, 2, 3, 4$.

$$P(0) = {}_4C_0\left(\frac{1}{6}\right)^0\left(\frac{5}{6}\right)^4 = 1 \cdot 1 \cdot \left(\frac{5}{6}\right)^4 = .4823$$

$$P(1) = {}_4C_1\left(\frac{1}{6}\right)^1\left(\frac{5}{6}\right)^3 = 4 \cdot \left(\frac{1}{6}\right)^1 \cdot \left(\frac{5}{6}\right)^3 = .3858$$

$$P(2) = {}_4C_2\left(\frac{1}{6}\right)^2\left(\frac{5}{6}\right)^2 = 6 \cdot \left(\frac{1}{6}\right)^2 \cdot \left(\frac{5}{6}\right)^2 = .1157$$

$$P(3) = {}_4C_3\left(\frac{1}{6}\right)^3\left(\frac{5}{6}\right)^1 = 4 \cdot \left(\frac{1}{6}\right)^3 \cdot \left(\frac{5}{6}\right)^1 = .0154$$

$$P(4) = {}_4C_4\left(\frac{1}{6}\right)^4\left(\frac{5}{6}\right)^0 = 1 \cdot \left(\frac{1}{6}\right)^4 \cdot \left(\frac{5}{6}\right)^0 = .0007$$

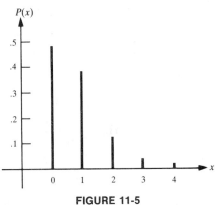

FIGURE 11-5

Figure 11-5 shows the results graphically.

PROBLEM 11-17   When dealt a five-card hand from an ordinary deck of cards, what is the probability of your getting (**a**) two pairs (**b**) a flush (all same suit) (**c**) a full house?

Solution   Recall that the probability for equally likely events is $P(A) = m/n$, where m is the number of outcomes in A and n is the number of outcomes in the sample space. The sample space for this problem is $_{52}C_5$, the number of ways of getting 5 cards from 52 cards:

$$\frac{52!}{5!(52-5)!} = \frac{52!}{5!47!} = \frac{52 \cdot 51 \cdot 50 \cdot 49 \cdot 48}{5!} = 2\,598\,960$$

For now, we'll leave this sample space value written as $_{52}C_5$ and place it in the denominator of $P(A) = m/n$.

(**a**) To figure out how many ways we can get 2 pairs, consider the patterns of the cards. There are $_{13}C_2$ ways to select any 2 values out of 13 card values. Then there are $_4C_2$ ways to have one of those cards pair off with a like-value card and $_4C_2$ ways for the second card to pair off with a like-value card. Then there are $_{44}C_1$ ways to insert a different fifth card. Thus

$$P(2\text{ pair}) = \frac{_4C_2 \cdot _4C_2 \cdot _{13}C_2 \cdot _{44}C_1}{_{52}C_5} = .0475$$

(**b**) For the flush we need to get any suit—there are $_4C_1$ ways to do that. Then we have to have all the cards in your hand be from the same suit—there are $_{13}C_5$ ways to do that. Thus

$$P(\text{flush}) = \frac{_4C_1 \cdot _{13}C_5}{_{52}C_5} = \frac{4 \cdot 1287}{2\,598\,960} = .00198$$

(**c**) For the full house we need three cards of one value and two of another. From part (**a**) we know that there are $_{13}C_2$ ways to select the two different values. Next, we find that there are $_2C_1$ ways of orienting the 3 cards versus 2 cards mix. (That is, once the two values have been chosen, we have two ways to produce a full house: either 3 cards of the first value and 2 cards of the second, or 2 cards of the first value and 3 cards of the second.) Finally, there are $_4C_3$ ways of getting three cards of one value and $_4C_2$ ways of getting 2 cards of the other value. Thus

$$P(\text{full house}) = \frac{_{13}C_2 \cdot _2C_1 \cdot _4C_3 \cdot _4C_2}{_{52}C_5} = .0014$$

PROBLEM 11-18   *The Birthday Problem.* In a group of r people, we want to calculate the probability that two people have the same birthday (day and month). Find a formula that will do this (ignore the possibility of leap years).

Solution   If r people are selected and their birthdates recorded, the sample space will have $(365)^r$ values. For example, $\{17, 3, 198, 246, 91, 311\}$ is one possible value of the birthdates of 6 people. Let $E = $ event that at least two people among the r people have the same birthday. Then $\bar{E}$ (the complement of E) stands for event that no two people have the same birthday. Thus we know that

$$|E| + |\bar{E}| = |\text{sample space}| = (365)^r$$

But $|\bar{E}|$ equals the number of ways of selecting r *different* numbers from 365 numbers; that is, $|\bar{E}| = 365 \cdot 364 \cdot 363 \cdot \cdots \cdot (365 - (r-1))$. Thus

$$P(\bar{E}) = \frac{|\bar{E}|}{(365)^r}$$

and finally

$$P(E) = 1 - P(\bar{E}) = 1 - \frac{365 \cdot 364 \cdot \cdots \cdot (365 - r + 1)}{(365)^r}$$

This is the formula that finds the probability we are seeking. Now to calculate. Use a pocket calculator to obtain

r	10	20	22	23	24	30	40	50	60
$P(E)$	.12	.41	.48	.51	.54	.71	.89	.97	.994

Conclusions: As soon as the group of people is larger than 22, the chances are better than 1/2 that two people have the same birthday. As the size of the group increases, the probability goes up dramatically. At size 60 it is almost an absolute surety! (If you work it right you could make some interesting bets on this one!)

PROBLEM 11-19   In one hour of duck hunting, the probability of bagging no ducks is .30, of bagging one duck is .35, of bagging two ducks is .22, of bagging three ducks is .12, and of bagging four ducks is .04. How many ducks can you expect to bag in one hour?

Solution   Use the expected value formula (eq. 11-7):

$$E(x) = \sum_{i=1}^{n} x_i P(x_i)$$

$$= 0(.30) + 1(.35) + 2(.22) + 3(.12) + 4(.04) = 1.31$$

PROBLEM 11-20   If you roll a pair of dice nine times, how many times would you expect to be successful in obtaining a total of 8 or more?

Solution   We need the probabilities of rolling 8 or more. These are easily obtained from the 6-by-6 grid of the dice sample space (see Example 11-24) and calculating:

$$P(8) = \frac{\text{number of possible pairs producing a sum of 8}}{36} = \frac{5}{36}$$

$$P(9) = \frac{4}{36}$$

$$P(10) = \frac{3}{36}$$

$$P(11) = \frac{2}{36}$$

$$P(12) = \frac{1}{36}$$

So $P(8 \text{ or more}) = P(8) + P(9) + P(10) + P(11) + P(12) = \frac{15}{36}$. Thus the expected value from nine rolls is

$$9 \cdot \left(\frac{15}{36}\right) = \frac{15}{4} = 3\frac{3}{4}$$

So less than half of the time you would achieve the desired result.

Supplementary Exercises

PROBLEM 11-21   If the probability that you win a hand of bridge is $\frac{3}{10}$ and you play 7 hands, what is the probability of winning (**a**) exactly 5 hands (**b**) 5 or more hands?

PROBLEM 11-22   The newlyweds plan to have 5 children. What is the probability that they will have 3 boys and 2 girls?

PROBLEM 11-23   You are dealt 5 cards from an ordinary deck. What is the probability that you will have (**a**) exactly one ace, (**b**) at least one ace, (**c**) at most one ace?

PROBLEM 11-24   Suppose that A and B are events such that $P(A) = .7$, $P(B) = .5$, and $P(A \cap B) = .28$. Find (**a**) $P(A \mid B)$, (**b**) $P(B \mid A)$, (**c**) $P(A \cup B)$, (**d**) $P(A - B)$.

PROBLEM 11-25 You roll two dice blindfolded. Someone tells you that one die shows a 4. What is the probability that the sum is 7?

PROBLEM 11-26 The probability that a TV set needs a repair within one year is 20%. You have six units in your apartment building. What is the probability that none of the sets will need repair within one year?

PROBLEM 11-27 An experiment consists of drawing one marble from a bag containing a mixture of orange, blue, and white marbles. (**a**) What is the sample space? (**b**) What is the sample space if two marbles are drawn?

PROBLEM 11-28 When two dice are rolled, what is the probability of doubles or a sum less than 4?

PROBLEM 11-29 If A and B are mutually exclusive events and $P(A) = .21$ and $P(B) = .34$, what is (**a**) $P(\overline{A})$, (**b**) $P(A \cup B)$, (**c**) $P(A \cap B)$?

Answers to Supplementary Exercises

11-21 (**a**) .025 (**b**) .029

11-22 $\frac{10}{32}$

11-23 (**a**) .2995 (**b**) .3412 (**c**) .9583

11-24 (**a**) .45 (**b**) .40 (**c**) .92 (**d**) .42

11-25 $\frac{2}{11}$

11-26 .262

11-27 (**a**) {O, B, W}
(**b**) {OO, OB, OW, BO, BB, BW, WO, WB, WW}

11-28 $\frac{2}{9}$

11-29 (**a**) .79 (**b**) 0 (**c**) .55

12 RELATIONS

THIS CHAPTER IS ABOUT

- ☑ **Relations Defined**
- ☑ **Graphical Representations of Relations**
- ☑ **Properties of Relations**
- ☑ **Matrix Representations of Relations**
- ☑ **Equivalence Relations**
- ☑ **Order Relations**

12-1. Relations Defined

A **relation** can be thought of as a table that shows the relationship of elements to other elements. If a set of students is matched with a set of jobs at the cafeteria, we might have the following table:

Student	Job
Alfie	Salad bar prep
Barb	Serving
Carlos	Dishwashing
Dianne	Dishwashing
Dianne	Serving
Ellis	Clean-up

We would say that Barb is related to Serving and that Dianne is related to both Dishwashing and Serving, etc.

A more compact way of representing a relation is to simply display a collection of ordered pairs. We actually define a relation to be a set of ordered pairs and the first and second members of the pairs are related. Thus the table above would produce the relation of ordered pairs {(Alfie, Salad bar prep), (Barb, Serving), (Carlos, Dishwashing), (Dianne, Dishwashing), (Dianne, Serving), (Ellis, Clean-up)}.

EXAMPLE 12-1 Express a relation between your three closest friends and the college courses they are taking.

Solution Say your three friends are Bob, Ed, and Jim and among them they take five classes. A possible relation might be: {(Bob, Biology), (Bob, Math), (Ed, Math), (Ed, Psychology), (Ed, Economics), (Jim, Psychology), (Jim, Biology), (Jim, Physics)}.

A **binary relation**, R, from a set A to a set B is a subset of the product set $A \times B$. If an ordered pair (a, b) is an element of R, we say that a is related to b and denote this relation by $a \, R \, b$. The set $\{a \in A \,|\, (a, b) \in R$ for some $b \in B\}$ is called the **domain** of relation R and the set $\{b \in B \,|\, (a, b) \in R$ for some $a \in A\}$ is called the **range** of relation R. If the relation is given as a table (as above) then the set of first elements (the left side) is the domain and the set of second elements (the right side) is the range.

EXAMPLE 12-2 Use the sets $A = \{a, b, c\}$ and $B = \{1, 2, 3\}$ to form a relation. Identify the domain and range.

Solution Match up elements of A with elements of B to form ordered pairs. Any combination of these pairs will produce a relation. One of the relations from A to B is the set $R = \{(a, 2), (a, 3), (c, 1), (c, 3)\}$. Here the domain is $\{a, c\}$ and the range is $\{1, 2, 3\}$.

EXAMPLE 12-3 Let $X = \{1, 3, 5\}$ and $Y = \{3, 4, 5, 6\}$. Form a relation, R, from X to Y which is defined by $(x, y) \in R$ if $y - x$ is even and non-negative. Find the set enumerating all the possible relation pairs, form the table of relation pairs, and identify the domain and range of the relation.

Solution Using the elements of sets X and Y, we want a set of all the ordered pairs whose elements satisfy three criteria:

(1) the x element is from X;
(2) the y element is from Y;
(3) $y - x$ is even and non-negative.

This gives us the set of relation pairs

x	y
1	3
1	5
3	3
3	5
5	5

The domain is $\{1, 3, 5\}$ and the range is $\{3, 5\}$.

Rather than describe a relation's elements as a set of ordered pairs or as a table, you can also describe its elements with words or symbols.

EXAMPLE 12-4 Define a binary relation on the set of positive integers.

Solution For positive integers a and b, let aRb if a is less than b. This binary relation is usually denoted by the symbol $<$ so we write $R = a < b$.

EXAMPLE 12-5 Define a binary relation on the set of all integers.

Solution Let m and n be integers. There are many relations we could create. For example, we can define mRn to be a relation if and only if $m - n$ is divisible by a specific integer t. This relation is important in mathematics and is denoted by $m = n \pmod{t}$. For $t = 5$, for example, some members of the relation are $(6, 1), (27, 7), (43, 13), (12, 2), \ldots$.

EXAMPLE 12-6 When a computer reads a program, it creates a table of symbol relations containing variable names and variable values. (That is, it matches a variable name with a current value.) Form such a relation.

Solution In a BASIC program, a relation table might be as follows:

Variable	Value
SUM	O
A$	"CAR"
X	100
B4	2
I	FALSE

12-2. Graphical Representations of Relations

We have already seen two ways of representing a relation: (1) in roster form, as a set of ordered pairs or as a table, and (2) in words, as a description such as "greater than" or "divides with no remainder." A third way to represent a relation is by means of a **directed graph**, or **digraph**, for short. These will be studied in greater detail in Chapter 14, but we use them here as a means of representing a relation.

If $A = \{x, y, z\}$ and $B = \{1, 2\}$ then some of the relations from A to B that may be formed include

$$R_1 = \{(x, 1), (y, 1), (y, 2), (z, 2)\}$$

$$R_2 = \{(y, 2), (z, 1)\}$$

$$R_3 = \{\quad\}$$

$$R_4 = A \times B$$

The directed graphs of these relations use dots (vertices) to represent the members of A and B; arrows are directed from a point a of A to a point b of B if aRb. Thus for our four relations above, we have the digraphs shown in Figure 12-1.

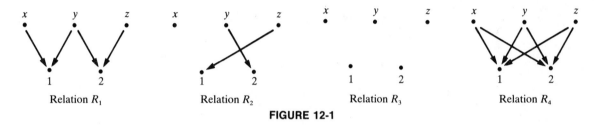

Relation R_1 Relation R_2 Relation R_3 Relation R_4

FIGURE 12-1

In a digraph an *edge* is formed if an arrow is directed from one vertex to another. Thus each pair in a relation provides an edge.

If we are using one set as both the domain set and the range set, it is not necessary to graph its vertices twice. The reordering of a finite set is an example of this. Let the ordering {apple, plum, cherry} become {cherry, apple, plum}. Then the relation matches apple → cherry, plum → apple, and cherry → plum, which allows us to use either diagram shown in Figure 12-2.

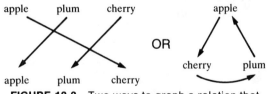

FIGURE 12-2 Two ways to graph a relation that has the same domain and range.

EXAMPLE 12-7 If a relation on set $A = \{a, b, c\}$ is given by $R = \{(a, b), (a, c), (b, a), (c, c)\}$, draw the directed graph.

Solution We do not need to duplicate the points of A; we use just one set for the digraph. There are four pairs in the relation, so the graph will have four arrows. For the pair (c, c), however, the arrow will point to itself. The complete digraph of R is shown in Figure 12-3.

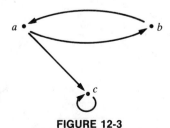

FIGURE 12-3

EXAMPLE 12-8   Create the directed graph using the relation "greater than or equal to" on the set $A = \{1, 2, 3, 4\}$.

Solution   Using the elements of A, we must first find all the ordered pairs whose elements satisfy the given definition of the relation. That is, we want to list all pairs whose x element is greater than or equal to its y element. We therefore have the pairs $(2, 1), (3, 1), (4, 1), (2, 2), (4, 2), (4, 3), (1, 1), (2, 2), (3, 3), (4, 4)$ as the relation set. Graphing these pairs produces the digraph in Figure 12-4.

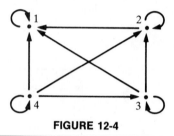

FIGURE 12-4

EXAMPLE 12-9   Find the relation displayed in the directed graph in Figure 12-5.

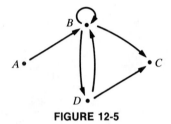

FIGURE 12-5

Solution   If there is a directed arrow (edge) either from one vertex to another or from a vertex to itself, then the vertices form an ordered pair of the relation. We have

$$R = \{(A, B), (B, B), (B, C), (B, D), (D, B), (D, C)\}$$

note: To be an ordered pair of a relation, the vertices must have an arrow directed from the x element to the y element of the pair. Thus (B, A) in Figure 12-5 is not an ordered pair of the relation, but (A, B) is.

EXAMPLE 12-10   Let $N = \{2, 3, 4, 6, 9, 10, 12\}$ and define a relation R on N by writing xRy if x divides y for $x, y \in N$. Create the digraph for R.

Solution   We must go through the list of all possible pairs and check to see if each pair satisfies the divisibility criterion of the relation. Once we find the set of all pairs of the relation, we draw the edges where the pairs so indicate. The relation is the set

$$R = \{(2, 2), (2, 4), (2, 6), (2, 10), (2, 12), (3, 3), (3, 6), (3, 9),$$
$$(3, 12), (4, 4), (4, 12), (6, 6) (6, 12), (9, 9), (10, 10), (12, 12)\}$$

Its digraph is shown in Figure 12-6.

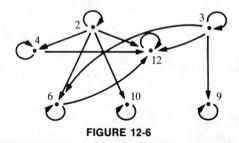

FIGURE 12-6

12-3. Properties of Relations

A. Symmetry

If R is a relation on a set S, then R is **symmetric** if whenever xRy, then also yRx. Geometrically, this implies that each edge of the digraph is bidirectional (has arrows going both ways) and that each vertex is related to itself. For example, the relations $R = \{1, 1), (1, 3), (3, 1), (3, 3)\}$ and $\{(1, 1), (3, 3)\}$ are symmetric, but $R = \{(1, 1), (1, 3), (3, 3)\}$ is not.

EXAMPLE 12-11 Consider the set of all straight lines in a plane. Let the relation R be defined by xRy if x is parallel to y. Show that this relation is symmetric.

Solution If s and t are two lines related by R, then s is parallel to t by their relation R. But t is also parallel to s by geometry rules, so we have sRt and tRs. In addition, we know by geometry that every line is parallel to itself, so the relation is indeed symmetric.

B. Reflexivity

If R is a relation on a set S, then R is **reflexive** if for each $x \in S$, xRx. Geometrically, each vertex must have a loop to itself. For example, the relation $R = \{(1, 1), (3, 3)\}$ is reflexive, as is the relation $R = \{(1, 1), (1, 3), (3, 3)\}$.

EXAMPLE 12-12 Show that on the set of integers, the relation xRy if x is less than or equal to y is reflexive.

Solution We need to pick any integer and test if xRx. Here we get xRx if x is less than or equal to x. This is clearly true, so the relation is reflexive.

C. Transitivity

If R is a relation on a set S, then R is **transitive** if whenever it is true that xRy and yRz, then it follows that xRz.

If a relation is given as a set of pairs, we can tell whether or not it is transitive by just looking at it. For example, if set $S = \{x, y, z, w\}$ and $R_1 = \{(x, y), (x, x), (y, w)\}$, then R_1 is not transitive since xRy and yRw does not have xRw. But if $R_2 = \{(x, y), (y, y), (x, z), (y, z), (w, z), (w, y), (w, x)\}$ then R_2 is transitive since xRy and yRz has xRz, and wRx and xRy has wRy, and wRx and xRz has wRz. If the relation is given by words or formulas, however, we may have to check it algebraically.

EXAMPLE 12-13 Show that on the set of integers the relation xRy if x is less than y is transitive.

Solution According to the algebraic laws for the set of integers, if $x < y$ and $y < z$, then $x < z$, so transitivity holds.

D. Irreflexivity

If R is a relation on a set S, then R is **irreflexive** if for each $x \in S$, we have $x\not Rx$. Geometrically, no vertex is related to itself. For example, $R = \{(1, 3), (3, 1), (4, 3)\}$ is irreflexive, but $R = \{(1, 1), (1, 3), (3, 1)\}$ is not.

EXAMPLE 12-14 Define a relation on $S = \{1, 2, 3\}$ by $R = \{(1, 2), (2, 3), (1, 3)\}$. Show that R is irreflexive.

Solution None of the elements satisfy xRx, so all elements satisfy $x\not Rx$. Thus it is irreflexive.

E. Antisymmetry

If R is a relation on a set S, then R is **antisymmetric** if for each pair of elements x and y, whenever xRy and yRx, then $x = y$. Geometrically, a vertex that is related to itself cannot also have bidirectional arrows. Thus $R = \{(1, 1), (1, 2), (1, 3)\}$ is antisymmetric, but $R = \{(1, 1), (1, 3), (3, 1)\}$ and $R = \{(1, 3), (3, 1)\}$ are not because in both cases $1R3$ and $3R1$, but $1 \neq 3$.

EXAMPLE 12-15   Show that the relation R: "greater than or equal to" is antisymmetric on the integers.

Solution   Let x and y be integers. If xRy, then x is greater than or equal to y. But if also yRx, then y is greater than or equal to x. This can be true only if x and y are actually equal, so this relation is antisymmetric.

EXAMPLE 12-16   On the directed graph in Figure 12-7, decide if the relation displayed is symmetric, reflexive, transitive, irreflexive, or antisymmetric.

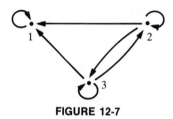

FIGURE 12-7

Solution   Designate the relation as R. Testing the various edges, we have

- $2R1$ but $1\not R2$, so it is not symmetric.
- $1R1$, $2R2$, and $3R3$, so it is reflexive.
- $2R3$ and $3R1$ and $2R1$.
  $3R2$ and $2R1$ and $3R1$. $\Big\}$ so it is transitive.
  No other cases in which xRy and yRz.
- $1R1$, etc., so it is not irreflexive.
- $2R3$ and $3R2$ but $3 \neq 2$, so it is not antisymmetric.

12-4. Matrix Representation of Relations

On a finite set $S = \{e_1, e_2, \ldots, e_n\}$ a relation R may be represented by an $n \times n$ matrix whose entries are zeroes and ones. To do so, we put a 1 in the j-th row and k-th column if $e_j Re_k$, and we put a 0 in the j-th row and k-th column if $e_j \not R e_k$.

EXAMPLE 12-17   Form the matrix representation of the relation given in Example 12-16.

Solution   From Figure 12-6, we have seven edges: $1R1$, $2R2$, $3R3$, $2R1$, $2R3$, $3R1$, $3R2$. Since S has 3 elements ($S = \{1, 2, 3\}$), we get a 3×3 matrix; each edge represented on the graph receives a corresponding value of 1 in the matrix:

$$\begin{array}{c} \\ 1 \\ 2 \\ 3 \end{array} \begin{array}{ccc} 1 & 2 & 3 \\ \left[\begin{array}{ccc} 1 & 0 & 0 \\ 1 & 1 & 1 \\ 1 & 1 & 1 \end{array}\right] \end{array}$$

EXAMPLE 12-18   Form the matrix representation of the directed graph in Figure 12-8.

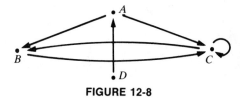

FIGURE 12-8

Solution   There are six edges and four vertices in the graph, so there will be six 1's in the representative 4×4 matrix:

$$
\begin{array}{c c c c c}
 & A & B & C & D \\
\begin{array}{c} A \\ B \\ C \\ D \end{array} &
\left[\begin{array}{c c c c}
0 & 1 & 1 & 0 \\
0 & 0 & 1 & 0 \\
0 & 1 & 1 & 0 \\
1 & 0 & 0 & 0
\end{array} \right]
\end{array}
$$

We can read some of the properties of a relation directly from its matrix representation.

(1) If the (i, j) entry always equals the (j, i) entry (that is, if the matrix is symmetric), then the relation is symmetric.
(2) If the matrix has a 1 in each position on the main diagonal, then the relation is reflexive.
(3) If the matrix has a 0 in each position on the main diagonal, then the relation is irreflexive.
(4) If the matrix has the property that the occurrence of a 1 in the (i, j) position and a 1 in the (j, k) position always implies a 1 in the (i, k) position, then the relation is transitive.

Every $n \times n$ matrix consisting of only 0's and 1's represents a relation on a set of n elements. Matrices therefore provide a convenient way to represent a relation on a computer.

EXAMPLE 12-19   Consider the 4×4 matrix

$$
M = \begin{bmatrix}
1 & 0 & 1 & 0 \\
0 & 0 & 1 & 0 \\
1 & 1 & 0 & 1 \\
0 & 0 & 1 & 1
\end{bmatrix}
$$

Create the directed graph of the relation that M represents.

Solution   We need four points; call them a, b, c, d. There are eight edges corresponding to the eight 1's in the matrix. They form the pairs $\{(a, a), (a, c), (b, c), (c, a), (c, b), (c, d), (d, c), (d, d)\}$. The digraph of the relation is shown in Figure 12-9.

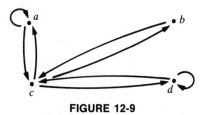

FIGURE 12-9

12-5. Equivalence Relations

Relations that satisfy the three properties of symmetry, reflexivity, and transitivity are called **equivalence relations**. These relations have a wide variety of applications in mathematics and computer science.

EXAMPLE 12-20 Consider the following relation: On set $S =$ the integers, let xRy if $x - y$ is divisible by 5. Show that this is an equivalence relation.

Solution We must show that the relation is symmetric, reflexive, and transitive. To do so, we use the laws of algebra:

- If 5 divides $x - y$, then 5 also divides $-(x - y) = y - x$, so the relation is symmetric.
- Since 5 divides $x - x = 0$, the relation is reflexive.
- If 5 divides $x - y$ and 5 divides $y - z$, then there are integers d_1 and d_2 such that $x - y = 5 \cdot d_1$ and $y - z = 5 \cdot d_2$. But $x - z = (x - y) + (y - z) = 5 \cdot d_1 + 5 \cdot d_2 = 5(d_1 + d_2)$, so $x - y$ and $y - z$ implies that $x - z$ is also divisible by 5. Thus the relation is transitive.

Since R is symmetric, reflexive, and transitive, it is an equivalence relation.

Let's make some observations now on the equivalence relation defined in Example 12-20. Note that the integers $\{\ldots, -15, -10, -5, 0, 5, 10, \ldots\}$ are all equivalent under this equivalence relation since their differences are divisible by 5. Thus each integer in such a collection is equivalent to every other integer in that collection. Note also that the integers in

$$\{\ldots, -14, -9, -4, 1, 6, 11, \ldots\}$$

are all equivalent, as are the integers in the collections

$$\{\ldots, -13, -8, -3, 2, 7, 12, \ldots\}$$

and

$$\{\ldots, -12, -7, -2, 3, 8, 13, \ldots\}$$

and

$$\{\ldots, -11, -6, -1, 4, 9, 14, \ldots\}$$

Every integer is in exactly one of these collections, which we call **equivalence classes**. For the equivalence relation described in Example 12-20, we have exactly five nonoverlapping equivalence classes. Their union is all of the integers. We say that the equivalence relation has **partitioned** the set of integers into *mutually exclusive* subsets (equivalence classes) which are *exhaustive* (everything covered). This kind of partitioning is a characteristic of an equivalence relation. An easy way to picture a partitioning is to consider a bag of marbles which are either red, white, or blue. If you were to divide that set into three subsets, each subset containing one color of marble, you would have a partition of the original set. In each subset, aRb if a and b have the same color.

EXAMPLE 12-21 Let $X =$ set of all automobiles, and define a relation R on X by xRy if x and y have the same number of doors. What are the equivalence classes?

Solution The set of automobiles is partitioned into those having one door (?), the two-door coupes, the three-door hatchbacks, the four-door sedans, the five-door wagons, the six-door limousines, and so on. Each automobile is in one and only one of the classes.

EXAMPLE 12-22 Let $S =$ set of all triangles in a plane. Let $t_1 R t_2$ if and only if t_1 is similar to t_2. Show that R is an equivalence relation.

Solution We must show that R is symmetric, reflexive, and transitive. According to the laws of geometry,

- R is symmetric since if t_1 is similar to t_2, then t_2 is similar to t_1.
- R is reflexive since every triangle is similar to itself.
- R is transitive since if t_1 is similar to t_2 and t_2 is similar to t_3, then t_1 is similar to t_3.

EXAMPLE 12-23 On the set of all rational numbers define a relation R so that xRy if $y - x$ is an integer. Is this an equivalence relation?

Solution Test the relation for symmetry, reflexivity, and transitivity.

- If $y - x$ is an integer then so is $x - y$; then $xRy \leftrightarrow yRx$. The relation is symmetric.
- Since $y - y$ is the integer zero, the relation is reflexive.
- If $y - x$ is an integer and $z - y$ is an integer, then $z - x = (z - y) + (y - x)$ is the sum of two integers and thus is an integer; the relation is transitive.

The relation is an equivalence relation.

12-6. Order Relations

A relation R defined on a set S is called a **partial order** on S when the relation R is reflexive, transitive, and antisymmetric. The set S is then called a **partially ordered set**, or **poset**.

EXAMPLE 12-24 Let S be the set of real numbers and define R by xRy if $x \leq y$. Show that S is a poset.

Solution We must show that the relation is reflexive, transitive, and antisymmetric.

- For every element $x \in S$, $x \leq x$, so the relation is reflexive.
- For every $x, y \in S$, if $x \leq y$ and $y \leq x$, then it is true that $x = y$; the relation is therefore antisymmetric.
- For $x, y, z \in S$, if $x \leq y$ and $y \leq z$, then $x \leq z$ is valid; the relation is transitive.

The relation is a partial order, so the set S is a poset.

Under the relation $\leq$ described in Example 12-24, the real numbers are reflexive, transitive, and antisymmetric. But there is one other property that the real numbers have in this relation: comparability. Two elements u and v in $\mathbb{R}$ are always **comparable** under the relation $\leq$; that is, it is true that either $u \leq v$ or $v \leq u$.

If a set S has a relation that is a partial order and if S also has the property that every pair of its elements is comparable under the relation, then S is a **totally ordered set** and the relation is a **total order**. Thus the set of real numbers under the relation $\leq$ is a totally ordered set.

EXAMPLE 12-25 Let T be a set of three elements $\{a, b, c\}$, and let $\mathscr{P} = \mathscr{P}(T)$ be the power set (the collection of all subsets of T). Let the relation R be defined by $\mathscr{P}_1 R \mathscr{P}_2$ if $\mathscr{P}_1 \subseteq \mathscr{P}_2$, (that is, $\mathscr{P}_1$ is a subset of $\mathscr{P}_2$). Show that $\mathscr{P}$ is a poset but not a totally ordered set.

Solution First show $\mathscr{P}$ is a poset by using set theory to show that $\mathscr{P}_1 R \mathscr{P}_2$ is a partial order (reflexive, transitive, and antisymmetric). Since $\mathscr{P}_1 \subseteq \mathscr{P}_1$ by definition, the relation is reflexive. For each pair of subsets, if $\mathscr{P}_1 \subseteq \mathscr{P}_2$ and $\mathscr{P}_2 \subseteq \mathscr{P}_1$, then $\mathscr{P}_1 = \mathscr{P}_2$ is valid; the relation is antisymmetric. Also, for any $\mathscr{P}_1, \mathscr{P}_2, \mathscr{P}_3 \in \mathscr{P}$, if $\mathscr{P}_1 \subseteq \mathscr{P}_2$ and $\mathscr{P}_2 \subseteq \mathscr{P}_3$, then $\mathscr{P}_1 \subseteq \mathscr{P}_3$ by set properties; the relation is transitive. Thus the relation is a partial order and $\mathscr{P}$ is a poset. Now, for $\mathscr{P}$ to be a totally ordered set, every pair of its elements must be comparable under the given relation. (That is, we must have $\mathscr{P}_1 \subseteq \mathscr{P}_2$ or $\mathscr{P}_2 \subseteq \mathscr{P}_1$ for every possible pair of subsets in $\mathscr{P}$.) But the subsets in $\mathscr{P}$ described by $\{a\}$ and $\{b\}$ are not comparable since they are distinct and neither is a subset of the other. Hence $\mathscr{P}$ is not a totally ordered set.

Pictures can often be an asset in understanding a concept. Consider the set $T = \{a, b, c\}$ with the relation $\subseteq$ from Example 12-25. We may picture this set and relation as a graph, called a **Hasse diagram**, in which all the possible subsets are the vertices and each subset relationship ($\subseteq$) is an arrow (edge) between two vertices; for example, an arrow from A to B means $A \subseteq B$ (Figure 12-10). Since T has 3 elements, there are $2^3 = 8$ subsets, which form the vertices.

In every Hasse diagram, we place a vertex U_2 higher than a vertex U_1 whenever U_2 is related to U_1 (in the case of Figure 12-10, whenever $U_1 \subseteq U_2$). We use a combination of two (or more) "upward" arrows to identify subsequent interrelationships among the vertices; this follows from the transitive property of the ordered relation. For example, in the case of Figure 12-10, vertex $\{c\}$ has no direct arrow to vertex

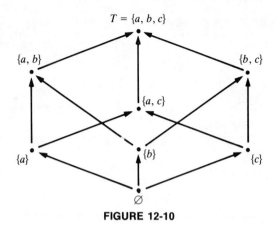

FIGURE 12-10

$\{a, b, c\}$, which it is obviously related to because $\{c\} \subseteq \{a, b, c\}$. Yet the Hasse diagram's upward arrows show the correct relation by building up to it: we see that subset $\{c\}$ is related to subset $\{b, c\}$, which in turn is related to subset $\{a, b, c\}$; by the property of transitivity, we can safely state that subset $\{c\}$ is also related to subset $\{a, b, c\}$.

EXAMPLE 12-26   Given the set $S = \{1, 2, 3, 4, 5, 6, 7, 8, 9\}$ under the relation xRy if x divides y, draw the Hasse diagram.

Solution   We must first determine the pairs in the relation. 1 is related to all of the elements; 2 is related to 4, 6, and 8; 3 is related to 6 and 9; 4 is related to 8. The rest of the elements are too large to meet the relation's criterion. (Of course each element is related to itself, but such vertex loops are not depicted in a Hasse diagram). Hence we get the diagram shown in Figure 12-11.

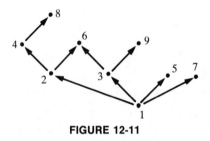

FIGURE 12-11

EXAMPLE 12-27   Use the set and relation of Example 12-26 and show that the set, which is a poset, is not a totally ordered set.

Solution   The elements 2 and 3 of set S are not comparable since 2 does not divide 3 and 3 does not divide 2. Since not every pair of elements in S is comparable, S is not totally ordered.

EXAMPLE 12-28   Given the set of integers and the relation xRy if $x < y$, show that the relation is not a partial order.

Solution   To be a partial order, a relation must be reflexive, transitive, and antisymmetric. This relation is not reflexive since if $a < b$, then $b \not< a$.

EXAMPLE 12-29   Let $S = \{2, 4, 8, 16\}$ and define xRy if x divides y (written $x \mid y$). Form the Hasse diagram.

Solution   We have $2 \mid 4, 2 \mid 8, 2 \mid 16, 4 \mid 8, 4 \mid 16$, and $8 \mid 16$, which can be simplified to $2 \mid 4, 4 \mid 8$, and $8 \mid 16$. The final diagram is easy (Figure 12-12).

$\bullet\ 16$

$\uparrow$

$\bullet\ 8$

$\uparrow$

$\bullet\ 4$

$\uparrow$

$\bullet\ 2$

FIGURE 12-12

EXAMPLE 12-30 Let $S = \{2, 3, 5, 6, 30, 60, 120, 180\}$ and define xRy if $x \mid y$. Form the Hasse diagram.

Solution The elements 2, 3, and 5 are primes and hence do not divide each other. They form the base of the diagram, and the others build on it (Figure 12-13).

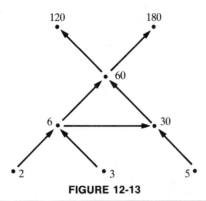

FIGURE 12-13

We close with three examples from computer science in which relations are important.

(1) A partial order is established when a large computer program contains a number of subprograms. If we say that subprogram P_1 is related to subprogram P_2, we can form a directed graph to indicate this with an arrow drawn from the calling program to the subprogram. The many arrows will form a hierarchy, which is the digraph of the calling procedure.

(2) Using strings of symbols (such as letters) created from a symbol set (such as an alphabet), we might use lexicographic ordering (see Problem 12-15) to list and/or manipulate the strings.

(3) If a system of programs for a large research project is being developed to interconnect several computer components, we could use a partial order to organize the various tasks involved to complete the project. Each task would be assigned a completion sequence position that is based on its position in the directed graph of the partial ordering created.

SUMMARY

1. A relation is a set of element pairs. The first members collectively form the domain. The second members collectively form the range.
2. A graphical representation of a relation is called a directed graph, or digraph.
3. There are five basic properties that a relation may possess: symmetry, reflexivity, transitivity, irreflexivity, and antisymmetry.
4. On a finite set S a relation R may be conveniently represented as an $n \times n$ matrix consisting of zeros and ones.
5. A relation that has the three properties of symmetry, reflexivity, and transitivity is called an equivalence relation.
6. An equivalence relation partitions the underlying set into mutually exclusive and exhaustive subsets called equivalence classes.
7. A relation defined on a set S is called a partial order if it is reflexive, transitive, and antisymmetric. The set then is called a partially ordered set, or poset.
8. If in a poset every pair of elements is comparable, then the set is also called a totally ordered set.

RAISE YOUR GRADES

Can you...?

☑ define a relation in two ways
☑ list the domain and range of a given relation
☑ provide the details on the construction of a directed graph

☑ convert a digraph to a relation listing and vice versa
☑ define the five properties of relations
☑ form a matrix representation of a given relation
☑ use the matrix representation to find the properties of the relation
☑ define an equivalence relation
☑ use an equivalence relation to partition a set
☑ define a partial order and a poset
☑ define a total order and a totally ordered set
☑ form the Hasse diagram for a given relation

SOLVED PROBLEMS

Relations and Graphical Representations of Relations

PROBLEM 12-1 Given set $P = \{1, 4, 6, 8\}$ and set $Q = \{2, 3, 6, 9, 10, 12\}$, form two relations using elements of P and Q.

Solution Just pair up the elements, one from P and one from Q, in any combination. You do not need to use all of them from either set. Lots of relations are possible, including

$$R_1 = \{(1, 2), (4, 3), (6, 6), (8, 12)\} \quad \text{and} \quad R_2 = \{(4, 10), (6, 3), (8, 9)\}$$

PROBLEM 12-2 Equality is a relation on a set A of rational members since we use the rules of arithmetic to determine when two given rationals are equal. If set $A = \{\frac{1}{2}, \frac{2}{3}, \frac{3}{4}\}$, list the number pairs in the relation and draw the directed graph.

Solution The number pairs must satisfy the requirement that $x = y$. There are only three ways the elements of set A can do so, so $R = \{(\frac{1}{2}, \frac{1}{2}), (\frac{2}{3}, \frac{2}{3}), (\frac{3}{4}, \frac{3}{4})\}$. The digraph will consist only of vertices pointing to themselves (Figure 12-14).

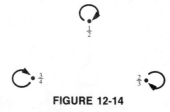

FIGURE 12-14

PROBLEM 12-3 Let R be the set of all real numbers. For $x, y \in R$ define a relation xRy if and only if $x^2 + y^2 = 9$. Define R and give its pictorial representation.

Solution The relation R consists of all points in a plane on the circle or radius 3 (Figure 12-15).

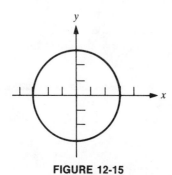

FIGURE 12-15

PROBLEM 12-4 Let a set S consist of $S = \{p, q\}$. Then the power set $\mathscr{P}(S)$ has four members. Let the relation be "subset of." Draw the directed graph of this relation.

Solution $\mathscr{P}(S) = \{\{\varnothing\}, \{q\}, \{p\}, \{p, q\}\}$, so let these four members be the four vertices of the diagram. Draw an arrow from one vertex to another if the first vertex is related to (is a subset of) the second. Be sure to cover all possible combinations and remember that a set is always a subset of itself. Figure 12-16 shows the final digraph.

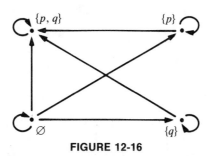

FIGURE 12-16

PROBLEM 12-5 Let set $S = \{1, 2, 3, 4\}$ and define a relation R on S as follows:

$$R = \{(1, 1), (1, 3), (1, 4), (2, 4), (3, 1), (4, 2), (4, 3)\}$$

Draw the directed graph.

Solution There are four elements and seven pairs, so our graph will have four vertices and seven edges. Be sure the arrows are directed from the x element to the y element (Figure 12-17).

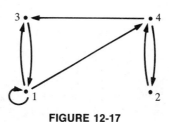

FIGURE 12-17

Properties of Relations

PROBLEM 12-6 Show that the relation R described on the set of positive integers as "being divisible by a positive integer" is reflexive.

Solution We need to show that for any positive integer x, xRx. But by the rules of arithmetic, every positive integer is divisible by itself, so reflexivity is obvious.

PROBLEM 12-7 We say that two integers have the same **parity** if both are odd or both are even. Show that on the set of integers that the relation R: "being of the same parity" is symmetric.

Solution If xRy because x and y have the same parity, then surely yRx since they have the same parity; thus R is symmetric.

PROBLEM 12-8 Discover the conditions necessary for a relation to be both symmetric and antisymmetric.

Solution If xRy then symmetry implies yRx but antisymmetry also implies $y = x$. Hence for both to apply no two distinct elements may be related; that is, we have no edges but only loops on the digraph. If $S = \{1, 2, 3\}$, for example, then $R = \{(1, 1), (2, 2), (3, 3)\}$ is both symmetric and antisymmetric.

PROBLEM 12-9 Construct a directed graph on $S = \{1, 2, 3\}$ that is reflexive but neither symmetric nor transitive. Will it be either irreflexive or antisymmetric?

Solution It is easy to make a relation reflexive: simply give each vertex a loop. To make the relation not symmetric, put an edge one way and not the other, such as $1R3$ but not $3R1$. To make it not transitive, keep $1R3$, put in the second requirement, such as $3R2$, but don't include the conclusion; that is, be sure $1 \not R 2$. The final relation is shown in Figure 12-18. Since this relation is already reflexive, it cannot also be irreflexive. And finally, since no pair (x, y) has xRy and tRx, we satisfy antisymmetry by default.

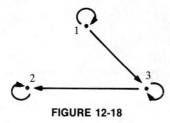

FIGURE 12-18

Matrix Representations of Relations

PROBLEM 12-10   Let $S = \{1, 2, 3\}$ and define relation $R = \{(1, 2), (1, 3), (2, 1), (3, 3)\}$. What is the matrix representation of this relation?

Solution   Since our original set S has three elements, the matrix will be 3×3. If a pair (i, j) of elements is represented in the relation R, place a 1 in the (i, j) position of the matrix. If a pair (i, j) is not represented in the relation, place a 0 in the (i, j) position of the matrix. The final representation of R will have four ones:

$$M_R = \begin{bmatrix} 0 & 1 & 1 \\ 1 & 0 & 0 \\ 0 & 0 & 1 \end{bmatrix}$$

PROBLEM 12-11   From the relation matrix $N = \begin{bmatrix} 1 & 0 & 1 \\ 0 & 1 & 0 \\ 1 & 1 & 1 \end{bmatrix}$ form the directed graph.

Solution   We need three points, call them A, B, C. There are six edges—corresponding to the six 1's in the matrix—that create the following pairs: $\{(A, A), (A, C), (B, B), (C, A), (C, B), (C, C)\}$. Use the pairs to draw the final digraph (Figure 12-19).

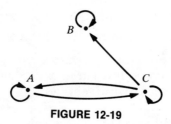

FIGURE 12-19

Equivalence Relations

PROBLEM 12-12   If the set S is the set of integers and R is defined by $(a, b) \in R$ if $a \le b$, show that R is not an equivalence relation.

Solution   Recall that to be an equivalence relation, R must be reflexive, transitive, and symmetric. Since $a \le a$ it is reflexive. If $a \le b$ and $b \le c$, then $a \le c$, so it is transitive. But if $a \le b$, it is not necessarily true that $b \le a$, so it is not symmetric. The relation is therefore not an equivalence relation.

PROBLEM 12-13   Suppose an equivalence relation on the set $S = \{1, 2, 3, 4\}$ produces the equivalence classes $E = \{\{1\}, \{2, 3, 4\}\}$. Display the entire relation.

Solution   Since each element in any equivalence class is related to every other element in its class, the entire relation consists of the listing of all the ordered pairs possible within all of its equivalence classes.

We therefore get

$$R = \{(1,1), (2,2), (2,3), (2,4), (3,2), (3,3), (3,4), (4,2), (4,3), (4,4)\}$$

PROBLEM 12-14 Let set $S = \{$Houston, San Diego, Los Angeles, Dallas, Atlanta$\}$. Let relation R be defined by aRb if a and b are cities in the same state. Show that R is an equivalence relation and find the equivalence classes.

Solution First prove R is symmetric, reflexive, and transitive.

- Let x and y be any two cities in S. Then if x and y are in the same state, so are y and x. Thus $xRy \Rightarrow yRx$ and the relation is symmetric.
- Certainly a city is in the same state as itself so xRx; it is reflexive.
- If x and y are in the same state and y and z are in the same state, then x and z are in the same state. Thus xRy and $yRz \Rightarrow xRz$; it is transitive.

The equivalence classes are:

$$E = \{\{\text{Houston, Dallas}\}, \{\text{San Diego, Los Angeles}\}, \{\text{Atlanta}\}\}$$

Order Relations

PROBLEM 12-15 The English alphabet in its usual order is an ordered set. The set of English words in alphabetical order is an ordered set. Are these partially ordered or totally ordered sets?

Solution First we need to define what the order for words is. (It is called *lexicographic order*.) We say x precedes y if any of the following hold: (**1**) if the first letter of x precedes the first letter of y; (**2**) if $x = ps$ and $y = pt$, where p is the longest common string of letters, then s precedes t. Now, the reflexive property holds; the antisymmetric property holds (if a precedes b and b precedes a, then $a = b$); and the transitive property holds (if a precedes b and b precedes c then a precedes c). We therefore have at least a partial order relation. In addition, every pair of letters or words is comparable: they are either the same letter (word) occupying the same position in the lexicographic order or they are different letters (words) and one is the predecessor of the other. Thus the English alphabet and the set of English words are both totally ordered sets.

PROBLEM 12-16 In the chapter on sets, (Chapter 8) we used the notation $|A|$ to denote the size of the set A. Let $S = \{1, 2, 3, 4\}$ and define a relation on the power set $\mathscr{P}(S)$ by indicating $\alpha R \beta$ if $|\alpha| \leq |\beta|$. Is R a partial order?

Solution To be a partial order R would need to be reflexive, antisymmetric, and transitive. This relation is not antisymmetric since for the sets $\alpha = \{1, 2\}$ and $\beta = \{3, 4\}$ we have $|\alpha| \leq |\beta|$ (that is, $2 \leq 2$), and we have $|\beta| \leq |\alpha|$ (that is, $2 \leq 2$), but $\alpha \neq \beta$.

Supplementary Exercises

PROBLEM 12-17 Decide if the following are relations:

(**a**) The set S_1 is the integers. For $x, y \in S_1$, let xRy if $x - y$ is divisible by 3.
(**b**) The set S_2 is the real number line. For $x, y \in S_2$, let xRy if $y - x \neq 0$.
(**c**) The set S_3 is the entire population of the U.S.A. For $x, y \in S_3$, let xRy if x is the mother of y.

PROBLEM 12-18 Which of the five properties of relations do the following possess?

(**a**) For $x, y \in R$, xRy if $x \geq y$.
(**b**) The definition in Problem 12-17a.

PROBLEM 12-19   Following Example 12-20, show that on the set of integers the definition "xRy if m divides $x - y$" holds for any positive integer m and thus produces an equivalence relation for each m.

PROBLEM 12-20   Determine if the relation represented by $\begin{bmatrix} 1 & 1 & 0 \\ 1 & 1 & 0 \\ 0 & 0 & 1 \end{bmatrix}$ is an equivalence relation.

PROBLEM 12-21   If $A = \{1, 2, 3, 4\}$ determine if $R = \{(1, 1),\ (2, 2),\ (3, 3),\ (4, 4),\ (1, 2),\ (3, 4)\}$ is an equivalence relation.

PROBLEM 12-22   If the set $S = \{1, 2, 3, 4, 5\}$ is partitioned into subsets $P = \{\{1, 3, 4\}, \{2\}, \{5\}\}$, what is the complete relation?

PROBLEM 12-23   Let the set S be all persons in the United States. Define xRy if x and y have the same father or mother. Is this relation an equivalence relation?

PROBLEM 12-24   Let the set $S = \mathbb{R}^2 = $ the entire xy plane. Define a relation R on S by $(a, b)R(c, d)$ if $a < c$ or if $a = c$, then $b \leq d$. Show that R is a partial order.

PROBLEM 12-25   Is the relation described in Problem 12-24 simply a partial order or is it also a total order?

PROBLEM 12-26   Let Y be the set of all women. Let R be defined by xRy if x is the sister of y. Is this relation reflexive, symmetric, antisymmetric, and/or transitive?

PROBLEM 12-27   Let $S = \mathbb{R}^2$ and define R by $(a, b)R(c, d)$ if $a \leq c$ and $b \leq d$. Is R a partial order or a total order?

PROBLEM 12-28   Let $S = \{3, 9, 27, 81\}$ and define R by xRy if $x \mid y$. Draw the Hasse diagram.

PROBLEM 12-29   Let S be the set of all students at University X. Define two students x and y to be related if x and y are in the same year in school. Is this relation an equivalence relation? If so, what are the equivalence classes?

Answers to Supplementary Exercises

12-17   (a) It is a relation. (b) It is a relation.
(c) It is a relation.

12-18   (a) Reflexive, transitive, antisymmetric
(b) Reflexive, symmetric, transitive

12-20   Yes, it is an equivalence relation.

12-21   No, it is not an equivalence relation.

12-22 $R = \{(1, 1),\ (1, 3),\ (1, 4),\ (3, 1),\ (3, 3),\ (3, 4),\ (4,1),$
$(4, 3),\ (4, 4),\ (2, 2),\ (5, 5)\}$.

12-23   No, it is not transitive.

12-25   It is also a total order.

12-26   It is symmetric only.

12-27   It is a partial order.

12-28
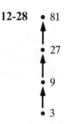

12-29   It is an equivalence relation; its equivalence classes are Freshmen, Sophomores, Juniors, and Seniors.

13 *FUNCTIONS*

THIS CHAPTER IS ABOUT

- ☑ **Functions and Basic Properties**
- ☑ **Additional Properties**
- ☑ **Inverses**
- ☑ **Composition of Functions**
- ☑ **Special Functions**

13-1. Functions and Basic Properties

Let X and Y be nonempty sets. The **function** f from X to Y, denoted as $f: X \to Y$ is a relation from X to Y satisfying two conditions:

(1) The domain of the function is the domain of the relation.
(2) If (x, y) and (x, z) are pairs in the relation, then $y = z$.

Less formally, we say that a function from X to Y is a "rule" that assigns to each element $x \in X$ exactly one element $y \in Y$.

We call X the **domain** of the function; we call Y the **codomain** of the function. If (x, y) is a pair in the function, we write $f(x) = y$. (Note the modification in notation from a relation, in which we write xRy—see Chapter 12.) A function is just a special case of a relation in which a given first element of any pair *cannot* have two different second elements among any of the pairs. Thus $R = \{(1, 1), (2, 3), (3, 1)\}$ is a function but $R = \{(1, 1), (1, 3), (3, 1)\}$ is not because the element 1 of the domain is assigned two different elements from the codomain: 1 and 3 in the first and second pairs, respectively.

Functions are sometimes called **mappings** or **transformations** since a picture of a function appears to map or transform some set of elements to another set of elements. If $f: X \to Y$ (that is, if $f(x) = y$), we say that f maps the set X to Y. To map a function means to draw directed segments between the elements of the function's pairs.

EXAMPLE 13-1   Let set $X = \{A, B, C\}$ and set $Y = \{1, 2, 3\}$. Create a function from X to Y and draw the mapping.

Solution   We will create a relation of pairs of elements that meet the conditions of a function. There are several ways to do this. For example, let $f = \{(A, 1), (B, 3), (C, 2)\}$. This is a function from X to Y because it assigns to each element $x \in X$ exactly one element $y \in Y$. Its mapping is a picture that shows its paired elements:

$$A \longrightarrow 1$$
$$B \quad \nearrow 2$$
$$C \quad \searrow 3$$

EXAMPLE 13-2   For sets $X = \{a, b, c, d\}$ and $Y = \{p, q, r\}$, say we have the pairing

$$G = \{(a, p), (c, r), (d, p), (b, q), (c, p), (a, r)\}$$

Is G a function?

Solution Drawing the map of the paired elements will help us to see if any domain element is paired up with more than one codomain element:

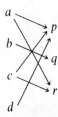

This is not a function since both a and c have two distinct mates, which violates condition (2) of the definition of a function.

---

note: The same codomain element can pair up with more than one domain element in a function, but not vice versa. Thus if $G = \{(a, p), (c, r), (d, p), (b, q)\}$, G would be a function even though both a and d of the domain pair up with p of the codomain.

EXAMPLE 13-3   Let $Z = \{s, t\}$. Find four functions that map Z to Z.

Solution   We must find four ways to map $\{s, t\}$ to $\{s, t\}$ without giving two unique codomain elements to the same domain element.

(1) The identity map $\{(s, s), (t, t)\}$, also written $f_1(s) = s$, $f_1(t) = t$, meets the criteria:

$$s \longrightarrow s$$
$$t \longrightarrow t$$

(2) The map $\{(s, t), (t, s)\}$, or $f_2(s) = t$, $f_2(t) = s$, is a function:

(3) The map $f_3(s) = t$, $f_3(t) = t$ is a function:

(4) The map $f_4(s) = s$, $f_4(t) = s$ is a function:

$$s \longrightarrow s$$
$$t \qquad t$$

The domain of a function does not have to be a finite set. If set X is the entire real line and the function f is defined as $f = \{(x, x^2) \mid x \in \mathbb{R}\}$, that is, $y = f(x) = x^2$, the function maps the real line to the non-negative real numbers in a graph (Figure 13-1). The individual values from the domain X that are

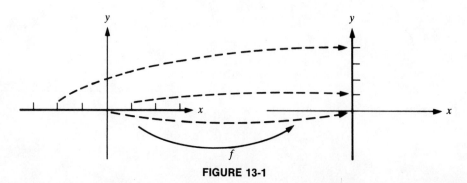

FIGURE 13-1

used are often called the **arguments** of the function, while the corresponding values in the codomain Y are called the **images** of the function. The **range** is the set of all possible images. Often the range is a proper subset of the codomain. Above, the arguments are real numbers and the images are non-negative real numbers.

EXAMPLE 13-4 Let set X be the positive integers from 1 through 4 and let set Y be the positive integers from 1 through 20. Define a relation of integer pairs (x, y) such that y is the square of x. Is this relation a function? Find the domain, codomain, and range and draw the mapping of X to Y.

Solution   The domain is the set $\{1, 2, 3, 4\}$. The codomain of the set is $\{1, 2, 3, \ldots, 18, 19, 20\}$. Since the square of each member of the domain is unique, this is a function. The range is the set of squares of the domain members and consists of $\{1, 4, 9, 16\}$ and is a proper subset of the codomain. The mapping is as follows:

```
Domain → 1          2              3                        4
         ↓          ↓              ↓                        ↓
Codomain →1  2  3   4  5  6  7  8  9  10  11  12  13  14  15  16  17  18  19
```

In computer science, as well as in mathematics, the use of the function is important. Many computers have built-in functions (called *library functions*) which perform the required operations. Examples include SQR, the function for squaring; ABS, the function for taking the absolute value; SIN, the function for computing the sine of an argument, etc.

13-2. Additional Properties

A function $f: X \to Y$ is said to be **one-to-one** (or **injective**) if for each two elements x_1 and x_2 in X such that $x_1 \neq x_2$, then $f(x_1) \neq f(x_2)$. Thus in a one-to-one mapping, we cannot have distinct elements of the domain corresponding to the same element in the range. The squaring function as defined in Example 13-4 is one-to-one.

note: If we were to expand the domain of Example 13-4 to include the negative integers $-1, -2, -3,$ -4, the new function would not be one-to-one. The new domain $\{-1, -2, -3, -4, 1, 2, 3, 4\}$ when mapped to the codomain $\{1, 2, 3, \ldots, 18, 19, 20\}$ produces the same range $\{1, 4, 9, 16\}$. We have two domain elements that share the same range element. For example, -3 and 3 both correspond to 9. Furthermore, on the set of real numbers the function $f(x) = x^2$ is not one-to-one.

A function $f: X \to Y$ is said to be **onto** (or **surjective**) if for each element $y \in Y$ there is at least one element $x \in X$ for which $f(x) = y$. For example, given the set $X = \{0, 1, 2, 3\}$ and $Y = \{a, b, c\}$, the function $f = \{(0, a), (1, a), (2, b), (3, c)\}$ is onto; the function $g = \{(0, a), (1, a), (2, b), (3, b)\}$ is not, however, because the element c of Y is not matched with an element of X.

Thus in an onto mapping the range is the codomain. A function that is both one-to-one and onto is called a **one-to-one correspondence** (or **bijection**).

EXAMPLE 13-5   Shown below are the maps of four functions. Determine if they are one-to-one, onto, or both.

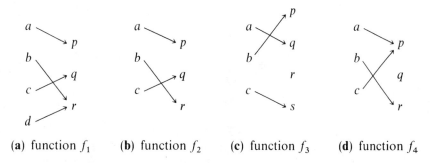

(**a**) function f_1 (**b**) function f_2 (**c**) function f_3 (**d**) function f_4

Solution

(a) f_1 is onto but not one-to-one since $f(b) = f(d) = r$.
(b) f_2 is both onto and one-to-one, so it is a bijection.
(c) f_3 is one-to-one but not onto since r is not an image of the function.
(d) f_4 is not one-to-one since $f(a) = f(c) = p$, and it is not onto since q is not an image of the function.

13-3. Inverses

If f is a one-to-one and onto function from X to Y, then the relation $\{(y, x) \mid (x, y) \in f\}$ is a function from Y to X. This function, denoted f^{-1}, is called the **inverse** of f. The domain of f becomes the codomain of f^{-1}, and the codomain of f becomes the domain of f^{-1}.

EXAMPLE 13-6   For function f_2 of Example 13-5, find f_2^{-1}.

Solution   Since f_2 is both one-to-one and onto, we know it has an inverse. To create it, we simply reverse the original element pairs $f_2 = \{(a, p), (c, q), (b, r)\}$ to get $f_2^{-1} = \{(p, a), (q, c), (r, b)\}$. The inverse's mapping is

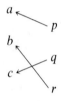

function f_2^{-1}

EXAMPLE 13-7   Let function f be described by $f = \{(1, b), (2, d), (3, b), (4, a)\}$. Does f have an inverse?

Solution   We may immediately create a potential inverse function by reversing the pairs:

$$f^{-1} \overset{?}{=} \{(b, 1), (d, 2), (b, 3), (a, 4)\}$$

However, this is not a function since the element b from the domain of this proposed function has two distinct second elements, 1 and 3. Therefore no inverse exists.

If the domain and codomain are not finite sets, the function f is usually given by a formula $y = f(x)$. To find the inverse in this case, this formula is solved for x as a function of y and then the roles of x and y are interchanged.

EXAMPLE 13-8   Let X be a set of real numbers and define a function f from X to X ($f: X \to X$), by $f(x) = 3x - 1$. Show that f is both one-to-one and onto. Then find f^{-1}.

Solution   To show f is one-to-one we must show that if $f(a) = f(b)$, then $a = b$; this is true:

$$f(a) = f(b) \Leftrightarrow 3a - 1 = 3b - 1$$

$$3a = 3b$$

$$a = b$$

To show f is onto we must show that if d is an element of the codomain (that is, if $d \in \mathbb{R}$), then there is an element c in the domain such that $f(c) = d$. From $y = 3x - 1$, the original formula for the function, solve for x:

$$y = 3x - 1$$

$$y + 1 = 3x$$

$$\tfrac{1}{3}(y + 1) = x$$

Now, if d is the given value for y, then the corresponding c value to make $f(c) = d$ is $c = \frac{1}{3}(d + 1)$ since

$$f(c) = f(\tfrac{1}{3}(d + 1)) = 3(\tfrac{1}{3}(d + 1)) - 1 = d + 1 - 1 = d$$

Thus the function f is both one-to-one and onto and it has an inverse. The inverse has already been calculated: We have solved the formula for x as a function of y, and now we simply reverse the roles of x and y in $\frac{1}{3}(y + 1) = x$ to get $f^{-1} = \frac{1}{3}(x + 1)$.

13-4. Composition of Functions

If f is a function from X to Y and g is a function from Y to Z, it is possible to combine them to obtain the **composition** of g and f, written $g \circ f$, from X to Z ($g \circ f$ is also called a *composite function*). That is, the composite $g \circ f$ has domain X and codomain Z. For $x \in X$ we define $(g \circ f)(x) = g(f(x))$. In this, we first apply f to $x \in X$ to get $f(x)$, an element of Y. Then we apply g to $f(x) \in Y$ to get $g(f(x)) \in Z$ (Figure 13-2). In order for $g \circ f$ to be defined, the codomain of f must be the domain of g.

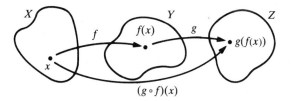

FIGURE 13-2   The composition of g and f, given $f: x \rightarrow y$ and $g: y \rightarrow z$; $g \circ f: x \rightarrow z$.

EXAMPLE 13-9   Let $f: \mathbb{R} \rightarrow \mathbb{R}$ be defined by $f(x) = x^2 + 1$ and let $g: \mathbb{R} \rightarrow \mathbb{R}$ be defined by $g(x) = 4x - 2$. Find $(g \circ f)(x)$.

Solution   Since the codomain of f is the domain of g, the composition of g and f exists.

$$\begin{aligned} (g \circ f)(x) = g(f(x)) &= g(x^2 + 1) \\ &= 4(x^2 + 1) - 2 \\ &= 4x^2 + 2 \end{aligned}$$

EXAMPLE 13-10   In Example 13-9 we could also define the composite function $(f \circ g)(x) = f(g(x))$ because the codomain of g is the domain of f. Find it.

Solution

$$\begin{aligned} (f \circ g)(x) = f(g(x)) &= f(4x - 2) \\ &= (4x - 2)^2 + 1 \\ &= 16x^2 - 16x + 5 \end{aligned}$$

In general, $(f \circ g)(x) \neq (g \circ f)(x)$.

If f is a bijective function (both onto and one-to-one) from X to Y, then we can define a function g from Y to X with the property that $(g \circ f)(x) = x$ for all $x \in X$ and that $(f \circ g)(y) = y$ for all $y \in Y$. Thus $g(f(x)) = x$ and $f(g(y)) = y$ and the functions f and g are **inverses of each other; that is,** $f^{-1} = g$ (Figure 13-3).

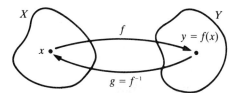

FIGURE 13-3   The inverse of a composite function, given $g(f(x)) = x$ and $f(g(x)) = y$.

EXAMPLE 13-11 Define a bijective function $f: \mathbb{R} \to \mathbb{R}$ by $f(x) = 2x + 1$ and let $g: \mathbb{R} \to \mathbb{R}$ be defined by $g(x) = \dfrac{1}{2}(x - 1)$. Show that these functions are inverses of each other.

Solution We must show that $g(f(x)) = x$ and that $f(g(y)) = y$. Since $f(x) = 2x + 1$ and $g(x) = \dfrac{1}{2}(x - 1)$, we have

$$g(f(x)) = g(2x + 1) = \frac{1}{2}((2x + 1) - 1) = \frac{1}{2}(2x + 1 - 1)$$

$$= \frac{1}{2}(2x) = x$$

Also,

$$f(g(y)) = f\left(\frac{1}{2}(y - 1)\right) = 2\left(\frac{1}{2}(y - 1) + 1\right)$$

$$= (y - 1) + 1 = y$$

Since $g(f(x)) = x$, then g is the inverse of f. And since $f(g(y)) = y$, then f is the inverse of g.

EXAMPLE 13-12 Let $f = \{(1, a), (2, b), (3, c)\}$ be a function from $X = \{1, 2, 3\}$ to $Y = \{a, b, c\}$. Let $g = \{(a, p), (b, q), (c, r)\}$ be a function from Y to $Z = \{p, q, r\}$. What is $(g \circ f)$? Is $f \circ g$ defined?

Solution The domain of g is the codomain of f, so $g \circ f$ is defined as the combination of the domain of f and the codomain of g:

$$(g \circ f)(x) = \{(1, p), (2, q), (3, r)\}$$

The composite $f \circ g = f(g(x))$ is not defined since the domain of f is not the codomain of g.

EXAMPLE 13-13 Let $f_1 = \{(a, c), (b, a), (c, b)\}$ be a function from $X = \{a, b, c\}$ to X and let $f_2 = \{(a, c), (b, c), (c, b)\}$ be another X to X function. Write **(a)** $f_1 \circ f_1$ **(b)** $f_2 \circ f_2$ and **(c)** $f_1 \circ f_2$ as sets of ordered pairs.

Solution We know that all of these composites exist because the functions in each pair meet the codomain and domain requirements (since all the elements come from the same set X). To find the ordered pairs of the composite, take the function of the given function.

(a) Given $f_1 = \{(a, c), (b, a), (c, b)\}$, we want $f_1 \circ f_1$, or $f_1(f_1(x))$. Since a maps to c, b maps to a, and c maps to b, the mapping of f_1 looks like this:

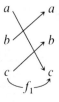

But that only represents putting the elements a, b, c through f_1 once; we must do it twice to get the composite $f_1(f_1(x))$. So we take the results of the first pass through f_1 and put them through f_1 again. We of course get a duplicate result:

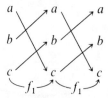

The first and third columns represent the composite pairs. In the first pass through f_1 [for $f_1(x)$], a is mapped to c, which in the second pass [for $f_1(f_1(x))$] is mapped to b; thus the composite $f_1(f_1(x))$ produces the pair (a, b). In the first pass, b maps to a, which in turn maps to c; thus the composite

$f_1(f_1(x))$ produces the pair (b, c). Finally, we see that c maps to b, which maps to a; thus the composite produces the final pair (c, a). Thus $f_1 \circ f_1 = \{(a, b), (b, c), (c, a)\}$.

(b) Here we want, $f_2 \circ f_2 = f_2(f_2(x))$ given that $f_2 = \{(a, c), (b, c), (c, b)\}$. Pass the values a, b, c twice through f_2 to get the mapping

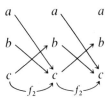

By examining the mapping we find that a maps to c, which maps to b; b maps to c, which maps to b; and c maps to b, which maps to c. The final set is $f_2 \circ f_2 = \{(a, b), (b, b), (c, c)\}$.

(c) For this one we need to send a, b, c first through f_2 then through f_1, which produces this mapping:

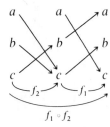

Thus $f_1 \circ f_2 = \{(a, b), (b, b), (c, a)\}$.

There are several interesting and useful results concerning compositions of functions. Suppose $f: X \to Y$ and $g: Y \to Z$ are functions. Then

(1) If f is one-to-one and g is one-to-one, then $g \circ f$ is one-to-one.
(2) If f is onto and g is onto, then $g \circ f$ is onto.
(3) If f is bijective and g is bijective, then $g \circ f$ is bijective.

13-5. Special Functions

In computer science and mathematics a number of special functions are used. We list here a few of them.

(1) Random functions   Often a computer is used to generate random numbers. Many methods exist but one of the popular ones uses the variable SEED and returns a random number via

$$\text{SEED} = (\text{SEED} * 13\,077 + 6925) \bmod 32\,768$$

$$\text{RANDOM} = \text{SEED}/32\,768.0$$

(2) Greatest integer function   Define $f: \mathbb{R} \to \mathbb{Z}$ by $y = f(x)$ where y is the greatest integer less than or equal to x and is denoted by $y = [x]$. For example, $[4.23] = 4$, $[-1.37] = -2$, $[12] = 12$ and $[.49] = 0$. The graph is interesting in that it gives rise to a new name: the staircase function (Figure 13-4).

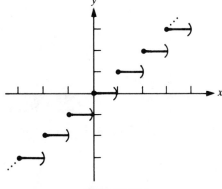

FIGURE 13-4

(3) Recursive function   One of the most popular recursive functions whose value is found via computers for large integer n is the factorial function:

$$\begin{cases} fac(0) = 1 \\ fac(n + 1) = (n + 1) * fac(n) & \text{for } n \geq 0 \end{cases}$$

Another recursive function studied earlier was the Fibonacci number generator:

$$\begin{cases} f(0) = f(1) = 1 \\ f(n) = f(n - 1) + f(n - 2) & \text{for } n > 1 \end{cases}$$

(4) Absolute value function

$$ABS(x) = \begin{cases} x \text{ if } x \geq 0 \\ -x \text{ if } x < 0 \end{cases}$$

(5) Exponential function   A very important mathematical function is $f: \mathbb{R} \to \mathbb{R}^+$ defined by $y = a^x$ for $a > 1$. This is called the exponential function with base a and is a bijection. Often a is an integer (especially $a = 2$ or $a = 10$), but the most convenient base is the irrational number e. The graphs of these three are shown in Figure 13-5.

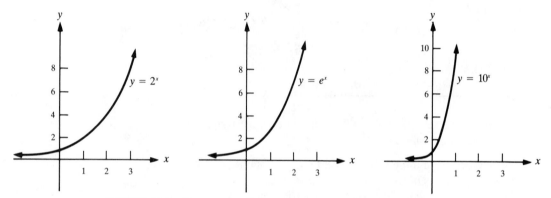

FIGURE 13-5   The graphs of three common exponential functions.

(6) Logarithmic function   The inverse of the exponential function is the logarithmic function with base $a > 1$. Figure 13-6 illustrates the inverse functions for the three exponential functions shown in Figure 13-5. For a thorough explanation of the exponential function and the logarithmic function, see the HBJ College Outline Series volume entitled *College Algebra*.

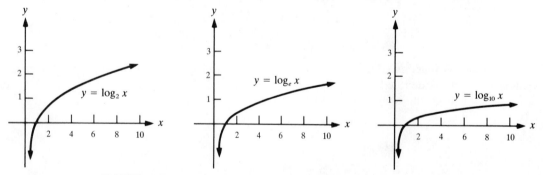

FIGURE 13-6   The graphs of three common logarithmic functions.

SUMMARY

1. A function is a special kind of relation in which no given first element may have two different corresponding second elements.
2. For finite sets, the mapping of the domain elements to the range elements is pictured with directed segments.

3. For infinite domains, the mapping is usually pictured as a graph.
4. A function is called one-to-one if distinct domain elements always have distinct range elements.
5. A function is called onto if the set of range elements coincides with the codomain.
6. If a function f is a bijection, an inverse function f^{-1} exists.
7. Two functions, f and g, may be composed to create new functions, denoted $f \circ g$ or $g \circ f$. Range and domain criteria must be met.
8. A number of special functions have been defined to make manipulation in mathematics and computer science easier.

RAISE YOUR GRADES

Can you . . . ?

☑ explain the difference between a relation and a function
☑ explain the concepts of domain, codomain, and range
☑ define a one-to-one function and give an example
☑ define an onto function and give an example
☑ explain how to find the inverse of a function defined on a finite domain
☑ explain how to find the inverse of a function defined on an infinite domain
☑ draw a mapping of the connections between domains and ranges when two functions are composed
☑ give examples to illustrate the special functions

SOLVED PROBLEMS

Functions and Properties

PROBLEM 13-1   Let $X = \{1, 2, 3, 4\}$ and $Y = \{a, b, c\}$.

(a) Find a function from X to Y.
(b) Can you find a function mapping X onto Y?
(c) Can you find a one-to-one function from X to Y?
(d) Can you find a one-to-one function from Y to X?

Solution

(a) There are many possibilities. Let $f = \{(1, a), (2, b), (3, c), (4, c)\}$. This is a function since no two pairs have the same first element and the entire domain X has been used.
(b) The function given in part (a) is from X onto Y since all of the codomain Y is in the range.
(c) It is impossible to make a one-to-one function from X to Y since four elements would have to match up with three elements; at least two domain elements would have to pair up with the same codomain element.
(d) From Y to X it is possible to have a one-to-one function. Let $f = \{(a, 1), (b, 2), (c, 3)\}$; this is one-to-one but not onto X.

PROBLEM 13-2   Suppose we have a listing of the states of the USA along with their respective populations. Create a function. What is its domain, codomain, and range?

Solution   We will associate fifty names in the set

$$\text{STATES} = \{\text{Alabama, Alaska, Arizona}, \ldots, \text{Wisconsin, Wyoming}\}$$

with fifty integers. The function will be:

$$f = \{(\text{Alabama}; 3{,}893{,}978), (\text{Alaska}; 401{,}851), \ldots, (\text{Wyoming}; 469{,}557)\}$$

Its domain is the set STATES; its codomain is the set of positive integers; its range is a subset of the positive integers consisting of the fifty population values.

PROBLEM 13-3 Let A be the set of 26 letters of the English alphabet; let B be the set of binary number strings of length 5. Create a one-to-one function $f: A \to B$.

Solution Let the j-th letter of the alphabet correspond to the 5-bit binary representation for the number j. A one-to-one function would be:

$$
\begin{array}{cccccccc}
a & b & c & d & \cdots & w & x & y & z \\
\updownarrow & \updownarrow & \updownarrow & \updownarrow & & \updownarrow & \updownarrow & \updownarrow & \updownarrow \\
00001 & 00010 & 00011 & 00100 & \cdots & 10111 & 11000 & 11001 & 11010
\end{array}
$$

PROBLEM 13-4 Define a function $f: N \to N$ by $f(n) = 3n$. Is f one-to-one? Is it onto?

Solution A function f is one-to-one if $f(n_1) = f(n_2)$ implies that $n_1 = n_2$. For our function, when $f(n_1) = f(n_2)$, we have $3n_1 = 3n_2$, or $n_1 = n_2$. Since it fulfills the requirement, the function is one-to-one. The function is not, however, onto since the range is only the multiples of 3; that is, the values of 2 and 7 and 20, ... from the codomain N are not in the range.

PROBLEM 13-5 Let X and Y be finite sets, with $|X| = |Y|$ (that is, they have the same number of elements). Let $f: X \to Y$ be defined such that the domain of f is all of X. Show that if f is one-to-one then it is onto, and then show the converse.

Solution Let $X = \{x_1, \ldots, x_n\}$ and $Y = \{y_1, \ldots, y_n\}$ (the same number of elements). If f is one-to-one, then for each member of X, an image exists and consists of $\{f(x_1), \ldots, f(x_n)\}$. But these elements are distinct members of Y, and include all of Y. Thus f is onto. Conversely, if f is onto, then the values $\{f(x_1), \ldots, f(x_n)\}$ form the entire set of Y. Since Y has exactly n elements, these values are all different. Hence each member of X has one and only one mate, and f is one-to-one.

Inverses

PROBLEM 13-6 Determine the inverse of the function given by $f = \{(1, 5), (2, 2), (3, -4)\}$.

Solution The inverse is the function given by $f^{-1} = \{(5, 1), (2, 2), (-4, 3)\}$. Note that the domain of f has become the codomain of f^{-1} and the codomain of f has become the domain of f^{-1}.

PROBLEM 13-7 Find the inverse function for $f(x) = \frac{1}{3}x + \frac{5}{3}$. Sketch f and f^{-1} on the same set of axes.

Solution Let $y = f(x) = \frac{1}{3}x + \frac{5}{3}$. Then solve the equation for x in terms of y and interchange the variables.

$$y = \tfrac{1}{3}x + \tfrac{5}{3}$$

$$\tfrac{1}{3}x = y - \tfrac{5}{3}$$

$$x = 3y - 5$$

$$y = 3x - 5$$

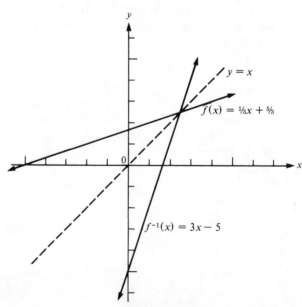

Thus

$$f(x) = \tfrac{1}{3}x + \tfrac{5}{3} \quad \text{and} \quad f^{-1}(x) = 3x - 5$$

You can see from Figure 13-7 that the graphs of the two functions are symmetric with respect to the line $y = x$.

FIGURE 13-7

PROBLEM 13-8 Find the expression for $f^{-1}(x)$ if $f(x) = \dfrac{1}{x+2}$

Solution Let $y = f(x) = \dfrac{1}{x+2}$. Then

$$y = \frac{1}{x+2} \qquad \text{[Solve the equation for } x.]$$

$$x + 2 = \frac{1}{y}$$

$$x = \frac{1}{y} - 2$$

$$y = \frac{1}{x} - 2 \qquad \text{[Interchange the variables.]}$$

Thus $\qquad\qquad f(x) = \dfrac{1}{x+2} \qquad \text{and} \qquad f^{-1}(x) = \dfrac{1}{x} - 2$

Composition of Functions

PROBLEM 13-9 Let $f: \mathbb{R}^+ \to \mathbb{R}^+$ be defined by $f(x) = \sqrt{x}$ and let $g: \mathbb{R}^+ \to \mathbb{R}^+$ be defined by $g(x) = \dfrac{1}{x}$. Find $g \circ f$ and $f \circ g$.

Solution Since the codomains and domains all belong to $\mathbb{R}^+$, they meet the criteria for composition of functions. The composite $(g \circ f)(x) = g(f(x)) = g(\sqrt{x}) = \dfrac{1}{\sqrt{x}}$. In turn, the composite $(f \circ g)(x) = f(g(x)) = f\left(\dfrac{1}{x}\right) = \sqrt{\dfrac{1}{x}} = \dfrac{1}{\sqrt{x}}$. In this rare instance $f \circ g = g \circ f$.

PROBLEM 13-10 Let $f(x) = x^2 + 4$ and $g(x) = x + 6$. Find the formula and domain for **(a)** $(f \circ g)(x)$ and **(b)** $(g \circ f)(x)$.

Solution

(a) $(f \circ g)(x) = f(g(x))$
$$= f(x + 6) = (x + 6)^2 + 4$$
$$= x^2 + 12x + 36 + 4$$
$$= x^2 + 12x + 40$$

The domain is the set of all real numbers.

(b) $(g \circ f)(x) = g[f(x)]$
$$= g[x^2 + 4]$$
$$= (x^2 + 4) + 6$$
$$= x^2 + 10$$

The domain is the set of all real numbers.

PROBLEM 13-11 Composition of functions may be extended to $f: X \to Y$, $g: Y \to Z$, and $h: Z \to W$, obtaining $(h \circ g \circ f)(x) = h(g(f(x)))$. Use $f(x) = x^3$, $g(y) = 2y + 5$, $h(z) = z^2 + 4$ to find $h \circ g \circ f$.

Solution

$$(h \circ g \circ f)(x) = h(g(f(x))) = h(g(x^3))$$
$$= h(2x^3 + 5)$$
$$= (2x^3 + 5)^2 + 4 = 4x^6 + 20x^3 + 29$$

PROBLEM 13-12 Let X be a set and let A be a subset of X. The **characteristic function** of A is the function $f: X \to \{0, 1\}$ defined for each $x \in X$ by

$$f(x) = \begin{cases} 0 & \text{if } x \notin A \\ 1 & \text{if } x \in A \end{cases}$$

If $X = \{1, 2, 3, 4, 5\}$ and $A = \{1, 3, 5\}$ write the characteristic function.

Solution   We have

$$f(x) = \begin{cases} 0 & \text{if } x = 2 \text{ or } 4 \\ 1 & \text{if } x = 1, 3, \text{ or } 5 \end{cases}$$

Supplementary Exercises

PROBLEM 13-13   Let $f: \mathbb{R} \to \mathbb{R}$ be given by $f(x) = 2x + \dfrac{1}{x}$. **(a)** Find the domain and range of f. **(b)** Find $f(-2)$, $f\left(\dfrac{1}{x}\right)$ and $\dfrac{1}{f(x)}$.

PROBLEM 13-14   Which of the following functions are one-to-one?

(a) $f: \mathbb{R} \to \mathbb{R}$ given by $f(x) = 5x - 2$
(b) $f: \mathbb{R}^+ \to \mathbb{R}^+$ given by $f(x) = \sqrt{x}$
(c) $f: \mathbb{R} \to \mathbb{R}$ given by $f(x) = 3x^2$

PROBLEM 13-15   Which of the following functions are onto?

(a) $f: \mathbb{R} \to \mathbb{R}$ given by $f(x) = 7x + 1$
(b) $f: X \to Y$ defined by $f(3) = b$, $f(5) = c$, $f(2) = b$, $f(7) = d$ where $X = \{2, 3, 5, 7\}$ and $Y = \{b, c, d\}$
(c) $f: \mathbb{R}^+ \to \mathbf{R}^+$ given by $f(x) = \sqrt{2x + 1}$
(d) $f: \mathbb{R} \to \mathbb{R}$ given by $f(x) = \sqrt[3]{x + 8}$

PROBLEM 13-16   A function $f: \mathbb{R} \to \mathbb{R}$ is called an **even function** if $f(x) = f(-x)$ for all x. The function is called an **odd function** if $f(x) = -f(-x)$ for all x. Show that $f(x) = x^2$ is even and $f(x) = x^3$ is odd.

PROBLEM 13-17   **(a)** Find another function besides $f(x) = x^2$ that is even. **(b)** Find another function besides $f(x) = x^3$ that is odd. **(c)** Find a function $f: \mathbb{R} \to \mathbb{R}$ that is neither even nor odd.

PROBLEM 13-18   Let $f: X \to Y$ and $g: Y \to Z$ and suppose $g \circ f$ is onto. Show that f does not have to be onto by finding an example.

PROBLEM 13-19   Let $f: X \to Y$ and $g: Y \to Z$ and suppose $g \circ f$ is one-to-one. Show that g does not have to be one-to-one by finding an example.

PROBLEM 13-20   Suppose R is a relation from X to Y where X and Y are the real numbers. To determine if the relation is actually a function we may use a geometric test called *the vertical line test*. After drawing a graph of the relation, determine if *any* vertical line drawn through the graph will touch the graph more than once. If yes, the graph is the graph of a relation; if no, the graph is a graph of a function. Use the vertical line test to determine which of the graphs in Figure 13-8 are graphs of functions and which are graphs of relations.

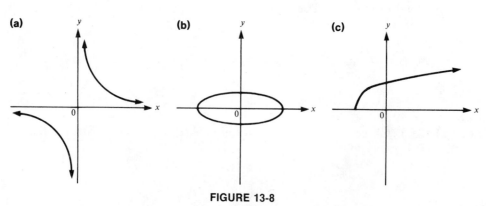

FIGURE 13-8

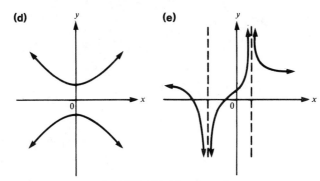

FIGURE 13-8 (*Continued*)

PROBLEM 13-21   Given $f(x) = x^2 + 5x$, find **(a)** $f(1)$, **(b)** $f(-5)$, **(c)** $f(a)$, **(d)** $f(a + h)$, and **(e)** $\dfrac{f(a + h) - f(a)}{h}$.

PROBLEM 13-22   Let $f(x) = 5 - 2x$. Find **(a)** $f(4)$, **(b)** $f(-4)$, **(c)** $f(t)$, **(d)** $f(-t)$, **(e)** $f(x + h)$, and **(f)** $\dfrac{f(x + h) - f(x)}{h}$.

PROBLEM 13-23   Let $g(x) = \dfrac{3x}{x^2 + 1}$. Find **(a)** $g(2)$, **(b)** $g(0)$, **(c)** $g(t^2)$, and **(d)** $[g(t)]^2$.

PROBLEM 13-24   Sketch the graphs of

(a) $f(x) = \begin{cases} -x \text{ if } x < -2 \\ 2 \text{ if } -2 \le x \le 3 \\ x - 1 \text{ if } x > 3 \end{cases}$ **(b)** $f(x) = \begin{cases} |y| \text{ if } x \le 4 \\ 8 - x \text{ if } x > 4 \end{cases}$

PROBLEM 13-25   For each pair of functions given, find the domain and the formula for $g \circ f$:

(a) $f(x) = -x + 4, g(x) = 2x^2 + 5$ **(b)** $f(x) = \dfrac{1}{3x}, g(x) = \sqrt{3x}$

(c) $f(x) = \dfrac{x^2 - 1}{x^2 + 1}, g(x) = \dfrac{1}{x}$

PROBLEM 13-26   Find the inverse of **(a)** $f(x) = \dfrac{1}{x}$ and **(b)** $f(x) = \dfrac{x}{x - 1}$.

PROBLEM 13-27   Given $f(x) = 2x^2 + x - 1$ and $g(x) = x^2 + 9$, find **(a)** $(f \circ g)(2)$ and **(b)** $(g \circ f)(4)$.

PROBLEM 13-28   Given $f(x) = 3x^3$, find $f^{-1}(x)$.

PROBLEM 13-29   Find the formula for the inverse of each function:

(a) $f(x) = -\dfrac{1}{3}(x + 2)$ **(b)** $g(t) = t^5$ **(c)** $h(x) = \dfrac{3 - x}{2x - 5}$

PROBLEM 13-30   On the same coordinate plane, sketch the graphs of the given function and its inverse:

(a) $f(x) = 4x$ **(b)** $g(x) = \sqrt{x + 3}$ **(c)** $h(x) = 10 - \tfrac{1}{2}x$

PROBLEM 13-31   If an equilateral triangle has side length x, then its area is given by $A = \tfrac{\sqrt{3}}{4}x^2$. **(a)** Derive a formula for the side length in terms of the area. **(b)** If the area is 24 in^2, what is the side length?

PROBLEM 13-32   Given $f(a) = a^3 + 3a$, compute $\dfrac{f(a + h) - f(a)}{h}$.

Answers to Supplementary Exercises

13-13 **(a)** The domain is all reals except $x = 0$; the range is $(-\infty < y \le 2\sqrt{2}) \cup (2\sqrt{2} \le y < \infty)$

(b) $f(-2) = -\dfrac{9}{2}, f\left(\dfrac{1}{x}\right) = x + \dfrac{2}{x}, \dfrac{1}{f(x)} = \dfrac{x}{2x^2 + 1}$

13-14 **(a)** and **(b)** are one-to-one

13-15 **(a)**, **(b)**, and **(d)** are onto

13-17 **(a)** $f(x) = x^4$ **(b)** $f(x) = x^5$ **(c)** $f(x) = x^2 + x^3$

13-18 Let $X = \{a, b\}$, $Y = \{c, d\}$, $Z = \{1\}$.
Define $f: X \to Y$ by $f = \{(a, c), (b, c)\}$.
Define $g: Y \to Z$ by $g = \{(c, 1), (d, 1)\}$.

13-19 Let $X = \{a, b\}$, $Y = \{c, d, e\}$, $Z = \{1, 2\}$.
Define $f: X \to Y$ by $f = \{(a, c), (b, c)\}$.
Define $g: Y \to Z$ by $g = \{(c, 1), (d, 1), (c, 2)\}$.

13-20 **(a)**, **(c)**, and **(e)** are functions; **(b)** and **(d)** are relations.

13-21 **(a)** 6 **(b)** 0 **(c)** $a^2 + 5a$ **(d)** $a^2 + 2ah + h^2 + 5a + 5h$ **(e)** $2a + 5 + h$

13-22 **(a)** -3 **(b)** 13 **(c)** $5 - 2t$ **(d)** $5 + 2t$ **(e)** $5 - 2x - 2h$ **(f)** -2

13-23 **(a)** $\dfrac{6}{5}$ **(b)** 0 **(c)** $\dfrac{3t^2}{t^4 + 1}$ **(d)** $\dfrac{9t^2}{(t^2 + 1)^2}$

13-24 **(a)**

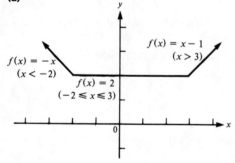

$f(x) = x - 1$
$(x > 3)$

$f(x) = -x$
$(x < -2)$

$f(x) = 2$
$(-2 \le x \le 3)$

(b)

$f(x) = |x|$
$(x \le 4)$

$f(x) = 8 - x$
$(x > 4)$

13-25 **(a)** $(g \circ f)(x) = 2x^2 - 16x + 37$; the domain is the set of all real numbers.

(b) $(g \circ f)(x) = \dfrac{1}{\sqrt{x}}$; the domain $= \{x \mid x > 0\}$.

(c) $(g \circ f)(x) = \dfrac{x^2 + 1}{x^2 - 1}$; the domain $= \{x \mid x \ne 1, x \ne -1\}$.

13-26 **(a)** $\dfrac{1}{x}$ **(b)** $\dfrac{x}{x - 1}$

13-27   (a) 350   (b) 1234

13-28 $\sqrt[3]{\frac{x}{3}}$

13-29   (a) $-3x - 2$ (b) $t^{1/5}$ (c) $\frac{5x+3}{2x+1}$

13-30   (a)

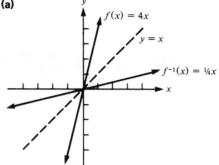

(b)

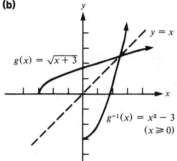

(c)

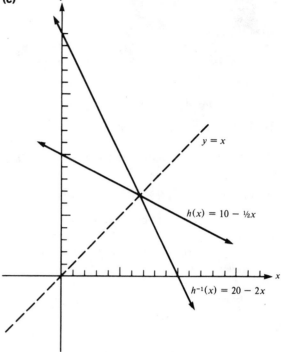

13-31   (a) $x = \sqrt{4a/\sqrt{3}}$ (b) $4\sqrt{2\sqrt{3}}$

13-32 $3a^2 + 3ah + h^2 + 3$

14 GRAPH THEORY

THIS CHAPTER IS ABOUT

☑ **Basic Ideas and Definitions**
☑ **Representations of Graphs**
☑ **Theory for Paths and Circuits**

14-1. Basic Ideas and Definitions

A **graph** is a collection of points, called **vertices** V, and a collection of lines or curves, called **edges** E, which connect some or all of the points. More specifically, if the graph is called G, then the nonempty, finite set of vertices is called V and the set of two-element subsets of V is called E. If the two-element subsets are ordered pairs, then the graph is a *directed graph* (see Section 12-2). If they are not ordered pairs, the graph is an **undirected graph**, and the edges have no direction to them.

If edge $e \in E$ has the representation $\{u, v\}$, then e is said to **connect** vertices u and v and vertices u and v are **adjacent**. The vertices are then the **endpoints** of the edge.

EXAMPLE 14-1 For the graph $G = (V, E)$ where $V = \{v_1, v_2, v_3, v_4\}$ and $E = \{\{v_1, v_2\}, \{v_2, v_3\}, \{v_3, v_4\}, \{v_4, v_2\}\}$, construct the diagram.

Solution The edges are undirected edges because the subsets are not ordered pairs. To draw the diagram, use dots for the four vertices then draw the edges by connecting the pairs of dots given in set E (Figure 14-1).

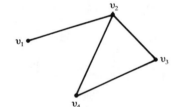

FIGURE 14-1

EXAMPLE 14-2 Given the locations of seven cities, create a roadmap to model a graph.

Solution Let capital letters denote the cities (vertices) and draw the roads (undirected edges) to connect them (Figure 14-2).

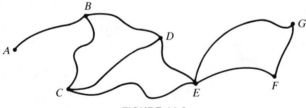

FIGURE 14-2

A roadmap as in Example 14-2 has its limitations as an illustration. In a graph we allow only one edge (road) to connect a pair of vertices, while a map could theoretically have several edges between the same two points. The edges may be curved or straight. Many different-looking diagrams may represent a given graph G, and any one of them is a valid representation of G.

EXAMPLE 14-3 Given the diagram of an undirected graph G_1 (Figure 14-3), write G_1 in set form, showing the vertices and edges.

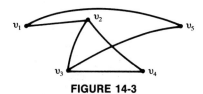

FIGURE 14-3

Solution The vertices are $V = \{v_1, v_2, v_3, v_4, v_5\}$. The edges are

$$E = \{\{v_1, v_2\}, \{v_2, v_3\}, \{v_3, v_4\}, \{v_1, v_5\}, \{v_2, v_4\}, \{v_3, v_5\}\}$$

EXAMPLE 14-4 Given the diagram of a directed graph G_2 (Figure 14-4), write G_2 in set form.

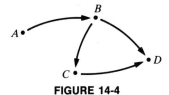

FIGURE 14-4

Solution The vertices are $V = \{A, B, C, D\}$. Since this is a directed graph, its edges will be ordered pairs.

$$E = \{(A, B), (B, C), (C, D), (B, D)\}$$

A few additional terms are helpful in discussing graphs. An edge $\{u, v\}$ (or (u, v)) in an undirected (or directed) graph is said to be **incident on u and v**, and the vertices are **incident on the edge**.

In a graph, an edge associated with a vertex pair $\{v, v\}$ or (v, v) is called a **loop**. If a graph has no loops, it is called a **simple graph**.

Two edges are said to be **parallel** if we have $\{u, v\}$ and $\{v, u\}$ or (u, v) and (v, u).

A graph H is called a **subgraph** of a graph G if the set of vertices of H is a subset of the set of vertices of G and if the set of edges of H is a subset of the set of edges of G.

agreement: In this chapter the graphs will be constructed with finite sets of vertices and edges.

EXAMPLE 14-5 Let $G = (V_G, E_G)$ be defined by $V_G = \{v_1, v_2, v_3, v_4\}$ and $E_G = \{\{v_1, v_2\}, \{v_1, v_3\}, \{v_2, v_3\}, \{v_2, v_4\}\}$. Let $H = (V_H, E_H)$ be defined by $V_H = \{v_1, v_2, v_3\}$ and $E_H = \{\{v_1, v_3\}, \{v_2, v_3\}$. Is H a subgraph of G?

Solution The diagrams of G and H should be helpful (Figure 14-5). Since graph H could be superimposed upon graph G, it is a subgraph. Its vertex set $V_H \subseteq V_G$ and its edge set $E_H \subseteq E_G$.

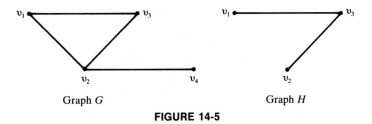

Graph G Graph H

FIGURE 14-5

If v is a vertex of graph G, then the **degree of v**, denoted $\delta(v)$, is the number of edges having v as an endpoint.

EXAMPLE 14-6 For the graph G of Example 14-5 (Figure 14-5) determine the degree of each vertex.

Solution We merely count the number of edges that are connected to each vertex: $\delta(v_1) = 2$, $\delta(v_2) = 3$, $\delta(v_3) = 2$, $\delta(v_4) = 1$.

In a graph G a **path** joining vertices A and B of G is a sequence of vertices of G: $v_0 (= A)$, v_1, v_2, $v_3, \ldots, v_m (= B)$ such that each pair $\{v_i, v_{i+1}\}$ is an edge of G. The path is said to have **length** m, since m edges are used to get from A to B. (Backtracking and repeating edges is permitted.)

EXAMPLE 14-7 Let graph G be diagrammed as in Figure 14-6. Find the paths from A to B and also find their lengths.

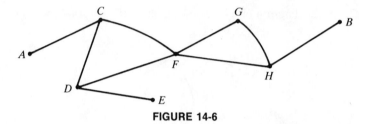

FIGURE 14-6

Solution This is a very intuitive concept. A path from A to B is a listing of the vertices that we visit as we journey from A to B. Thus $ACFHB$ is a path of length 4. And $ACFGHB$ has length 5. And $ACDFGHB$ has length 6. There are many other paths that could be written that would seem ridiculous: by backtracking, crossing over, going in circles a while, etc. Thus a total count of the number of paths is not possible to compute.

A graph G is said to be **connected** if every pair of vertices in G may be joined by a path. Intuitively we see that the graph is all joined together; it is one piece. If there is a pair of vertices not joined by a path, the graph is **disconnected**.

EXAMPLE 14-8 Construct a graph that is connected; then one that is disconnected.

Solution See Figure 14-7.

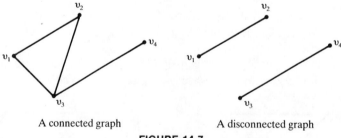

A connected graph   A disconnected graph

FIGURE 14-7

A path $v_0, v_1, v_2, \ldots, v_m$ in a graph G is an **Euler path** (pronounced "Oiler") if every edge of G appears once and only once in the path. If the Euler path begins and ends at the same vertex—that is, if it's **closed**—it is called an **Euler circuit**. And if a route in a graph G begins and ends at the same vertex and passes through each vertex of G exactly once, it is called a **Hamiltonian circuit**.

note: In an Euler circuit, every *edge* of G appears once and only once in the path, while in a Hamiltonian circuit, every *vertex* of G appears once and only once in the path.

EXAMPLE 14-9 Construct **(a)** a graph containing an Euler path, **(b)** a graph containing an Euler circuit, and **(c)** a graph containing a Hamiltonian circuit.

Solution We will work with five vertices.

(a) The Euler path must link all the vertices without repeating edges. One such path would be v_0, v_1, v_2, v_4, v_3 (Figure 14-8).

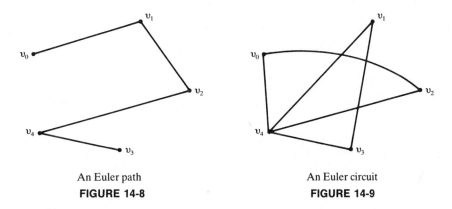

An Euler path

FIGURE 14-8

An Euler circuit

FIGURE 14-9

(b) It would be easy to add to the Euler path of Figure 14-8 the additional edge $\{v_3, v_0\}$ to make it into an Euler circuit. It would then also be a Hamiltonian circuit. But let's offer a little more general situation of $v_0, v_4, v_1, v_3, v_4, v_2, v_0$ (Figure 14-9).

(c) In the Euler circuit we are concerned with including each edge once and only once. Now we are concerned with including each vertex once and only once. Thus we may skip some edges if necessary. A route around the outside does it quite nicely (Figure 14-10a). There are other possibilities, including $v_0, v_1, v_3, v_2, v_4, v_0$, etc. (Figure 14-10b).

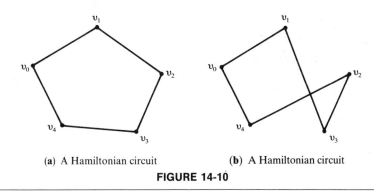

(a) A Hamiltonian circuit   **(b)** A Hamiltonian circuit

FIGURE 14-10

EXAMPLE 14-10 Consider the diagrams (a) and (b) in Figure 14-11. Decide if the diagram contains an Euler path, an Euler circuit, or a Hamiltonian circuit.

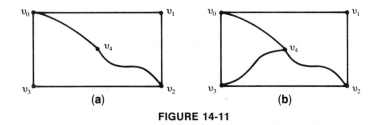

(a)   **(b)**

FIGURE 14-11

Solution

(a) We have an Euler path. It is a sequence of vertices that include every edge once and only once: $v_0, v_1, v_2, v_3, v_0, v_4, v_2$.

We do not have an Euler circuit since once we get around any closed subgraph there are edges that have not been covered. Try this with your pencil.

This does not have a Hamiltonian circuit either since any route around must include a closed subgraph with at least one vertex not yet visited. Thus we could not begin and end at the same vertex and pass through each vertex exactly once. Use your pencil to trace out some possible solutions. See that they do not work.

(b) There is no Euler path since the new edge now in the diagram cannot be covered when the other edges are covered exactly once. Try it with your pencil.

Since an Euler circuit is an Euler path with a common beginning–ending point, we obviously don't have this one either.

We do have Hamiltonian circuits. One of them is $v_0, v_1, v_2, v_3, v_4, v_0$. Use your pencil to find others.

We will return to the study of these special paths in the Solved Problems.

14-2. Representations of Graphs

Besides using a diagram to picture a graph, we have a more formal mathematical method of representing it. This is especially important in computer applications, where we can analyze graphs with algorithms.

Let G be a graph with vertex set $V = \{v_1, v_2, \ldots v_n\}$. Then the **adjacency matrix** of the graph G is the $n \times n$ matrix

$$A = [a_{ij}]$$

where

$$a_{ij} = \begin{cases} 1 & \text{if vertices } v_i \text{ and } v_j \text{ are adjacent} \\ 0 & \text{if vertices } v_i \text{ and } v_j \text{ are not adjacent} \end{cases}$$

Remember that vertices v_i and v_j are adjacent if the edge set E contains the edge $\{v_i, v_j\}$.

EXAMPLE 14-11   Suppose a graph G has vertex set $V = \{v_1, v_2, v_3, v_4\}$ and an adjacency matrix

$$A = \begin{bmatrix} 0 & 1 & 1 & 1 \\ 1 & 0 & 0 & 0 \\ 1 & 0 & 0 & 1 \\ 1 & 0 & 1 & 0 \end{bmatrix}$$

(a) What is the edge set? **(b)** Draw the graph.

Solution

(a) Since matrix element a_{12} is 1, the vertices v_1 and v_2 are adjacent, so we have an edge $\{v_1, v_2\}$. Similarly, since a_{31} is 1, vertices v_3 and v_1 are adjacent, and we have edge $\{v_3, v_1\}$. Continuing this analysis provides the edge set

$$E = \{\{v_1, v_2\}, \{v_1, v_3\}(v_1, v_4\}, \{v_3, v_4\}\}$$

(b) Draw the diagram of G from edge set E (Figure 14-12).

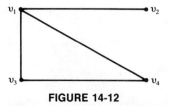

FIGURE 14-12

Example 14-11 reveals some important properties of the adjacency matrix.

(1) The adjacency matrix is a square $n \times n$ matrix.
(2) Since edge $\{v_i, v_j\}$ is the same as edge $\{v_j, v_i\}$, we will have $a_{ij} = a_{ji}$ in the adjacency matrix. In other words, the adjacency matrix is **symmetric**.
(3) If, in a graph, an edge never joins a vertex to itself, the main diagonal entries a_{ii} are all zero.
(4) The degree of a vertex, $\delta(v)$, is the sum of the entries in the row (or column) for that vertex.
(5) If $n(E)$ denotes the number of edges in a graph, then the adjacency matrix contains $2n(E)$ entries that are 1's. This follows since each edge is represented twice: once as $a_{ij} = 1$ and again as $a_{ji} = 1$. Thus the sum of the entries in any adjacency matrix is twice the number of edges in the edge set of the graph.

EXAMPLE 14-12 Find the adjacency matrix for the graph in Figure 14-13.

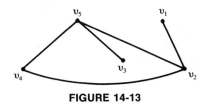

FIGURE 14-13

Solution Since there are five vertices, the adjacency matrix A will be 5×5. Since the edge set is

$$E = \{\{v_5, v_2\}, \{v_5, v_3\}, \{v_5, v_4\}, \{v_1, v_2\}, \{v_2, v_4\}\}$$

the entries in A that will be 1 are a_{52} and a_{25}, a_{53} and a_{35}, a_{54} and a_{45}, a_{12} and a_{21}, a_{24} and a_{42}. Thus the matrix is

$$A = \begin{array}{c} \\ v_1 \\ v_2 \\ v_3 \\ v_4 \\ v_5 \end{array} \begin{array}{ccccc} v_1 & v_2 & v_3 & v_4 & v_5 \\ \begin{bmatrix} 0 & 1 & 0 & 0 & 0 \\ 1 & 0 & 0 & 1 & 1 \\ 0 & 0 & 0 & 0 & 1 \\ 0 & 1 & 0 & 0 & 1 \\ 0 & 1 & 1 & 1 & 0 \end{bmatrix} \end{array}$$

The sum of the entries is 10, twice the number of edges ($2n(E) = 2 \times 5 = 10$). Note also that the row (column) sums agree with the degree of the vertices from the graph.

EXAMPLE 14-13 Diagram the graph G represented by the adjacency matrix

$$A = \begin{bmatrix} 0 & 1 & 0 & 1 \\ 1 & 0 & 1 & 1 \\ 0 & 1 & 0 & 1 \\ 1 & 1 & 1 & 0 \end{bmatrix}$$

Solution There will be four vertices and the edges will join v_1 to v_2, v_1 to v_4, v_2 to v_3, v_2 to v_4, v_3 to v_4 (and vice versa). Thus the graph is as in Figure 14-14.

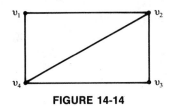

FIGURE 14-14

The adjacency matrix is an efficient and effective way of describing a graph in a computer. Analysis of the graph with respect to Euler paths, Euler circuits, etc., may then be accomplished by establishing investigative algorithms.

14-3. Theory for Paths and Circuits

In Section 14-1 we introduced definitions for a path (a sequence of vertices), a connected graph (a graph in which every pair of vertices can be joined by a path), an Euler path (a path in which each edge appears only once), and an Euler circuit (an Euler path that is closed). We now state some important theoretical results that use all three of these concepts.

THEOREM 1   A graph has an Euler circuit if and only if it is connected and all vertices have even degree.

EXAMPLE 14-14   The famous Königsberg Bridge problem solved by Leonhard Euler in 1736 illustrates this theorem. Königsberg was an eighteenth-century city situated at the confluence of two rivers. In the river was an island, and seven bridges crossed the rivers at various spots to join four land masses *A*, *B*, *C*, *D* (Figure 14-15). The challenge to Euler was to take a walk which crossed each bridge once and returned him to his starting point. Show that his Theorem 1 (above) was an ingenious solution.

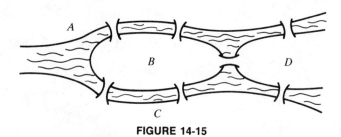

FIGURE 14-15

Solution   The graph of the bridge walking problem uses vertices as the land masses and edges as the bridges that join them (Figure 14-16). After modeling the problem as in Figure 14-16, Euler reasoned that if the walk got *to* vertex *D* it would also have to *leave* vertex *D*. If *D* was not the starting point (or even if it was!), one extra edge was incident at *D*. That edge could not, then, be covered and the proposed walk could not be done. The degree of *D* (as well as of *A* and *C*) was 3, an odd number. No Euler circuit exists!

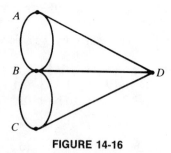

FIGURE 14-16

THEOREM 2   The sum of the degrees of all the vertices of any graph is an even number.

THEOREM 3   For any graph, the number of vertices that have odd degree is an even number.

THEOREM 4   An Euler path will exist from vertex *V* to vertex *W* in a graph *G* if and only if *V* and *W* are the only vertices of *G* having odd degree.

EXAMPLE 14-15   Examine the graph in Figure 14-17. **(a)** Does it have an Euler path? **(b)** Does it have an Euler circuit?

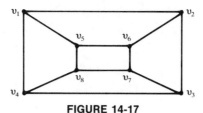

FIGURE 14-17

Solution Every vertex has degree 3. The total degree is 24—an even number (see Theorem 2). There are eight vertices with odd degree (see Theorem 3).

(a) No Euler paths exist (by Theorem 4).
(b) No Euler circuit exists (by Theorem 1).

EXAMPLE 14-16 Examine the graph in Figure 14-18. **(a)** Does it have an Euler path? **(b)** Does it have an Euler circuit?

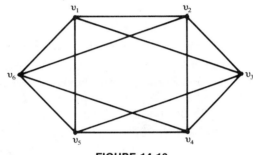

FIGURE 14-18

Solution Every vertex has degree 4. The total degree for the six vertices is 24 (see Theorem 2). None have odd degree (see Theorem 3).

(a) No Euler path from distinct vertices V and W exist (by Theorem 4).
(b) Euler circuits certainly do exist (by Theorem 1). An Euler circuit that begins and ends at v_1 and that covers each edge exactly once is

$$v_1, v_3, v_5, v_1, v_2, v_4, v_6, v_2, v_3, v_4, v_5, v_6, v_1$$

Trace it with your pencil.

Note that a systematic approach was used to find the Euler circuit in Example 14-16. The subgraph v_1, v_3, v_5, v_1 was traversed with a closed path (Figure 14-19), which also was, by itself in its graph, an Euler circuit. Moving from v_1 around the outside of the entire graph would again get us back to v_1, BUT such a circuit would not allow us to pick up the remaining edges. These leftover edges made another subgraph v_2, v_4, v_6, v_2 (Figure 14-20) that was also a closed Euler circuit, so when we got to vertex v_2, we needed to insert this subgraph. Note now that the Euler circuit in the answer to Example 14-16 can be broken down into three subgraphs but with the insertions in the right places (Figure 14-21).

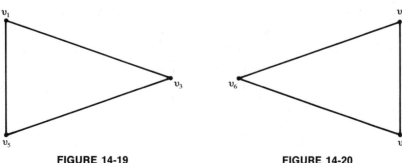

FIGURE 14-19   **FIGURE 14-20**

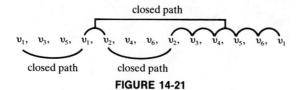

FIGURE 14-21

Put another way, if the theorems tell us that an Euler circuit exists, we can find it systematically by these steps:

(1) Identify a closed subgraph and write the sequence of vertices, beginning at any one of them (call it the "chosen one").
(2) Redraw the original graph with the subgraph edges removed.
(3) Redo steps 1 and 2 until no further subgraph exists.
(4) Write the sequence of vertices for the remaining closed path.
(5) In this sequence, wherever a "chosen one" appears, chop the sequence open and insert the sequence obtained in step 1.

The final sequence will be the Euler circuit.

EXAMPLE 14-17 You are given a graph G (for "gargantuan") that represents a road map used by a traveling milk delivery person (Figure 14-22). He leaves his home H in the morning, visits clients along each stretch of road, never wants to pass by a client twice in a day (saves on gasoline), and returns home at night. Find an Euler circuit. (Note: His clients are on the edges, not at the vertices. Treat the vertices as "intersections" where he may turn onto another road.)

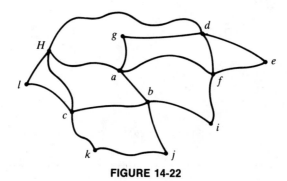

FIGURE 14-22

Solution Every vertex has even degree (they are all twos and fours) so we do have an Euler circuit. We find it systematically. We begin at H and find a closed subgraph and write the sequence of vertices: H, a, b, c, H. The "chosen one" will be H. Now remove these edges from the original graph (Figure 14-23).

Another closed subgraph of the remaining graph is d, f, a, g, d. Let the "chosen one" for this subgraph be d. Remove these edges in turn, and we are left with Figure 14-24.

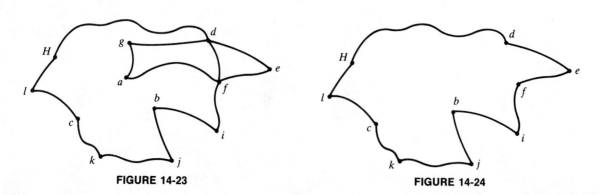

FIGURE 14-23   **FIGURE 14-24**

Now write a sequence for the lone remaining closed subgraph (we could start anywhere, but we want the milk delivery person home so we pick H):

$$H, d, e, f, i, b, j, k, c, l, H$$

At d in this sequence, we insert the second subsequence found above, and at the final H we insert the first subsequence we found. (We could have put that second one in front also—does he want a morning or afternoon coffee break at Home?) We finally obtain

$$H, d, f, a, g, d, e, f, i, b, j, k, c, l, H, a, b, c, H$$

as the Euler circuit. Trace it with your pencil.

SUMMARY

1. A graph is a collection of points, called vertices, and a collection of lines or curves, called edges, that join pairs of vertices.
2. In a directed graph, the edges are ordered pairs (u, v) with the edge directed from vertex u to vertex v.
3. In an undirected graph, the edges have no direction to them—they only connect two vertices.
4. A convenient way to represent a graph with n vertices is by constructing an $n \times n$ adjacency matrix. This allows computer analysis and manipulation of the graph.
5. The degree of a vertex v is the number of edges that have an end at v. The sum of the degrees of all the vertices is twice the number of edges.
6. In a graph, a path from vertex A to vertex B is a sequence of edges connected one to another that begins at A and ends at B.
7. If every edge in the graph appears once and only once in a path, the path is an Euler path. If this Euler path begins and ends at the same vertex, it is called an Euler circuit.
8. A route in a graph that begins and ends at the same vertex and passes through each vertex exactly once is called a Hamiltonian circuit.

RAISE YOUR GRADES

Can you...?

☑ change the set form of a graph into a diagram
☑ form a subgraph of a given graph
☑ find the degree of a vertex of a graph
☑ explain the difference between a path and an Euler path
☑ explain how you determine if a graph is connected
☑ define how to construct the adjacency matrix for a graph
☑ recite the theorems on paths and circuits
☑ describe the steps needed to systematically find an Euler circuit for a graph

SOLVED PROBLEMS

Basic Ideas and Definitions

PROBLEM 14-1 Construct a graph (**a**) with four vertices and three edges that is simple, (**b**) with four vertices and ten edges, (**c**) with four vertices and seven edges that is simple.

Solution Recall that a simple graph has no loops or parallel edges.

(a) See Figure 14-25a

(b) This one will require loops; see Figure 14-25b.

(c) This is impossible. With four vertices, there are only $_4C_2 = 6$ distinct connections without using loops or parallel connections, which would prevent the graph from being simple (Figure 14-25c).

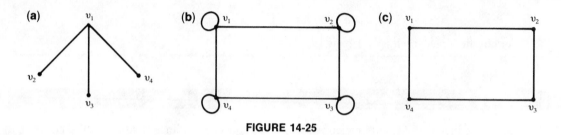

FIGURE 14-25

PROBLEM 14-2 Let graph G have vertices $V = \{A, B, C, D, E\}$ and edges $E = \{\{B, C\}, \{C, D\}, \{D, B\}, \{E, A\}\}$. Represent G by a diagram.

Solution You may draw the vertices anywhere, then just add in the edges between the vertex pairs that define the edges (Figure 14-26). This is a disconnected graph.

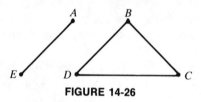

FIGURE 14-26

PROBLEM 14-3 An undirected graph is shown in Figure 14-27. Define this graph in set form.

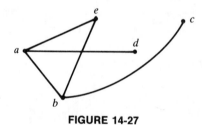

FIGURE 14-27

Solution The vertex set is $V = \{a, b, c, d, e\}$ and the edge set is $E = \{\{a, b\}, \{b, c\}, \{a, d\}, \{a, e\}, \{b, e\}\}$.

PROBLEM 14-4 A directed graph has vertices $V = \{v_1, v_2, v_3\}$ and edges $E = \{(v_1, v_2), (v_1, v_3), (v_3, v_1), (v_3, v_2)\}$. Draw the diagram.

Solution Remember to use directed arrows instead of line segments (Figure 14-28).

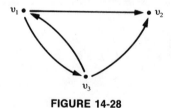

FIGURE 14-28

PROBLEM 14-5 Find the degree of each vertex of the graph in Figure 14-29.

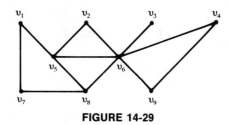

FIGURE 14-29

Solution The degree of a vertex is the number of edges that are incident upon it (connected to it). Thus $\delta(v_1) = 2$, $\delta(v_2) = 2$, $\delta(v_3) = 1$, $\delta(v_4) = 2$, $\delta(v_5) = 4$, $\delta(v_6) = 6$, $\delta(v_7) = 2$, $\delta(v_8) = 3$, $\delta(v_9) = 2$.

PROBLEM 14-6 In the graph in Figure 14-30, find two different paths from A to D.

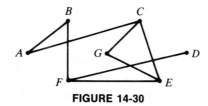

FIGURE 14-30

Solution Beginning at A we can go to B or to C. From B we can go only to F. Then directly to D gives the path $ABFD$ of length 3. If we had gone from A to C, we could go next to E. From E we back up to F and then go to D. This path then is $ACEFD$ of length 4. There are an infinite number of others.

PROBLEM 14-7 Given the graph in Figure 14-31, determine if it has an Euler path, an Euler circuit, and/or a Hamiltonian circuit.

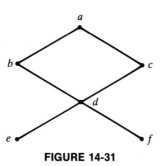

FIGURE 14-31

Solution An Euler path is e, d, c, a, b, d, f. There is no Euler circuit since vertices e and f have odd degree (1). Since vertex d is passed through twice, we have no chance of a path's being Hamiltonian.

note: If edge $\{e, f\}$ were added to Figure 14-31, we would have an Euler circuit but still not a Hamiltonian one.

PROBLEM 14-8 For the graph in Figure 14-32, determine if it has an Euler path, an Euler circuit, and/or a Hamiltonian circuit.

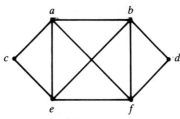

FIGURE 14-32

Solution   All vertices have even degree, so (by Theorem 4) no Euler path between distinct vertices exists. But an Euler circuit is $a, b, d, f, b, e, c, a, f, e, a$. A Hamiltonian circuit simply skirts the outside of the graph: a, b, d, f, e, c, a.

Representations of graphs

PROBLEM 14-9   For the graph in Figure 14-33, construct the adjacency matrix.

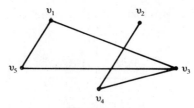

FIGURE 14-33

Solution   In this matrix we insert a 1 in row i and column j if an edge connects vertex v_i with vertex v_j and insert a 0 otherwise. The matrix is symmetric and is 5×5 because there are five vertices. Since the graph is simple (no loops or parallel edges) the main diagonal will contain only zeros:

$$A = \begin{array}{c} \\ v_1 \\ v_2 \\ v_3 \\ v_4 \\ v_5 \end{array} \begin{array}{c} \begin{matrix} v_1 & v_2 & v_3 & v_4 & v_5 \end{matrix} \\ \begin{bmatrix} 0 & 0 & 1 & 0 & 1 \\ 0 & 0 & 0 & 1 & 0 \\ 1 & 0 & 0 & 1 & 1 \\ 0 & 1 & 1 & 0 & 0 \\ 1 & 0 & 1 & 0 & 0 \end{bmatrix} \end{array}$$

PROBLEM 14-10   Given the adjacency matrix

$$A = \begin{array}{c} \\ a \\ b \\ c \\ d \end{array} \begin{array}{c} \begin{matrix} a & b & c & d \end{matrix} \\ \begin{bmatrix} 0 & 1 & 1 & 0 \\ 1 & 0 & 1 & 1 \\ 1 & 1 & 0 & 0 \\ 0 & 1 & 0 & 0 \end{bmatrix} \end{array}$$

(a) What is the edge set?
(b) Draw the graph.

Solution   Since the vertices are labeled a, b, c, d, the edge set will be $E = \{\{a,b\}, \{a,c\}, \{b,c\}, \{b,d\}\}$ and the graph is as in Figure 14-34.

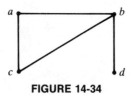

FIGURE 14-34

PROBLEM 14-11   Find the adjacency matrix for the graph in Figure 14-35.

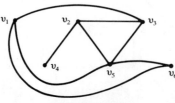

FIGURE 14-35

Solution Since vertex v_1 is connected to v_3, v_5, v_6, we insert a 1 at positions a_{13}, a_{15}, and a_{16}. Since vertex v_2 is connected to v_3, v_4, v_5 we insert a 1 at a_{23}, a_{24}, and a_{25}. Similarly, we get a 1 in positions a_{35} and a_{56}. Of course we get the symmetric counterparts. Thus the adjacency matrix is:

$$A = \begin{array}{c} \\ v_1 \\ v_2 \\ v_3 \\ v_4 \\ v_5 \\ v_6 \end{array} \begin{array}{cccccc} v_1 & v_2 & v_3 & v_4 & v_5 & v_6 \\ \left[\begin{array}{cccccc} 0 & 0 & 1 & 0 & 1 & 1 \\ 0 & 0 & 1 & 1 & 1 & 0 \\ 1 & 1 & 0 & 0 & 1 & 0 \\ 0 & 1 & 0 & 0 & 0 & 0 \\ 1 & 1 & 1 & 0 & 0 & 1 \\ 1 & 0 & 0 & 0 & 1 & 0 \end{array}\right] \end{array}$$

PROBLEM 14-12 Suppose G is the graph in Figure 14-36. Construct the adjacency matrix for G. Check the row sums and the degree(s) of the vertices.

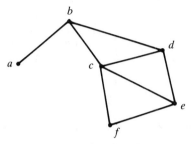

FIGURE 14-36

Solution Since there are six vertices, the matrix will be 6×6.

$$M_G = \begin{array}{c} \\ a \\ b \\ c \\ d \\ e \\ f \end{array} \begin{array}{cccccc} a & b & c & d & e & f \\ \left[\begin{array}{cccccc} 0 & 1 & 0 & 0 & 0 & 0 \\ 1 & 0 & 1 & 1 & 0 & 0 \\ 0 & 1 & 0 & 1 & 1 & 1 \\ 0 & 1 & 1 & 0 & 1 & 0 \\ 0 & 0 & 1 & 1 & 0 & 1 \\ 0 & 0 & 1 & 0 & 1 & 0 \end{array}\right] \end{array} \quad \begin{array}{cc} \text{row sums} & \delta(v) \\ 1 & 1 \\ 3 & 3 \\ 4 & 4 \\ 3 & 3 \\ 3 & 3 \\ 2 & 2 \end{array}$$

PROBLEM 14-13 Suppose G has the adjacency matrix

$$M_G = \begin{array}{c} \\ p \\ q \\ r \\ s \\ t \end{array} \begin{array}{ccccc} p & q & r & s & t \\ \left[\begin{array}{ccccc} 0 & 1 & 0 & 1 & 0 \\ 1 & 0 & 1 & 1 & 0 \\ 0 & 1 & 0 & 0 & 1 \\ 1 & 1 & 0 & 0 & 1 \\ 0 & 0 & 1 & 1 & 0 \end{array}\right] \end{array}$$

Determine the degree of each vertex and draw the graph.

Solution Using the row sums gives us the degrees $\delta(p) = 2$, $\delta(q) = 3$, $\delta(r) = 2$, $\delta(s) = 3$, $\delta(t) = 2$. The graph is shown in Figure 14-37.

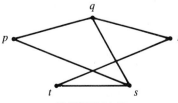

FIGURE 14-37

Theory for Paths and Circuits

PROBLEM 14-14   Prove Theorem 2: The sum of the degrees of all the vertices of any graph is an even number.

Solution   Each edge of a graph is incident upon two vertices; that is, edge $\{v_1, v_2\}$ contributes 1 to the degree of vertex v_1 and 1 to the degree of vertex v_2. Thus when we add up the degrees of the entire set of vertices we will have a number that is twice as large as the number of edges. Since the number of edges is an integer, twice this integer is an even number.

PROBLEM 14-15   Prove Theorem 3: For any graph, the number of vertices that have odd degree is an even number.

Solution   By Theorem 2, the sum of the degrees of all the vertices is an even number, call it $2n$. If we subtract from $2n$ the sum of the degrees of all the vertices with even degrees, we still have an even number of degrees, call it $2m$. This $2m$ is the sum of the degrees of the vertices that have odd degree. There must be an even number of such vertices for the sum of these odd numbers to be even (since odd + odd = even, and odd + even = odd).

PROBLEM 14-16   Determine if the graph in Figure 14-38 has an Euler path and/or an Euler circuit.

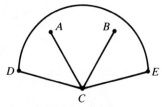

FIGURE 14-38

Solution   Use Theorem 4: An Euler path will exist from vertex V to vertex W in a graph G if and only if V and W are the only vertices of G having odd degree. We note that vertices A and B have degree 1 while D and E have degree two and C has degree four. Thus an Euler path exists from A to B (by Theorem 4); one such path is $ACEDB$. By theorem 1, no Euler circuit exists since we have vertices with odd degree.

PROBLEM 14-17   Using Theorem 4, show that in the graph of Problem 14-5 (Figure 14-29) there is an Euler path between two of its vertices. Find it.

Solution   If two specific vertices of a graph are the only vertices with odd degree, then an Euler path exists from one of them to the other one. In the Figure 14-29, $\delta(v_3) = 1$ and $\delta(v_8) = 3$; all of the other vertices have even degree. We know an Euler path exists from v_3 to v_8. In this path each edge must appear only once. One orientation to get the Euler path is $v_3, v_6, v_4, v_9, v_6, v_2, v_5, v_6, v_8, v_7, v_1, v_5, v_8$. Use your pencil to draw this path and verify that it does include each edge exactly once.

PROBLEM 14-18   Verify that the graph in Figure 14-39 has an Euler circuit and find one beginning and ending at A.

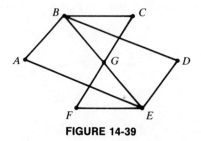

FIGURE 14-39

Solution   The degree of each vertex is even, so Theorem 1 guarantees that an Euler circuit exists. We now need to find a path that begins and ends at *A* and includes each edge exactly once. One such path is *A, B, C, G, B, D, E, F, G, E, A*. Trace over it with your pencil to see its construction.

PROBLEM 14-19   In the graph in Figure 14-40, find three different Hamiltonian circuits.

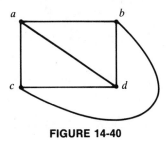

FIGURE 14-40

Solution   We need to begin and end at the same vertex, but we may pass through any given vertex only once. The circuits are *a, b, c, d, a; a, b, d, c, a;* and *a, c, b, d, a*.

PROBLEM 14-20   There are six stations in an assembly plant that are to be connected by an intercom link. We have enough wire to connect ten pairs of stations. Construct the graph of the connections so any pair of stations can talk to each other (possibly using intermediaries) and so as to minimize failure of communication due to lines being accidentally cut.

Solution   We have six vertices and ten edges. Since a cut line would prevent a station from talking if only two lines were connected to it, we want the minimum degree of each vertex to be three. (Figure 14-41). Note that if any one line is cut, all of the stations can still communicate using only one or two intermediary lines.

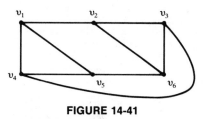

FIGURE 14-41

PROBLEM 14-21   A **complete graph on *n* vertices**, called K_n, is made up of *n* vertices, all of which are connected to each other with no loops and no parallel edges. Form the graphs of K_3 and K_4.

Solution   To form K_3, we will place three vertices and connect them all (Figure 14-42a). To form K_4, we place four vertices and connect them all (Figure 14-42b).

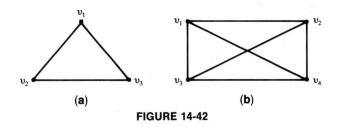

FIGURE 14-42

PROBLEM 14-22   Form the graph of K_5 and find a formula for the number of edges in the graph of K_n.

Solution   To form K_5, we place five vertices and connect them all (Figure 14-43).

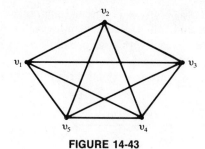

FIGURE 14-43

Looking at the graphs to Problem 14-21 (Figure 14-42) and Figure 14-43, we have three examples to work from and we can chart their data:

Number of vertices in K_n	Number of edges in K_n
3	3
4	6
5	10

In forming a complete graph on n vertices we draw in the edges as follows: From v_1 draw edges to $v_2, v_3, v_4, \ldots, v_n$. There are $n-1$ of these. From v_2 draw edges to $v_3, v_4, \ldots, v_n$. There are $n-2$ of these. From v_3 draw edges to $v_4, v_5, \ldots, v_n$. There are $n-3$ of these. Continue this until you finally draw the edge from v_{n-1} to v_n. The total number of edges is

$$(n-1) + (n-2) + (n-3) + (n-4) + \cdots + 3 + 2 + 1$$

This is the sum of all the integers from 1 to $n-1$. The algebraic formula for this sum (the sum of all edges of K_n) is

$$S_{n-1} = \frac{(n-1)(n)}{2}$$

Test this on the above examples. For K_3, $S_2 = \dfrac{2 \times 3}{2} = 3 \checkmark$. For K_4, $S_3 = \dfrac{3 \times 4}{2} = 6 \checkmark$. For K_5, $S_4 = \dfrac{4 \times 5}{2} = 10 \checkmark$.

PROBLEM 14-23 Suppose the set of vertices of a graph is divided into two disjoint subsets V_1 and V_2. If every edge of the graph is incident upon one vertex in V_1 and one vertex in V_2, then the graph is called **bipartite**. Construct two examples of bipartite graphs.

Solution

(a) Let $V_1 = \{A, B, C\}$, $V_2 = \{D, E, F\}$. Then Figure 14-44a is a bipartite graph.
(b) Let $V_1 = \{a, b\}$, $V_2 = \{c, d, e, f\}$. Then Figure 14-44b is a bipartite graph.

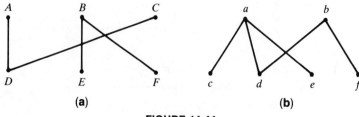

(a)   **(b)**

FIGURE 14-44

There are many other possibilities. Constructing the bipartite graph is easy. Construct set V_1 at the top in a row. Construct set V_2 at the bottom in a row. Draw in whatever edges you want but DON'T draw a "horizontal edge." Note that if V_1 has m vertices and V_2 has n vertices, then the maximum number of edges is $m \times n$; that is, from each of the m vertices in V_1 you could draw an edge to each of the n vertices in V_2.

PROBLEM 12-24 Suppose we have a set V_1 of m vertices and a disjoint set V_2 of n vertices. If we draw all possible edges that are incident upon one vertex in V_1 and one vertex in V_2, the graph is called the **complete bipartite graph** with parts V_1 and V_2, and is denoted $K_{m,n}$ (and we always put the smaller number first in the subscript). Construct (a) $K_{2,2}$ (b) $K_{2,3}$ (c) $K_{2,4}$.

Solution

(a) Let $v_1 = \{a_1, a_2\}$ and let $V_2 = \{b_1, b_2\}$. Draw all possible edges from points in V_1 to points in V_2 (Figure 14-45a).

(b) Let $V_1 = \{a_1, a_2\}$, $V_2 = \{b_1, b_2, b_3\}$, we have Figure 14-45b.

(c) Let $V_1 = \{a_1, a_2\}$, $V_2 = \{b_1, b_2, b_3, b_4\}$. This produces the graph in Figure 14-45c.

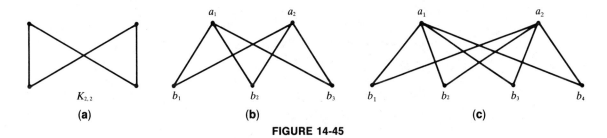

FIGURE 14-45

Note that in a complete bipartite graph the number of edges is always equal to $m \times n$, which is the maximum number of edges possible.

PROBLEM 14-25 A graph G is said to be **planar** if its diagram can be drawn in a two-dimensional plane and no two edges intersect. Show that K_4 is planar but that K_5 is not planar.

Solution To show a graph is planar we need only to show its diagram. K_4 may be drawn as in Figure 14-46a.

K_5 may tried in a similar way, as in Figure 14-46b. Note, however, that vertices v_3 and v_5 are not yet connected. But to do so is impossible. Vertex v_3 is "closed in" by existing edges and v_5 is on the outside. Vertex v_5 must intersect another path or be on another plane to get to v_3.

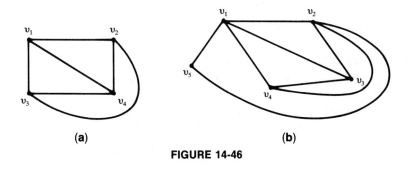

FIGURE 14-46

PROBLEM 14-26 Show that $K_{2,2}$ and $K_{2,3}$ are planar, but $K_{3,3}$ is not planar.

Solution

The diagram of $K_{2,2}$ is easy to reconstruct to be planar (Figure 14-47a).

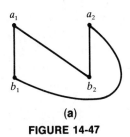

FIGURE 14-47

$K_{2,3}$ may also be reconstructed (Figure 14-47b).

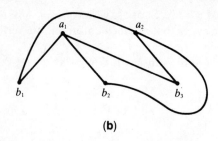

(b)

Now try $K_{3,3}$ (Figure 14-47c).

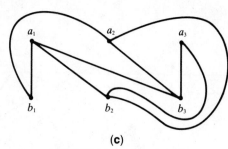

(c)

FIGURE 14-47 *(Continued)*

We first connect a_1 to all the b's, then a_2 to all the b's (follow the plan in Figure 14-47b but loop around a_3). Next, we try to connect a_3 to the b's. We can get from a_3 to b_2 and from a_3 to b_3, but we cannot get from a_3 to b_1 since vertex a_3 is now "closed in" by existing edges. Thus $K_{3,3}$ is not planar.

Two important theoretical results that illustrate how our body of knowledge may be extended are verified by Problems 14-25 and 14-26:

THEOREM 5 K_n is planar for $n \leq 4$. K_n is not planar when $n \geq 5$.

THEOREM 6 $K_{2,n}$ is planar for $n \geq 2$. $K_{m,n}$ is not planar when $3 \leq m \leq n$.

Supplementary Exercises

PROBLEM 14-27   (a) Draw a simple graph with four vertices and five edges. (b) Draw a simple graph with four vertices and six edges. (c) Draw a simple graph with four vertices and seven edges.

PROBLEM 14-28   In the graph in Figure 14-48 determine the degree of each vertex. Show that
$$\sum_{i=1}^{6} = \delta(v_i) = 2(n(E)).$$

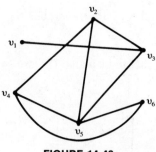

FIGURE 14-48

PROBLEM 14-29   If graph G is as in Figure 14-49, find three distinct subgraphs.

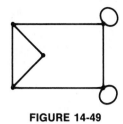

FIGURE 14-49

PROBLEM 14-30 Redraw the graph in Figure 14-50 so that it is planar (no crossings).

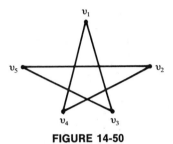

FIGURE 14-50

PROBLEM 14-31 An art gallery has the floorplan shown in Figure 14-51. Is it possible to begin in the hallway at BEGIN and walk thru each door exactly once and end up back in the hallway? *Hint*: Redraw using rooms (and hallway) as vertices and doors as edges. Look for an Euler circuit. If it is possible, draw one.

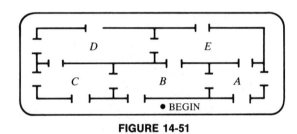

FIGURE 14-51

PROBLEM 14-32 Form the adjacency matrix for the graph in Figure 14-52.

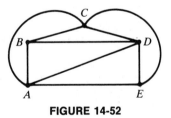

FIGURE 14-52

PROBLEM 14-33 (**a**) Find the degree of each vertex in the graph in Figure 14-53. (**b**) Without actually counting, determine the number of edges in this graph.

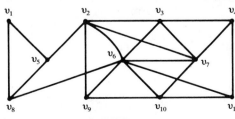

FIGURE 14-53

PROBLEM 14-34   In a newspaper puzzle posed in 1859 by Sir William Rowan Hamilton, the map shown in Figure 14-54 represents 20 cities around the world. The connections between cities represent the possible travel routes. Show that a person could begin at any city, visit each city only once keeping on the established routes, and return to the home city. (You are finding a Hamiltonian circuit.)

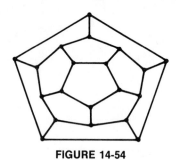

FIGURE 14-54

PROBLEM 14-35   There are five teams in the Red division and five in the Blue division. Find a schedule that will have each team play two games with teams in its own division and two games with teams in the other division.

PROBLEM 14-36   You hold a summer job as a road inspector for County Cork, whose "map" is shown in Figure 14-55. One of your responsibilities is to drive along a given collection of highways to inspect the roads for possible repairs. (a) If you live in town T, is it possible to make a round trip which takes you over each section of highway exactly once? (b) If you moved to town W, would such a round trip (beginning and ending at home) still be possible? Use your pencil to construct a possible trip.

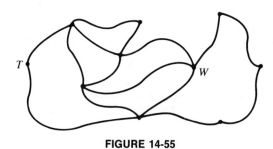

FIGURE 14-55

PROBLEM 14-37   A home that is for sale has the floor plan shown in Figure 14-56. Could the realtor form a path starting from the front door and passing through each room so that each doorway is used once in the tour and an exit is made into the backyard? Try to draw the path with your pencil.

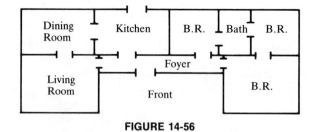

FIGURE 14-56

PROBLEM 14-38   If K_n denotes the complete graph of order n, find the number of edges in (a) K_6 (b) K_{10}.

PROBLEM 14-39   By redrawing the orientation of the vertices show that the graph in Figure 14-57 is bipartite.

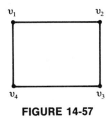

FIGURE 14-57

PROBLEM 14-40 Try to redesign the graph in Figure 14-58, which is bipartite, so that it is planar.

FIGURE 14-58

Answers to Supplementary Exercises

14-27 **(a)** **(b)** 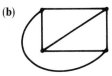 **(c)** Impossible

14-28 $\delta(v_1) = 1$, $\delta(v_2) = 3$, $\delta(v_3) = 3$, $\delta(v_4) = 3$, $\delta(v_5) = 4$, $\delta(v_6) = 2$. The number of edges is 8. $\sum \delta(v_i) = 16$.

14-29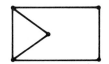

14-30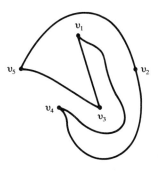

14-31 Yes, it is possible: $\delta(\text{BEGIN}) = 8$, $\delta(A) = 4$, $\delta(B) = 4$, $\delta(C) = 4$, $\delta(D) = 4$, $\delta(E) = 4$.

14-32

	A	B	C	D	E
A	0	1	1	1	1
B	1	0	1	1	0
C	1	1	0	1	1
D	1	1	1	0	1
E	1	0	1	1	0

14-33 **(a)** $\delta(v_1) = 2$, $\delta(v_2) = 5$, $\delta(v_3) = 4$, $\delta(v_4) = 3$, $\delta(v_5) = 3$, $\delta(v_6) = 7$, $\delta(v_7) = 5$, $\delta(v_8) = 3$, $\delta(v_9) = 3$, $\delta(v_{10}) = 4$, $\delta(v_{11}) = 3$

(b) $\sum_1^{11} \delta(v_i) = 42 = 2n(E)$, so $n(E) = 21$

14-35

14-36 (a) Yes, each vertex has even degree so an Euler circuit exists.
(b) Yes, if you moved to W, it is still possible and would include an intermediate pass through town during the trip.

14-37 No. If each room is a vertex and each door is an edge, then the odd degree of vertex Foyer and vertex Kitchen do not allow such a path.

14-38 $n(E) = \dfrac{n(n-1)}{2} = \dbinom{n}{2} = {}_nC_2$. So (a) In K_6, $n(E) = \dfrac{6 \times 5}{2} = 15$; (b) In K_{10}, $n(E) = \dfrac{10 \times 9}{2} = 45$.

14-39

14-40 Impossible.

15 TREES

15-1. Definitions and Properties

A path in an undirected graph (see Chapter 14) is called a **cycle** if it begins and ends at the same vertex and no edges in the graph are repeated. A graph that does not contain any cycles is called *acyclic*. Also, a graph is said to be *connected* if every pair of vertices can be joined by a path.

- A graph G with vertex set V and edge set E is a **tree** if the graph is connected and acyclic.

EXAMPLE 15-1 Construct a tree with (**a**) three vertices, (**b**) five vertices, (**c**) thirteen vertices.

Solution Be sure all the vertices are connected and that there are no cycles (Figure 15-1).

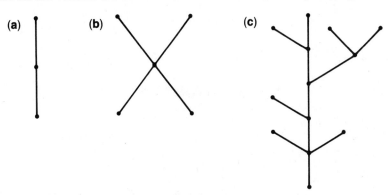

FIGURE 15-1

EXAMPLE 15-2 Create three graphs that are not trees.

Solution To *not* be a tree, a graph may contain a cycle (Figure 15-2a), have loops or duplicate edges (Figure 15-2b), or be disconnected (Figure 15-2c).

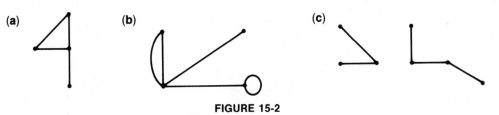

FIGURE 15-2

- An acyclic graph—whether or not it's connected—is usually called a **forest**. Hence each component of a forest is a tree.

EXAMPLE 15-3 Construct a forest.

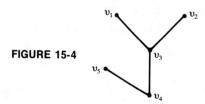

Solution Figure 15-3 shows a forest of three trees.

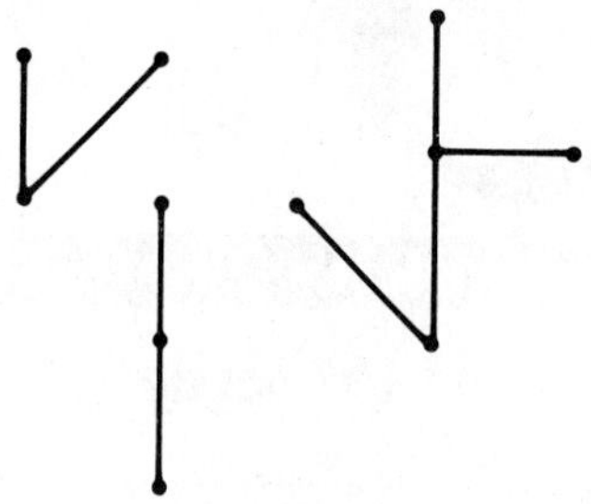

FIGURE 15-3

We can immediately observe several properties of trees:

(1) In a tree, there is exactly one path that connects vertex v_i to vertex v_j.
(2) If m is the number of edges and n is the number of vertices, then $m = n - 1$.
(3) Removing any edge from a tree makes a disconnected graph, which is not a tree.
(4) Adding an extra edge to a tree creates a cycle in the graph.
(5) Every tree with more than one edge has at least two vertices of degree 1.

EXAMPLE 15-4 Illustrate the five properties of trees.

Solution First form a tree having five vertices (Figure 15-4).

FIGURE 15-4

(1) Observe that for any two vertices chosen, there is exactly one path that joins them.
(2) For the five vertices, we do have four edges.
(3) Removing, say, edge $\{v_3, v_4\}$ does disconnect the graph.
(4) Adding, say, $\{v_1, v_5\}$ does create a cycle $(v_1, v_3, v_4, v_5, v_1)$.
(5) Vertices v_1, v_2, and v_5 all have degree 1.

There are many ways in which trees may be used to help analyze mathematical problems. For example, a tree may be formed to illustrate the possible outcomes when a fair coin is tossed twice (Figure 15-5). Because a tree offers exactly one path from the beginning of a test to each of the final outcomes, a tree may be used to help in calculating probabilities (see Chapter 11).

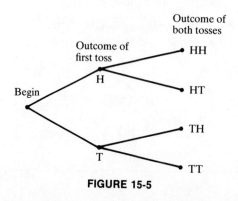

FIGURE 15-5

EXAMPLE 15-5 You are given three containers numbered I, II, and III. In container I there are 2 red balls and 1 blue ball. Container II has 3 red and 2 blue balls, while container III has 1 red and 2 blue balls. You select a container at random, then select one ball at random from that container. What is the probability that the ball is red?

Solution Form the tree, and on each edge (branch) indicate the probability of that event's happening given the number of choices available in each container (Figure 15-6). The probability that a red ball is chosen is the sum of the probabilities on the "red branches." This is

$$\frac{1}{3} \times \frac{2}{3} + \frac{1}{3} \times \frac{3}{5} + \frac{1}{3} \times \frac{1}{3} = \frac{2}{9} + \frac{3}{15} + \frac{1}{9} = \frac{8}{15}$$

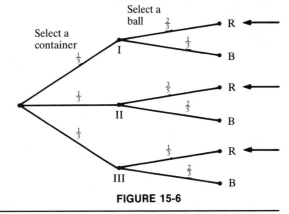

FIGURE 15-6

Example 15-5 shows that any sequential decision-making process is apt to form a tree.

15-2. Spanning Trees

Suppose that G is a connected graph with vertex set V and edge set E. A **spanning set** for G is a subset $\bar{E}$ of E such that another graph H with vertex set V and edge set $\bar{E}$ is connected. That is, a spanning set H is a connected subset of G that still covers all the vertices of G. If the graph $H = \{V, \bar{E}\}$ is a tree, then it is called a **spanning tree** for G. In other words, a spanning tree of G is an acyclic, connected subset of G that covers all the vertices of G.

To create a spanning tree of a given graph G, we remove all edges of G that create cycles, being careful to keep the graph connected. The spanning tree of a graph with n vertices will have $n - 1$ edges.

EXAMPLE 15-6 For the graph in Figure 15-7, find a spanning tree.

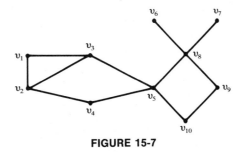

FIGURE 15-7

Solution If the graph were a tree, it would be its own spanning tree. But since this graph has cycles, it is not a tree; and thus we must remove edges in order to get a tree, which will also be a spanning tree. The edges to be removed are those that are part of cycles. We remove edges $\{v_1, v_3\}$, $\{v_4, v_5\}$, $\{v_5, v_{10}\}$ to obtain the spanning tree in Figure 15-8. To verify it's a tree, note that it is connected and has no cycles,

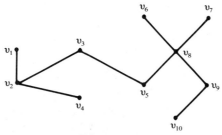

FIGURE 15-8

that each pair of vertices is joined by a path, and that if any additional edge were to be removed it would no longer be a tree. It's a spanning tree of the graph in Figure 15-7 because it spans all the vertices of the original graph. Can you create another spanning tree of this graph?

- Every connected graph has at least one spanning tree.

EXAMPLE 15-7   Find two different spanning trees for the graph in Figure 15-9.

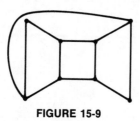

FIGURE 15-9

Solution   Remove edges so that there are no cycles. Two possible trees are shown in Figure 15-10.

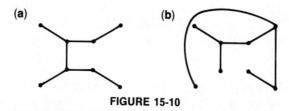

FIGURE 15-10

EXAMPLE 15-8   On the two spanning trees in the solution to Example 15-7 (Figure 15-10), put letter labels on the vertices and show that they are distinctly different spanning trees.

Solution

Label the one in Figure 15-10a arbitrarily (see Figure 15-11a). Note that vertices f and g have degree three and that they have an edge joining them. Only two vertices in Figure 15-11b have degree three and they are not adjacent. Thus these are distinctly different spanning trees.

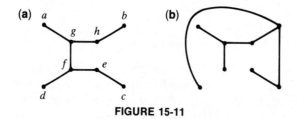

FIGURE 15-11

15-3. Minimal Spanning Trees

Say we have a graph G whose edges are **weighted**; that is, they are given different values to represent, say, the cost of various parts of a construction project. We want to form a **minimal spanning tree**, a tree that spans G by using the combination of edges that has the least total weight value.

An algorithm may be constructed to find the minimal spanning tree T that begins at a prechosen vertex. Let the vertices be denoted as $v_1, v_2, v_3, \ldots, v_n$. Here's the general form of the algorithm:

MINIMAL SPANNING TREE ALGORITHM

1. Select vertex v_1 to be part of tree T.

2. If T has $n - 1$ edges, the tree is complete.

3. Survey all of the edges not yet chosen for inclusion into T that are incident on a vertex in T and that do not complete a cycle if added to T.

4. Pick the edge with minimum weight and add it to T.

5. If more than one edge has this minimum weight, select the edge $\{v_i, v_j\}$ with the smallest i. Refer to this as $\{v_{i_m}, v_j\}$.

6. If two or more edges $\{v_{i_m}, v_j\}$ have equal minimum weights, select the one with smallest j.

7. Return to Step 2.

EXAMPLE 15-9 You have recently purchased an island and wish to develop a resort complex that includes six towns. You need to connect these towns with paved roads but you must keep road construction costs to a minimum. On the map (graph) of your island (Figure 15-12), the construction engineer has drawn all the possible roads linking the towns, and has weighted them with the costs involved in building each section (in thousands of dollars). Create a minimal (cost) spanning tree that links all the towns with the least construction expense.

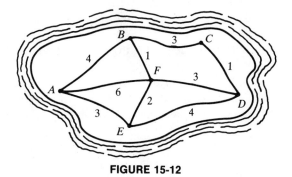

FIGURE 15-12

Solution To show that different costs may be incurred, we give two distinct spanning trees and compute their total costs (Figure 15-13). To find the *minimal* spanning tree, we will follow the algorithm to see if a still lower total cost than that in Figure 15-13b could be obtained. Let's use A as the beginning point. Examining Figure 15-12, choose the edge incident at A with smallest weight. This is $\{A, E\}$, with weight 3.

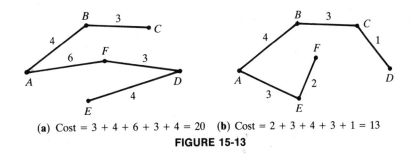

(a) Cost $= 3 + 4 + 6 + 3 + 4 = 20$ (b) Cost $= 2 + 3 + 4 + 3 + 1 = 13$

FIGURE 15-13

Now look at edges incident on A or on E that do not complete a cycle. The candidates are $\{A, B\}$, $\{A, F\}$, $\{E, F\}$, $\{E, D\}$. The one with the minimum weight is $\{E, F\}$, with weight 2.

Now look at edges incident on A or on E or on F that do not complete a cycle. They are $\{A, B\}$, $\{E, D\}$, $\{F, D\}$, $\{F, B\}$. The one with the minimum weight is $\{F, B\}$, with weight 1.

Now look at all edges incident on A or E or F or B that do not complete a cycle. They are $\{E, D\}$, $\{F, D\}$, and $\{B, C\}$. Two of these have weight 3. Choose the one with "smaller" first letter, $\{B, C\}$.

Now look at all edges incident on A or E or F or B or C that do not complete a cycle. They are $\{E, D\}, \{F, D\}, \{C, D\}$. The one with smallest weight is $\{C, D\}$, with weight 1.

We now have edges $\{A, E\}, \{E, F\}, \{F, B\}, \{B, C\}, \{C, D\}$—five edges for six vertices. Step 2 of the algorithm now holds, and we have a complete minimal spanning tree. The total cost is 10.

Figure 15-13b, with cost 13, may be modified to become the minimal spanning tree. Just replace edge AB by edge BF and the "cost-10 tree" is obtained.

EXAMPLE 15-10 An oil company has eight storage tanks which need to be interconnected with pipelines. The only requirement is that the company must have a way to move oil from any tank to tank A. And, of course, the installation costs must be kept to a minimum. The costs of installing the connections are shown on the diagram in Figure 15-14. Find the minimal (cost) spanning tree having vertex A as the root vertex. (*Hint:* Use a red pencil to color in the edges as they are added. We need seven edges.)

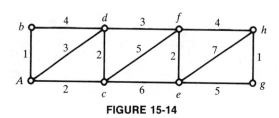

FIGURE 15-14

Solution From A, find the minimum cost edge. It is $\{A, b\} = 1$.

From vertices A and b find the incident edge that does not complete a cycle but that has the minimum cost. It is $\{A, c\} = 2$.

From vertices A, b, and c, find the incident edge that does not complete a cycle but that has the minimum cost. It is $\{c, d\} = 2$.

From vertices A, b, c, and d repeat the analysis. We pick up $\{d, f\} = 3$.

From vertices A, b, c, d, and f repeat the analysis. We pick up $\{f, e\} = 2$.

From vertices A, b, c, d, f, and e repeat the analysis. We pick up $\{f, h\} = 4$.

The final analysis picks up $\{h, g\} = 1$.

The minimal (cost) spanning tree (Figure 15-15) has a total cost $C = 1 + 2 + 2 + 3 + 2 + 4 + 1 = 15$.

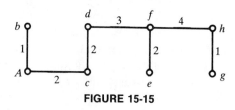

FIGURE 15-15

EXAMPLE 15-11 A graph G has seven vertices and pairs of these are connected by edges whose weights are shown in the following matrix. If two vertices are not connected, the entry is designated with a 0.

$$
G = \begin{array}{c|ccccccc}
 & A & B & C & D & E & F & G \\
\hline
A & 0 & 3 & 1 & 2 & 0 & 6 & 4 \\
B & 3 & 0 & 2 & 0 & 1 & 2 & 3 \\
C & 1 & 2 & 0 & 5 & 3 & 2 & 0 \\
D & 2 & 0 & 5 & 0 & 4 & 1 & 3 \\
E & 0 & 1 & 3 & 4 & 0 & 2 & 5 \\
F & 6 & 2 & 2 & 1 & 2 & 0 & 3 \\
G & 4 & 3 & 0 & 3 & 5 & 3 & 0
\end{array}
$$

(a) Draw graph G as a diagram.

(b) Beginning at vertex A, use the algorithm to find a minimal spanning tree for graph G.

Solution   The graph is shown in Figure 15-16 (see Section 14-2 for how to convert a matrix into a diagram). We have seven vertices so we need to find six edges for the minimum spanning tree.

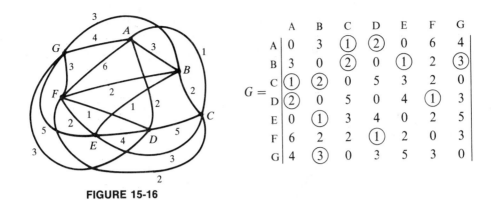

$$G = \begin{array}{c|ccccccc} & A & B & C & D & E & F & G \\ \hline A & 0 & 3 & ① & ② & 0 & 6 & 4 \\ B & 3 & 0 & ② & 0 & ① & 2 & ③ \\ C & ① & ② & 0 & 5 & 3 & 2 & 0 \\ D & ② & 0 & 5 & 0 & 4 & ① & 3 \\ E & 0 & ① & 3 & 4 & 0 & 2 & 5 \\ F & 6 & 2 & 2 & ① & 2 & 0 & 3 \\ G & 4 & ③ & 0 & 3 & 5 & 3 & 0 \end{array}$$

FIGURE 15-16

Begin in row A of the matrix and find the smallest weighted value. This is edge $\{A, C\}$. Trace over the edge $\{A, C\}$ on Figure 15-20 in red pencil. Circle the matrix entries as the edge is added, two circles for each edge. Now in rows A and C find the smallest number (not one already used and not one that completes a cycle). There are three 2's. Picking the one with "smallest" first letter, we include edge $\{A, D\}$. Now in rows A, C, and D find the smallest number (not already circled and not one that completes a cycle). A 1 occurs for edge $\{D, F\}$, so we include edge $\{D, F\}$. Now in rows A, C, D, and F find the smallest number. There are five 2's not yet used. Pick the one with the "smallest" first letter, $\{C, B\}$. (Notice that we couldn't use $\{C, F\}$ because it would create a cycle.) Looking in rows A, B, C, D, and F we find a 1 for edge $\{B, E\}$. Include this.

We now have five edges. One more is needed so that we can include vertex G in the spanning tree. Looking in rows A, B, C, D, E, and F we find lots of 2's. The "smallest" first letter edge is $\{B, F\}$, but it forms a cycle. Don't use it. Next "smallest" is $\{C, F\}$, it too forms a cycle. Next is $\{E, F\}$. It forms a cycle. So none of the edges of weight 2 are usable. Go on to the edges of weight 3. We finally find that edge $\{B, G\}$ has minimal weight and does not form a cycle. Adding the circled weights of the matrix entries above the main diagonal gives the total weight value 10 for the minimal spanning tree. A pictorial of the minimal spanning tree makes the construction clear (Figure 15-17). Note that all the vertices are included and there are no cycles.

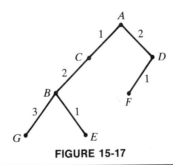

FIGURE 15-17

15-4. Rooted Trees

In a directed graph, the **indegree** of a vertex is the number of edges that terminate at that vertex. The **outdegree** is the number of edges that begin or emanate from the vertex. A **rooted tree** is a tree with a unique vertex r, called the **root vertex**, which has zero indegree. In simple terms, all line segments (edges) in a rooted tree are directed away from the root. The root is the only vertex with no edges coming to it. When all of the directed edges point down, we actually may draw the rooted tree without the directing arrowheads.

agreement: In this text, all rooted trees will be drawn "upside down." The root will be at the top and the branches will point down.

EXAMPLE 15-12   Create a digraph of rooted tree with 13 vertices. Then reproduce it as an undirected graph.

Solution   Select an arbitrary point, say, *A*, as the root vertex, and place other vertices so that the edges point downward and all of the vertices are reached by some path (Figure 15-18a). To convert it to an undirected graph, simply replace each directed edge with an undirected edge (Figure 15-18b).

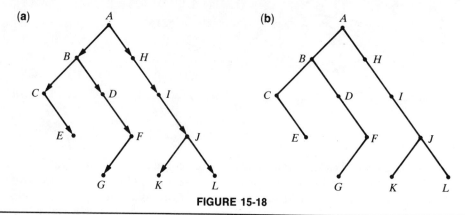

FIGURE 15-18

In a rooted tree, if there is an edge from vertex *v* to vertex *w*, we say that *v* is the **parent** of *w* and *w* is the **child** (son) of *v*. All vertices that can be reached by some "downward" directed path from vertex *v* are the **descendents** of *v*. The vertices on the path from the root to a vertex *v* are the **ancestors** of *v*. If a vertex has outdegree zero, it is a **leaf** vertex.

In a tree, all vertices that are not leaves are called **interior vertices**. If every interior vertex has outdegree 2, the tree is a **binary tree**. If every interior outdegree is *m*, it is called an ***m*-ary tree**.

If a path of length *n* exists from the root to a vertex *v*, then vertex *v* is at **level *n***. For example, the root is at level zero. Its immediate descendents (sons) are at level 1. The "grandsons" are at level 2. We usually draw a tree so that vertices of a given level are horizontally aligned.

EXAMPLE 15-13   Examine the trees in Figure 15-19. Describe their properties.

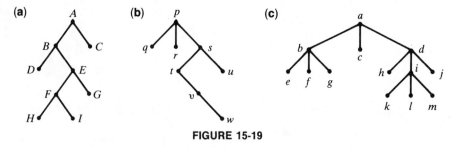

FIGURE 15-19

Solution

(a) This is a rooted tree with its root at vertex *A*. It is a binary tree. It has nine vertices, of which five are leaves and four are interior. There are four levels in this tree (below the root). Paths exist and have lengths 1, 2, 3, and/or 4.
(b) This is a tree rooted at *p*. Every interior root does not have the same outdegree *m*, so it is not *m*-ary. It has eight vertices, of which four are leaves and four are interior. There are four levels below the root.
(c) This is a tree rooted at *a* that is a 3-ary (ternary) tree. It has nine leaves and four interior vertices and has three levels. The maximum path length is 3.

The **height** of a rooted tree is the integer representing the maximum path length in the tree. Tree (a) in Example 15-13 has height 4, tree (b) has height 4, and tree (c) has height 3.

15-5. Traversing Binary Trees

One of the important applications of trees occurs in computer science and involves **traversing** a binary tree, that is, listing all of its vertices in a certain order. There are three kinds of traversals: pre-order, in-order, and post-order. Each has its own algorithm describing its procedure. Two of these lead us to advantageous, parentheses-free methods for evaluating algebraic expressions. The third corresponds to ordinary algebraic evaluation methods.

An arithmetic expression may be represented as a binary tree with the operations as the internal vertices and the numbers or symbols as the leaves. The final operation performed in the expression is represented by the root. The left number or symbol in the operation is on the left branch, and the right number is on the right branch. Thus Figure 15-20 is the arithmetic expression tree for $a + b$.

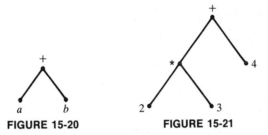

FIGURE 15-20   FIGURE 15-21

If the arithmetic expression is more complicated—that is, if one of the numbers or symbols is replaced by another subexpression—we proceed in the same way. Thus Figure 15-21 is the arithmetic expression tree for $(2 * 3) + 4$. We will refer to the left branch of a root as the **left subtree** of that root and the right branch as the **right subtree** of that root.

The entire left (right) subtree coming from the tree's main root forms a collective expression. If a subtree has subtrees, as in Figure 15-21, each of them is an expression.

Examining the subtrees in a tree reveals the structure of the expression the tree represents. Because each subtree stands for a parenthetical expression, we build the expression as a series of subexpressions nested by parentheses. Thus the expression represented by Figure 15-22 is

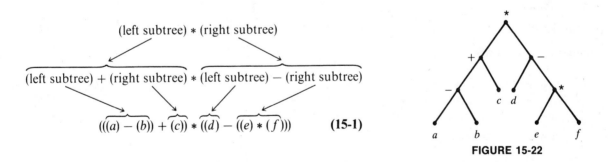

$$((\overline{(a) - (b)}) + \overline{(c)}) * (\overline{(d)} - \overline{(e) * (f)})) \qquad \textbf{(15-1)}$$

FIGURE 15-22

Since most of the parentheses in eq. (15-1) are unnecessary, we actually would have

$$((a - b) + c) * (d - (e * f))$$

as the final expression the tree represents.

If a tree has height zero, its only vertex is its root r. The listing of the vertices would be $\{r\}$. If the binary tree has height 1, then it has three vertices (Figure 15-23).

FIGURE 15-23

There are three listing orders of the vertices in Figure 15-23:

- in-order listing $= \{a, r, b\}$
- post-order listing $= \{a, b, r\}$
- pre-order listing $= \{r, a, b\}$

In all three traversal methods, left always comes before right (note that vertex *a* is before vertex *b* in all the listings), but the position of the root varies.

For trees of higher height we proceed inductively.

A. In-order traversal

To list the vertices of binary trees in an in-order listing, we use this algorithm:

IN-ORDER LISTING ALGORITHM

1. Traverse the left subtree.
2. Include the root.
3. Traverse the right subtree.

EXAMPLE 15-14   Use-in-order traversal to list the vertices of the arithmetic expression tree in Figure 15-24.

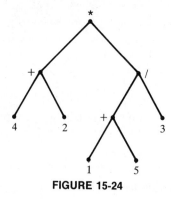

FIGURE 15-24

Solution   Following the algorithm,

1. First list the vertices of the left subtree (Figure 15-25).
   To do so by in-order traversal, list its left vertex first, then the root, then its right vertex. This would be $\{4, +, 2\}$.
2. Now add the root of the whole tree $\{*\}$.
3. Now go through the right subtree by working left, root, right.

FIGURE 15-25

The vertex listing order would be

$$\overbrace{\{4, +, 2,}^{\text{left subtree}} \underset{\text{root}}{*,} \overbrace{1, +, 5, /, 3\}}^{\text{right subtree}}$$

If we now include all of the appropriate parentheses to designate the arithmetic operation gleaned from the diagram, we would have

$$(4 + 2) * ((1 + 5)/3)$$

Without the parentheses in place, the value of the expression in this order is up for grabs—what operation would you do first?)

From Example 15-14 we can see that the operations appear on the interior vertices and on the root and that the variables appear on the leaves of the tree.

The disadvantage of in-order traversal is that it depends on parentheses being appropriately placed in the arithmetic expression. If they are not, we receive an ambiguous description of the tree and the expression it contains.

EXAMPLE 15-15   Form two expression trees for the in-order traversal of the expression $3 - 4 * 2 + 1 * 5 - 2$.

Solution   We may put any operation at the root, but to have it somewhat "balanced" we pick one near the middle of the expression. Then we put everything to the left of the root operation into the left branch and everything to the right of the root operation into the right branch. Keep everything in the in-order mode. Two possible trees are shown in Figure 15-26.

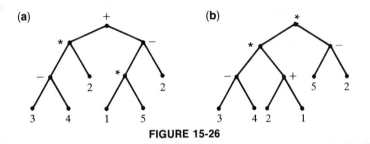

FIGURE 15-26

If we are just given the expression, then parentheses are needed to correctly evaluate it. The relationships indicated by the given parentheses will then determine the tree structure. Conversely, the given tree structure will dictate where the parentheses belong in the final expression. (Unless, of course, the omission of parentheses was deliberate and we are meant to apply the order of operations on the expression. Then this would generate yet another expression tree.)

EXAMPLE 15-16   Following the tree structures for in-order traversal, evaluate the two expressions in the solution to Example 15-15.

Solution   In the tree in Figure 15-26a we get $(3 - 4) * 2$ from the left subtree. We get $(1 * 5) - 2$ from the right subtree. We then put these together by the root operation to get

$$((3 - 4) * 2) + ((1 * 5) - 2) = (-1 * 2) + (5 - 2) = -2 + 3 = 1$$

In the tree in Figure 15-26b we get $(3 - 4) * (2 + 1)$ from the left subtree and $(5 - 2)$ from the right. The whole expression now is

$$((3 - 4) * (2 + 1)) * (5 - 2) = (-1 * 3) * (3) = (-3) * 3 = -9$$

We receive two different trees—and thus two different answers—because the original expression did not have parentheses to guide us. Without parentheses, and assuming the usual order of arithmetic operations—multiplication, division, addition, subtraction, and left to right—the "correct" tree for the expression as given in Example 15-15 would be Figure 15-27.

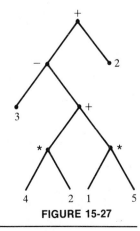

FIGURE 15-27

EXAMPLE 15-17 Given the in-order expression (which includes all of the required parentheses) $((2 + 5) - (3 - (2 * 4))) * (1 - (6 * 2))$, form its binary tree.

Solution Note that two main subexpressions are connected by *. Let x denote $(2 + 5) - (3 - (2 * 4))$ and y denote $(1 - (6 * 2))$. We then have the tree in Figure 15-28a. But expression x is the difference of two subexpressions. Let $s = (2 + 5)$ and $t = (3 - (2 * 4))$. Thus expression x is as in Figure 15-28b, and the total expression now is as in Figure 15-28c. Expression s is easily graphed (Figure 15-28d), and expression t is graphed as in Figure 15-28e. Finally, expression y is graphed as in Figure 15-28f. We put everything together and get Figure 15-28g as the graph of the entire expression.

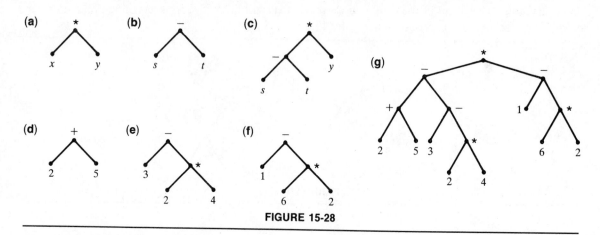

FIGURE 15-28

B. Post-order traversal

To list the vertices of a binary tree in post-order listing we use this algorithm:

POST-ORDER TRAVERSAL ALGORITHM

1. Traverse the left subtree.
2. Traverse the right subtree.
3. Include the root.

EXAMPLE 15-18 Use post-order traversal to list the vertices of the arithmetic expression represented by the trees in Figure 15-29.

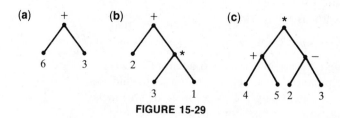

FIGURE 15-29

Solution Remember to go {left, right, root}.

(a) The vertex listing would be $\{6, 3, +\}$
(b) The vertex listing would be $\{2, 3, 1, *, +\}$
(c) The vertex listing would be $\{4, 5, +, 2, 3, -, *\}$

EXAMPLE 15-19 Give the in-order listing and the post-order listing and their respective arithmetic expressions for the binary rooted tree in Figure 15-30.

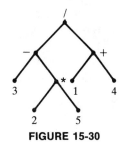

FIGURE 15-30

Solution The in-order listing is $\{3, -, 2, *, 5, /, 1, +, 4\}$. The arithmetic expression is $3 - 2 * 5/1 + 4$. Its value is not determined until parentheses are inserted. We get them from the tree structure

$$(3 - (2 * 5))/(1 + 4) = (3 - 10)/(5) = -\frac{7}{5}$$

The post-order listing is $\{3, 2, 5, *, -, 1, 4, +, /\}$. The arithmetic expression is

$$3\ 2\ 5\ *\ -\ 1\ 4\ +\ /$$

We have no parentheses in this expression. They are not needed. We will find out why shortly.

A post-order listing of the vertices of a binary tree that represents an algebraic expression without commas or parentheses is called the **Reverse Polish Notation** (**RPN**) for the algebraic expression. If we are given a string of symbols that is the RPN for an expression, we can re-create the *unique* algebraic expression that it represents.

Let the elements representing variables be denoted as being from set X. Let the elements representing binary operations be denoted as being from set B. Then the RPN is taken from the set $X \cup B$. The RPN has the following properties: (**1**) The number n of elements (variables and operators) is odd. (**2**) If $n = 2m + 1$, then m elements are from B and $m + 1$ elements are from X. [Remember that we always have one more leaf vertex (variable) than we have interior vertices (operators)]. We follow this algorithm:

REVERSE POLISH NOTATION ALGORITHM

1. If $n = 3$ then the RPN is of the form pq $\square$ where p and q are from X (that is, they are variables) and $\square$ is from B (that is, it's an operator). This is easily converted to the unique algebraic expression (p $\square$ q).

2. If $n > 3$, find the leftmost element in the RPN that is a member of B (that is, find the leftmost operator); call it e_i. Using it and the two members of X that immediately precede it, form the short RPN subexpression $e_{i-2}e_{i-1}e_i$ and replace it with a new variable x. The RPN is now two elements shorter since a group of three is replaced by one. (When you make the final conversion, this shorter RPN subexpression will be represented as $x = e_{i-2}e_ie_{i-1}$.)

3. Repeat Step 2 until the expression has length 3. Then do as in Step 1.

EXAMPLE 15-20 Convert the RPN expression $3\ 2\ 5\ *\ -\ 1\ 4\ +\ /$ from Example 15-19 to its unique arithmetic expression.

Solution The RPN expression is longer than 3 elements, so we move to Step 2 of the algorithm. The leftmost binary operator in the expression is *. Combine it with its two predecessors 2 and 5 to form $x = (2 * 5)$. The "shorter" RPN expression is now $3\ x\ -\ 1\ 4\ +\ /$. It is still longer than 3, so go back to Step 2. The leftmost binary operator is $-$. Combine it with its two predecessors 3 and x to get $y = (3 - x) = (3 - (2 * 5))$. The RPN expression is now y 1 4 $+$ $/$. It is longer than 3. The leftmost binary operator is $+$. Combine it with its two predecessors 1 and 4 to get $z = (1 + 4)$. The expression is now y z $/$. This has length 3. By Step 1 of the algorithm we get (y/z). Substituting back, we have

$$(y/z) = (3 - (2 * 5))/(1 + 4)$$

which is precisely the arithmetic expression described by the binary rooted tree of Example 15-19 (Figure 15-30).

EXAMPLE 15-21 Evaluate the arithmetic expression for the graph in Figure 15-31 using the reverse Polish notation and its algorithm.

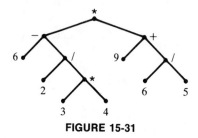

FIGURE 15-31

Solution We must first get the post-order listing, then go to the RPN expression, then use the algorithm to evaluate it. Begin in the left sub-tree:

$$\{6, 2, 3, 4, *, /, -\}$$

Then in the right sub-tree:

$$\{9, 6, 5, /, +\}$$

Then add in the root to get the final post-order listing:

$$\{6, 2, 3, 4, *, /, -, 9, 6, 5, /, +, *\}$$

The RPN expression is obtained by removing the commas and braces:

$$6\ 2\ 3\ 4\ *\ /\ -\ 9\ 6\ 5\ /\ +\ *$$

We now follow the algorithm steps:

Group:	6 2 (3 4 *) / − 9 6 5 / + *			
Replace:	6 2 x / − 9 6 5 / + *	where $x = (3 * 4)$
Group:	6 (2 x /) − 9 6 5 / + *	
Replace:	6 y − 9 6 5 / + *	where $y = (2/x) = (2/(3 * 4))$
Group:	(6 y −) 9 6 5 / + *	
Replace:	z 9 6 5 / + *	where $z = (6 - y) = 6 - (2/(3 * 4))$
Group:	z 9 (6 5 /) + *	
Replace:	z 9 w + *	where $w = (6/5)$
Group:	z (9 w +) *	
Replace:	z v *	where $v = (9 + 2) = (9 + (6/5))$
Finalize:	f	where $f = z * v$

We finally obtained

$$f = (6 - (2/(3 * 4))) * (9 + (6/5))$$

The value of f is now calculated as

$$f = \left(6 - \frac{2}{12}\right) * \left(\frac{51}{5}\right) = \frac{35}{6} * \frac{51}{5} = \frac{119}{2}$$

If the chip in a hand-held calculator were constructed to do the operations following the RPN evaluation algorithm, then you would just punch calculator buttons in the order of the RPN expression and the unique value would be displayed. You would nevery worry about adding parentheses and/or considering the order of operations when using the calculator.

C. Pre-order traversal

We don't need to present any new ideas for this third method. Just remember that here the algorithm is:

PRE-ORDER TRAVERSAL ALGORITHM

1. List the root first.
2. Traverse the left subtree.
3. Traverse the right subtree.

This operation is not an exact reversal of the post-order of traversal although it has many of the same basic properties. The notation for this pre-order traversal without commas and brackets is called **Polish Prefix Notation** or PPN. This particular form is useful in the theory of mathematical logic.

EXAMPLE 15-22 Find the listing of vertices in pre-order traversal for the binary tree in Figure 15-32.

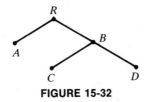

FIGURE 15-32

Solution Follow the order: {root, left, right} throughout. The listing is $\{R, A, B, C, D\}$.

Here's the algorithm for converting a tree to PPN form:

THE POLISH PREFIX NOTATION CONVERSION ALGORITHM

1. If there are three elements, the PPN is of the form $\Box pq$, where p and q are variables and $\Box$ is an operator.
2. For more than three elements, find the rightmost element that is an operator, call it e_i. Using it and the two variables that immediately follow it, form the short PPN expression $e_i e_{i+1} e_{1+2}$ and replace it with the variable x. The PPN is now two elements shorter since three elements have been replaced by one.
3. Repeat Step 2 until the expression has exactly three elements remaining. Then do as in Step 1.

EXAMPLE 15-23 The following is a PPN arithmetic expression. Evaluate it and create the tree for the expression:

$$+ * - 5 1 + 4 3 2$$

Solution We apply algorithmic steps similar to the RPN process except we find the *rightmost* operator and combine it with the two *following* variables.

$$
\begin{array}{llllllll}
 & + & * & - & 5 & 1 & + & 4 & 3 & 2 \\
\text{Group:} & + & * & - & 5 & 1 & (+ & 4 & 3) & 2 \\
\text{Replace:} & + & * & - & 5 & 1 & & x & & 2 \\
\text{Group:} & + & * & (- & 5 & 1) & & x & & 2 \\
\text{Replace:} & + & * & & y & & & x & & 2 \\
\text{Group:} & + & (* & & y & & & x) & & 2 \\
\text{Replace:} & + & & & z & & & & & 2 \\
\text{Group:} & (+ & & & z & & & & & 2) \\
\text{Replace:} & & & & t
\end{array}
$$

where $x = (4 + 3)$

where $y = (5 - 1)$

where $z = (y * x) = ((5 - 1) * (4 + 3))$

where $t = (z + 2)$

Thus

$$t = ((5 - 1) * (4 + 3)) + 2 = (4 * y) + 2 = 28 + 2 = 30$$

The binary tree is shown in Figure 15-33.

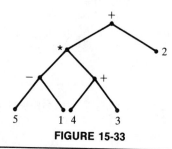

FIGURE 15-33

15-6. Applications

We close with two interesting applications of trees. The first involves alphabetizing a collection of words, possibly those in a sentence or in a list of names. This is an application of a *sorting process*, which is very common in computer science applications.

THIS IS A COLLECTION OF WORDS THAT NEEDS TO BE ALPHABETIZED

We create the following rules.

1. The first word becomes the root word.
2. Each succeeding word W goes to the left branch if it occurs before the root alphabetically and to the right branch if it occurs after the root alphabetically.
3. If another word is encountered on the left or right branch at a lower level, then the test is applied again using the word on the new level as the "root word."
4. Only leaves may be added.
5. An in-order traversal produces the alphabetized list.

Figure 15-34 illustrates the sequence of building the alphabet tree for this sentence.

note: In the final tree, we insert "ghost branches" (dotted edges) to complete the binary nature of the tree. The in-order traversal now provides the following listing of the vertices (ellipses stand in for the ghost branches):

{..., A, ALPHABETIZED, BE,..., COLLECTION,..., IS, NEEDS, OF, THAT, THIS, TO, WORDS,...}

Removing the commas, brackets, and "ghosts" gives the complete alphabetized list..

The second application involves a **decision tree**, which is an *n*-ary tree whose branches form the bases for decisions at various steps of a process.

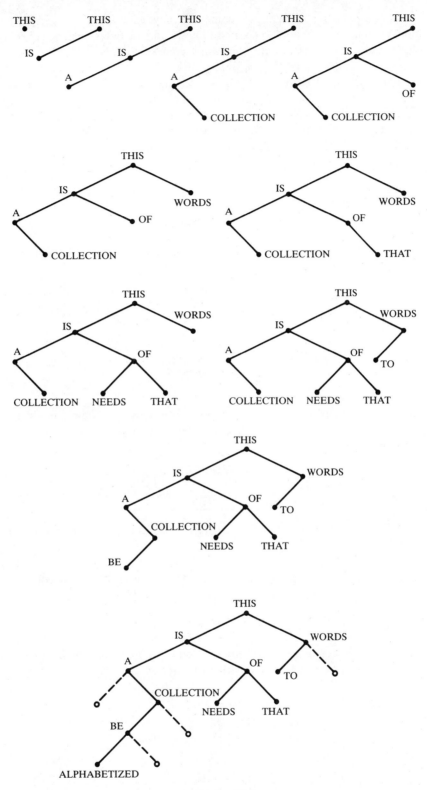

FIGURE 15-34

Suppose we have a collection of eight gold coins, one of which is known to be a lighter, counterfeit one. With a simple balance we are able to compare the weights of two stacks of coins and determine which stack is lighter. Our problem is to determine which coin is the counterfeit one in the fewest number of weighings.

We divide our collection into three equal (or nearly equal) stacks. Each coin is identifiable, and to keep track of individual members, we will number them. Let the stacks be $\{1, 2, 3\}$, $\{4, 5\}$, $\{6, 7, 8\}$. Place the two equal stacks on the balance. If one stack is lighter, it contains the counterfeit coin. If they weigh the same, the omitted stack contains the bad one. Repeat the process. The decision tree for eight coins is shown in Figure 15-35.

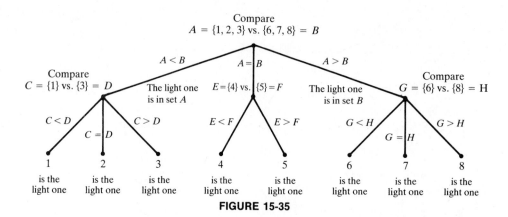

FIGURE 15-35

We need at most two weighings to determine the lighter coin. This is the height of the tree.

With n coins in the collection and one lighter one amongst them, the number of weighings w is established by the inequality string

$$3^{w-1} < n \leq 3^w$$

EXAMPLE 15-24 Form the decision tree for locating a light counterfeit coin in a collection of twelve coins.

Solution Assume they are numbered 1 through 12. Make three stacks of 4 coins each. Set up a systematic approach (Figure 15-36). Since

$$3^2 < 12 \leq 3^3$$

we have a maximum of three weighings and a minimum of two weighings. Note that one counterfeit coin out of a collection of 27 coins could be found with only three weighings.

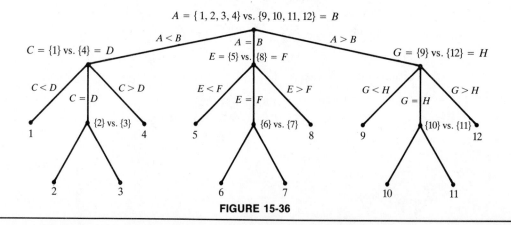

FIGURE 15-36

SUMMARY

1. A tree is a special kind of graph. It is connected and contains no cycles. The number of edges is one less than the number of vertices. There is exactly one path that connects any pair of vertices.
2. Trees have applications in diverse areas. These include probability, chemistry, mathematical logic, sorting and searching, computer design, and decision making.
3. In a graph, a spanning tree is made up of all of the vertices of the graph and a subset of the edges of the graph, such that these two sets form a tree.
4. If the edges of a graph have weights attached to them, then a minimum spanning tree is a spanning tree whose edges have the minimum possible total weight.
5. A rooted tree is simply a tree in which one vertex is designated as the root. This root vertex must have indegree zero. A rooted tree is drawn in "inverted position," with the root at the top and the branches below. The leaves are at the ends of branches and they have outdegree zero.
6. The height of a tree is the length of the longest path from the root to a vertex below it.
7. Methods of listing the vertices of a binary tree produce important applications in mathematics and computer science. The three main listing methods are pre-order, in-order, and post-order.

RAISE YOUR GRADES

Can you . . . ?

☑ form a tree with n vertices and $n - 1$ edges
☑ demonstrate four distinct properties of a tree
☑ follow an algorithm to find a minimum spanning tree in a graph
☑ recite the definitions for at least eight different properties of a rooted tree
☑ distinguish between the three main methods of vertex traversal for binary trees
☑ convert an arithmetic expression tree to its corresponding PPN and RPN forms
☑ form a tree from either a PPN or RPN expression
☑ list at least four different areas of application where trees are important

SOLVED PROBLEMS

Definitions and Properties

PROBLEM 15-1 Construct a tree with 13 vertices and 12 edges.

Solution Remember that a tree must have no cycles and must be a complete graph. Figure 15-37 is one of many possibilities.

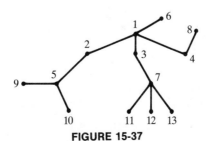

FIGURE 15-37

PROBLEM 15-2 Construct a forest having four trees in it.

Solution Recall that a forest is simply a collection of trees, as in Figure 15-38.

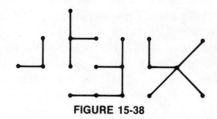

FIGURE 15-38

PROBLEM 15-3 Construct a binary tree to form the matches for an eight-person single elimination tennis tournament.

Solution See Figure 15-39.

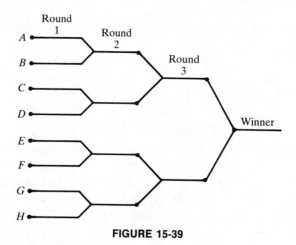

FIGURE 15-39

PROBLEM 15-4 Construct all possible trees having five vertices.

Solution There are only three possible ways to connect 5 vertices without cycles (Figure 15-40). Any other construction would be a modified copy of one of these.

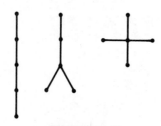

FIGURE 15-40

PROBLEM 15-5 Given six vertices with degrees $\delta(v_1) = 2$, $\delta(v_2) = 1$, $\delta(v_3) = 2$, $\delta(v_4) = 3$, $\delta(v_5) = 1$, and $\delta(v_6) = 1$ is it possible to construct a tree?

Solution Since the sum of the degrees is twice the number of edges (Section 14-2), we have

$$\sum_{i=1}^{6} \delta(v_1) = 2n(E)2 + 1 + 2 + 3 + 1 + 1 = 10$$

So $n(E) = 5$ (that is, a graph with 10 degrees will have 5 edges). But, for a graph to be a tree, the number of its edges must be one less than the number of its vertices. This holds true for our given vertices [$n(v) = 6$ and $n(E) = 5$], so it *is* possible to construct a tree with the given vertices. Figure 15-41 shows one such diagram.

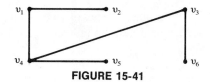

FIGURE 15-41

Spanning Trees

PROBLEM 15-6   For the graph in Figure 15-42 find a spanning tree.

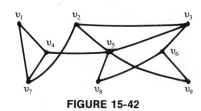

FIGURE 15-42

Solution   We must remove edges that help form cycles but we must leave the graph connected. We remove edges $\{v_1, v_7\}$, $\{v_2, v_7\}$, $\{v_7, v_8\}$, $\{v_5, v_9\}$, $\{v_5, v_3\}$, $\{v_8, v_6\}$. We are left with the graph in Figure 15-43, which has the same nine vertices and eight remaining edges.

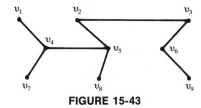

FIGURE 15-43

PROBLEM 15-7   Find two distinct spanning trees for the graph in Figure 15-44.

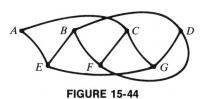

FIGURE 15-44

Solution   The given graph has seven vertices and ten edges. The spanning tree will have the same seven vertices but six edges. We therefore need to remove four edges and leave a connected, acyclic graph. Figure 15-45 shows two possibilities.

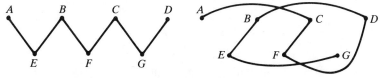

FIGURE 15-45

Minimal Spanning Trees

PROBLEM 15-8   In Western County, Alaska, there are five towns that must be accessible during the winter by having their connecting roads plowed. To save money, the county road commissioners agree to plow a set of roads that allow travel between any two towns—they will not necessarily plow the most direct route, but they will plow only a minimum number of miles. Using the map in Figure 15-46, find the network of roads to be plowed based on a minimum mileage constraint. Begin at the county seat *S*.

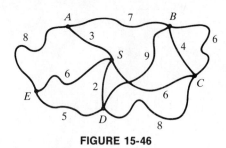

FIGURE 15-46

Solution There are six vertices so we want five edges in the final tree *T*. We begin at vertex *S* and include it in the tree. Of all the edges incident at *S* the one with smallest weight is edge $\{S, D\}$. It is included into the tree *T*. Of all edges incident on *S* or *D*, the one with minimum weight is edge $\{S, A\}$. Include it into *T*. Of all edges incident upon *S*, *D*, or *A*, the minimum weight edge is $\{D, E\}$. Include it into *T*. Of all edges incident at *S*, *D*, *A*, or *E* that do not also complete a cycle, the one with minimum weight is $\{S, C\}$. Include it into *T*. We now have four edges. Survey vertices *S*, *D*, *A*, *E*, and *C* and find the minimal weight, noncycle-forming edge to complete the tree. It is edge $\{C, B\}$. The tree now is as shown in Figure 15-47, and the total miles to be plowed is $5 + 2 + 3 + 6 + 4 = 20$ miles.

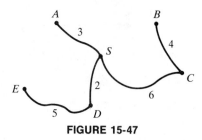

FIGURE 15-47

Rooted Trees

PROBLEM 15-9 Prove the following proposition: An *m*-ary tree which has *j* interior vertices has $n = mj + 1$ vertices in all.

Solution Each of the *j* interior vertices has outdegree *m* (i.e., the vertices each have *m* sons). The graph thus has *mj* sons in all. The root is not a son. Thus we have a total of $mj + 1$ vertices.

PROBLEM 15-10 A binary tree has 18 leaves. How many vertices does it have?

Solution Let *n* be the number of vertices. With *j* of these as interior vertices and 18 as leaves, we have $n = j + 18$. But by the proposition in Problem 15-9 this binary ($m = 2$) tree has $n = 2j + 1$. Thus $j + 18 = 2j + 1$ or $j = 17$. We have 17 interior vertices and 18 leaves for a total of 35 vertices.

PROBLEM 15-11 Draw all binary rooted trees of height 2.

Solution Since every interior vertex has outdegree 2, there is only one tree whose leaves are all at level two (Figure 15-48a). But we can find two more binary trees whose leaves are at other levels (Figure 15-48b and 15-48c).

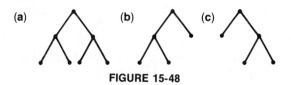

FIGURE 15-48

PROBLEM 15-12   For the tree in Figure 15-49,

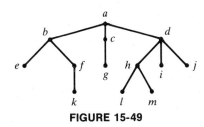

FIGURE 15-49

(a) Find all the descendents of *b*.   (c) Find all the ancestors of *m*.

(b) Find the parent of *i*.   (d) Find all the leaf vertices and interior vertices.

Solution

(a) The descendents of *b* are all those vertices that may be reached by "downward" paths from *b*. They are *e*, *f*, and *k*.

(b) The parent of *i* is *d*. The grandparent is *a*.

(c) The ancestors of *m* are the vertices on the path from the root to vertex *m*. They are *h*, *d*, and *a*.

(d) The leaf vertices are those with zero outdegree. These are *e*, *k*, *g*, *l*, *m*, *i*, and *j*. All the rest of the vertices are interior vertices. These are *a*, *b*, *c*, *d*, *f*, and *h*.

Traversing Binary Trees

PROBLEM 15-13   Construct an expression tree for $(5 + 3) * (6 - (3/2))$.

Solution   Remember that a subtree represents a parenthetical expression and that variables go on the leaf vertices, operators on the internal vertices. The operation $*$ will be the root. Put $(5 + 3)$ in the left branch and $(6 - (3/2))$ in the right branch (Figure 15-50).

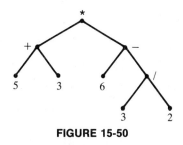

FIGURE 15-50

PROBLEM 15-1   Give the pre-order listing of the vertices for the tree of Problem 15-13 (Figure 15-50) and the PPN expression.

Solution   In pre-order we list {root, left, right}. The root is $\{*\}$. The left branch in pre-order is $\{+, 5, 3\}$. The right branch in pre-order is $\{-, 6, /, 3, 2\}$. The whole tree then is $\{*, +, 5, 3, -, 6, /, 3, 2\}$. The PPN expression then is $*\ +\ 5\ 3\ -\ 6\ /\ 3\ 2$.

PROBLEM 15-15   Form the post-order listing of the vertices for the tree of Problem 15-13 (Figure 15-50) and the RPN expression.

Solution   This time we list {left, right, root}. The left branch in post-order is $\{5, 3, +\}$. The right branch in post-order is $\{6, 3, 2, /, -\}$. The whole tree is $\{5, 3, +, 6, 3, 2, /, -, *\}$ and the RPN expression is then $5\ 3\ +\ 6\ 3\ 2\ /\ -\ *$.

PROBLEM 15-16   The value of the expression in Problem 15-13 is easily found to be $8 * (9/2) = 36$. (a) Verify that the PPN expression of Problem 15-14 gives the same value. (b) Verify that the RPN expression of Problem 15-15 gives the same value.

Solution

(a) Work from right to left:

$$* + 5\ 3\ -\ 6\ /\ 3\ 2$$
$$* + 5\ 3\ -\ 6(/\ 3\ 2) \qquad \text{Let } x = (3\ /\ 2)$$
$$* + 5\ 3\ (-\ 6\ x) \qquad \text{Let } y = (6 - x) = 6 - 3/2 = 9/2$$
$$*(+\ 5\ 3)\ y \qquad \text{Let } z = (5 + 3) = 8$$
$$(*\quad z\quad y \qquad \text{Let } t = z * y = 8 * 9/2$$
$$t \qquad\qquad = 36$$

(b) Work from left to right:

$$5\ 3\ +\ 6\ 3\ 2\ /\ -\ *$$
$$(5\ 3\ +)6\ 3\ 2\ /\ -\ * \qquad \text{Let } x = (5 + 3) = 8$$
$$x\qquad 6(3\ 2\ /)-\ * \qquad \text{Let } y = (3\ /\ 2) = 3/2$$
$$x\qquad (6\quad y\quad -)* \qquad \text{Let } z = (6 - y) = 6 - 3/2 = 9/2$$
$$(x\qquad\quad z\qquad *) \qquad \text{Let } t = x * z = 8 * 9/2$$
$$t \qquad\qquad\qquad = 36$$

PROBLEM 15-17   Construct an expression tree for the RPN expression

$$a\ b\ -\ c\ +\ d\ e\ -\ *$$

Solution   We must break this down using replacement. Begin at the leftmost operator and group the operator with its two predecessors.

$$(a\ b\ -)\ c\ +\ d\ e\ -\ *$$
$$(x\qquad\ c\ +)d\ e\ -\ *$$
$$y\qquad (d\ e\ -)*$$
$$(y\qquad z\qquad *)$$
$$t$$

Now we build the tree in steps by reversing the above actions (Figure 15-51).

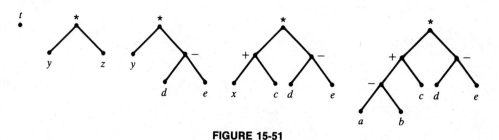

FIGURE 15-51

PROBLEM 15-18   Construct an expression tree for the PPN expression

$$+\ -\ A\ B\ /\ *\ +\ E\ F\ D\ C$$

Solution   Break down the expression using replacements. Begin at the rightmost operation and group the operation with its two successors.

$$+\ -\ A\ B\ /\ *(+\ E\ F)D\ C$$
$$+\ -\ A\ B\ /(*\qquad x\qquad D)C$$
$$+\ -\ A\ B(/\qquad y\qquad C)$$
$$+\ (-\ A\ B)\qquad z$$
$$(+\qquad w\qquad z)$$
$$t$$

Build the tree from its root using the replacement groups as branches (Figure 15-52).

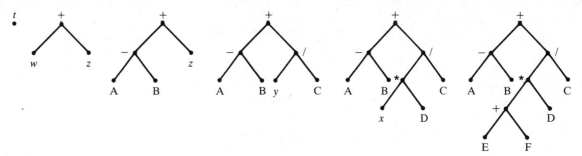

FIGURE 15-52

Applications

PROBLEM 15-19 The customers of XYZ Corp are kept in an alphabetical file by last name. The present tree structure is as shown in Figure 15-53, with an in-order traversal producing the alphabetized list. Add the new customers ADAMS, CAMP, GAGE, ORK, TUNE, SILL, and WILLS to the tree. Do an in-order traversal of the new tree and thus obtain the new alphabetical listing.

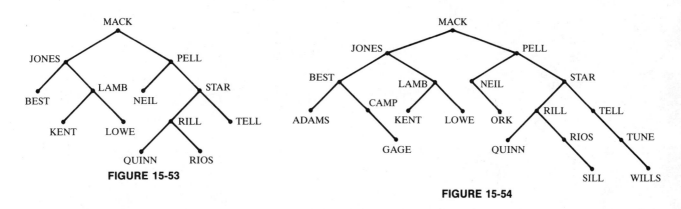

FIGURE 15-53

FIGURE 15-54

Solution Before you look at the answer in Figure 15-54, cover up the figure, copy the tree in Figure 15-53, and put the new customers on it. Remember that you may add only at leaves and that in-order listing goes {left, root, right}. Use alphabetical comparisons for each new customer: Go left if she comes before the "root matrix"; go right if she comes after. Compare your tree with the one we created in Figure 15-54.

PROBLEM 15-20 Find a rooted tree that (**1**) will exhibit all paths through the maze in Figure 15-55 from *X* to *Y*, (**2**) that goes through no gate twice, and (**3**) that does not involve backtracking.

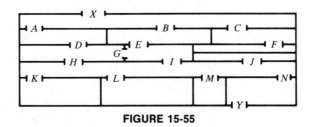

FIGURE 15-55

Solution Begin with *X* as the root and add branches as the paths to the set of potential next gates—the new vertices (Figure 15-56).

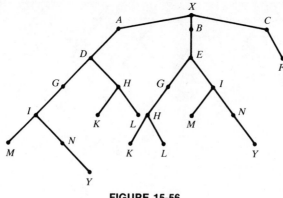

FIGURE 15-56

PROBLEM 15-21 In a collection of five gold commemorative medals, one is deemed to be counterfeit. Using a balance scale, show that the light one of the group may always be found in only two weighings. Could it be done in only one weighing?

Solution Number of medals 1 through 5 and form the three groups $\{1, 2\}$, $\{3\}$, $\{4, 5\}$. Figure 15-57 shows the decision tree. We could find the counterfeit coin in one weighing only if it is the third coin.

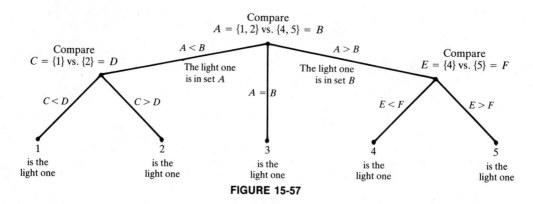

FIGURE 15-57

PROBLEM 15-22 Form the decision tree for showing that one counterfeit coin that is heavier than all the rest may be detected in three weighings of 17 coins.

Solution Number the coins 1 through 17 and make three stacks. The decision tree (Figure 15-58) requires only 3 levels, thus only 3 weighings are necessary.

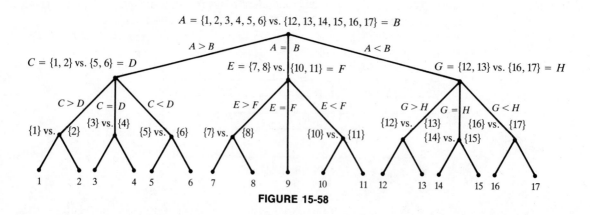

FIGURE 15-58

PROBLEM 15-23 A path through a maze may be modeled by a rooted tree. If the maze begins at B and ends at E, we need to find the paths on the tree whose root is at B that lead to leaf E. Suppose a rat needs to execute the maze shown in Figure 15-59. What are its possible paths?

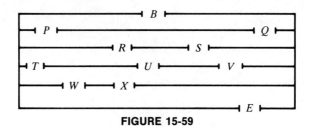

FIGURE 15-59

Solution Each vertex of the tree will be a gate of the maze. Draw an edge from vertex v to vertex w if an open path is available from gate v to gate w (Figure 15-60). There are four possible solution paths: *BPRTWE*, *BPRUXE*, *BQRTWE*, and *BQRUXE*. Using your pencil, draw these paths in the maze.

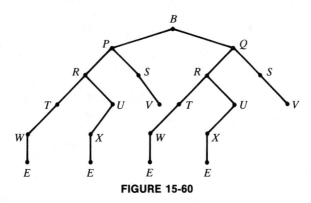

FIGURE 15-60

Supplementary Exercises

PROBLEM 15-24 Construct an expression tree for $(((5-2)*4)+3)/((6-3)+(7-(6/2)))$.

PROBLEM 15-25 Construct an expression tree for $(p+q/r)*((s-t)/v)$.

PROBLEM 15-26 Convert the expressions of **(a)** Problem 15-24 and **(b)** Problem 15-25 to reverse Polish notation.

PROBLEM 15-27 Given the tree in Figure 15-61, find **(a)** the left subtree of vertex D **(b)** the right subtree of vertex F **(c)** the right subtree of vertex B **(d)** the left subtree of vertex C **(e)** the left subtree of vertex F.

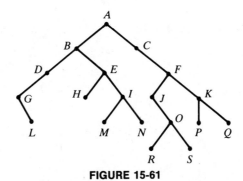

FIGURE 15-61

PROBLEM 15-28   Find the value of the RPN expression 5 2 − 6 * 3 4 + +.

PROBLEM 15-29   Find the value of the RPN expression 3 6 3 2 2 / + * −.

PROBLEM 15-30   Find the value of the PPN expression + * 4 / 8 2 − + 3 1 5.

PROBLEM 15-31   Find the value of the PPN expression * + − +7 3 5 2 4.

PROBLEM 15-32   Construct an expression tree for the PPN expression + * A C − B D.

PROBLEM 15-33   Construct an expression tree for the RPN expression P Q − R + S T * U + −.

PROBLEM 15-34   Construct a tree whose in-order listing is *RVPSMTQWU* and whose post-order listing is *VRSPTWUQM*.

PROBLEM 15-35   Construct a tree with 5 vertices for which the pre-order listing is the same as the in-order listing.

PROBLEM 15-36   For each of the three trees in Figure 15-62, find the pre-order listing of the vertices, the in-order listing of the vertices, and the post-order listing of the vertices.

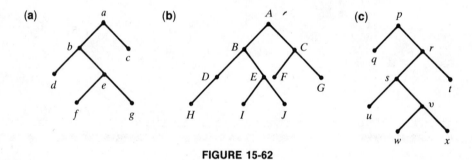

FIGURE 15-62

PROBLEM 15-37   On a piece of paper construct any binary tree. Beginning at the root, draw a curved path around the perimeter of the tree that moves counterclockwise and stays as close to the tree as possible. Make a list of the vertices as they are passed (no repeats). Show that this list is the pre-order listing.

PROBLEM 15-38   Form a method of moving around the perimeter of a tree similar to the method described in Problem 15-37 but which lists the vertices in post-order listing.

PROBLEM 15-39   Form the decision tree for finding, using only a balance scale, one light baseball from a collection of 7 identical looking balls. What is the minimum and maximum number of weighings needed?

Answers to Supplementary Exercises

15-24   **15-25**

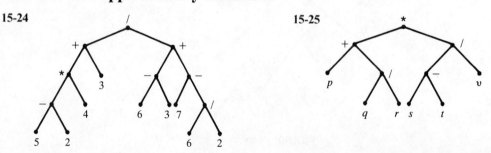

15-26 (a) $5\ 2\ -\ 4\ *\ 3\ +\ 6\ 3\ -\ 7\ 6\ 2\ /\ -\ +\ /$

(b) $p\ q\ r\ /\ +\ s\ t\ -\ v\ /\ *$

15-27 (a)   (b)

(c)   (d) None

(e)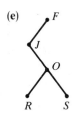

15-28 25

15-29 -21

15-30 15

15-31 28

15-32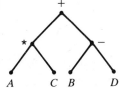

15-33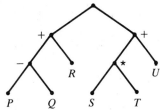

15-34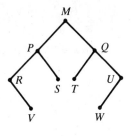

15-35

15-36 (a) Pre-order: *abdefgc*; In-order: *abfegac*;
Post-Order: *dfgebca*.
(b) Pre-order: *ABDHEIJCFG*;
In-order: *HDBIEJAFCG*;
Post-order: *HDIJEBFGCA*.
(c) Pre-order: *pqrsuvwxt*;
In-order: *qpuswvxrt*;
Post-order: *quwxvstrp*.

15-37 *ABDEHJKICFG*

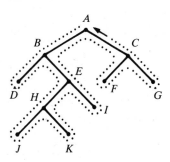

15-39 Minimum is 1. Maximum is 2.

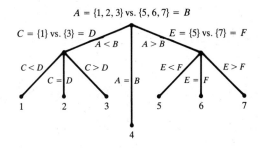

EXAM 3 (Chapters 11–15)

1. **(a)** Construct the sample space for tossing three honest coins. **(b)** Use this to find the probability of obtaining two heads when the three coins are tossed.

2. When rolling a pair of dice, what is the probability of obtaining **(a)** a sum of exactly 3, **(b)** a sum of more than 9, **(c)** a sum of at least 4?

3. Suppose you roll a single die five times. You win a point if the die shows one spot (call it an ace). What is the probability you will get more than 3 points?

4. If a relation on a set $A = \{a, b, c, d\}$ is given by $\mathbf{R} = \{(a, d), (b, b), (c, a), (b, d)\}$, **(a)** find the domain and the range; **(b)** draw the directed graph for $\mathbf{R}$

5. Begin with a set of four elements a, b, c, and d. Construct the subsets that have exactly one element and those that have exactly three elements. Now let X be this collection of sets, with ϕ added in. Form the Hasse diagram of X using the relation $\subseteq$ (proper subset).

6. Let Z denote the set of integers $\{\ldots, -2, -1, 0, 1, 2, 3, \ldots\}$. Form a partition of the integers by constructing the sets $\{5Z, 1 + 5Z, 2 + 5Z, 3 + 5Z, 4 + 5Z\}$. In which set is 46?

7. **(a)** Show that the function $f: \mathbb{R} \to \mathbb{R}$ defined by

$$f(x) = \frac{5x - 2}{3}$$

is one-to-one and onto. **(b)** Find its inverse.

8. Given $f: \mathbb{R} \to \mathbb{R}$ defined by $f(x) = 2x^2 - 3$ and $g: \mathbb{R} \to \mathbb{R}$ defined by $g(x) = 8x + 1$, find **(a)** $(f \circ g)(x)$ and **(b)** $(g \circ f)x$.

9. Consider the graph in Figure E3-1:

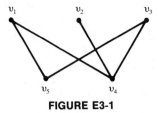

FIGURE E3-1

(a) Form the adjacency matrix. **(b)** Does this graph have an Euler circuit? **(c)** Does it have an Euler path?

10. In Figure E3-2, find, if possible, **(a)** an Euler circuit and **(b)** a Hamiltonian circuit.

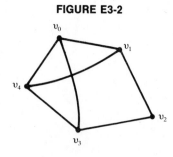

FIGURE E3-2

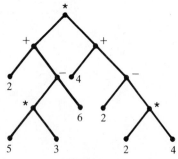

FIGURE E3-3

11. Given the arithmetic expression tree in Figure E3-3,

   (a) find its in-order listing, the arithmetic expression, and its value.
   (b) find its pre-order listing, arithmetic expression, and its value.

12. Find the minimum spanning tree for Figure E3-4 using vertex A as the root vertex.

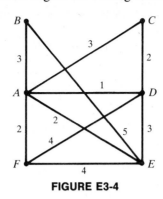

FIGURE E3-4

Solutions to Exam 3

---

1. (a) Since each coin has two possible outcomes, H or T, we have $2^3 = 8$ possible outcomes for the sample space. They are

$$S = \{HHH, HHT, HTH, HTT, THH, THT, TTH, TTT\}$$

(b) The probability of getting two heads is the number of outcomes favoring this result divided by the number of outcomes in the sample space. The favorable results are HHT, HTH, and THH. The probability, then, is 3/8.

2. The sample space is the 6×6 grid of possible outcomes (see Example 11-24). You need to find the favorable ones.

(a) There are only two outcomes of exactly 3: the outcomes $(1, 2)$ and $(2, 1)$. Thus the probability of getting a sum of 3 is 2/36.

(b) The outcome "sum of more than 9" includes all the sums 10, 11, and 12. The three diagonals containing these on the sample space include $(4, 6), (5, 5), (6, 4), (5, 6), (6, 5)$ and $(6, 6)$. This probability is 6/36.

(c) The outcome "sum of at least 4" includes all the sums from 4 through 12. It's easier to find the complement— "sum of 2 or 3." There are three outcomes in the complement and hence 33 outcomes in the favorable space. This probability is 33/36.

3. The is a binomial experiment with the probability of success in a single trial $p = \frac{1}{6}$. We have five tries, so $n = 5$, and we need $P(4) + P(5)$ (the probability of getting 4 points or 5 points). Use eq. (11-5).

$$P(4) = {_5}C_4(\tfrac{1}{6})^4(1 - \tfrac{1}{6})^1 = 5 \cdot (\tfrac{1}{6})^4(\tfrac{5}{6}) = \tfrac{25}{7776}$$

$$P(5) = {_5}C_5(\tfrac{1}{6})^5(1 - \tfrac{1}{6})^0 = 1 \cdot (\tfrac{1}{6})^5 \cdot 1 = \tfrac{1}{7776}$$

$$P(\text{more than 3 points}) = P(4) + P(5) = \tfrac{26}{7776}$$

4. (a) The domain is the set of first elements of the pairs: Domain $= \{a, b, c\}$. The range is the set of second elements of the pairs: Range $= \{a, b, d\}$.

(b) The graph is shown in Figure E3-5.

FIGURE E3-5

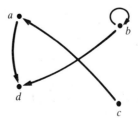

5. The sets are $A = \{a\}, B = \{b\}, C = \{c\}, D = \{d\}, E = \{a, b, c\}, F = \{a, c, d\}, G = \{b, c, d\}$ and, of course, ϕ. Set X will be this collection of sets. The Hasse diagram is shown in Figure E3-6.

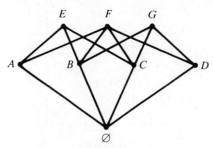

FIGURE E3-6

6. First form $5Z = \{\ldots, -10, -5, 0, 5, 10, 15, \ldots\}$. Then our other sets will be

$$1 + 5Z = \{\ldots, -9, -4, 1, 6, 11, 16, \ldots\}$$

$$2 + 5Z = \{\ldots, -8, -3, 2, 7, 12, 17, \ldots\}$$

$$3 + 5Z = \{\ldots, -7, -2, 3, 8, 13, 18, \ldots\}$$

$$4 + 5Z = \{\ldots, -6, -1, 4, 9, 14, 19, \ldots\}$$

Note that the union of these five sets is the entire set of integers Z. Thus 46 is in only one of them. To find out which one, deduct from 46 enough multiples of 5 to find it in one of the sets above: $46 - 40 = 46 - 8 \cdot 5 = 6$; it is in set $1 + 5Z$.

7. (a) It is one-to-one since if $x_1 \neq x_2$, then

$$f(x_1) = \frac{5x_1 - 2}{3} \quad \text{and} \quad f(x_2) = \frac{5x_2 - 2}{3}$$

and they're not equal. This is apparent when we set

$$\frac{5_1 - 2}{3} = \frac{5x_2 - 2}{3}$$

and multiply by 3, add 2, and divide by 5 on both sides. The function is onto since it maps all of $\mathbf{R}$ onto all of $\mathbf{R}$.

(b) You find the inverse by solving the function

$$y = \frac{5x - 2}{3}$$

for x and interchanging x and y:

$$3y = 5x - 2$$

$$3y + 2 = 5x$$

$$\frac{3y + 2}{5} = x \Rightarrow y = f^{-1}(x) = \frac{3x + 2}{5}$$

8. (a) $(f \circ g)(x) = f(g(x)) = f(8x + 1) = 2(8x + 1)^2 - 3 = 2(64x^2 + 16x + 1) - 3 = 128x^2 + 32x - 1$
(b) $(g \circ f)(x) = g(f(x)) = g(2x^2 - 3) = 8(2x^2 - 3) + 1 = 16x^2 - 23$

9. (a) There are five edges so we will have ten 1's in a 5×5 matrix:

$$
\begin{array}{c c}
 & \begin{array}{c c c c c} v_1 & v_2 & v_3 & v_4 & v_5 \end{array} \\
\begin{array}{c} v_1 \\ v_2 \\ v_3 \\ v_4 \\ v_5 \end{array} &
\left[\begin{array}{c c c c c}
0 & 0 & 0 & 1 & 1 \\
0 & 0 & 0 & 1 & 0 \\
0 & 0 & 0 & 1 & 1 \\
1 & 1 & 1 & 0 & 0 \\
1 & 0 & 1 & 0 & 0
\end{array}\right]
\end{array}
$$

(b) Since both vertex v_2 vertex v_4 have odd degrees (1 and 3, respectively), no Euler circuit exists.

(c) An Euler path does exist from vertex v_2 to vertex v_4 since they are the only two vertices with odd degree.

10. (a) There are four vertices of odd degree, so no Euler circuit exists. **(b)** There are many Hamiltonian circuits. One of them is $v_0, v_3, v_2, v_1, v_4, v_0$.

11. (a) The in-order listing is $\{2, +, 5, *, 3, -, 6, *, 4, +, 2, -, 2, *, 4\}$. The arithmetic expression without parentheses is

$$2 + 5 * 3 - 6 * 4 + 2 - 2 * 4$$

but with appropriate parentheses added by examining the form of the tree, the arithmetic expression becomes

$$(2 + ((5 * 3) - 6)) * (4 + (2 - (2 * 4)))$$

whose value is

$$(2 + 9) * (4 + (-6)) = 11 * (-2) = -22$$

(b) The pre-order listing and arithmetic expression in PPN is $* + 2 - * 5\,3\,6 + 4 - 2 * 2\,4$

To evaluate this expression we insert the appropriate parentheses according to the form of the tree:

$$* + 2 - * 5\,3\,6 + 4 - 2(*\,2\,4)$$
$$* + 2 - * 5\,3\,6 + 4(-2\,8)$$
$$* + 2 - * 5\,3\,6(+ 4\{-6\})$$
$$* + 2 - (* 5\,3)\,6\{-2\}$$
$$* + 2(- 15\,6)\{-2\}$$
$$*(+ 2\,9)\{-2\}$$
$$(* 11\{-2\}$$
$$-22$$

12. The root vertex A is chosen. Since we need 5 edges to complete the spanning tree, we begin by surveying all edges incident on A. The one with the lowest (minimal) value is $\{A, D\}$. We include this value in our tree.

Of all remaining edges incident on A or D, we have a lowest weight 2 for $\{A, F\}$, $\{A, E\}$, and $\{D, C\}$. From these we select the edge with smallest first letter (A), then the one with the smallest second letter (E); so we include $\{A, E\}$.

Of all the remaining edges incident on A, D, or E, the only edges with lowest weight 2 are $\{A, F\}$ and $\{D, C\}$. Since edge $\{A, F\}$ has the smaller first letter, we include $\{A, F\}$ in our tree.

Of all the remaining edges incident on A, D, E, or F, the only edge with weight 2 is $\{D, C\}$, which we include as our fourth edge.

Now we survey all edges incident on A, C, D, E, or F and find weight 3 on $\{A, B\}$, $\{A, C\}$, $\{D, E\}$, and $\{D, F\}$. The latter three all form cycles, but edge $\{A, B\}$ is a likely candidate. Include it as the last edge. The minimal spanning tree (Figure E3-7) has total weight 10.

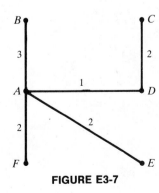

FIGURE E3-7

FINAL EXAM

1. Design an algorithm to calculate the area of an equilateral triangle whose side length is x.
2. Convert the base 2 number 11010110110001_2 to (a) base 8, (b) base 10, and (c) base 16.
3. Evaluate the expression $5 * (-4) \uparrow 4 + 8 \uparrow (0.5 \uparrow -2) + 9/2 \uparrow 0$ and then write its result in scientific notation.
4. Solve the equations:

$$\text{(a) } 4(2t - 6) + 3(5 - t) = 5(4t + 1) \qquad \text{(b) } 2x^2 + x - 6 = 0 \qquad \text{(c) } \begin{cases} x + 3y = 8 \\ -2x + 5y = 6 \end{cases}$$

5. For the matrix $\mathbf{A} = \begin{bmatrix} 5 & 4 \\ -3 & -2 \end{bmatrix}$ find (a) $\mathbf{A}^T$, (b) $\det \mathbf{A}$, (c) $\mathbf{A}^2$, and (d) $\mathbf{A}^{-1}$.

6. (a) Show that the expression $(p \vee q) \wedge [(\sim p) \wedge (\sim q)]$ is a contradiction.
   (b) Show that the expression $(\sim p \wedge q) \vee p \vee [\sim(p \vee q)]$ is a tautology.

7. Prove the following conjecture:

   "I am not going swimming or I will play racquetball. I will play basketball or I will go swimming. But I will not play racquetball. Therefore I will play basketball."

8. Let set $S = \{A, B, C\}$, $T = \{+, *\}$. Find (a) $S \times T$, (b) $|S \times T|$, and (c) the power set of S.
9. Suppose we are given the circuit diagram in Figure EF-1.

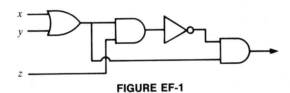

FIGURE EF-1

   (a) Find the Boolean function $f(x, y, z)$.
   (b) Simplify $f(x, y, z)$ algebraically.
   (c) Construct the truth table for the simplified $f(x, y, z)$.
   (d) Draw the circuit diagram for the simplified case.

10. (a) A child has a penny, nickel, dime, quarter, and half-dollar. How many different two-coin payouts can be made?
    (b) On a special shelf, a bookstore owner wants to display five out of eight new books just received. In how many unique ways can this be done?

11. Find the probability of

    (a) obtaining an 8 or a Club on a single draw from an ordinary 52-card deck.
    (b) being dealt all four Aces in a five-card poker hand.
    (c) selecting two red chips from a box containing five red, four white, and three blue chips if the two chips are drawn simultaneously.

12. Let $X = \{a, b, c\}$. Define a relation on X by

$$R = \{(a, a), (b, b), (b, c), (c, b), (a, c)\}$$

    (a) Construct the matrix representation of R.
    (b) Is the relation reflexive, symmetric, and/or transitive?

13. Define a function $f: \mathbb{R} \to \mathbb{R}$ by $f(x) = \sqrt{x - 2}$.

    (a) Is f one-to-one? Onto?
    (b) Find f^{-1} on a restricted domain.

14. If we have the graph adjacency matrix
$$\begin{array}{c} \\ a \\ b \\ c \\ d \end{array} \begin{array}{c} \begin{array}{cccc} a & b & c & d \end{array} \\ \begin{bmatrix} 0 & 1 & 0 & 1 \\ 1 & 0 & 1 & 1 \\ 0 & 1 & 0 & 1 \\ 1 & 1 & 1 & 0 \end{bmatrix} \end{array},$$

(a) What is its edge set?

(b) Draw the graph.

(c) Does it have an Euler path?

15. For the tree in Figure EF-2,

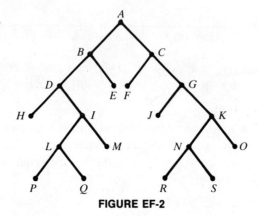

FIGURE EF-2

(a) Find the left subtree of vertex D.

(b) Find the post-order listing of the vertices.

Solutions to Final Exam

1. A natural language algorithm would be

   1. Input x.
   2. Calculate $h = $ SQRT(3) times $x/2$.
   3. Calculate $A = \frac{1}{2}xh$.
   4. Output A.

2. (a)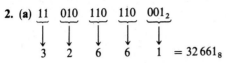
   $$11 \quad 010 \quad 110 \quad 110 \quad 001_2$$
   $$\downarrow \quad \downarrow \quad \downarrow \quad \downarrow \quad \downarrow$$
   $$3 \quad 2 \quad 6 \quad 6 \quad 1 \quad = 32661_8$$

   (b)
   $$1 \quad 1 \; 0 \; 1 \; 0 \; 1 \quad 1 \; 0 \; 1 \quad 1 \; 000 \; 1$$
   $$\downarrow \quad \downarrow \quad \downarrow \quad \downarrow \quad \downarrow \quad \downarrow \quad \downarrow \quad \downarrow$$
   $$2^{13} + 2^{12} + 2^{10} + 2^8 + 2^7 + 2^5 + 2^4 + 2^0$$
   $$= 8192 + 4096 + 1024 + 256 + 128 + 32 + 16 + 1 = 13745_{10}$$

   (c)
   $$11 \quad 0101 \quad 1011 \quad 0001$$
   $$\downarrow \quad \downarrow \quad \downarrow \quad \downarrow$$
   $$3 \quad 5 \quad B \quad 1 \quad \rightarrow 35B1_{16}$$

3. Use the PEMDAS order—work parentheses first, then exponentiation, then multiplication, division, addition, and subtraction, always working left to right.
   $$5 * (-4) \uparrow 4 + 8 \uparrow (0.5 \uparrow -2) + 9/2 \uparrow 0$$
   $$5 * (-4) \uparrow 4 + \quad 8 \uparrow (4) \quad + 9/2 \uparrow 0$$
   $$5 * \quad (256) \quad + \quad 4096 \quad + \quad 9/1$$
   $$1280 \quad + \quad 4096 \quad + \quad 9 \quad = 5385 = 5.385 \times 10^3$$

4. (a) $4(2t - 6) + 3(5 - t) = 5(4t + 1)$
   $$(8t - 24) + (15 - 3t) = 20t + 5$$
   $$5t - 9 = 20t + 5$$
   $$-14 = 15t$$
   $$t = -14/15$$

   (b) Use the quadratic formula:
   $$x = \frac{-1 \pm \sqrt{1^2 - 4(2)(-6)}}{2 \times 2} = \frac{-1 \pm \sqrt{1 + 48}}{4} = \frac{-1 \pm 7}{4}$$

   So $x = \frac{3}{2}$ or $x = -2$.

   (c) Multiply the first equation by 2 and then add it to the second:
   $$11y = 22$$
   $$y = 2$$

   Then plug y into the first equation and solve for x:
   $$x = 8 - 3y = 8 - 6 = 2$$

   The solution is $(2, 2)$.

5. (a) $\mathbf{A}^T = \begin{bmatrix} 5 & -3 \\ 4 & -2 \end{bmatrix}$

   (b) $\det(\mathbf{A}) = 5(-2) - (4)(-3) = -10 + 12 = 2$

   (c) $\mathbf{A}^2 = \mathbf{A} \times \mathbf{A} = \begin{bmatrix} 5 & 4 \\ -3 & -2 \end{bmatrix}\begin{bmatrix} 5 & 4 \\ -3 & -2 \end{bmatrix} = \begin{bmatrix} 13 & 12 \\ -9 & -8 \end{bmatrix}$

(d) $A^{-1} = \frac{1}{2}\begin{bmatrix} -2 & -4 \\ 3 & 5 \end{bmatrix} = \begin{bmatrix} -1 & -2 \\ \frac{3}{2} & \frac{5}{2} \end{bmatrix}$

6. (a) If the truth table provides all F's, we have a contradiction:

p	q	$p \vee q$	$\sim p$	$\sim q$	$(\sim p) \wedge (\sim q)$	$(p \vee q) \wedge [(\sim p) \wedge (\sim q)]$
T	T	T	F	F	F	F
T	F	T	F	T	F	F
F	T	T	T	F	F	F
F	F	F	T	T	T	F

(b) If the truth table provides all T's, we have a tautology:

p	q	$\sim p$	$\sim p \wedge q$	$(\sim p \wedge q) \vee p$	$p \vee q$	$\sim (p \vee q)$	$(\sim p \wedge q) \vee p \vee [\sim(p \vee q)]$
T	T	F	F	T	T	F	T
T	F	F	F	T	T	F	T
F	T	T	T	T	T	F	T
F	F	T	F	F	F	T	T

7. Put the three activities into symbolic notation

   p: I will go swimming.
   q: I will play racquetball.
   s: I will play basketball.

The conjecture now is $[\sim p \vee q] \vee [s \vee p] \vee (\sim q) \rightarrow s$, and a direct proof is as follows:

   (1) $\sim p \vee q$ Hypothesis
   (2) $\sim q$ Hypothesis
   (3) $\sim p$ Simplification of (1) and (2)
   (4) $s \vee p$ Hypothesis
   (5) s Simplification of (3) and (4)

8. (a) $S \times T = \{(x,y) \mid x \in S \text{ and } y \in T\} = \{(A,+),(B,+),(C,+),(A,*),(B,*),(C,*)\}$
   (b) $|S \times T| = $ number of elements in $S \times T = 6$
   (c) $\mathcal{P}(S) = $ set of all subsets of $S = \{\phi, \{A\}, \{B\}, \{C\}, \{A,B\}, \{A,C\}, \{B,C\}, S\}$.

9. (a) At the first gate we have $x + y$. This proceeds to become $(x + y) \cdot z$. This is negated to $\overline{(x+y) \cdot z}$. Then the final result becomes $\overline{(x+y) \cdot z} \cdot (x + y)$.

   (b)
$$[\overline{(x+y) \cdot z}] \cdot (x+y) = [\overline{x+y} + \overline{z}] \cdot (x+y) \quad \text{DeMorgan's Law}$$
$$= [\overline{x+y} \cdot (x+y) + \overline{z} \cdot (x+y) \quad \text{Distributive Law}$$
$$= 0 + \overline{z} \cdot (x+y) \quad \text{Complementation Axiom}$$
$$= (x+y) \cdot \overline{z} \quad \text{Commutative Law}$$

   (c)

x	y	z	$x+y$	$\overline{z}$	$(x+y) \cdot \overline{z}$
0	0	0	0	1	0
0	1	0	1	1	1
1	0	0	1	1	1
1	1	0	1	1	1
0	0	1	0	0	0
0	1	1	1	0	0
1	0	1	1	0	0
1	1	1	1	0	0

   (d) Current will flow in the circuit only if one of the switches x or y is on and z is simultaneously off. Figure EF-3 shows the circuit diagram.

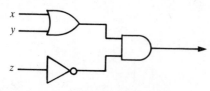

FIGURE EF-3

10. **(a)** This is $_5C_2 = \dfrac{5!}{2!3!} = 10$

    (b) This is $_8P_5 = \dfrac{8!}{3!} = 6720$

11. **(a)** There are 52 cards, of which the possible successful draws would be the 8 of Hearts, the 8 of Diamonds, the 8 of Spades, the 8 of Clubs, or any one of the other twelve Clubs. Thus we have 16 possible successes, and $P(8 \text{ or Club}) = 16/52$. An alternative way is to use the special addition rule of mutually exclusive events, remembering to subtract the event that is not mutually exclusive (the probability of getting the 8 of Clubs):

$$P(8 \text{ or Club}) = P(8) + P(\text{Club}) - P(8 \text{ and Club})$$

$$= \frac{4}{52} + \frac{13}{52} - \frac{1}{52} = \frac{16}{52}$$

    (b) This is conditional probability:

$$P(A \text{ and } A \text{ and } A \text{ and } A) = \tfrac{4}{52} \cdot \tfrac{3}{51} \cdot \tfrac{2}{50} \cdot \tfrac{1}{49} = \tfrac{1}{13} \cdot \tfrac{1}{17} \cdot \tfrac{1}{25} \cdot \tfrac{1}{49} = \tfrac{1}{270\,725}$$

    The fifth card is immaterial. Its probability is 1.

    (c) This is conditional probability:

$$P(\text{red and red}) = P(\text{red}) \cdot P(\text{red} \,|\, \text{red}) = \tfrac{5}{12} \cdot \tfrac{4}{11} = \tfrac{5}{33}$$

12. **(a)** $R = \begin{array}{c} \\ a \\ b \\ c \end{array} \begin{array}{ccc} a & b & c \\ \left[\begin{array}{ccc} 1 & 0 & 1 \\ 0 & 1 & 1 \\ 0 & 1 & 0 \end{array}\right] \end{array}$

    (b) It's not reflexive since (c,c) is missing. It's not symmetric since we have (a,c) but not (c,a). And it's not transitive since we have (a,c) and (c,b) but not (a,b).

13. **(a)** We are given that f maps the real numbers to the real numbers. Since the domain of f consists only of x values ≥ 2, and the range of f consists only of positive real numbers, we find f to be one-to-one but not onto.

    (b) On the domain $D = \{x \,|\, x \geq 2\}$, we have

$$y = \sqrt{x - 2}$$
$$y^2 = x - 2$$
$$x = y^2 + 2$$

Thus $f^{-1}(x) = x^2 + 2$.

14. **(a)** $E = \{\{a,b\}, \{a,d\}, \{b,c\}, \{b,d\}, \{c,d\}\}$

    (b) The graph of this edge set will be as in Figure EF-4.

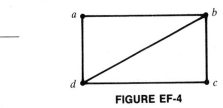

FIGURE EF-4

    (c) An Euler path exists from vertex b to vertex d since these are the only vertices with odd degree.

15. **(a)** The left subtree of D consists only of the vertex H.

    (b) In post-order listing we traverse left, right, root. Thus we get *HPQLMIDEBFJRSNOKGCA*.

INDEX

Entries in italics indicate solved problems.